INTERNATIONAL ASTRONOMICAL UNION

SYMPOSIUM No. 104

EARLY EVOLUTION OF THE UNIVERSE AND ITS PRESENT STRUCTURE

Edited by G. O. ABELL and G. CHINCARINI

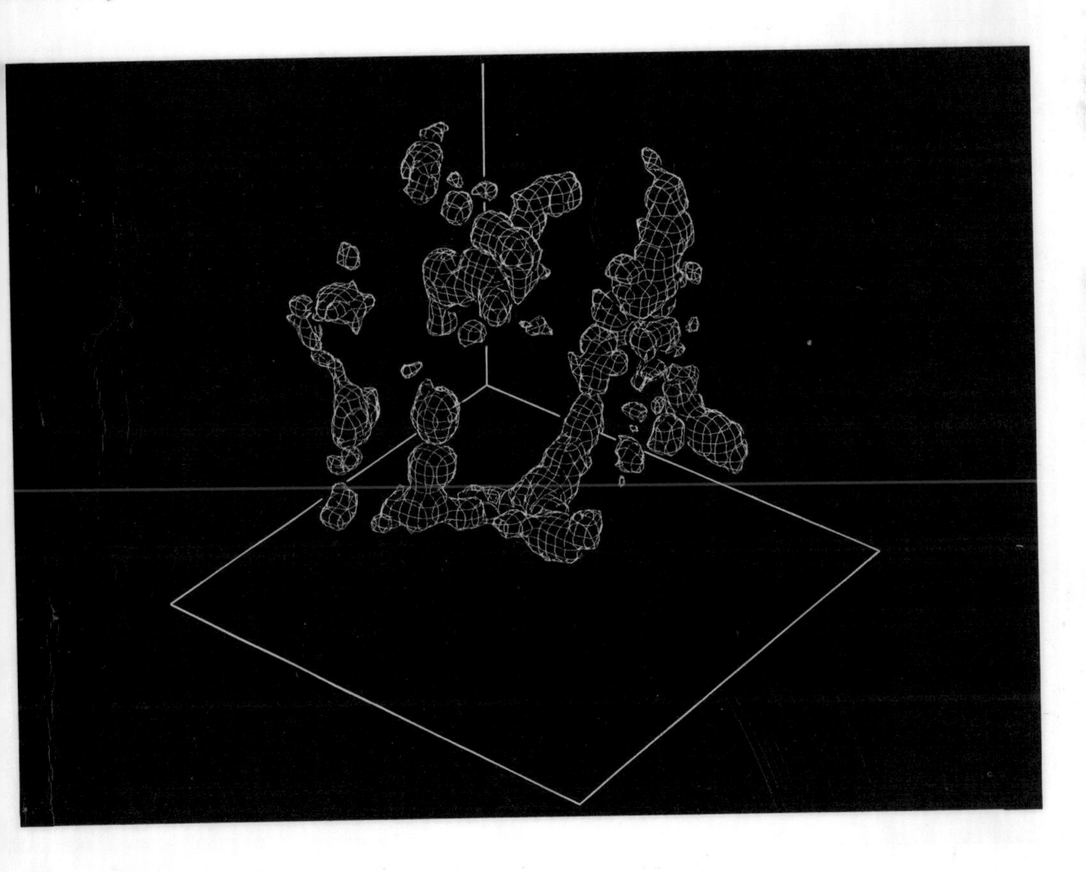

INTERNATIONAL ASTRONOMICAL UNION

D. REIDEL PUBLISHING COMPANY

DORDRECHT / BOSTON / LANCASTER

EARLY EVOLUTION OF THE UNIVERSE AND ITS PRESENT STRUCTURE

Charles Donald Shane (1895-1983), whose extensive study of the distribution of galaxies laid the observational foundation for a good part of the more recent work described in these Proceedings.

INTERNATIONAL ASTRONOMICAL UNION
UNION ASTRONOMIQUE INTERNATIONALE

SYMPOSIUM No. 104

HELD IN KOLYMBARI, CRETE, AUGUST 30 – SEPTEMBER 2, 1982

EARLY EVOLUTION OF THE UNIVERSE AND ITS PRESENT STRUCTURE

EDITED BY

G. O. ABELL
University of California, Los Angeles, U.S.A.

and

G. CHINCARINI
University of Oklahoma, Norman, U.S.A. and
European Southern Observatory, Garching bei Munchen, F.R.G.

D. REIDEL PUBLISHING COMPANY
A MEMBER OF THE KLUWER ACADEMIC PUBLISHERS GROUP
DORDRECHT / BOSTON / LANCASTER

Library of Congress Cataloging in Publication Data

Main entry under title:

Early evolution of the universe and its present structure.

At head of title: International Astronomical Union.
Includes indexes.
1. Cosmology–Congresses. 2. Astronomy–Congresses.
I. Abell, George Ogden, 1927- II. Chincarini, Guido L.
III. International Astronomical Union.
QB981.E15 1983 523 83-17641
ISBN 90-277-1653-6
ISBN 90-277-1662-5 (pbk.)

Published on behalf of
the International Astronomical Union
by
D. Reidel Publishing Company, P. O. Box 17, 3300 AA Dordrecht, Holland

Sold and distributed in the U.S.A. and Canada
by Kluwer Academic Publishers,
190 Old Derby Street, Hingham, MA 02043, U.S.A.

In all other countries, sold and distributed
by Kluwer Academic Publishers Group,
P. O. Box 322, 3300 AH Dordrecht, Holland

Printed in The Netherlands

TABLE OF CONTENTS

SCIENTIFIC ORGANIZING COMMITTEE

Preface

Since the last International Astronomical Union Symposium that dealt with matters cosmological, there have been dramatic advances, both on the observational and theoretical fronts.

Modern high-efficiency detectors have made possible extensive magnitude-limited redshift surveys, which have permitted observational cosmologists to construct three-dimensional maps of large regions of space. What seems to emerge is a distribution of matter in extensive, flat, but probably filamentary, and possibly interconnected, superclusters, serving as interstices between vast voids in space. Meanwhile, theoretical ideas that were highly speculative a few years ago have begun to be taken seriously as possibly describing conditions in the very early universe. And brand new ideas, such as that of the inflationary universe, hold promise of solving outstanding observational, theoretical, and philosophical problems in cosmology. A new look at grand unified theories and concepts of supersymmetry have brought observational and theoretical cosmologists to a common meeting ground with modern particle physicists.

What better place to explore that figurative "meeting ground" than at a real one in the Aegean Islands, where our science of astronomy was born two millenia ago? A symposium on the early universe and its present structure held on the Island of Crete had the appeal of a classical Greek drama: unity of time (when these new ideas are coming to focus); unity of place (the seat of the beginning of modern cosmology); and unity of action (immediately following the 1982 General Assembly at Patras). I.A.U. Commissions 28 (Galaxies), 40 (Radio Astronomy), and 47 (Cosmology) were therefore delighted to accept the invitation of the Orthodox Academy of Crete to hold I.A.U. Symposium 104 at the Academy's splendid facilities at Kolymbari, on the northern shore of the west end of Crete, from 30 August through 2 September 1982.

The symposium was attended by 192 astronomers from 23 nations. There were 78 invited and contributed papers presented orally, and 43 as posters. We received manuscripts for 33 of the poster papers and for all but one of the oral papers (we deeply regret that one invited

G. O. Abell and G. Chincarini (eds.), Early Evolution of the Universe and Its Present Structure, xiii–xv.

speaker failed to submit a manuscript for his important contribution). The topics covered included a review of surveys of the universe at all wavelengths, the latest data on the spectrum and large- and small-scale isotropy (and anisotropy) of the cosmic background radiation, the observed parameters of superclusters and the large-scale structure of the distribution of matter in space, the possible nature of unseen matter (including neutrinos of non-zero rest mass), the kinematics and dynamics of matter in the universe, and theoretical considerations of the early universe, including the inflationary universe and phase transitions, predictions of grand unified and supersymmetry theories, and some comments on unconventional cosmologies.

As is usual with the Proceedings of I.A.U. symposia, the plan was to publish camera-ready manuscripts submitted by the authors. For the most part, this has been done. Many authors, however, do not have access to appropriate typewriters, and we have retyped their contributions. We also retyped other manuscripts that we felt could be improved in appearance or, especially, in English grammar and spelling (which, we acknowledge, present a formidable challenge to those for whom English is not a native language--and, alas, to many for whom it is!). Moreover, we have typed all the questions and answers from written versions received from questioners and authors after the actual exchanges (thereby giving them the opportunity to word their queries and responses in a form suitable for publication). We are happy to report that we have been able to collect 81 percent of the discussion in this written format.

We assumed responsibility for correcting typographical errors where we caught them in otherwise presentable manuscripts, but only if the original manuscript was in a type font we could match. Where we retyped authors' papers, and in typing the discussion, we are aware of the risk of introducing errors of our own (typographical and otherwise). There has been no opportunity to submit the retyped manuscripts to authors for approval, so where we have erred we can only apologize and take full blame. We have made a careful effort to avoid altering the meaning of any author, and hope that we have not done so.

No scientific meeting can be successful unless the local arrangements are comfortable and convivial. We believe from many comments received that Symposium 104 was highly successful, and can lay full credit for that achievement on the Local Organizing Committee, under the superb guidance of Professor George Contopoulos. First, the choice of the Orthodox Academy as a meeting place was ideal (despite the summer heat!), in a beautiful setting on the beach of the Sea of Crete. All sessions were in the morning (8:00 A.M. to noon) and evening (5:00 to 8:00 P.M.), leaving afternoon time for informal discussions, relaxation, and swimming--an opportunity fully exploited. Indeed, significant scientific discussions were held between delegates treading water in the sea! Many participants, while en route from Patras to Kolymbari, took advantage of a splendid tour of mainland Greece and the Island of Crete arranged by the Local Organizing Committee. Many more took lodging at the Academy itself, and still more took meals there; these arrangements were marvelous. There were tours to nearby points of interest for spouses, and a special tour and swimming outing for all at a nearby beach, where some of us enjoyed (in retrospect!) a vigorous bout with jellyfish. None of us will forget the "Crete Night," which featured a sumptuous feast of Cretan specialties as well as folk songs and dancers.

We are extremely grateful to Bishop Irineos, Metropolitan of Kissamos and Selinos, and President of the Academy of Crete, for his hospitality and for inviting us to the Academy, and for coming in person to welcome us. We are also most grateful to the Academy and its Director, Mr. A. Papaderos for providing greatly reduced prices for room and board. In particular, Drs. S. Persides and S. Bonanos worked extremely hard before, during, and after the symposium to make it a success. We thank Olympic Airways for donating 15 free tickets from Athens to Crete, the Director of Museums for providing free admission to our tours, and the National Observatory of Athens for providing free lodging and transportation for several participants en route to the symposium. All other members of the Local Organizing Committee and volunteers, and especially Mr. M. Sfakianakis, were absolutely marvelous in their warmth, helpfulness, and in the generosity of their time. Also, we thank Mr. K. Dimopoulos, President of Balkania Travel Agency, for attaining really bargain prices in transportation and lodging, and for his special and personal concern that each of us received proper and efficient travel arrangements.

Finally, we thank the unsung secretaries who have worked so hard, for no extra compensation, both in planning the meeting and in helping with the preparation of these Proceedings. A partial list includes Mr. Forrest Barger and Mrs. Marietta Stevens, of the Department of Astronomy, UCLA, Miss Deborah MacArthur, of the Department of Physics and Astronomy, University of Oklahoma, and for the lion's share of the typing, cutting and pasting, correcting of spelling and other errors, and general help in editing, Mr. Robert L. O'Daniel, at UCLA.

G.O. Abell
G. Chincarini

Participants

Aaronson, M., University of Arizona, Tucson, U.S.A.
Abell, G.O., University of California, Los Angeles, U.S.A.
Aguilar, L.A., University of California, Berkeley, U.S.A.
Baldwin, J.E., Cavendish Laboratory, Cambridge, U.K.
Barnothy, J.M., 833 Lincoln St., Evanston, Illinois 60201, U.S.A.
Barnothy, M.F., 833 Lincoln St., Evanston, Illinois 60201, U.S.A.
Beckman, J.E., Queen Mary College, London, U.K.
Bekenstein, J.D., Ben Gurion University, Beer Sheva, Israel
Benn, C., Sterrewacht, Leiden, Netherlands
Bhavsar, S.P., University of Sussex, Falmer, Brighton, U.K.
Bludman, S.A., University of Pennsylvania, Philadelphia, U.S.A.
Boksenberg, A., Royal Greenwich Observatory, U.K.
Bonanos, S., N.R.C. Democritos, Aghia Paraskevi, Attiki, Greece
Bond, R., Institute of Astronomy, Cambridge, U.K.
Bonometto, S.A., Istituto di Fisica G. Galilei, Padova, Italy
Braccesi, A., Istituto di Astronomia, Bologna, Italy
Brecher, K., Boston University, U.S.A.
Brosch, N., Huygens Laboratorium, Rijksuniversiteit Leiden, Netherlands
Bruzual, G., C.I.D.A., Mérida, Venezuela
Burbidge, G., Kitt Peak National Observatory, Tucson, Arizona, U.S.A.
Burbidge, E.M., University of California, San Diego, U.S.A.
Carr, B.J., Institute of Astronomy, Cambridge, U.K.
Chen, J.-S., Peking National Observatory, China
Chincarini, G., University of Oklahoma, Norman, U.S.A.
Ciovanardi, C., Italy
Clavel, J., Observatoire, Meudon, France
Clowes, R., University of Durham, U.K.
Code, A.D., University of Wisconsin, Madison, U.S.A.
Contopoulos, G., University of Athens, Greece
Couch, W.J., University of Durham, U.K.
Cristiani, S., Istituto di Astronomia dell'Università di Roma, Italy
Danese, L., Istituto di Astronomia, Padova, Italy
Davies, R.D., University of Manchester, Cheshire, U.K.
de Barnardis, P., Italy
Dekel, A., Yale University, New Haven, Connecticut, U.S.A.
Dent, W.A., University of Massachusetts, Amherst, U.S.A.
de Zotti, G., Istituto di Astronomia, Padova, Italy
di Serego, S., ESTEC, Noordwijik, Netherlands
Djorgovski, S., University of California, Berkeley, U.S.A.
Dressler, A., Mount Wilson and Las Campanas Observatories, Pasadena, U.S.A.
Durret, F., Institut d'Astrophysique, Paris, France
Dyer, C.C., Scarborough College, University of Toronto, West Hill, Ontario, Canada
Efstathiou, G., Institute of Astronomy, Cambridge, U.K.
Einasto, J., Tartu Observatory, Estonia, U.S.S.R.
Ekers, R.D., NRAO, Socorro, New Mexico, U.S.A.
Ellis, G.F.R., University of Cape Town, Cape, South Africa
Ellis, R.S., University of Durham, U.K.

Feast, M.W., South African Astronomical Observatory, Observatory, South Africa
Fisher, P.C., 2401 Sharon Oaks Drive, Menlo Park, California, U.S.A.
Fitton, B., ESA, Noordwijk, Netherlands
Freeman, K.C., Mount Stromlo Observatory, A.C.T., Australia
Geller, M.J., Center for Astrophysics, Cambridge, Massachusetts, U.S.A.
Gerbal, D., Observatoire, Meudon, France
Gioia, I.M. Center for Astrophysics, Cambridge, Massachusetts, U.S.A.
Giovanelli, R.G., Arecibo Observatory, Puerto Rico, U.S.A.
Giuricin, G., Osservatorio Astronomico, Trieste, Italy
Goldman, I., NASA Goddard Institute for Space Studies, New York, U.S.A.
Gorenstein, M., Massachusetts Institute of Technology, Cambridge, U.S.A.
Gott, J.R., III, Princeton University, New Jersey, U.S.A.
Haddock, F.T., University of Michigan, Ann Arbor, U.S.A.
Hanes, D.A., Anglo-Australian Observatory, Epping, N.S.W., Australia
Hara, T., Kyoto Sangyo University, Kyoto, Japan
Hardy, E., Université Laval, Quebec, Canada
Harms, R.J., University of California, San Diego, U.S.A.
Harrison, E.R., University of Massachusetts, Amherst, U.S.A.
Hartwick, F.D.A., University of Victoria, Victoria, B.C., Canada
Haubold, H.J., Zentralinstitut für Astrophysik, Sternwarte, Babelsberg, Pottsdam, Germany D.R.
Heeschen, D.S., NRAO, Charlottesville, Virginia, U.S.A.
Heyvaerts, J., Institut d'Astrophysique, Paris, France
Hoffman, Y., Tel-Aviv University, Israel
Hogan, C., Institute of Astronomy, Cambridge, U.K.
Huchra, J.P., Center for Astrophysics, Cambridge, Massachusetts, U.S.A.
Hunstead, R.W., University of Sydney, Australia
Inagaki, S., Institute of Astronomy, Cambridge, U.K.
Irwin, J.B., 2744 N. Tyndall Ave., Tucson, Arizona 85729, U.S.A.
Jones, B., Observatoire, Meudon, France
Jones, J., Institut d'Astrophysique, Paris, France
Jugaku, J., Tokyo Astronomical Observatory, Japan
Kafatos, M., George Mason University, Fairfax, Virginia, U.S.A.
Kalinkov, M., Bulgarian Academy of Sciences, Sofia, Bulgaria
Kapahi, V.K., Tata Institute of Fundamental Research, Bangalore, India
Kardašhev, N., Space Research Institute, Moscow, U.S.S.R.
Katgert, P., Sterrewacht, Leiden, Netherlands
Katgert-Merkelijn, J.K., Sterrewacht, Leiden, Netherlands
Kazanas, D., NASA Goddard Space Flight Center, Greenbelt, Maryland, U.S.A.
Kellerman, K.I., NRAO, Greenbank, West Virginia, U.S.A.
Kiang, T., Dunsink Observatory, Castlenock, Ireland
Kokkotas, K., Greece
Kolbenstvedt, H., University of Trondheim, Dragvoll, Norway
Koo, D.C., Carnegie-DTM, Washington, D.C., U.S.A.
Kristian, J., Mount Wilson and Las Campañas Observatories, Pasadena, U.S.A.
Kron, R.G., Yerkes Observatory, Williams Bay, Wisconsin, U.S.A.
Kunth, D., Institut d'Astrophysique, Paris, France

Kviz, Z., University of New South Wales, Broken Hill, Australia
Laing R.A., NRAO, Charlottesville, Virginia, U.S.A.
Lasenby, A.N., University of Manchester, Cheshire, U.K.
Latham, D.W., Center for Astrophysics, Cambridge, Massachusetts, U.S.A.
Lazarides, G., The Rockefeller University, New York, U.S.A.
Lindley, D., Institute of Astronomy, Cambridge, U.K.
Lucchin, F., Istituto di Fisica Galileo Galilei, Padova, Italy
Lukash, V.N., Space Research Institute, Moscow, U.S.S.R.
Maccacaro, T., Center for Astrophysics, Cambridge, Massachusetts, U.S.A.
McCrea, W.H., 87 Houndean Rise, Lewes, Sussex BN7 1EJ, U.K.
MacGillivray, H.T., Royal Observatory, Edinburgh, U.K.
McHardy, I.M., Leicester University, Leicester, U.K.
Mandolesi, N., Istituto Te.S.R.E., Bologna, Italy
Marano, B., Istituto di Astronomia dell'Università, Bologna, Italy
Maraschi, L., Istituto di Fisica, Milano, Italy
Maucherat-Joubert, M., Laboratoire d'Astronomie Spatiale, Marseille, France
Mavrides, S., Institut Henri Poincaré, Paris, France
Mazure, A., Observatoire, Meudon, France
Melchiorri, F., Istituto di Fisica, Roma, Italy
Melnick, J., University of Chile, Santiago, Chile
Menon, T.K., University of British Columbia, Vancouver, Canada
Michaelidis, P., Greece
Miller, R.H., University of Chicago, U.S.A.
Mould, J., Kitt Peak National Observatory, Tucson, Arizona, U.S.A.
Murdoch, H.S., University of Sydney, Australia
Natale, V., Istituto di Ricerca sulle Onde Elettromagnetiche del CNR, Firenze, Italy
Nieto, J.-L., Observatoire du Pic-du-Midi, France
Norman, C.A., Sterrewacht, Huygens Laboratorium, Leiden, Netherlands
Novikov, I.D., Space Research Institute, Moscow, U.S.S.R.
Occhionero, F., Laboratorio Astrofisica Spaziale, Roma, Italy
Olivo-Melchiorri, B., Istituto di Fisica, Roma, Italy
Oort, J.H., Sterrewacht Leiden, Netherlands
Osmer, P.S., Cerro Tololo Inter-American Observatory, La Serena, Chile
Paal, G., Konkoly Observatory, Budapest, Hungary
Palmer, P., University of Chicago, U.S.A.
Parijskij, Yu., Special Astrophysical Observatory, Pulkovo, U.S.S.R.
Peacock, J., Royal Observatory Edinburgh, U.K.
Perola, G.C., Istituto di Fisica, Roma, Italy
Persides, S., University of Thessaloniki, Greece
Peterson, B.A., Mount Stromlo Observatory, A.C.T., Australia
Petrosian, V., Stanford University, Stanford, California, U.S.A.
Petrou, M., Greece
Puget, J.L., Institut d'Astrophysique, Paris, France
Quintana, H., Universidad Católica, Santiago, Chile
Rees, M., Institute of Astronomy, Cambridge, U.K.
Roberts, M.S., NRAO, Charlottesville, Virginia, U.S.A.
Robertson, J.G., Anglo-Australian Observatory, Epping, N.S.W., Australia

Rowan-Robinson, M., Queen Mary College, London, U.K.
Salpeter, E.E., Cornell University, Ithaca, New York, U.S.A.
Savage, A., U.K. Schmidt Telescope Unit, Coonabarabran, N.S.W., Australia
Schallwich, D., Ruhr-Universität Bochum, D.F.R.
Schatzman, E.L., Observatoire de Nice, France
Schmid-Burgk, J., Max-Planck-Institut für Radioastronomie, Bonn, D.F.R.
Schmidt, M., California Institute of Technology, Pasadena, California, U.S.A.
Schuch, N.J., Observatorio Nacional, Rio de Janeiro, Brazil
Sciama, D., University Observatory, Oxford, U.K.
Scott, E.L., University of California, Berkeley, U.S.A.
Segal, I., Massachusetts Institute of Technology, Cambridge, U.S.A.
Setti, G., Università di Bologna, Italy
Sfakianakis, M., Holargos Athens, Greece
Shafer, R.A., University of Maryland, College Park, U.S.A.
Shandarin, S.F., USSR Academy of Sciences, Moscow, U.S.S.R.
Shaver, P.A., European Southern Observatory, Garching bei München, D.F.R.
Shaya, E., 2501 Coyne Street, No. 201, Honolulu, Hawaii, U.S.A.
Shectman, S.A., Mount Wilson and Las Campañas Observatories, Pasadena, U.S.A.
Shklovsky, I.S., Space Research Institute, Moscow, U.S.S.R.
Signore, M., Observatoire de Paris, France
Smith, H.J., University of Texas, Austin, U.S.A.
Smith, H.A., Naval Research Laboratory, Washington, D.C., U.S.A.
Smoot, G.F., Lawrence Berkeley Laboratory, Berkeley, California, U.S.A.
Spyrou, N., University of Thessaloniki, Greece
Stabell, R., University of Tromso, Norway
Stecker, F.W., NASA Goddard Space Flight Center, Greenbelt, Maryland, U.S.A.
Szalay, A., Eotvos University, Budapest, Hungary
Tarenghi, M., European Southern Observatory, Garching bei München, D.F.R.
Thompson, L.A., University of Hawaii, Honolulu, U.S.A.
Trevese, D., Osservatorio Astronomico di Roma, Italy
Trimble, V., University of California, Irvine, and University of Maryland, College Park, U.S.A.
Tully, B., University of Hawaii, Honolulu, U.S.A.
Tyson, J.A., Bell Laboratories, Murray Hill, New Jersey, U.S.A.
Urry, M., NASA Goddard Space Flight Center, Greenbelt, Maryland, U.S.A.
Vagnetti, F., Istituto di Fisica, Roma, Italy
van der Laan, H., Sterrewacht, Leiden, Netherlands
van Woerden, H., Rijksuniversiteit te Gronigen, Netherlands
Veron, M., European Southern Observatory, Garching bei München, D.F.R.
Veron, P., European Southern Observatory, Garching bei München, D.F.R.
Vignato, A., Osservatorio Astronomico di Roma, Italy
Vittorio, N., Istituto di Fisica, Roma, Italy
Voglis, N., Greece
Wall, J.V., Royal Greenwich Observatory, U.K.

Wampler, E.J., European Southern Observatory, Garching bei München, D.F.R.
Wehinger, P.A., Arizona State University, Tempe, U.S.A.
Wilkinson, D.T., Princeton University, New Jersey, U.S.A.
Windhorst, R.A., Sterrewacht Leiden, Netherlands
Wolfe, A.M., University of Pittsburgh, Pennsylvania, U.S.A.
Wright, E.L., University of California, Los Angeles, U.S.A.
Wright, J.P., National Science Foundation, Washington, D.C., U.S.A.
Wyckoff, S., Arizona State University, Tempe, U.S.A.
Xanthopoulos, B., University of Thessaloniki, Greece
Zamorani, G., Istituto di Radioastronomia, Bologna, Italy

KEYNOTE: STRUCTURE OF THE UNIVERSE

J.H. Oort
Sterrewacht, Leiden

A clear keynote for our conference is given by the diagram shown in Figure 1. This year we can celebrate that it is just half a century ago that it was produced by Harlow Shapley and Adelaide Ames. This marvellous picture illustrates nearly all we know about the properties of the Universe, except its expansion. It can teach us practically all the lessons we are still learning today. Most prominent is the Virgo cluster. Also evident are structures on larger scale, such as the appendages on both sides of the cluster, spanning a total length of 20 to 30 Mpc.* They are the kernel of the Local, or Virgo, supercluster. The supercluster has a complicated structure, evident from the clumpiness in the distribution along its axis, but likewise from the arrangement of the galaxies in its environment. The centre of the local supercluster (the Virgo cluster) lies at a distance of about 20 Mpc. The supercluster probably extends somewhat beyond us. However, everything in the Shapley-Ames picture should be considered as connected with this one supercluster. Features marked A, B and C may be independent structures; A lies at a distance of roughly 50 Mpc; C, at a distance of ∿ 70 Mpc, lies similarly well outside the Local Supercluster. The strongly elongated feature B, with an average velocity of 1400 km s^{-1} and a dispersion of only 300 km s^{-1}, appears also well isolated. All three are roughly 30 Mpc long, and may be considered as separate superclusters.

Look once more at the Figure. I want to comment on two other phenomena. The first: the occurrence of several large near-empty regions which cannot be ascribed to extinction in our Galaxy. To a certain extent the voids are a natural complement to the superclustering, but it is interesting to see the large contrast in density between "empty" and populated regions of space. The second phenomenon which I must refer to

*Throughout my lecture I shall use a Hubble constant of 50 km s^{-1} Mpc^{-1} for deriving distances and dimensions. Discussion of this constant is outside the programme of this symposium, and would be beyond the scope of my lecture. There are weighty reasons, principally based on the ages of globular clusters, to believe that H_0 lies between roughly 45 and 60 km s^{-1} Mpc^{-1}.

G. O. Abell and G. Chincarini (eds.), Early Evolution of the Universe and Its Present Structure, 1–6.

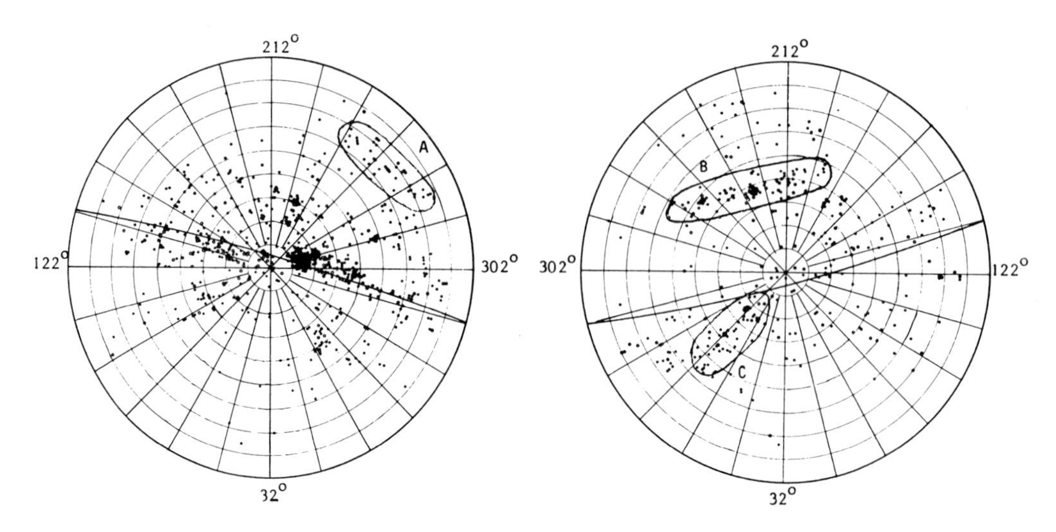

Figure 1. Distribution of galaxies brighter than the 13th magnitude in galactic coordinates; latitude circles are drawn every 10°. The north galactic hemisphere is on the left (Shapley & Ames 1938).

is the ever-present small-scale structure, the "clumping." The clumps come in all sizes: doubles, triples, multiples, small and large groups; they come also in different shapes, and seem to defy an encompassing description. However, some very useful characteristics have been brought out through a correlation description, worked out in great detail by Peebles. His two-point correlation, or covariance, function, $\xi(r)$, has over a large range of r the remarkably simple form $(r/r_o)^{-\gamma}$, with $\gamma = 1.8$; r_o, the "clustering length," is about 10 Mpc.

Perhaps the most interesting thing in the Shapley-Ames picture is the almost complete lack of equilibrium. Even the Virgo cluster itself, which is the only large feature showing signs of mixing, lacks the strong central concentration typical of a relaxed cluster. Every large feature in the picture appears cosmogonically young, and seems to bear the stamp of its origin. This impression is confirmed by the better insight afforded by the now almost complete radial velocity measurements: the smallness of the velocity dispersion indicates that the crossing times along the major axes of the Virgo supercluster are of the order of the age of the Universe, so that hardly any mixing can have occurred in these directions. It is only in the direction of the short axes that there are distinct signs of evolution. Similar conditions seem to hold for feature B, and possibly also for the two other superstructures, A and C.

For the small structures conditions are different. The "clumps" can all have been formed by gravitational clustering in an initially homogeneous Universe, provided that protogalaxies originated at a sufficiently early epoch. This has been shown convincingly by N-body simulations. Such simulations can reproduce in a striking manner the observerved covariance function. The marvellous outcome of these calculations naturally prompts one to ask whether also the larger structures might be explained this way. On the basis of our present knowledge the answer must probably be "No, at least not all such structures." In particular, it seems doubtful whether, starting from a random distribution, the strongly flattened, or elongated, shapes and the sometimes enormous dimensions can be produced that are characteristic of many superclusters. I mention, however, that interesting simulations have recently been made that do produce supercluster-like formations, by choosing initial conditions in which fluctuations beneath a certain, rather large, scale are suppressed or by having galaxies formed at a very early epoch.

We have arrived here at one of the crucial questions for this symposium, viz., do the large structures we observe around us teach us something about the Universe before decoupling?

Before we can answer this question we must scrutinize more closely whether there is any possibility that structures of 50-100 Mpc, with the characteristics of superclusters, can have originated in the era after decoupling. An original scenario has been suggested by Ostriker and Cowie. They considered the chain reaction which might be started by the blast waves in the intergalactic medium produced by the burst of radiation and supernovae expected to accompany the birth of a galaxy. The ensuing shocks may lead to the birth of other galaxies, and under suitable conditions might produce a chain reaction by which large, and possibly flat, structures could be built up. But it is still quite uncertain whether the process could lead to structures having such large dimensions as the largest superstructures that have been observed. Unsatisfactory aspects of the scenario are that it does not explain the origin of the seed galaxies from which the chain reactions must be started, and that, further, it is not clear how it could explain the apparent prevalence of elongated structures among the superclusters. A test of its applicability to superclusters might be obtained from a study of the systematic motions in their surroundings.

If the origin of the large superclusters has to be sought in the earlier Universe, it is evidently of very great interest to study their properties. Several communications will deal with observations tending to this purpose. The investigations made so far have indicated the following:

1. Superclusters fill perhaps 10% of the Universe; the space between them seems to contain comparatively little luminous matter. Most galaxy clusters, and the majority of galaxies, may belong to superclusters.
2. They show no signs of equilibrium or even mixing.
3. Traversal times along long axes are probably as long as the age of the Universe.
4. The largest dimensions reach up to about 100 Mpc, perhaps even 150 Mpc. There are indications that small-amplitude density fluctuations may exist up to 300 Mpc.
5. On scales beyond this the Universe appears essentially homogeneous.
6. The average separation between superclusters along a line of sight is of the order of 200 Mpc.

7. Mass estimates range from $\sim 10^{15}$ to perhaps 10^{17} $M_\odot$.
8. The majority of superclusters appears strongly elongated on the sky; many are probably elongated in space as well.
9. Some contain two or more "chains," at large angles to each other.
10. The major axes of rich clusters are -- at least in some cases -- aligned with the supercluster in which they are situated. The major axes of all rich clusters are strongly correlated with the direction toward the nearest neighbouring rich cluster, up to distances of 50-100 Mpc.
11. The orientation of double galaxies may be connected with that of the superstructure in which they lie.
12. In the outskirts of some superclusters indications of a deceleration of the Hubble expansion have been observed.
13. Superclusters may form a more or less connected cell structure, or network, throughout the Universe. This very important aspect, which was first suggested by Einasto, requires much more study.

In order that condensations be formed, the Universe must have carried within it deviations from homogeneity. The most massive structures probably existed before the decoupling of matter and radiation. In the so-called "adiabatic" theory, such structures should have had masses exceeding about 10^{14} $M_\odot$, smaller condensations having been destroyed by radiation drag in the fireball stage. It is interesting that all structures having the characteristics of superclusters (such as lack of equilibrium, or mixing, very long crossing times, and sizes such as could hardly have been produced in the time elapsed since the decoupling) appear to have just the sort of masses that could survive the radiation era.

In an alternative -- more controversial -- model of the Universe, the so-called isothermal model, fluctuations existed on all scales, and galaxies could presumably have begun to form soon after decoupling. The formation of clusters and superclusters would then have been non-dissipative; it could nevertheless have produced flat and oblong structures. It is evidently of interest to search for observational tests which can decide between the two models. A non-dissipative collapse cannot produce as steep a density gradient as a dissipative one. A close study of the galaxy distribution perpendicular to the long axes of edge-on, or cigar-like, superclusters might therefore provide such a test. Is, for instance, the steepness of the density drop toward positive supergalactic Z in Tully's Fig. 4 (Tully 1982) compatible with a non-dissipative formation?

Let us now consider the adiabatic picture. In this scenario clusters of galaxies, as well as detached galaxies, including quasars, would have formed only after the collapse of the superstructures. This should thus have taken place well before the birth of the oldest quasars, say at $z \sim 5$ or earlier. The well-known theory, that has been worked out so extensively by the Moscow school, is that at decoupling there were fluctuations of large scale, but small amplitude, in the universal medium, which at first expanded at the same rate as the Universe. The expansion of regions of excess density gradually lagged behind, and they finally collapsed. It was realized in an early stage, by Zeldovich and co-workers, as well as independently by Icke in Leiden, that unless the shapes of the fluctuations would have been implausibly close to spheres they would always collapse first along their smaller diameters, and form either "pancakes," "cigars," or more complicated structures. This yielded a natural explanation of some of the striking properties of superclusters.

What amplitudes must the initial fluctuations (at $z = 1000$) have had in order to collapse at $z = 3$, corresponding to an age of 1.63×10^9 years in an Einstein-de Sitter Universe, or 3.45×10^9 years in a Universe with $\Omega = \rho/\rho_{crit} = 0.1$? For a spherical fluctuation we have the following simple relation between the fractional density excess η at $z = 1000$ above the average density in an Einstein-de Sitter Universe, and the time of collapse t in units of 10^9 years: $t = 0.00019\ \eta^{-3/2}$. For the Einstein-de Sitter case this gives $\eta = 0.011$; for the open Universe $\eta = 0.007$. In the latter case the excess over the mean density in the Universe considered would be 0.016. For a collapse at $z = 10$ instead of 3, the density excesses should have been 0.030 and 0.025 for the two cases respectively.

Fluctuations with amplitudes as large as these are improbable on theoretical grounds. Expected amplitudes for supercluster masses are an order of magnitude smaller. The amplitudes quoted are likewise excluded by observations of the background radiation. A mass of $10^{16}\ M_\odot$ subtends at $z = 1000$ an angle of $19'.2$ in an Einstein-de Sitter Universe, and $4'.85$ in a Universe with $\Omega = 0.1$. On such scales upper limits of roughly 0.0004 have been found for fluctuations in the background radiation, again much smaller than the figures given above.

There may be various ways out of this apparent discrepancy. One such way is to assume that the initial fluctuations deviated considerably from a spherical shape. The collapse times along the short axes can then be substantially shortened, while the structures continue to expand along their long axes. The collapse times can further be reduced by the "overshoot" mechanism suggested by the Soviet astronomers. An earlier collapse could also occur if the greater part of the density consists of particles such as heavy neutrinos that decoupled from radiation at a much earlier epoch. Small density fluctuations in the neutrino gas could then recollapse at an earlier time, and form potential wells in which baryons would collect. A combination of these various possibilities may have led to a sufficiently early formation of the Zeldovich "pancakes" (between about $z = 5$ and 10) without requiring fluctuations in the cosmic background temperature that conflict with observations.

It is evidently worth a great effort to obtain more accurate measurements on the fluctuation of the background on scales from 2 to 20'. Expected amplitudes of fluctuations corresponding to protosuperclusters are not _very_ far below the upper limits already established. Discovery of these fluctuations, and determination of their brightness, diameters and shapes, might be the greatest contribution to cosmology that can present be made.

REFERENCES

Shapley, H., and Ames, Adelaide: 1938, Harvard Obs. Ann. 88, No. 2.
Tully, R.B.: 1982, Ap. J. 257, p. 389.

DISCUSSION

Inagaki: Are superclusters joined together, or isolated from one another?

Oort: In some cases superclusters seem to be interconnected, but more redshifts are required before such connections can definitely be outlined. It is unknown whether there are superclusters which are truly isolated.

Miller: What is the distinction between clusters and superclusters? Is the distinction clearly defined observationally? Are clusters anything more than irregularities along a supercluster chain or blobs at the junction of two or more chains?

Oort: A rich cluster is usually fairly well defined and distinguished from the surrounding supercluster (if it lies in a supercluster). A useful distinction between clusters and superclusters is that the first are generally relaxed, while the second are not. The Virgo cluster is an intermediate case: reasonably well-mixed, but lacking the central concentration of an equilibrium structure.

X-RAY SOURCES OF COSMOLOGICAL RELEVANCE

T. Maccacaro and I.M. Gioia
Harvard/Smithsonian Center for Astrophysics, Cambridge, U.S.A.
and Istituto di Radioastronomia del CNR, Bologna, Italy.

1. INTRODUCTION

The imaging and spectroscopic instruments onboard the Einstein Observatory (Giacconi et al. 1979) have been extensively used to study in detail the X-ray properties of a large variety of astronomical objects. In this paper we will briefly discuss some of the most relevant results on extragalactic astronomy obtained mainly with the Imaging Proportional Counter (IPC).

About 2000 different IPC pointings were carried out at high galactic latitude, to study the X-ray behavior of preselected objects or classes of objects. In the great majority of cases, the IPC has provided us with much more information than was asked for. The answer to the observer's enquiry on the target of the observation in fact, can usually be found in a small central region of the IPC image, while the remaining square degree contains important unsolicited information on the soft X-ray universe, through the presence of serendipitous sources. The importance of a systematic search for these sources was immediately realized by several groups, and programs were undertaken to count, identify and study a large number of serendipitous sources.

For each class of objects, we shall consider the results derived from the targeted observations as well as those derived from the survey of serendipitous sources. In Section 2 the X-ray data on Active Galactic Nuclei (AGNs: quasars and Seyfert galaxies) are discussed, BL Lacs are discussed in Section 3, and clusters of galaxies in Section 4. A brief summary is given in Section 5.
A Hubble constant of 50 km/(s Mpc) is used througout the paper.

2. QUASARS AND SEYFERT GALAXIES (AGNs)

From analyses of samples of optically selected and radio selected QSOs (Tananbaum et al. 1979, Ku et al. 1980, Zamorani et al. 1981) it has been shown that:

G. O. Abell and G. Chincarini (eds.), Early Evolution of the Universe and Its Present Structure, 7–18.

1) Quasars are, as a class, powerful X-ray sources.
2) On average, the X-ray luminosity correlates with optical luminosity.
3) Radio loud QSOs show a correlation between radio and X-ray emission.
4) For a given optical luminosity the average X-ray emission of radio loud quasars is $\simeq 3$ times higher than that of radio quiet quasars.
5) The ratio between X-ray and optical luminosity, Lx/Lo, is possibly a function of Lo and/or redshift.

This last point is extremely important in its implications. It is a key point for the understanding of many QSOs properties and, in particular, the link between optically selected and X-ray selected QSOs. Zamorani (1982) has recently reanalyzed the available X-ray data on optically selected QSOs and Seyfert galaxies, assuming that the dependence of the Lx/Lo is exclusively on Lo and not on redshift. He found that the absence of correlation between Lx/Lo and Lo is rejected at the 6σ level and that the data are best described by:

$$Lx \propto Lo^{\beta} \qquad \text{with } \beta = .66 \pm .06 \qquad (1)$$

Namely, Lx/Lo decreases with increasing optical luminosity. More recently Avni and Tananbaum (1982) have analyzed the dependence of Lx/Lo on Lo and redshift. They conclude that the explicit dependence of this ratio is predominantly on Lo.

We shall now summarize the relevant results on X-ray selected AGNs and than show how the above relation between Lx and Lo nicely links the properties of X-ray selected AGNs to those of optically selected AGNs.

The many works devoted to serendipitous X-ray sources (Grindlay et al. 1980, Kriss and Canizares 1982, Maccacaro et al. 1982a, Margon et al. 1982, Reichert et al. 1982, Stocke et al. 1982a) have shown that:

1) The large majority of serendipitous sources are optically identified with AGNs.
2) The increasing detection rate of AGNs as function of the decreasing limiting X-ray flux implies that AGNs, even when selected in the X-rays, show evidence of cosmological evolution.
3) There is a continuity between

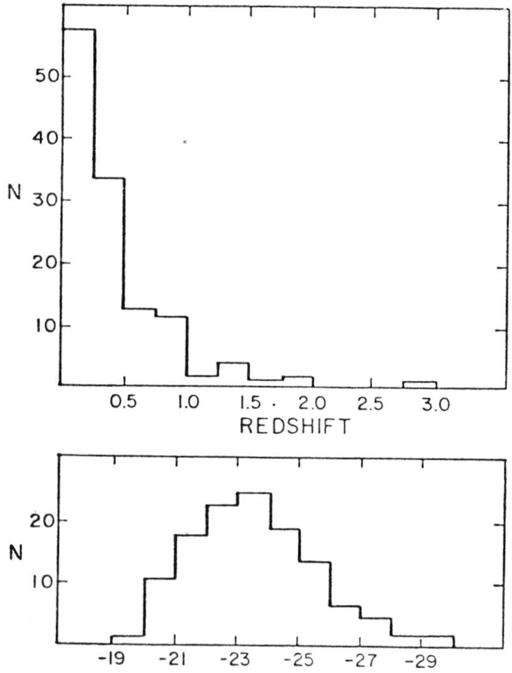

Figure 1. Redshift and absolute magnitude distribution for a sample of 127 X-ray selected AGNs.

the X-ray properties of Seyfert galaxies and quasars.
4) Most of these newly discovered AGNs are characterized by low redshift and low absolute luminosity.

Figure 1 shows the redshift and absolute magnitude distributions of a sample of 127 AGNs taken from the work of Grindlay et al. 1980, Kriss and Canizares 1982, Gioia et al. 1982a, Margon et al. 1982, and Stocke et al. 1982a. As pointed out by Margon et al. (1982), high redshift and high optical luminosity objects are almost totally absent from X-ray selected samples. About 90% of these objects in fact, have a redshift smaller than 1 and an absolute magnitude fainter than -26.

Recently the first complete sample of AGNs extracted from a fully identified sample of X-ray sources and thus exclusively defined by its X-ray properties became available. It consists of 31 AGNs extracted from the Einstein Observatory Medium Sensitivity Survey (Maccacaro et al. 1982a, Stocke et al. 1982a). The analysis of this sample has confirmed the previous findings and has further shown that X-ray selected AGNs do indeed evolve (Maccacaro et al. 1982b). In the framework of pure luminosity evolution of the form:

$$L(z) = L(0)\exp(Cz/(1+z)) \quad (2)$$

their evolution is best described by the parameter C = 5.1. Figure 2 shows the result of the Ve/Va test (a generalized V/Vmax test) used to determine the best value of C.

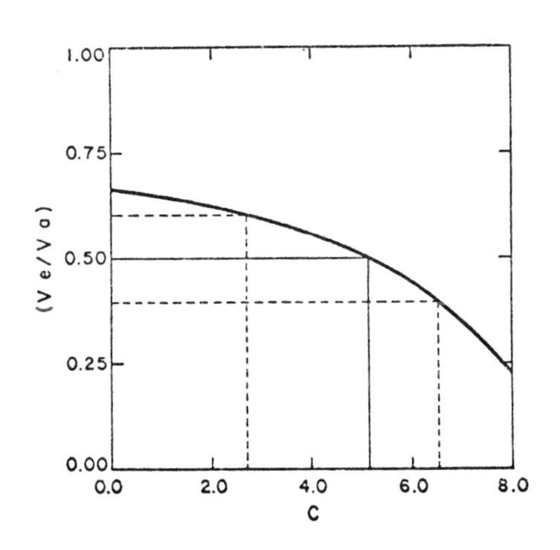

Figure 2. The resulting <Ve/Va> for different trial values of the evolution rate. Dashed lines: 95% confidence region, thin line: best fit value (adapted from Maccacaro et al. 1982b).

A significant difference exists between the X-ray and optical evolution rates. For optically selected samples, in fact, higher values are derived (cf. Schmidt 1978, Marshall et al. 1982).
It can be shown that the dependence of Lx/Lo on the optical luminosity can explain the difference in the evolution rate and in the redshift distribution between optically selected and X-ray selected AGNs.

In the framework of pure luminosity evolution, and with the very important assumption that X-ray surveys and optical surveys select from the same parent population of objects, one can easily derive the amount of evolution expected to characterize X-ray selected AGNs from the amount of evolution required to describe optically selected QSOs, if the relation between their optical and X-ray properties is known.
Combining an optical evolution law of the form of Eq. (2) with the relation between X-ray and optical luminosities of Eq. (1), one obtains:

$$Lx(z) = Lx(0)\exp(\beta Co\cdot z/(1+z)) \qquad (3)$$

where β is from Eq. (1) and Co is the parameter which describe the optical evolution. We can define $Cx = \beta\cdot Co$ and therefore, since β is smaller than 1, Cx is smaller than Co. In particular, for values of Co between 7 and 8, one expects Cx to be between 4.7 and 5.3 since $\beta = .66$, (cf. Avni and Tananbaum 1982 for a more detailed discussion of the link between optical and X-ray evolution rate).
The lack of high redshift, high luminosity objects among X-ray selected AGNs detected in a flux-limited survey, can also be explained by the dependence of Lx/Lo on Lo.
Figure 3a shows the expected redshift distribution of X-ray selected AGNs in a survey with limiting sensitivity of 10^{-14}ergs/(cm^{2}s), a few times better than the Einstein Deep Surveys. The expected redshift distribution is very similar to the redshift distribution of optically selected objects. Such a survey in fact is deep enough to allow an almost uniform sampling, at all redshift, of the optical luminosity function. If we now increase the limiting X-ray flux of the survey to 10^{-13}ergs/(cm^{2}s) we see that we loose objects at high redshifts (Figure 3b). Again the reason for this is easily understood: in a flux limited sample, objects at high redshifts can be detected only if they have high luminosities. When the limiting flux is increased the objects which disappear first are the high luminosity objects since for them the production of X-rays is relatively less efficient as indicated by the decreasing dependence of Lx/Lo on Lo.

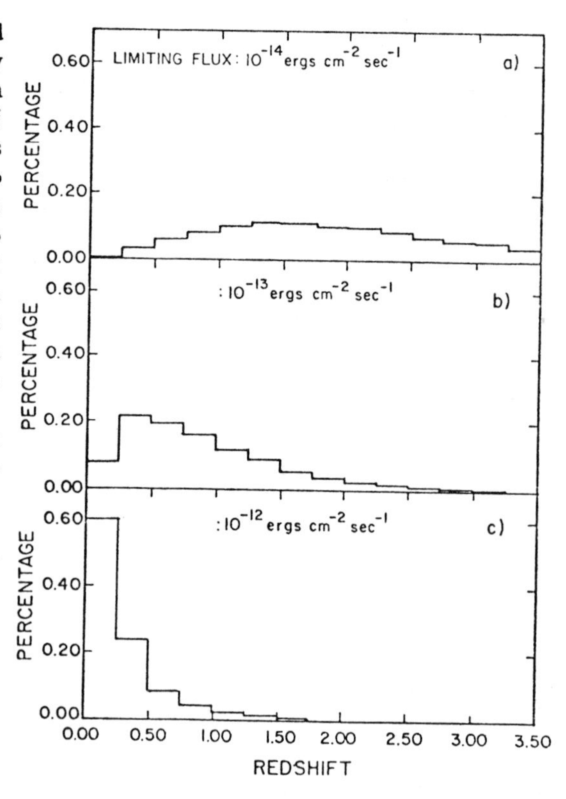

Figure 3. Expected redshift distribution for a sample of X-ray selected AGNs (see text for details).

If we further increase the limiting flux by another factor of 10, (we are now dealing with a sensitivity comparable to that of a 1000 seconds IPC observation), the redshift distribution is remarkably weighted toward very low redshifts (Figure 3c), and is now similar to the observed redshift distribution (Figure 1). These expected redshift distributions (done in collaboration with P. Giommi) have been derived integrating the optical luminosity function of quasars, assuming pure luminosity evolution and converting optical luminosities into X-ray luminosities according to Eq. (1). In a recent work, Zamorani (1982) has independently reached these same conclusions.

It is worth mentioning that if instead of the Einstein Observatory, AXAF (the Advanced X-ray Astronomical Facility, a 1.2 meter X-ray telescope under study) had been launched, the whole issue of the apparent anomalous redshift distribution of X-ray selected AGNs would probably never have arisen. The AXAF capability to detect sources at 10^{-14}ergs/(cm^{2}s) in a few thousand seconds (Murray 1980, Zombeck 1982) would have in fact allowed us to easily detect virtually all QSOs, including the high redshift, high luminosity ones.

3. BL LACS OBJECTS

The Hewitt and Burbidge (1980) catalogue of quasi stellar objects contains about 1500 QSOs but only 60 BL Lacs. Of these not more than a third have a measured redshift. This huge difference between the number of QSOs and BL Lacs known, reflects the elusive nature of BL Lacs and limits our capability of a systematic study of their properties. Nonetheless a number of BL Lacs have been observed with the Einstein Observatory (e.g. Ku 1979, Schwartz and Ku, 1980, Maccagni and Tarenghi 1981). From the results of these observations we know that BL Lacs, in many respects, are similar to QSOs in the X-ray band. They are, as a class, powerful X-ray emitters with an X-ray to optical flux ratio similar to those of QSOs.

BL Lacs, however, show a more frequent and more pronounced X-ray variability. If one believes that optical, radio and X-ray emission are related, this is not surprising since BL Lacs are, by definition, characterized by strong flux variations. More noticeable is the difference between the X-ray spectra of AGNs and BL Lacs.
Figure 4 shows a histogram of X-ray (2-10 keV) spectral indexes for a sample of 23 AGNs, and for 7 BL Lacs. The data on AGNs are taken from Halpern and Grindlay (1982), those on BL Lacs from Worral et al. (1981), Maccagni and Tarenghi (1982) and Maccagni et al. (1982). The distribution is very narrow for the AGNs suggesting that these objects may be characterized, in the X-rays, by a unique power low spectrum with an energy slope of 0.7. BL Lacs instead do not show any preference for a particular spectral index and they seem to be characterized by a steeper spectrum with no evidence for a low energy cut-off. In several cases variations of the slope of the spectrum have been reported (e.g. Worral et al. 1981, Kondo et al. 1981, Maccagni et al. 1982).

The most surprising result however, is probably the almost total absence of BL Lacs from samples of serendipitous sources. Only one source, 1E1402+0416 has been reported in the literature as positively identified with a BL Lac object (Maccacaro et al. 1982a, Stocke et al. 1982b), despite the many serendipitous X-ray sources which have been identified. This result implies a further significant difference between the X-ray properties of BL Lacs and QSOs.

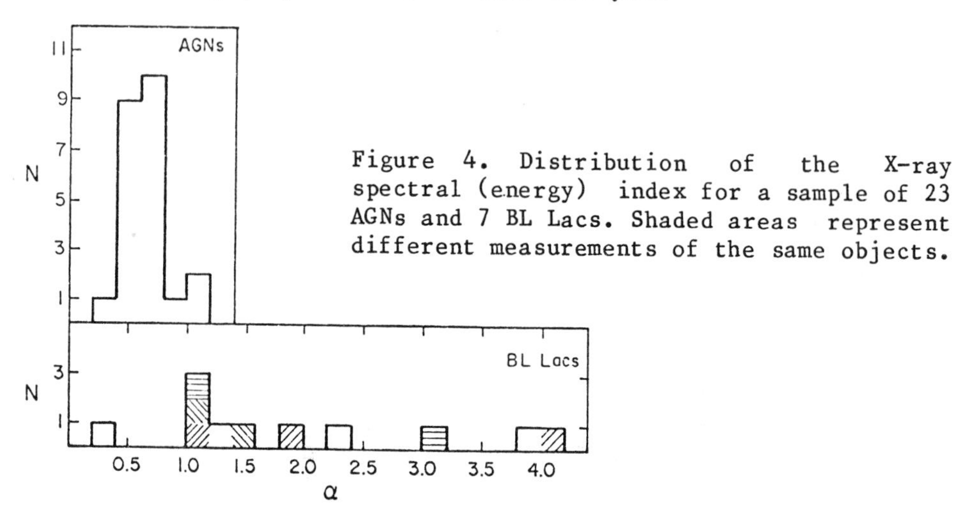

Figure 4. Distribution of the X-ray spectral (energy) index for a sample of 23 AGNs and 7 BL Lacs. Shaded areas represent different measurements of the same objects.

Let us compare, as a function of the X-ray flux, the detection rate of these two classes of objects. For the comparison we are considering two X-ray surveys: the HEA01-A2 all sky survey (Piccinotti et al. 1982) and the Einstein Observatory Medium Survey (Maccacaro et al. 1982a). The former covers the entire high galactic latitude sky with a sensitivity of $\simeq 3 \times 10^{-11}$ ergs/(cm^{2}s) (2-10 keV). The latter covers 50 square degrees of high galactic latitude sky with sensitivities in the range 10^{-13} - 10^{-12} ergs/(cm^{2}s) in a softer (0.3 - 3.5 keV) band. The numbers of extragalactic sources detected in the two surveys are similar, 61 in the HEA01-A2 survey and 49 in the Medium Survey.

If we now define as a QSO any AGN with absolute optical magnitude brighter than -24 we have 3 QSO and 4 BL Lacs detected in the HEA01-A2 survey. The ratio of detections between QSOs and BL Lacs is therefore of the order of the unity. In the Medium Survey, about 100 times more sensitive, 16 QSOs, but only 1 BL Lac, are detected.

Why do we find so few BL Lac objects? Since their spectra are predominantly steeper than those of QSOs the different energy band of the two surveys cannot account for the lack of objects among serendipitous Einstein sources. Just the the opposite would be expected, a steep spectrum should favour their discovery in soft X-ray surveys. The explanation we favour is that BL Lacs do not show, at least at X-ray wavelengths, the same amount of cosmological evolution observed for QSOs.

4. CLUSTER OF GALAXIES

The IPC, with its moderate angular resolution is an extremely powerful instrument to study the X-ray emission from cluster of galaxies. Observations with the Einstein Observatory have shown a large variety of structures in the X-ray surface brightness distribution of cluster of galaxies, from smooth and centrally peaked to highly clumped emission. An extensive review of the X-ray properties of clusters is given in Forman and Jones (1982). Here, we shall very briefly summarize some of the results on a specific class of clusters, those which show double or multiple structures.
The issue of the presence of sub-structure in clusters of galaxies was addressed by Abell and coworkers as early as in 1964 and then investigated by others. Recently, however, interest on this topic has very much revived, due largely to the X-ray images taken with the Einstein Observatory (Forman et al. 1981, Henry et al. 1981, Beers et al. 1982, Gioia et al. 1982b)

Since the X-rays mainly come from diffuse intracluster gas and not from individual galaxies, an X-ray image of the cluster can give us immediate information on the overall cluster morphology. At optical wavelengths, the capability to easily recognize substructures is often limited by the presence of background and foreground galaxies and, in the case of distant clusters even by stellar contamination. Figure 5 shows the X-ray isointensity contours for the clusters of galaxies A98, A115, A1750, and SC0627-54, superposed on optical photographs. The double structure of these cluster can be immediately seen.

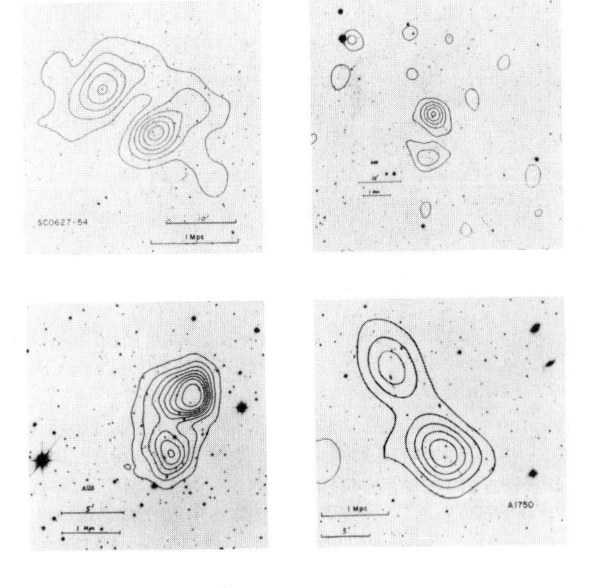

Figure 5. X-ray isointensity contours for the four cluster of galaxies A98, A115, A1750 and SC0627-54 (adapted from Forman et al. 1981).

Optical contour maps for these 4 clusters, are presented by Geller and Beers (this volume) along with a discussion of the comparison between X-ray and optical data.

A cluster which shows multiple subcondensations both in the X-rays and in the optical is shown in Figure 6. The cluster is A2069 and was discovered as an X-ray source by Gioia et al. (1982b) while extending the CFA survey of serendipitous sources. The similarity between the X-ray and the optical contours is striking. This close correspondence between the gas and the galaxy distribution indicates that the galaxies in this system do indeed map the mass distribution.

Obvious questions are whether these subsystems are gravitationally bound and whether these clusters represent an intermediate stage of cluster evolution in the sense that the two or more subcondensations will eventually merge. The answer comes only from a detailed dynamical analysis of the cluster itself.

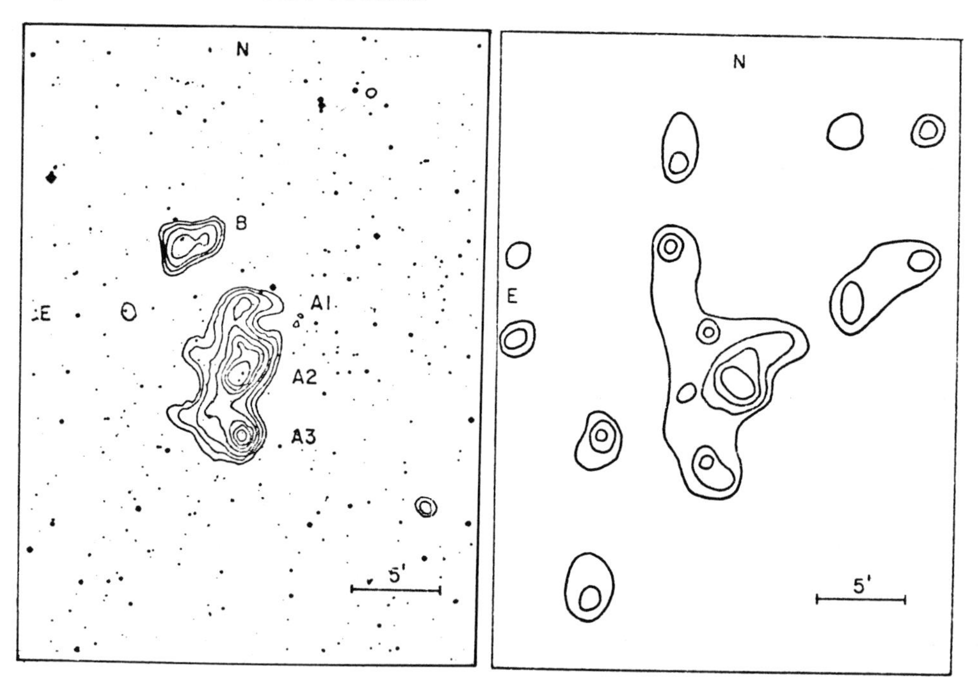

Figure 6. X-ray (left) and optical (right) contours for the cluster of galaxies A2069 (adapted from Gioia et al. 1982b).

In the case of A98 such analysis has recently been done by Beers et al. (1982). They conclude that at the 98% confidence level the two components are bound, and they estimate that in about 3 billion years the two subclusters will finally merge. A similar analysis for other clusters is already in progress (Beers et al. 1983).

5. SUMMARY

We have seen that for optically selected QSOs Lx/Lo decreases with increasing Lo, that X-ray selected AGNs evolve cosmologically and that the rate of evolution required by the X-ray data is smaller than the rate determined from optically selected samples. We have shown that both the difference in the evolution rate between optically selected and X-ray selected AGNs and the anomalous redshift and absolute magnitude distribution of X-ray selected AGNs are possibly induced by the discussed relation between Lx/Lo and Lo.
We have also shown that BL Lacs, in the X-rays, do not evolve as AGNs do. These two classes of objects differs also in their X-ray spectrum. While AGNs are characterized by a very narrow distribution of spectral indexes, BL Lacs do not show any preference for a particular spectral index. Moreover BL Lacs have, on average, a steeper X-ray spectrum and in several cases variation of the shape of the spectrum have been reported.

X-ray images of clusters of galaxies are a powerful tool for the study of their morphological structure and have revealed that most clusters are not symmetric relaxed systems. Clusters of galaxies characterized by double or multiple structure have been found, which may represent an intermediate evolutionary stage in the process of the cluster formation. Dynamical studies have revealed, at least in some cases, that the sub-systems in these clusters are gravitationally bound and will eventually merge.

We thank S. Murray for many useful comments and M. Elvis for a careful reading of the manuscript. This work has received financial support from NASA contract NAS8-30751 and from the Italian SAS (Servizio Attivita´ Spaziali).

REFERENCES

Abell, G.O., Neyman, J. and Scott, E.L. 1964, A. J. 69, 529.
Avni, Y., and Tananbaum, H. 1982, Ap. J. (Letters), submitted.
Beers, T., Geller, M., and Huchra, J., 1982, Ap. J., 257, 23.
Beers, T., Huchra, J., and Geller, M., 1983, Ap. J., in press (Jan 15).
Forman, W., Bechtold, J., Blair, W., Giacconi, R., van Speybroeck, L., and Jones, C. 1981, Ap. J. (Letters), 243, L133.
Forman, W. and Jones, C. 1982, Annual Rev. Astron. and Astrophys., in press.
Giacconi, R., et al. 1979, Ap. J., 230, 540.
Gioia I.M. et al. 1982a, in preparation.
Gioia, I.M., Geller, M.J., Huchra, J.P., Maccacaro, T., Steiner, J.E. and Stocke, J. 1982b, Ap. J. (Letters) 255, L17.
Grindlay, J.E., Steiner, J.E., Forman, W.R., Canizares, C.R., and McClintock, J.E. 1980, Ap. J. (Letters), 239, L43.
Halpern, J. and Grindlay, J. 1982, Nature, submitted.
Henry, J. P., Henrikson, M., Charles, P., and Thorstensen, J. 1981, Ap. J. (Letters), 243, L137.

Hewitt, A. and Burbidge, G. 1980, Ap. J. Suppl. 43, 57.
Kondo, Y., et al. 1981, Ap. J. 243, 690.
Kriss, G.A., and Canizares, C.R., 1982, Ap. J. in press.
Ku, W.H.M., 1979, in IAU Joint Discussion, Extragalactic High Energy Astrophysics, ed. H. van der Laan.
Ku, W.H.M., Helfand, D.J. and Lucy, L.B. 1980, Nature, 288, 323.
Maccacaro, T., et al. 1982a, Ap. J., 253, 504.
Maccacaro, T., Avni, Y., Gioia, I.M., Giommi, P., Liebert, J., Stocke, J.. and Danziger, J. 1982b, Ap. J.. submitted.
Maccagni, D. et al. 1982, in preparation.
Maccagni, D. and Tarenghi, M. 1981, Ap. J. 243, 42.
Margon, B., Chanan, G.A., and Downes, R.A. 1982, Ap. J. (Letters), 253, L7.
Marshall, H.L., Avni, Y., Tananbaum, H. and Zamorani, G. 1982 Ap. J. submitted.
Murray, S.S., 1980, Physica Scripta, 21, 684.
Piccinotti, G. et al. 1982, Ap. J. 253, 485.
Reichert, G.A., Mason, K.O., Throstensen, J.R., and Boyer, S. 1982, Ap. J., in press.
Schmidt, M. 1978, Physica Scripta, 17, 329.
Schwartz, D.A., and Ku, W.H.M., 1980, BAAS, 12, 873.
Stocke, J.. Liebert, J., Stockman, H., Maccacaro, T., Griffiths, R.E., Giommi, P., Danziger, J. and Lub, J. 1982b, Mon. Nat. R. Astr. Soc. 200, 27P.
Stocke, J.T., Liebert, J.W.. Griffiths, R.E., Maccacaro, T., Gioia, I.M., and Danziger, J.W., 1982a, in preparation.
Tananbaum, H. et al. 1979, Ap. J. (Letters), 234, L9.
Worral, D.M., Bold, E.A., Holt, S.S., Mushotzky, R.F., and Serlemitsos 1981, Ap. J. 243, 53.
Zamorani, G. 1982 Ap. J. (Letters), in press.
Zamorani, G. et al. 1981 Ap. J. 245, 357.
Zombeck, M.V., 1982 to be published in the Cospar proceedings.

DISCUSSION

Peacock: Kembhavi and Fabian have concluded that Lx/Lo for quasars increases with Lo. Why does your result differ from theirs? Might it be that by including Seyferts in your Lx-Lo plot you are obtaining a spuriously low slope for your regression line?

Maccacaro: The available data do not allow Lx/Lo to increase with increasing optical luminosity. There is no doubt about it. Data on Seyfert galaxies are in excellent agreement with data on low-luminosity quasars. I believe you have misinterpreted Kembhavi and Fabian's paper.

Zamorani: Kembhavi and Fabian's results were based on an analysis of X-ray and optical properties of QSOs as a function of "observed" optical magnitude, instead of intrinsic optical luminosity. Their conclusion was that the ratio of X-ray to optical luminosity is a decreasing function of "observed" optical magnitude.

Rowan-Robinson: The study by Janet Cheney and me (Monthly Notices 1981) of the optical evolution of quasars gave an evolution factor, C_{opt}, of 5 or 6, with pure exponential luminosity evolution.

Maccacaro: Other works (e.g., Marshall et al. 1982) give a higher value. It will be interesting to compare the samples used in the two studies.

Segal: Published studies (circa 1980-1981) have shown that the non-parametric chronometric cosmology fits very well the QSO X-ray luminosity redshift relation, with optimal statistical treatment of the cut-off bias, without evolution. Later V/V_{max} studies have indicated agreement with spatial homogeneity, i.e., lack of number evolution.

Therefore, is not your finding of strong evolution for X-ray sources:

a) Entirely an artifact of the use of a Friedmann model as the theoretical basis of your analysis;

b) Scientifically dubious in view of the total lack of predictive power of the Friedmann model as regards quasars and other large-redshift objects?

Maccacaro: Not being familiar with the chronometric cosmology, I cannot comment on your question.

Windhorst: Would not you have missed the BL Lacs at high redshifts in the Medium Sensitivity Survey, compared to the BL Lacs in the Bright Survey? (I presume the latter are at low-to-moderate redshifts.) You said that most BL Lacs have very steep X-ray spectra ($1 \lesssim \alpha_X \lesssim 4$). So BL Lacs at higher redshift must have an enormous dimming due to the redshift and will drop below the unit of the Medium Survey.

Maccacaro: The lack of BL Lacs in The Einstein Observatory Medium Survey cannot be easily accounted for even by the K correction. Besides, why are the low redshift BL Lacs unaffected?

Wampler: Complete samples of radio quasars show that the ratio of radio quasars to radio BL Lacs is 10:1. Your statement that there are too few BL Lacs discovered by Einstein is based on a much lower ratio from a small sample of X-ray quasars. Isn't it possible that the true ratio is closer to 10:1 than to 1:1?

Maccacaro: We have shown that the ratio of detection of QSOs and BL Lacs is a function of the X-ray flux. Moreover, the radio behaviour may differ from the X-ray behaviour.

Ekers: What is your definition of AGN?

Maccacaro: With the generic term of AGN (active galactic nuclei) we refer to quasars and Seyfert galaxies.

P. Veron: A few years ago, Setti and Woltjer showed that, optically, BL Lac objects are evolving less strongly than quasars (if they evolve at all); therefore, it is not surprising to see that you find the same to be true in the X-ray domain.

Maccacaro: Setti and Woltjer indeed suggested that BL Lacs are evolving less strongly than quasars. Our results give further support to this picture.

M. Burbidge: In the viewgraph you showed of the L_{opt}, L_x ratio against L_{opt}, to the eye it seemed that the points for QSOs and for Seyferts would be fitted better by two parallel lines, of steeper slope than the single line which you showed. Have you tried making such a fit?

Maccacaro: The eye can be misled by the presence of many upper limits. Points for quasars and for Seyfert galaxies lie, however, on the same correlation; the proof is that the same correlation is found if only QSOs are considered.

(The figure Prof. Burbidge asked about, which was displayed by Dr. Maccacaro, is from Zamorani 1982, and is not reproduced here. Ed.)

The e^+-e^- Annihilation Line and The Cosmic X-Ray Background

Demosthenes Kazanas and Richard A. Shafer
NASA/Goddard Space Flight Center, Code 665, Greenbelt, MD. 20771 and University of Maryland

Abstract

The possibility that the processes responsible for the Cosmic X-ray Background (CXB) would also produce an e^--e^+ annihilation feature is examined. Under the assumption that these processes are thermal, the absence of a strong e^--e^+ annihilation feature places constraints on the compactness (L/R ratio) of these sources. Observations favor sources of small compactness ratio.

1. INTRODUCTION

The fact that the X-ray sky is dominated by an isotropic component (the so called Cosmic X-ray Background, hereafter CXB) had been established by the earliest X-ray astronomy observations (Giacconi et al 1962). The subsequent sattelite X-ray observations, especially by the A-2 and A-4 experiments on HEAO 1 (Marshall et al. 1980), allowed the detailed spectral determination of CXB. The observed spectrum in the region 5-150 KeV, along with the higher energy data is shown in fig 1. Marshall et al. (1980) find a remarkably good fit of this spectrum to that expected from thermal bremsstrahlung from

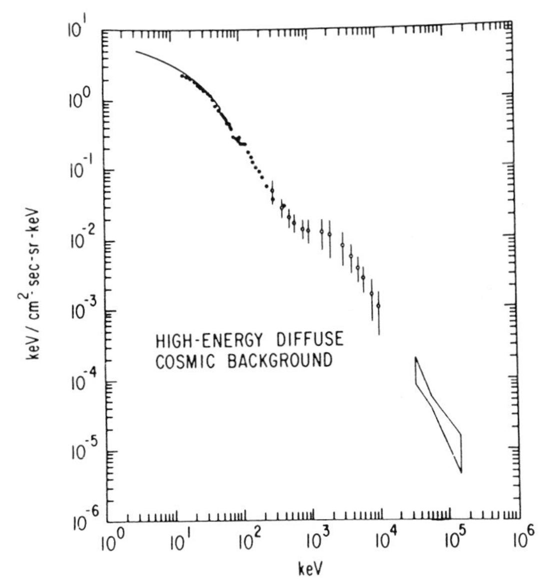

Fig. 1 - The diffuse X and gamma-ray background. (From Ramaty and Lingenfelter 1982).

G. O. Abell and G. Chincarini (eds.), Early Evolution of the Universe and Its Present Structure, 19–22.

an optically thin plasma of temperature 40 ± 5 keV. Interestingly enough, no studied population of sources is known to have a thermal spectrum with the required properties. One can of course contrive to combine sources with a variety of spectra emitting over a range of red shifts to produce the observed total background spectrum, even if the individual spectra are different than that of CXB (DeZotti 1982). The shape of of the spectrum however clearly suggests a thermal distribution of rather specific temperature and it would be more "natural" if the CXB could be explained as such. A thin thermal bremsstrahlung from a heated intergalactic medium will provide such a spectrum (Cowsik and Kobetich 1972; Field and Perrenod 1975).

However, the energy required to heat a diffuse uniform medium to such a temperature is quite large. Clumping of the medium however may provide a solution to the problem since it would reduce the input energy requirements. Taken to its extremes, the thin thermal bremsstrahlung clumps might be reduced to a size comparable to galaxies, or smaller, becoming equivalent to the known "compact" sources of x-rays.

For either the heated intergalactic medium or the "compact source" models for the CXB, the bulk of the emission originates at redshifts $\gtrsim$ 2-3 (or maybe even larger) so that the corresponding source temperature would be $kT \gtrsim$ 100-200 keV. For these temperatures a sufficiently compact source would thermally produce electron-positron pairs from the tails of the photon and particle distributions. Under certain conditions the positron abundance would be sufficient to produce an observable e^+-e^- annihilation feature in the CXB.

2. THE POSITRON ABUNDANCE

In a thermal plasma of temperature $kT \gtrsim$ 100 KeV it is possible to produce positrons at significant abundances by ee, eγ or $\gamma\gamma$ collisions since a non-negligible fraction of the particles and photons at the tails of the distributions fulfills the pair production threshold condition. Their steady state abundance is determined by the balance between pair production and annihilation reactions (Lightman 1982).

In the cases of interest for the CXB the dominant production is due to $\gamma\gamma$ reactions so the ee, eγ will not be currently considered. The approximate expressions for the relevant rates then are:

$$R_{\gamma\gamma} \simeq \frac{3}{8} \sigma_T \, c n_\gamma^2 \, \phi^3 \, e^{-2\phi} \quad \text{(Weaver 1976)}$$
$$R_{+-} \simeq \frac{3}{16} \sigma_T \langle v \rangle n_+ n_- \qquad (1)$$

where η_+, η_-, η_γ are the positron, electron and photon number densities, σ_T the Thomson cross section, $<v> \simeq (2kT/m_e)^{1/2}$ the mean electron velocity, and $\phi = m_e c^2/kT$. The balance equation $R_{\gamma\gamma} = R_{+-}$ then gives:

$$(e^{-2\phi}\phi^3 (\frac{\eta_\gamma}{\eta_+ + \eta_-}))^2 = 1/2 \frac{1}{\phi^{1/2}} \frac{\eta_+\eta_-}{(\eta_+ + \eta_-)^2} \tag{2}$$

If $m = M/M_0$ is the mass of the typical source in solar masses, then its luminosity L and radius R can be expressed in terms of the solar mass Eddington luminosity ($\simeq 1.3 \times 10^{38}$ erg s^{-1}) and Schwarzschild radius ($\simeq 3 \times 10^5$ cm) as $L = 1.3 \times 10^{38}$ mF erg s^{-1} and $R = 3 \times 10^5$ m/f cm where F, f are numerical factors between zero and one ($0 < F,f \lesssim 1$). Then

$$\eta_\gamma/(\eta_+ + \eta_-) \simeq Ff\ 1.2 \times 10^4/(1 + Z)\ T_{40}(1 + \tau)\ \tau \tag{3}$$

Where τ is the optical depth of the sources and T_{40} the observed CXB temperature in units of 40 keV. The balance equation then reads:

$$e^{-\phi}\ \phi^{7/4} \frac{F,f\ 1.7 \times 10^4}{(1 + Z)\ T_{40}(\tau + 1)\tau} = \frac{\lambda^{1/2}}{1 + \lambda} \tag{4}$$

where $\lambda = \eta_+/\eta_-$ is the positron abundance at the sources.

3. COMPARISON TO OBSERVATION

Eq (4) provides an estimate of the positron abundance λ and can be directly related to observation since no obvious annihilation feature is observed in the CXB. The corresponding condition is that the annihilation spectral luminosity (Ramaty and Meszaros 1981) be smaller than that of thermal bremsstrahlung. The latter condition gives

$$\frac{\lambda^{1/2}}{1+\lambda} \lesssim 6\ 10^{-2}\ e^{-\phi/2} \left[(1+Z)\ T_{40}\right]^{1/4} \tag{5}$$

Elimination of λ between (4) and (5) and taking into account that $\phi \simeq 12.8/(1+Z)T_{40}$ gives

$$Ff \lesssim 4.10^{-8} \exp\left[6.4/(1+Z)T_{40}\right] \left[(1+Z)\ T_{40}\right]^3 (\tau+1)\ \tau \tag{6}$$

Eq (6) is independent of the mass of the sources and determines (as a function of the redshift z at which the CXB was produced) their L/R ratio so it is compatible with the absence of a prominent annihilation feature in the CXB. The optical depth τ of the sources is of course unknown, however it is constrainted to be $\tau \lesssim 3$ otherwise the corresponding Comptonization Wien peak should be apparent in the spectrum (Lightman and Band 1981). For compact sources ($f \simeq 0.1 - 0.01$) eq (6) constraints the sources to emit at a small fraction of their Eddington luminosity ($10^{-3} - 10^{-2}$), even for $\tau=3$ and $Z=0$, an important constraint in understanding the nature of the sources of CXB.

4. REFERENCES

Cowsik, R. and Kobetich, E. J., 1972, Ap. J., 177, 585.

Dezotti, G., 1982, Astron. and Astrophys.

Field, G. B. and Perrenod, S., 1977, Ap. J., 215, 717.

Giacconi, R., Gursky, J., Paolini, F. and Rossi, B., 1962, Phys. Rev. Lett., 9, 439.

Lightman, A. D. and Band, D. L., 1981, Ap. J., 251, 715.

Lightman, A. D., 1982, Ap. J., 253, 842.

Marshall, F., Boldt, E., Holt, S., Miller, R., Mushotzky, R., Rose, L. A., Rothchild, R. and Serlemitsos, P., 1980, Ap. J., 235, 377.

Ramaty, R. and Meszaros, P., 1981, Ap. J., 250, 384.

Ramaty, R. and Lingenfelter, R. E., 1982, Annual Reviews of Nuclear and Particle Science.

Weaver, T. A., 1976, Phys. Rev. A, 13, 1536.

DISCUSSION

G. Burbidge: How did you choose the epoch at which the sources generated the original photons?

Kazanas: The epoch is not chosen. The constraints on the L/R (Ff here) ratio of the sources are given as a function of the redshift z. So once a redshift is chosen, the absence of the line limits the ratio L/R.

QSO SURVEYS AND QUASAR EVOLUTION

A. Braccesi
Istituto di Astronomia, Bologna, Italy

Summary - Recent results of QSO surveys are reviewed and QSO's cosmic evolution reconsidered.

1 - INTRODUCTION

For the sake of conciseness, let me refer to the paper I wrote a few years ago together with some Bologna colleagues in which we analyzed the available evidence concerning the space distribution and evolutionary properties of QSOs (Braccesi et al. 1980).

Since that time extremely valuable new information has been gathered, thanks to the work of many researchers. This information, which is only partially published, is summarized below.

1 - Almost complete information on the Green and Schmidt bright quasar search is now available in the synthetic form of a Hubble diagram (Schmidt and Green, 1982). The sample includes 108 quasars, down to an average magnitude B=16.2, found in a 10700 sq.deg. area.

2 - Combining the results by Steppe (1978) and Berger and Frignant (1977), Steppe, Véron and Véron (1979) have produced a very reliable list of UVX quasar candidates which includes 21 objects down to B=18.9 found in a 20.6 sq.deg. area centered on S.A.57.

3 - Marshall et al. (1982a) published spectral information on a complete sample of BFG objects (Braccesi et al. 1970) which includes 22 quasars down to B=18.25 in a 37.2 sq.deg. area and on a complete sample of 13h+36° vf UVX objects (Formiggini et al. 1980). This second sample contains 10 quasars down to B=19.2 in a 1.72 sq.deg. region.

4 - Extending the previous work, Marshall (1982) and Kron (1982) have completed the spectral observations of the 13h+36° vf UVX quasar candidates down to B=19.75 finding that most of the observed objects were quasars. No systematic spectral difference has been found between the objects classified as "stellar" and "extended" by Bonoli

G. O. Abell and G. Chincarini (eds.), Early Evolution of the Universe and Its Present Structure, 23–29.

et al. (1980), a result which has been question by Véron and Véron (1982). Further work on P.S. IIIa-J plates is going on in Bologna on this important point. It is not irrelevant to add that the fuzziness of a number of UVX objects found by Bolton and Savage (1978) in the 2204-1855 field is now ascribed to "an artifact of the singlet corrector" then in use on the U.K. Schmidt (Savage, 1982).

5 - Four fields, 25 sq.deg. each, have been searched for quasar candidates by the U.K. Schmidt Telescope Unit. For a substantial fraction of the objects redshifts have been determined mainly from objective prism plates, which did permit extending the search even to the higher z's. The detailed results are being prepared for publication. The authors and the most relevant data are listed in the following table:

Field	Authors	quasars/sq.deg.	
		B<19.5	B<19.75
0200-50	Savage,A. and Bolton,J.B.	2.8	4.0
2203-1855	Savage,A. and Bolton,J.B.	2.2	3.6
0053-2803	Clowes,R.G. and Savage,A.	4.3	5.1
0112-35	Savage,A.,Trew,A.,Chen,J. and Weston,T.	5.5^{x}	6.9^{x}

(x) J magnitudes.

The large difference in the number of quasars found in the first and last two fields is attributed by Savage (1982) to the improved observing set-up and the better seeing conditions existing when the last two fields were observed. In the following discussion only the data concerning the last two fields will thus be considered.

6 - Kron and Chiu (1981) have examined with a variety of techniques the stellar objects in a 0.1 sq.deg. area near S.A.57 . In this area they found 8 quasars down to J=21 .

7 - Koo and Kron (1982) have made extensive multicolour studies of all the objects with B<23 in a 0.3 sq.deg. field centered on S.A.68 . In this area they found the following numbers of UVX quasar candidates: 24 at B<21.5 , 40 at B<22 , 65 at B<22.5 , 105 at B<23 .

8 - CTIO spectral surveys of high-z objects have been thoroughly discussed by Osmer (1980). The Curtis Schmidt survey covers 375 sq. deg., the 4 m survey 5.1 sq.deg. In spite of the many efforts made (see also Clowes 1981 and Woltjer and Setti 1982) the inconsistency between the C.S. and the 4 m results is not fully understood.

This wealth of data improve considerably the factual knowledge with respect to 1979, when the paper I mentioned at the beginning was written. Not only the number of available z's has substantially increased but, mainly thanks to the work of Green and Schmidt, the coverage of the Hubble plane has been considerably extended toward the brighter magnitudes. Furthermore, a sufficient number of complete samples of UVX quasar candidates has been observed spectroscopically to allow one to base the study of the number-magnitude relation on sounder grounds.

2 - THE NUMBER-MAGNITUDE RELATION FOR UVX QUASARS

Figure 1 shows the number-magnitude relation for UVX, for spectroscopically confirmed samples of quasars. Continuous spectrum objects and extragalactic H II regions have been excluded. Also shown in the figure are the Koo and Kron (1981) UVX object counts. The results of the Palomar Bright Quasar Search have been corrected for 20% losses (Green and Schmidt 1978), those for the BFG objects for a 23% loss (Braccesi et al. 1980).

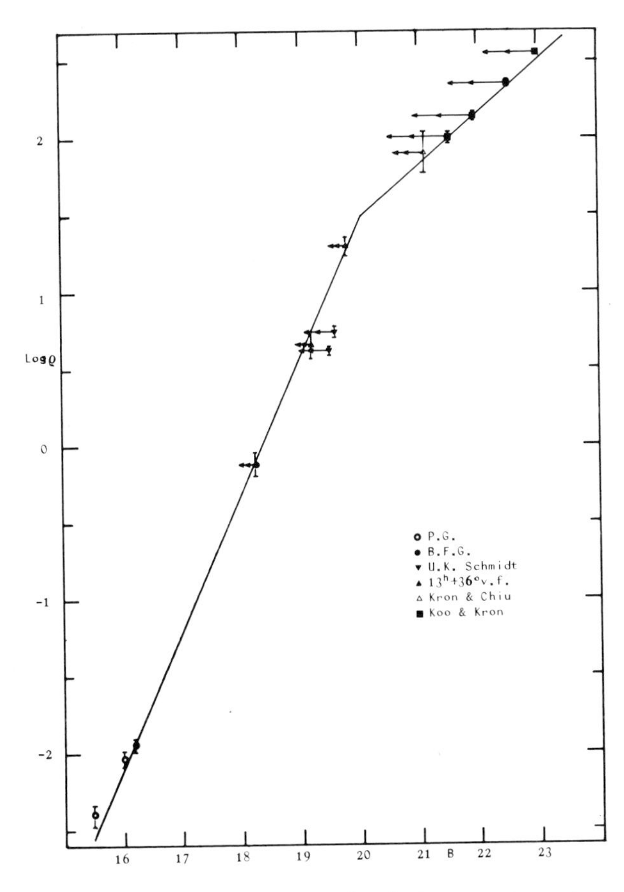

Fig. 1 - The number--magnitude relation for UVX quasars, objects/sq.deg. Error bars indicate 1σ statistical uncertainty. Horizontal arrows show tentative correction for galactic absorption based on neutral hydrogen columns.

Up to B=19.75 the experimental points, except those from the U.K. Schmidt, are nice fitted by the relation:

$$\text{Log } N(B) = 2.20\ ((B-18.30)/2.5) \quad \text{objects/sq.deg.} \qquad (1)$$

which is nearly coicident with that given by Braccesi et al. (1980). The main difference with our previous work is that there we assumed a careful colour selection could lead to the inclusion among UVX quasar candidates of objects with z up to 2.5, whereas we now know that the real limit is about 2.15.

With regard to the two apparently discordant points from the U.K. Schmidt, it is interesting to see the results which are obtained when one applies the corrections for interstellar absorption based on the galactic neutral hydrogen column densities, as suggested by Teerikorpi (1981). These corrections are shown by the horizontal arrows in fig.1, the two end points of the arrows correspond to the double valued relation found between the hydrogen column density and absorption.

One can see that, because of the larger corrections found for the U.K. Schmidt points, all the values in the $18.25 < B < 19.75$ range become, when corrected, consistent with each other. Also the points of Koo and Kron (1981) seem to require a large correction; this will not change the very different slope for the counts found at these very faint magnitudes, 1.00 instead of 2.20, but the magnitude at which the slope changes will be moved from B=20.0 to B=20.4. To apply this correction to the Palomar Bright Quasar Survey, objects should be considered one by one. Certainly, a somehow less steep slope in the range $15.5 < B < 18$ will result.

This discussion of the possible effects of galactic absorption has been introduced mainly to remind us that these effects are large enough to require to be taken properly into account before definitive results on the number-magnitude relation for quasars could be obtained.

3 - THE SPACE AND ABSOLUTE MAGNITUDE DISTRIBUTION OF QUASARS

While the number of quasars increases very rapidly with apparent magnitude, and thus only carefully selected samples with accurate magnitudes can be used in the study of the number-magnitude relation, the fractional distribution in z changes only very slowly. This justifies the use of miscellaneous data to provide a better and more statistically significant coverage of the Hubble plane.

As shown by Braccesi et al. (1980), one can take advantage in this way of a larger sample of objects, using the previously established number-magnitude relation to obtain the normalizations needed to derive the surface densities as a function of apparent magnitude and redshift. When this has been done, in order to compute the space densities as a function of z and absolute magnitudes, one only needs to choose a cosmological model.

These computations have been repeated, strictly following the procedures outlined in the above mentioned paper, taking, as previously, H=50 and $q_0=0$. The number of objects included in the analysis has been increased from 164 to 263, but even more important is the fact that, as stressed before, the coverage of the Hubble plane is now more extended in the magnitude domain.

The results of the analysis are given in Fig. 2 which shows the run of the isodensity lines for UVX quasars -- i.e., quasars with $z < 2.15$ -- as a function of absolute magnitude and look-back time. The points in the plot are statistically independent, one from the other, and the overall consistency of the picture appears very rewarding. Also shown in the figure is the local density of Seyfert I nuclei, taken from Véron 1979, and the density of high-z ($2.5 < z < 3.5$) quasars derived from Table 4 of Osmer (1980) paper. It should be added that

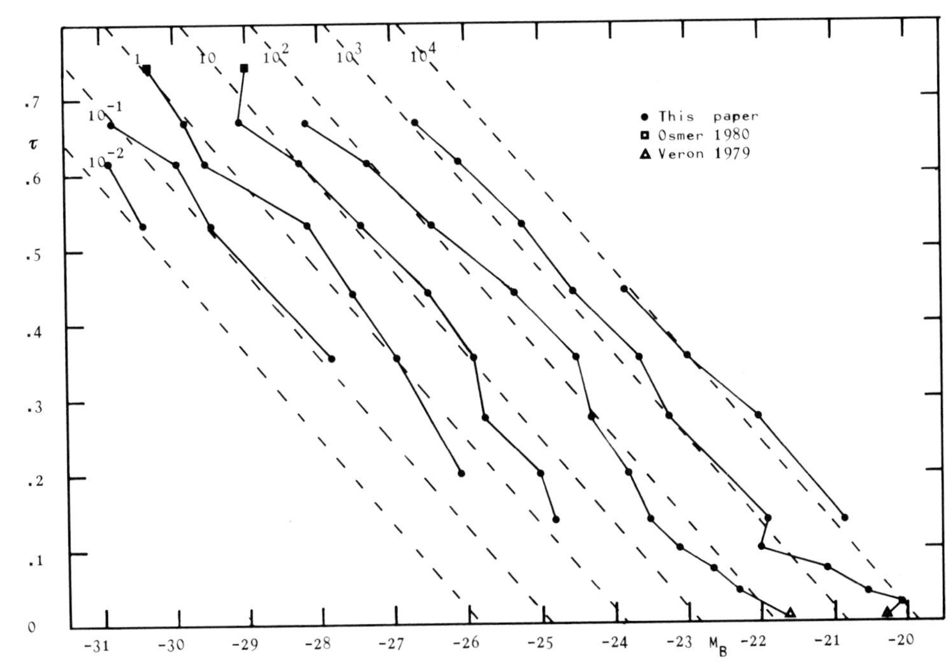

Fig. 2 - Quasars isodensity lines (objects/Gpc^{3}magn.) as a function of absolute magnitude and look-back time. Broken lines show the predictions from the evolutionary model summarized by relation 2.

the densities we have derived for the UVX quasars in the highest redshift interval ($1.85 < z < 2.5$) are in good agreement with those computed from Osmer's table for the $1.8 < z < 2.5$ domain.

The isodensity lines in Fig. 2, in spite of the very large interval in z and absolute magnitude they cover, are, on the whole, surprisingly straight, parallel and equidistant. The space density of quasars, as a function of absolute magnitude and look-back time, may thus be represented by:

$$\text{Log } \rho = 9.0\ \tau + (M_B + 23.75) \quad \text{quasars/Gpc}^3\text{magn.} \tag{2}$$

namely by a luminosity independent exponential increase of density with look-back time and a straight luminosity function. Nothing peculiar seems to happen at $M_B=-24$ where a tentative division between quasars and Seyferts is suggested by Woltjer and Setti (1982). These authors, however, noticed "a strong evolution" even for the $M_B>-24$ objects, an evolution which may now look less "surprising." Schmidt and Green (1982) have proposed a luminosity dependent density evolution, but no evidence of this emerges from the data summarized in Fig. 2.

Relation (2) can be reconciled with the bending of the number-magnitude relation at $B_L\sim 20$ only if we consider the highest isodensity line shown in Fig. 2, 10^4 quasars/Gpc^{3}mag., as the point where the very steep luminosity function we have found should undergo a very substantial flattening, and not an artifact of selection effects. This point has been verified by comparison of the model with the original data and unmistakably found to be so. With this further specification, relation (2) describes a pure luminosity evolution model in which a constant number of objects are dimming with a time constant of 0.12 the age of the universe or 2.2 billion years independently from the epoch or the absolute magnitudes.

The property of the luminosity evolution models of predicting in a straightforward way a flattening of the number-magnitude relation has been recently discussed by Bònoli et al. (1980), Cheney and Rowan-Robinson (1981), Mathez and Nottale (1982) and Marshall et al. (1982b) mainly in connection with Koo and Kron (1982) results on the flattening of the number-magnitude relation of quasars at the very faint magnitudes or with the problem of the quasar contribution to the X-ray background. It must be said that the models which have been proposed are considerably different from each other. Let us hope that the model presented here might represent a stable first-order approximation to the real cosmological evolution of the quasar population.

I must thank Ann Savage, Jiansheng Chen, Roger Clowes, Richard Kron and Herman Marshall for the generous information they provided me on the progress of their work.

REFERENCES

Berger,J.and Frignant,A.M.1977, Astron.& Astrophys.Suppl.28,123
Bolton,J.B.and Savage,A.1978,in I.A.U.Symp.79,p.125
Bònoli,F.,Braccesi,A.,Marano,B.,Merighi,R.and Zitelli,V.1980,Astron.& Astrophys.90,L10
Braccesi,A.,Formiggini,L.and Gandolfi,E.1970,Astron.& Astophys.5,264
Braccesi,A.,Zitelli,V.,Bònoli,F. and Formiggini,L.1980,Astron.& Astrophys 85,80
Cheney,J.E.and Rowan-Robinson,M.1981,M.N.RAS 195,497
Clowes,R.G.1981,M.N.RAS 197,731
Formiggini,L.,Zitelli,V.,Bònoli,F.and Braccesi,A.1980,Astron.& Astrophys. 85,80
Green,R.F.and Schmidt,M.1978,Astrophys.J.220,L1
Koo,D.C. and Kron,G.R.1982,Astron.& Astrophys.105,107
Kron,G.R. and Chiu,L.J.1981,P.A.S.P.93,397
Kron,G.R.1982, private communication
Marshall,H.L., Tananbaum,H.,Zamorani,G.,Hurchra,J.P.,Braccesi,A.and Zitelli,V.1982a,Astrophys.J.(submitted)
Marshall,H.L.,Avni,Y.,Tananbaum,H.and Zamorani,G.1982b (preprint)
Marshall,H.L.1982,private communication
Mathez,G. and Nottale,L.1982Astron.& Astrophys.(in press)
Osmer,P.S. 1980,Astrophys.J.Suppl.42,523
Savage,A.1982,private communication
Schmidt,M.and Green,R.F.1982,Pont.Ac.Scient.Scripa Varia,48,293
Steppe,H.1978,Astron.& Astrophys.Suppl.31,209
Steppe,H.Véron,P.andVéron,M.P.1979,Astron.&Astrophys.78,125
Teerikorpi,P.1981,Astron.&Astrophys.98,300
Véron,P.1979,Astron.&Astrophy.78,46
Véron,P.and Véron,M.P.1982,Astron.&Astrophys.105,405
Woltjer,L. and Setti,G.1982,Pont.Ac.Scient.Scripta Varia,48,281

DISCUSSION

Segal: In the 1980 Ap. J. we published an analysis of your earlier complete optical quasar sample within the non-evolutionary, non-parametric, chronometric cosmology, showing consistency with absence of evolution and a turnover in the $N(< m)$ relation slightly faintward of 20th magnitude. This analysis made a statistically optimal, non-parametric allowance for the observational magnitude cutoff. Is there any reason to doubt that the more recent observations that you cite are also consistent with the chronometric cosmology, and thus with the real physical absence of evolution in the quasar population?

Braccesi: I worked in the usual cosmological frame. I must add that I was not able to get a personal opinion on chronometric cosmology because of the mathematical difficulties of your book, which are too great for me.

Kron: You are advocating a large total absorption in the direction of SA 68 ($b = -46°$). The galaxy counts in the direction of SA 57 ($b = +86°$) and SA 68 are similar, indicating that absorptions in the two directions are not greatly different.

Braccesi: The values I presented for the absorption are possibly wrong. What I wanted to underline is the need to get a better knowledge of the phenomenon.

(Kiang asked if the data permit one to distinguish between a density evolution and a luminosity evolution. Ed.)

Braccesi: Relation 2 may describe both a density or a luminosity evolution. The relevant point is whether one limits its domain to a given lowest absolute luminosity or a given highest constant space density. As I said, to fit the counts, and to get correct fractional distributions in z, one is forced to choose a limiting space density, i.e., a luminosity evolution model.

Schmidt: The Palomar Bright Quasar Survey contains 114 objects in its final form. Green and I have submitted a paper describing the Survey for publication and preprints will be distributed soon. In our derivation of evolution, we find that the increase in space density is very strongly dependent on optical absolute luminosity and that the luminosity function is curved at all redshifts. Both of these results appear to be at variance with Dr. Braccesi's results.

Braccesi: I hope the results presented here will make the subject of a more extended paper in which all the details of the analysis will be presented. We will then be able to compare the procedure and understand the origin of the different conclusions.

AUTOMATED QUASAR DETECTION

Roger G. Clowes
Department of Physics, University of Durham

John A. Cooke and Steven M. Beard
Department of Astronomy, University of Edinburgh

The existing spectral searches for quasars have increased the number of quasars known very substantially but have not contributed proportionately to an understanding of the collective properties because of the selection effects. To fully exploit the spectral searches we have developed the technique of automated quasar detection (AQD) using objective-prism plates from the UK Schmidt Telescope, the COSMOS measuring machine at the Royal Observatory Edinburgh, and the STARLINK nodes at Durham and Edinburgh.

AQD has the following advantages: (i) the selection criteria and consequently the selection effects are known, pre-defined and rigidly maintained, (ii) large and complete samples may be obtained from several plates, (iii) emulsion responses are taken into account, (iv) selected objects emerge with low-resolution spectrophotometry and values for equivalent widths and line widths, (v) repeat searches using new plates need not be avoided, (vi) non-spectral searches (eg selection by ultraviolet excess) can be run simultaneously, (vii) measuring machines and computers do all the repetitive work.

Preliminary results for a field that has already been searched by eye indicate that AQD is very successful: essentially all of the eyeball quasars that satisfied the selection criteria were re-discovered (a few were mistakenly rejected during quality testing) and new quasar candidates were discovered in significant numbers.

G. O. Abell and G. Chincarini (eds.), Early Evolution of the Universe and Its Present Structure, 31.

THE ASIAGO CATALOGUE OF QUASI STELLAR OBJECTS

S. di Serego Alighieri
Space Science Department of ESA
ESTEC, Noordwijk, The Netherlands.

A catalogue containing the QSOs known with reasonable reliability from literature published before the end of 1981 has been compiled at the Asiago Observatory by C. Barbieri, M. Capaccioli, S. Cristiani, G. Nardon and A. Omizzolo, and will be published before the end of 1982 by the Memorie della Societá Astronomica Italiana.
The present release is an enlarged and updated version of that published by Barbieri et al. (1975). It provides basic data (names, equatorial and galactic coordinates at 1950.0, photometry, redshifts and information on radio emission, variability, morphology, presence of absorption lines, X-ray emission and finding charts) for 2004 QSOs, together with references, individual notes and cross reference tables. Two separate lists contain X-ray data for 271 QSOs (X-ray luminosity, optical-X and optical-radio spectral indexes) and absorption line data for 243 QSOs (absorption systems with their redshift, identified ions and wavelengths, spectral range, resolution and dispertion of the spectra).
The catalogue has been carefully checked for internal consistency and several discrepancies with the lists by Hewitt and Burbidge (1980) and by Véron and Véron (1975) have been resolved. The authors intend to provide a magnetic tape version of the catalogue to the Centre de Données Stellaires of the Strasbourg Observatory for distribution to the interested community.

REFERENCES

Barbieri,C., Capaccioli,M. and Zambon,M.: 1975, Mem.S.A.It., **46**, 461.
Hewitt,A. and Burbidge,G.: 1980, Astrophys. J. Suppl., **43**, 57
Erratum: 1981, Astrophys. J. Suppl., **46**, 113.
Véron,P. and Véron,M.P.: 1981, private communication.

G. O. Abell and G. Chincarini (eds.), Early Evolution of the Universe and Its Present Structure, 33.

THE QUASAR REDSHIFT LIMIT

Patrick S. Osmer
Cerro Tololo Inter-American Observatory
Casilla 603
La Serena, Chile

The concept of a redshift limit for quasars was first mentioned a decade ago by Schmidt (1970, 1972) and Sandage (1972) who independently realized that faint, high redshift quasars were not being found in the numbers expected from the redshift and magnitude distribution of bright quasars. Subsequently the limit moved out in redshift to $z \sim 3.5$ from $z \sim 2.5$ after the discovery of OQ172 ($z = 3.53$, Wampler et al. 1973) and numerous other quasars with $z > 3$. It should be said at the outset that the limit concept is in need of definition. While it would be most interesting if an absolute limit existed, i.e., no quasars beyond a certain redshift, in reality the limit value is more likely to be the redshift at which the quasar space density turns down significantly.

What is the case for a quasar redshift limit? On purely observational grounds there has been great difficulty in finding quasars with $z > 3.5$. In the nine years since the discovery of OQ172, the number of known quasars has increased by more than a factor of 5, yet only in March of this year was a larger redshift found, PKS 2000-330, with $z = 3.78$ (Peterson et al. 1982). In the meantime more than 20 quasars with $3.0 < z < 3.5$ were discovered. While there was formerly a justified concern about observational biases against the discovery of quasars with $z > 2.5$ because of their loss of a characteristic ultraviolet excess, such concern is no longer warranted. The development and success of the objective-prism technique for finding quasars has shown that a color-independent approach can be very effective for $z > 3$ (see review by Smith 1978). Similarly, the improvement of radio positions means that sources now can be identified only from positional coincidences. Indeed there has probably been a bias in favor of large redshifts in recent work on radio catalogs as a result of researchers looking for neutral or red stellar objects near radio positions.

Specific searches for quasars with $z > 3.5$ with the objective-prism technique by Koo and Kron (1980) and Osmer (1982) have failed to find any despite limiting magnitudes that should have been more than faint enough. Osmer addressed the space denity question in particular. Because he did discover 15 emission-line quasars and galaxies with $0.03 < z < 3.36$ in his survey, he could make a good estimate of the

G. O. Abell and G. Chincarini (eds.), *Early Evolution of the Universe and Its Present Structure*, 35–38.

limiting sensitivity attained and show that from 9 to 22 quasars were expected with $3.7 < z < 4.7$ if the space density were constant or followed the exp $(10\ \tau)$ form of density evolution. As none were found, he concluded that the space density must decrease significantly for $z \gtrsim 3.5$. Note that a density law increasing as steeply as $(1+z)^6$ had previously been ruled out (Carswell and Smith 1978).

Obviously the discovery of PKS 2000-330 shows that the redshift cutoff is not absolute at $z \sim 3.5$, but does not in itself affect the conclusion that the space density begins to decrease near $z = 3.5$. In the fact the spectrum of PKS 2000-330 shows such a strong Lyα emission line that it would have been easily detected with the grating-prism technique, more easily than many of the objects that were found. It is most important to continue the search for quasars with $z > 3.5$ to improve our knowledge of the space density in this critical redshift region.

A decrease in the quasar density for $z > 3.5$ is related to the number counts of faint stellar objects and to the X-ray background. It was clear from Schmidt's redshift magnitude tables in 1972 that a redshift cutoff would produce a substantial flattening in the apparent luminosity function for quasars fainter than 20th magnitude. More recently Bohuski and Weedman (1979) and Koo and Kron (1982), to name just two examples, have commented on how star counts at faint magnitudes provide useful limits on the number of faint quasars, and the latter paper shows evidence for a flattening of the quasar counts at 21st magnitude. Similarly Setti and Woltjer (1979) have discussed how the quasar counts must flatten if the integrated X-ray flux of quasars is not to exceed that of the observed X-ray background radiation. Thus, integrated number counts of faint stellar objects and the X-ray background data are consistent with the concept of a redshift limit, although it should be emphasized that they are by no means a proof.

On the theoretical side, the redshift limit may be interpreted as the formation epoch of quasars, which is rather later in the evolution of the universe than most galaxies are believed to have formed. Alternatively, the limit may be due to intergalactic absorption, or the time at which the intergalactic medium becomes ionized and transparent at the wavelength of Lyα and below. Already the limit has been mentioned in connection with the idea that massive neutrino decay photoionizes the intergalactic medium (Sciama 1982).

Whatever the outcome, the quasar redshift limit will be a lively topic for the next several years.

REFERENCES

Bohuski, T.J., and Weedman, D.W. 1979, Ap. J., 231, 653.
Carswell, R.F., and Smith, M.G. 1978, M.N.R.A.S., 185, 381.
Koo, D.C., and Kron, R.G. 1980, P.A.S.P., 92, 537.
Koo, D.C., and Kron, R.G. 1982, Astron. Astrophys., 105, 107.
Osmer, P.S. 1982, Ap. J., 253, 28.
Peterson, B.A., Savage, A., Jauncey, D.L., and Wright, A.E. 1982, Ap. J. (Letters) in press.
Schmidt, M. 1970, Ap. J., 162, 371.

_____. 1972, Ap. J., 176, 273.
;andage, A. 1972, Ap. J., 178, 25.
;ciama, D.W. 1982, M.N.R.A.S., 198, 25.
;etti, G., and Woltjer, L. 1979, Astron. and Astrophys., 76, L1.
;mith, M.G. 1978, Vistas in Astr., 22, 321.
Jampler, E.J., Baldwin, J.A., Burke, W.L., Robinson, L.B., and Hazard, C. 1973, Nature, 243, 336.

DISCUSSION

3raccesi: I think very high-z objects (z > 3.5) should be searched for at relatively bright magnitudes ($\sim$ 18), allowing for a sufficient sky coverage, and not trying to go extremely faint in very limited areas. This suggestion comes from the evolutionary model I presented before.

Osmer: This is a valuable suggestion. I can add that a number of plates have been taken to date with the Curtis Schmidt by Smith, and I believe with the Burrell Schmidt by Pesch, to look for bright quasars with 7 > 3.5. The results have all been negative.

P. Verón: In his survey for high redshift quasars, Osmer has found 12 quasars (and three low redshift emission line objects). The observed wavelength of the emission lines (one per object) seen on the grens plate falls into two narrow wavelength windows which correspond to the two sensitivity maxima of the IIIa-F emulsions in the used range λλ5700 - 7000 (see figure). These windows cover one-third or one-quarter

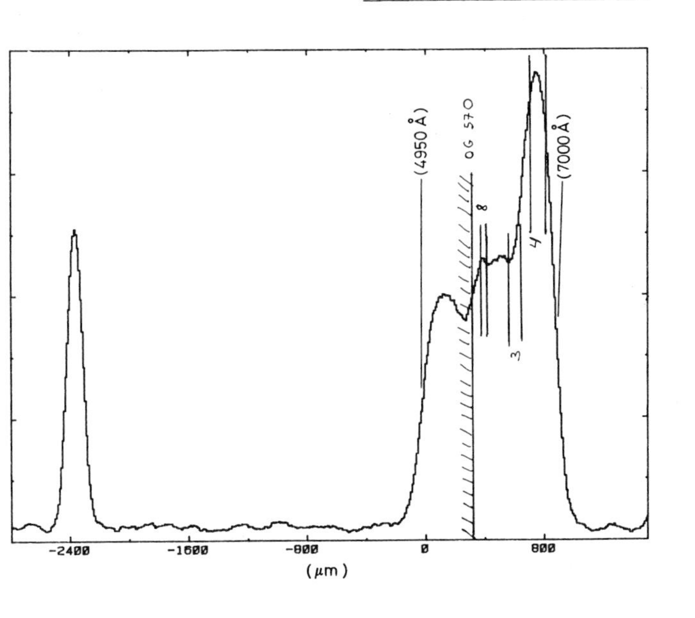

Spectrum of a red star from a grens plate taken on a IIIa-F emulsion at the prime focus of the CFH telescope in Hawaii. The cut-off due to the OG570 used by Osmer is shown as well as the three narrow wavelength windows which contain all of the emission lines found by Osmer on his own plate.

The three lines detected on a sensitivity minimum of the F emulsion are all [O III] λ5007 from low redshift galaxies; this line is probably easier to detect than broad quasar emission lines because it is usually strong and narrow. All quasar emission lines have been detected on the sensitivity maxima.

---

of the total range available. This suggests that Osmer's survey could be incomplete by as much as a factor of four, this being due to the very wavy shape of the spectral response of the IIIa-F emulsion. If this is the case, Osmer's estimate of about eighth high redshift, Lyman α quasar in his field (assuming constant space density between redshift 3.5 and 5 is reduced to two and the fact that none has been found is not significant.

Osmer: I agree that the wavy nature of the IIIa-F wavelength respons makes it a difficult emulsion to use. While it is true that the quasar emission lines were found in narrow wavelength regions, some of the galaxy emission lines (note that these particular galaxies look very similar to quasars on the grism plates) do not fall in between the main peaks. Nonetheless, Dr. Verón makes a very good point.

If we continue his argument, however, it is remarkable that so many quasars were found in the C IV line in the narrow redshift range As Drs. Peterson and Savage have pointed out, this means that the number may be higher than previously thought, which still suggests that quasars with $3.7 < z < 4.7$ should have been found. Their thesis is that the Lα emission is weak at $z > 3.5$ due to absorption cutting into the line.

Peterson: A new QSO with a z of 3.6 has been found last week in Australia by Shanks and Clowes from the University of Durham

Osmer: This reminds me of 1973 when OH 471 ($z = 3.40$) and OQ 172 ($z = 3.53$) were found within months of each other.

Savage: Have you taken a spectrum of PKS 2000 - 330 with your grism system? And can you see the Lyman α line?

Osmer: No, I have not yet been able to obtain the grism spectrum.

ENERGY DISTRIBUTION AND VARIABILITY OF BL LAC OBJECTS. THE CASES OF PKS 2155-304 AND 3C 66A.

L. Maraschi[1,2], D. Maccagni[1], E.G. Tanzi[1], M. Tarenghi[3] and A. Treves[1,2].

1) Istituto di Fisica Cosmica, CNR, Milano, Italy
2) Istituto di Fisica dell'Università, Milano, Italy
3) European Southern Observatory, Garching bei Muenchen, FRG

PKS 2155-304 was repeatedly observed in 1979 and 1980 with the International Ultraviolet Explorer. Variations up to a factor of 2 in one year and by 20% in a day are found. The maximum amplitude of variation in X-rays is similar but the timescales are much shorter (a factor of 2 in one day; Urry and Mushotzky, 1982). In all cases the 1200-3100 A continuum is well fitted by a power law with frequency spectral index α_{UV} between -0.7 ± 0.03 and -0.9 ± 0.03. Optical and ultraviolet observations taken within one day show different spectral slopes (Fig. 1). Separate power law fits in the two bands yield $\alpha_{opt} = -0.46\pm0.01$ and $\alpha_{UV} = -0.80\pm0.02$. The observations by Urry and Mushotzky indicate that the energy distribution steepens further in the soft X-ray region.

3C 66A is characterized by violent X-ray activity, its intensity varying by a factor of 10 with time-scale shorter than 6 months, possibly uncorrelated with optical variations (Fig. 2). An ultraviolet spectrum taken in 1981 August shows a featureless continuum which is best-fitted by a power law with $\alpha_{UV} = -1.79\pm0.03$, steeper than that derived from (non-simultaneous) infrared photometry, thus suggesting a special "break" in the region 10^{14} - 10^{15} Hz. The X-rays measured during the maximum intensity peak exhibit a spectral slope similar to the UV continuum, although lying a factor of 100 above its extrapolation (Fig. 3). There is an indication of a spectral steepening when the X-ray flux is low.

Two points should be stressed: i) The interpretation of the observed steepening at 10^{14} - 10^{15} Hz as due to radiative losses implies radiative lifetimes much shorter than the variability time-scale in the optical-UV band. Allowing for relativistic motion introduces a dependence of the break frequency, calculated as in Maraschi et al (1980) on δ^5 where δ is the usual Doppler factor (Konigl, 1981). High values of δ ($25<\delta<10$ for PKS 2155-304 and $\delta\simeq40$ for 3C 66A) can obviate the above difficulty. However, it is intriguing that the spectral feature under discussion occurs at a frequency similar to that of the so-called 3000 A bump in QSOs.
ii) Variability in the optical and ultraviolet is much smaller than in the X-rays over comparable time-scales, indicating that the emission regions and/or mechanisms in the two frequency ranges are not strictly related.

G. O. Abell and G. Chincarini (eds.), Early Evolution of the Universe and Its Present Structure, 39–40.

REFERENCES

Konigl, A., 1981, Ap.J., 243, 700.
Maraschi,L., Tanzi,E.G., Tarenghi,M., Treves, A., 1980, Nature, 285, 555.
Urry, C.M., and Mushotzky, R.F., 1982, Ap. J., 253, 38.

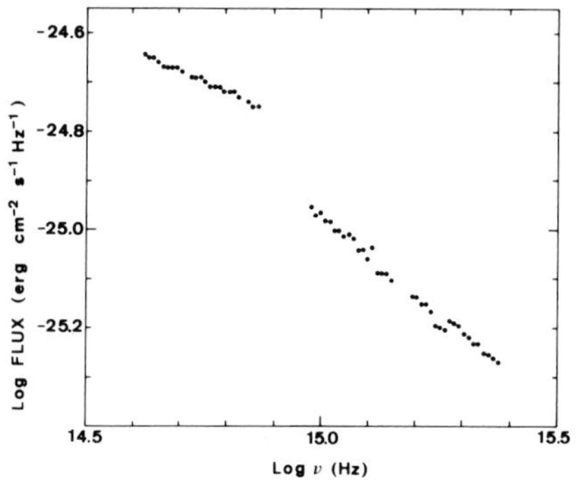

Fig. 1 - Ultraviolet and optical observations of PKS 2155-304, taken one day apart (on November 14 and 15, 1979, respectively).

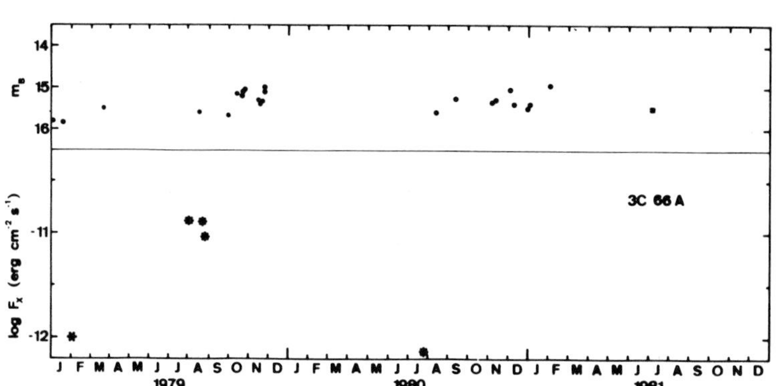

Fig. 2 - Light curve of 3C 66A in the X-rays (lower panel) and in the B passband (upper panel) (Barbieri, private communication; Pica, private communication).

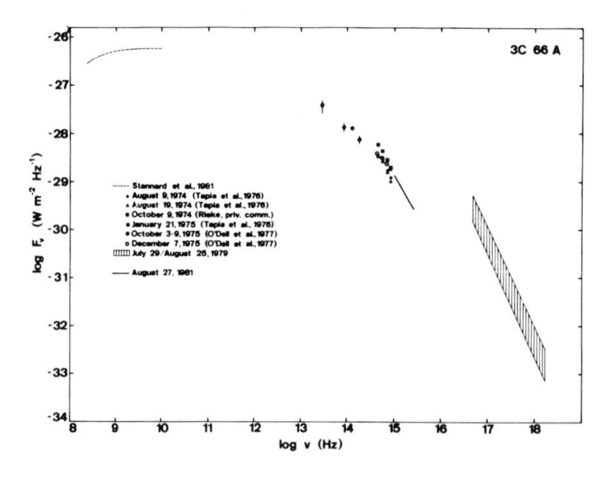

Fig. 3 - Overall energy distribution of 3C 66A obtained combining non simultaneous observations. The X-ray spectrum refers to the highest state recorded during our observations.

LONG-TERM VARIABILITY OF QUASI-STELLAR OBJECTS, AND THEIR DISTRIBUTION IN THE HUBBLE DIAGRAM

John E. Beckman and Mark R. Kidger,
Department of Physics, Queen Mary College,
Mile End Road,
London E1 4NS.
England.

Summary

A stochastic model for the energy source of QSO's is used to fit the light curves of 43 objects taken from long-period photometry (minimum duration of observations, 8 years per object). The model fits are encouraging enough to allow us to derive absolute luminosities for individual QSO's and to re-plot the Hubble Diagram with the values thus computed. We find a significantly improved fit to the expected unit slope in the plot of log z against $1/5(m_B - M_B)$, and a best fit value of $q_o = 0.1$ (± 0.4).

G. O. Abell and G. Chincarini (eds.), Early Evolution of the Universe and Its Present Structure, 41.

RADIO-SOURCE EVOLUTION AND THE REDSHIFT CUT-OFF

J.A. Peacock
Royal Observatory, Edinburgh

1. INVESTIGATING HIGH-REDSHIFT SPACE

The idea that there may be a cut-off in the distribution of quasars at high redshifts ($z \sim 4$) has been of some recent interest through the work of Osmer (1982). The observation of such an epoch of quasar creation is potentially of great importance in relation to theories of galaxy formation, but the evidence from optically-selected quasar samples remains uncertain: quite apart from the notorious problems in achieving quantifiable completeness in objective-prism surveys, any observed lack of high-redshift quasars may always be attributed to absorption either by a neutral IGM or by dust in intervening galaxies. Radio-selected samples, however, do not suffer from these problems, and this paper aims to review what studies of extragalactic radio sources can tell us about the numbers of objects at the highest redshifts.

The problem with radio cosmology is that complete redshift information is available only for sources of high flux density - there is thus an uncertainty in constructing the radio luminosity function (RLF) and its epoch dependence. However, the partial data which are available may be combined to yield self-consistent RLFs which explore the uncertainties allowed by the observational constraints. Initial efforts of this sort (e.g. Wall, Pearson & Longair 1980) assumed that the evolution with redshift had a form similar to $\rho \propto \exp(mt)$, where $m=0$ for weak sources and $m \sim 10$ for the most powerful sources (t is look-back time in units of the age of the Universe). This differential evolution fitted low-frequency source counts quite well, but was unsatisfactory in general, because the assumed arbitrary form was hard to extend to incorporate additional observations.

2. FREE-FORM EVOLUTION

A scheme to account for all flux-density/redshift data at all frequencies was produced by Peacock & Gull (1981). This assumed smooth RLFs (expanded as free series expansions) for flat-spectrum and steep-spectrum sources separately; different expansions consistent with the data were used to map out the features of the RLF which were well-constrained. This study demonstrated that strong differential evolution

G. O. Abell and G. Chincarini (eds.), Early Evolution of the Universe and Its Present Structure, 43–45.

applied for both spectral classes. To study the behaviour of these RLFs at high redshift, Figure 1 shows the analogue of the simple exp(mt) law - cuts through the RLFs at P(2.7 GHz) = 10^{27} $WHz^{-1} sr^{-1}$, a typical quasar luminosity. These are plotted against t for 4 models - two different values of q_o, with and without an imposed cut-off at z=5. In each case the upper lines are for the steep-spectrum RLF, the lower for flat-spectrum.

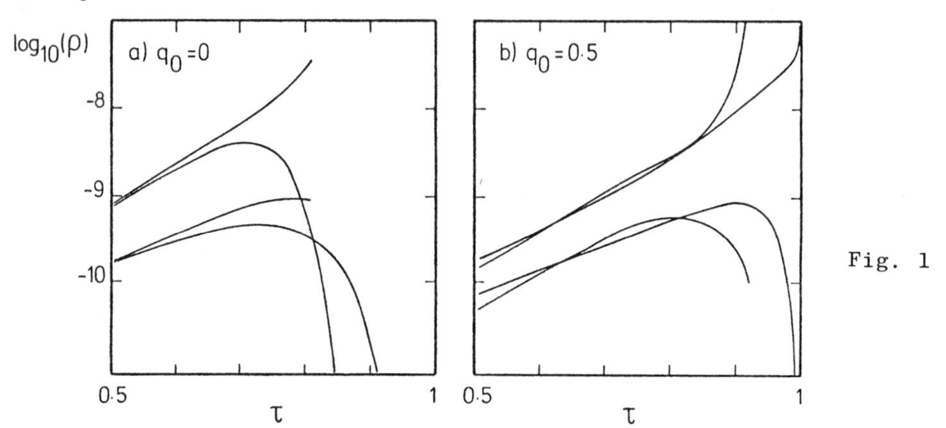

Fig. 1

We see from this that although there is no evidence for a decrease in the rate of evolution at high redshift for the steep-spectrum population in every case the flat-spectrum RLF has ceased to evolve by z ∿ 3. This result may seem uncertain due to the difference between various RLFs, but there is support for its reality from data on flat-spectrum quasars from the Parkes ±4° sample studied by Wills & Lynds (1978). In applying the V/V_m test to quasars with limiting redshifts less than and greater than 1.8, they found $\langle V/V_m \rangle = 0.65$ for the nearer subset, but $\langle V/V_m \rangle = 0.52$ for the more distant one - confirming the slackening of evolution at high redshift.

3. FUTURE PROSPECTS

Our knowledge of radio-source evolution is advancing on two fronts - through the gathering of new redshift data on faint sources, and through the synthesis of all new results to produce new evolving RLFs. The Peacock & Gull analysis is presently being extended and we expect on the basis of the results given here that the redshift cut-off for flat-spectrum sources will be confirmed. As yet, the high-redshift evolution of steep-spectrum sources is uncertain, but it is simply a matter of time before this is resolved.

REFERENCES

Osmer, P.S., 1982. Astrophys. J., 253, 28.
Peacock, J.A. & Gull, S.F., 1981. Mon.Not.R.astr.Soc., 196, 611.
Wall, J.V., Pearson, T.J. & Longair, M.S., 1980. Mon.Not.R.astr.Soc., 193,683.
Wills, D. & Lynds, R., 1978. Astrophys. J. Suppl., 36, 317.

DISCUSSION

Peterson: Do you use simple power-law spectra for the radio sources in your models?

Peacock: Yes, we do. A spread of spectral indices makes no important difference to the predictions. Also, I do not believe that spectral curvature matters unless you want to go to $z \sim 10$, which is highly uncertain anyway.

Peterson: How well can your models predict counts at $z > 3.5$ when they are based on samples with maximum redshifts of 2.6 that show evolution?

Peacock: The whole point is that a luminosity function which evolves as fast at $z \sim 3$ as at $z \sim 2$ would predict many ultrahigh redshift sources which are not observed, so that the evolution must have turned off by $z \sim 3$. Higher redshifts are more uncertain; obviously, it will be a long while before we can delineate accurately any actual turn-over in the luminosity function.

Segal: The chronometric cosmology predicts that objects of spectral index α will appear relatively abundant at redshifts $\sim(1-\alpha)^{-1}$ but cut off observationally at somewhat larger redshifts. Isn't it possible that the cut-off you observe in flat-spectrum sources derives in part from this effect and in part from the general cut-off at larger redshifts predicted by the chronometric cosmology, and that physically there is in reality no evolution of these sources?

Peacock: It is clear that studies like this cannot tell you about the geometry of the universe -- it must be assumed in order to obtain an evolving luminosity function. Now, Occam's razor tells me that there is no point in working with any geometry but the simplest one which satisfies all known tests -- to me this is general relativity. I agree that this would not be so if the chronometric cosmology could explain the data without evolution -- but it cannot. If the Hubble D-Z relation is taken, low luminosity sources do not evolve, while they would evolve if the chronometric D-Z relation were taken.

A MODEL FOR THE COSMOLOGICAL EVOLUTION OF RADIO SOURCES

C.R. Subrahmanya and V.K. Kapahi
Tata Institute of Fundamental Research
Post Box 1234, Bangalore 560012, India.

The existing schemes to express the evolution of radio sources are either too simple to fit all the data (multi-frequency counts, identifications, spectra etc.) or have too many parameters and little predictive value. They do not also fit the identification statistics at different flux-densities (Swarup, Subrahmanya & Kapahi 1982, in 'Astrophysical Cosmology', Proc. Vatican Study Week) and the counts from recent deep surveys (van der Laan 1982, in 'Astrophysical Cosmology'). Here, we report a preliminary attempt to derive a model that retains the simplicity of a small number of parameters and ease of refinement when new data are added.

Following Peacock & Gull (1981, MNRAS 196, 611), we consider two spectral populations - 'flat' and 'steep' - taking $\alpha = 0$ for the former and a P-α correlation for the latter. For both types, evolution is introduced only for $P_{0.4} > 10^{24.5}$ W/Hz/sr (H_o = 50 km/s/Mpc) and is expressed as $\exp[\beta(P)(1-t/t_o)]$, where $\beta \propto \log P$ with slopes β_1, β_2 and β_3 in the $\log P_{0.4}$ ranges 24.5-26, 26-27 and 27-28 respectively, and $1.5\beta_1+\beta_2+\beta_3$ for $\log P \geq 28$. The luminosity ranges were chosen considering that the correlation between radio and optical luminosities is important only for $\log P \lesssim 24.5$ (Auriemma et al. 1977, A&A 57, 41) and quasars generally have $\log P \gtrsim 26$.

In our method, the radio luminosity function (RLF) is first determined for 'flat' sources using flat-spectrum counts at 2.7 and 5 GHz and the observed luminosity distribution(LD) for the $S_{2.7} \geq 1.5$ Jy sample of Peacock & Wall (1981, MNRAS 194, 331). An initial set of parameters is then found for 'steep' sources to fit the total counts at 0.4 GHz and LD for $S_{0.4} \geq 10$ Jy, which is well determined in the log P range 24.5 to 28. Observational estimates of local luminosity function (LLF) are available for lower luminosities (Fig. 1) but, for very low luminosities (spirals and irregulars), LLF is very uncertain.

After determining the LLF and an initial RLF, the parameters are refined to simultaneously fit (i) total counts at 0.4, 1.4, 2.7 and 5 GHz; (ii) spectrally separated counts at 2.7 and 5 GHz; and

G. O. Abell and G. Chincarini (eds.), Early Evolution of the Universe and Its Present Structure, 47–48.

(iii) percentage identification - flux density relation for galaxies on the PSS prints (predicted using the bivariate luminosity function as in Swarup, Subrahmanya & Venkatakrishna 1982, A&A 107, 190). We could obtain satisfactory models for $q_o = 0$ and 0.5 both with and without a cutoff in z, although the fit was considerably poorer for $q_o = 0$ in the absence of a redshift cutoff. The predicted counts are compared with observations in Fig. 2 for the model with $q_o = 0.5$ and $z_c = 3.5$ with the following parameters:

'Flat': β_1=4.0, β_2=1.5, β_3=0.0, $\alpha = 0$
'Steep': β_1=3.8, β_2=4.5, β_3=4.0, $\alpha = 0.7$ for $\log P \leq 24.5$ and $0.7+0.08(\log P-24.5)$ otherwise.

The fit with all the data is remarkably good in spite of the limited number of parameters.

Our results show that it is necessary to assume evolution for both flat and steep spectrum sources, although evolution for 'flat' sources could be somewhat milder. The recently observed flattening of the normalised counts at $S_{1.4} \lesssim 1$ mJy is well fitted by the adopted LLF at $\log P \leq 22$, which implies that most of the sources at these flux levels are local ($z < 0.1$). Should the redshifts turn out to be larger, evolution with epoch would be indicated for a source population of low luminosities.

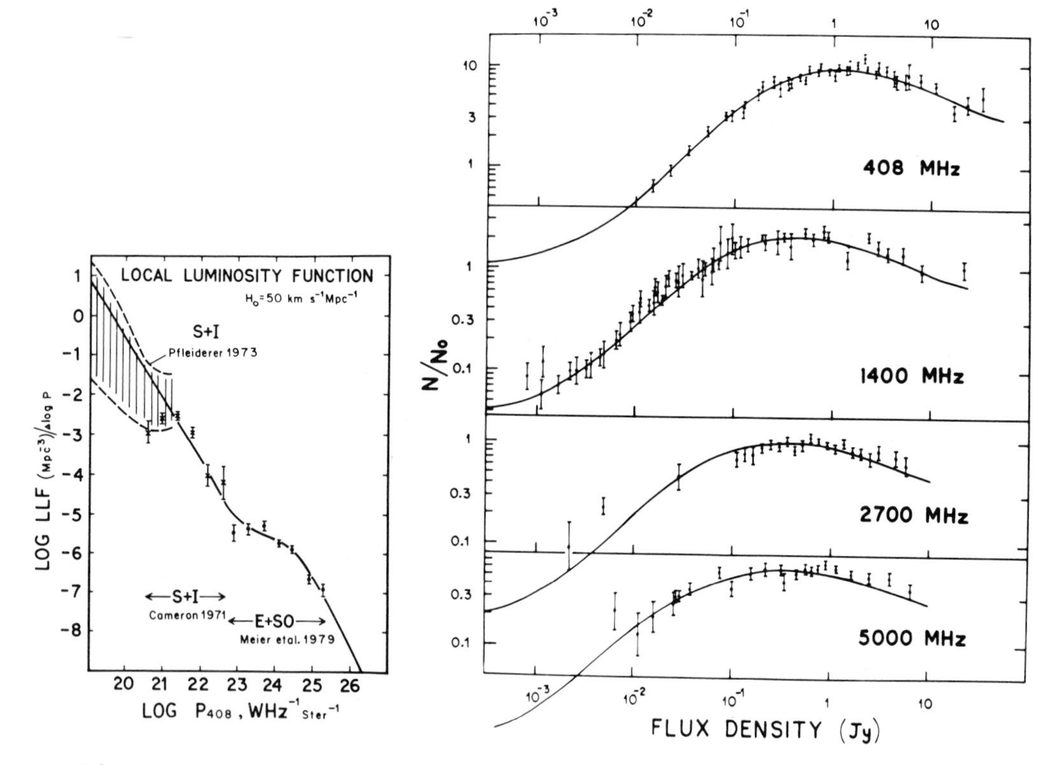

Figure 1.

Figure 2. Source counts.

QUASAR GALAXIES: TWO-DIMENSIONAL IMAGE DECONVOLUTIONS

P. A. Wehinger and S. Wyckoff
Physics Department, Arizona State University
T. Gehren
Max Planck Institut für Astronomie, Heidelberg
H. Spinrad
Astronomy Department, University of California, Berkeley

Studies of quasar images which have adequate spatial resolution and reach sufficiently faint surface brightness levels indicate that virtually all low redshift ($z \lesssim 0.6$) quasars are surrounded by faint nebulosities extending $\sim$ 3-20 arcsec from the quasar nucleus (at 26 R mag arcsec^{-2}) (Wyckoff *et al.* 1980, 1981, Hutchings *et al.* 1981, Wehinger *et al.* 1983). Furthermore, the average integrated absolute magnitude and average metric diameter of the quasar nebulosities (quasar nucleus removed) are roughly those expected for galaxies at the corresponding (cosmological) quasar distances. Moreover, statistical support for the cosmological interpretation of the redshifts as well as the galaxy interpretation of the fuzz was found in correlations between the angular isophotal diameters of the quasar nebulosities and the redshifts, and between the integrated apparent magnitudes and the angular isophotal diameters (Wyckoff *et al.* 1981). Spectroscopic observations of quasar fuzz now convincingly support the galaxy interpretation for the quasar nebulosities (Boroson and Oke 1982, Oke *et al.* 1983).

Calibrated surface photometry of 40 quasars (34 radio and 6 optically selected) and 7 BL Lac objects have been analyzed (Wehinger *et al.* 1983). Isophotal maps to $\sim$ one percent the surface brightness of the night sky ($\sim$ 26-27 R mag arcsec^{-2}) were derived for each quasar image, a star in each field (defining the point spread function), and for the quasar nebulosities after a two-dimensional deconvolution of the nucleus. Essentially all quasars with $z \lesssim 0.4$ were found to be resolved for these prime-focus photographic observations which were obtained with the 3.6-m ESO and the 4-m CTIO telescopes. The angular diameters and integrated magnitudes of the deconvolved images are for this larger sample of objects again entirely consistent with the hypothesis that quasars are the luminous nuclei of distant galaxies.

The present sample includes quasars in the redshift range $0.1 < z < 2.5$ with three quasars having $z \gtrsim 1$. None of these high redshift quasars was resolved to the detection levels achieved. This

G. O. Abell and G. Chincarini (eds.), Early Evolution of the Universe and Its Present Structure, 49–50.

result is also consistent with the galaxy nucleus interpretation of quasars. A detailed report of this work will appear in *The Astrophysical Journal*.

REFERENCES

Boroson, T. and Oke, J.B. 1982, *Nature*, *296*, 397.
Hutchings, J., Crampton, D., Campbell, B. and Pritchet, C. 1981, *Ap.J.*, *247*, 743.
Oke, J.B., Boroson, T. and Green, R. 1983, *Ap.J.*, (in press).
Wehinger, P.A., Wyckoff, S., Gehren, T. and Spinrad, H. 1983, *Ap.J.*, (in press).
Wyckoff, S., Wehinger, P.A., Spinrad, H. and Boksenberg, A. 1980, *Ap.J.*, *240*, 25.
Wyckoff, S., Wehinger, P.A. and Gehren, T. 1981, *Ap.J.*, *247*, 750.

QSO LUMINOSITIES AT λ 1 MM

W.A. Sherwood
Max Planck Institute for Radio Astronomy

The results discussed here are based on observations at λ 1 mm of 36 optically selected quasars (Sherwood et al., 1982; in preparation) and of 24 flat spectrum radio sources with S_ν (5 GHz) $\geq$ 1 Jy (Kühr et al., 1979, 1981). H = 50 km/(s Mpc) and q_o = 0 are used.

In Figure 1 we compare the power for sources detected at $\geq$ 3σ in the raw data at λ 1 mm (dots) with other objects similarly detected in X-rays (crosses) (Ku et al., 1980; Zamorani et al., 1981). The millimeter and X-ray samples are based on optical and radio selection criteria. Radio selected quasars tend to have intermediate redshifts and luminosities relative to the optical samples. X-ray selected quasars have a mean redshift of only 0.42 (Margon et al., 1982). There may be low mm luminosity QSOs at high redshifts but there appear to be no high luminosity QSOs at low redshift. Nearly three-quarters of the sources (28) in

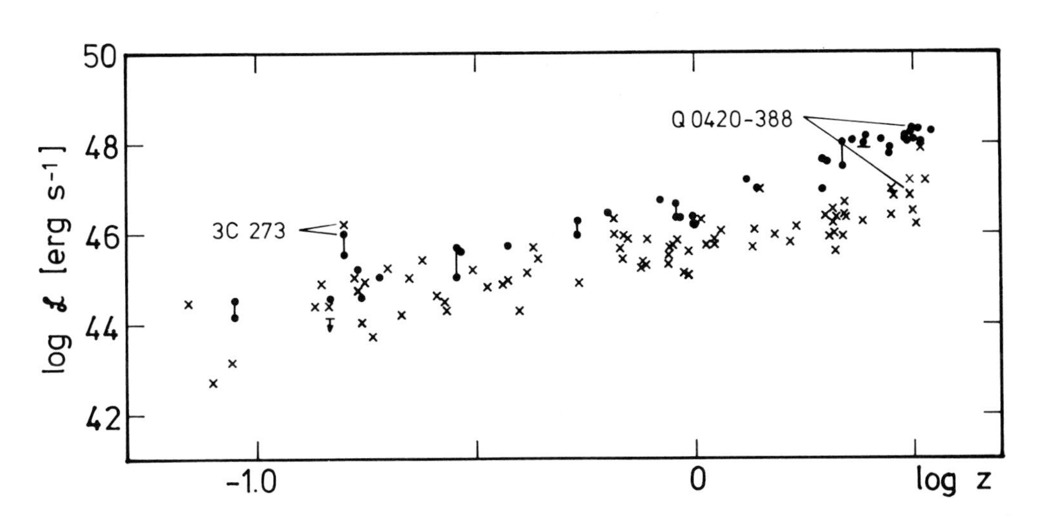

Figure 1. The power of quasars at λ 1 mm (dots) and in X-rays (crosses) as a function of redshift, z.

G. O. Abell and G. Chincarini (eds.), Early Evolution of the Universe and Its Present Structure, 51–54.

this millimeter sample are more luminous than 3C 273. At high redshifts QSOs emit more power near λ 1 mm than they do at X-ray wavelengths. The difference decreases toward lower redshifts and eventually reverses, i.e., more power in X-rays than at λ 1 mm. A millimeter survey will be of great interest to see how frequently such high luminosity objects occur.

The rate of detection of quasars at λ 1 mm is a function of selection criteria and redshift. As seen in Table 1 the detection rate at the 3σ flux density level of optically selected quasars increases with redshift while that for radio selection declines. This is also seen for the independently selected radio sample observed by Jones et al. (1981). Both radio and optical samples are flux density (apparent magnitude) limited.

Finally, the millimeter to X-ray luminosity ratio may be directly compared in Figure 2. The ratio $\mathcal{L}(300\ \mathrm{GHz})/\mathcal{L}(\mathrm{XR})$ is nearly independent of z except for a K-term of the form

$$(1+z)^{(\alpha(\mathrm{XR})\ -\ \alpha(300))}$$

if $\alpha(\mathrm{XR}) \neq \alpha(300)$. For the present data reasonable errors in the spectral indices do not explain the appearance of Figure 2. The complete lack of simultaneous observations of these sources, some of which are obviously variable at one or both frequencies, is the major source of error.

Nevertheless, for a ratio which ought to be nearly independent of z there is a remarkably good regression. Selection effects cannot be ruled out until a millimeter survey has been undertaken. The distribution may exist only among the most luminous quasars or there may be parallel sequences for quasars of lower luminosity at any epoch.

For comparison the data for four BL Lac objects having some kind of redshift estimates are represented in the figure by triangles. These

Table 1: Statistics of Detections (3σ Flux Density Level) at λ 1 mm

$z < 1.00$ (assuming all BL Lac objects have $z < 1.00$)		
20 of 30 sources:		
selection:	optical (mag)	7 of 12
	radio	13 of 18
(cf Jones et al. (1981):	radio	9 of 14)
$1 \leq z < 3.00$		
11 of 20 sources:		
selection:	optical (mag)	11 of 14
	radio	0 of 6
(cf Jones et al.:	radio	0 of 1)
$z > 3.00$		
7 of 10 sources:		
(The other three have $\geq$ 3σ detections in the raw data)		
(selection: optical emission line redshift)		

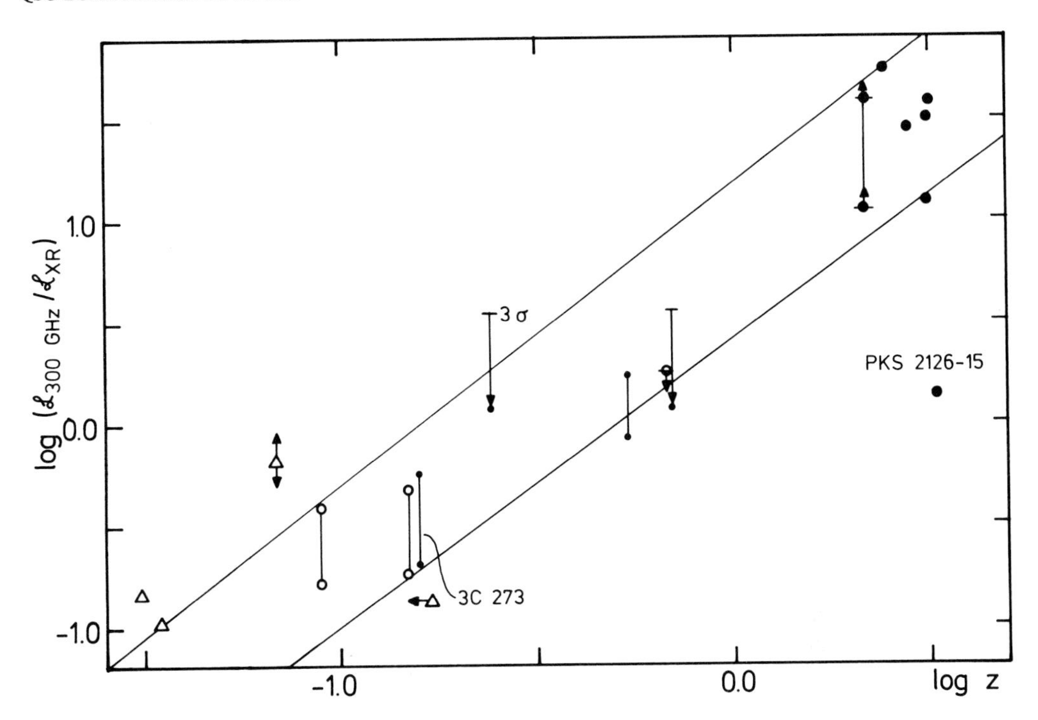

Figure 2. The ratio of millimeter to X-ray power vs. z
The large filled and open circles represent quasars selected for the presence of emission lines and UV excess respectively. The points are for radio selected quasars except the one marked by the 3σ upper limit-it represents an X-ray discovered quasar (Chanan et al., 1981) with a 1σ detection at λ 1 mm.

redshifts appear to agree with the mean relation to within a factor of ∿ 2.

Summary

1) the luminosity of quasars observed at λ 1 mm can exceed 10^{48} erg s^{-1} ($H = 50$ km/(s Mpc), $q_o = 0$
2) at redshifts $z > 0.5$ the upper envelope of the millimeter luminosity distribution exceeds that of the X-ray distribution
3) a millimeter survey may lead to the detection of very large redshift quasars
4) there is a good correlation between redshift and the ratio of 1 millimeter luminosity to X-ray luminosity. There is certainly no global millimeter to X-ray luminosity ratio.

This work has been supported by DFG-SFB 131, Radioastronomie. The observations at λ 1 mm were made at the European Southern Observatory. I would like to thank all who helped with the observing, especially Drs. E. Kreysa and H.-P. Gemünd. Drs. P. Biermann and J. Schmid-Burgk suggested improvements to the paper.

REFERENCES

1) Chanan, G.A., Margon, B., Downes, R.A.:
1981, Astrophys. J. 243, L5

2) Jones, T.W., Rudnick, L., Owen, F.N., Puschell, J.J., Ennis, D.S., Werner, M.W.:
1981, Astrophys. J. 243, 97

3) Ku, W.H.-M., Helfand, D.J., Lucy, L.B.:
1980, Nature, 288, 323

4) Kühr, H., Nauber, U., Pauliny-Toth, I.I.K., Witzel, A.:
1979, M.P.I.f.R. preprint No. 55

5) Kühr, H., Witzel, A., Pauliny-Toth, I.I.K., Nauber, U.:
1981, Astron. Astrophys. Suppl. Ser. 45, 367

6) Margon, B., Chanan, G.A., Downes, R.A.:
1982, Astrophys. J. 253, L7

7) Sherwood, W.A., Schultz, G.V., Kreysa, E., Gemünd, H.-P.:
1982, Extragalactic Radio Sources
D.S. Heeschen and C.M. Wade (eds.), (Reidel, Dordrecht) pg 305

8) Zamorani, G., Henry, J.P., Maccacaro, T., Tananbaum, H., Soltan, A., Avni, Y., Liebert, J., Stocke, J., Strittmatter, P.A., Weymann, R. J., Smith, M.G., Condon, J.J.:
1981, Astrophys. J. 245, 357

QSO REDSHIFT LIMIT AND PERIODICITY IN A FIB UNIVERSE

Jeno M. Barnothy and Madeleine F. Barnothy
833 Lincoln St., Evanston, IL 60201

The FIB cosmology was conceived half a century ago, to explain the origin and spectrum of the ultrahigh energy cosmic ray particles we observed in deep mines. It succeeded in explaining quantitatively the 3°K microwave background radiation (IAU Symposium No. 44), the redshift periodicity (PASP 1976), and the cutoff of QSO's beyond z=3.5 (AAS 1982).

The FIB universe corresponds to the $\dot{R}=\ddot{R}=0$ solution of the Friedman equations, with radiant energy in the form of neutrinos dominating over matter. The perfect cosmological principle is valid. Length, time and energy quantities of signal carriers increase relative to the same quantities of elementary particles. The perfect cosmological principle demands that all three quantities change according to the same exponential function of travel time: $\omega/2H_o$. The spin of photons and neutrinos changes as $s=s_oe^{\omega}$; hence, $\omega=2\ln(1+z)$.

When the photon spin approaches 1.5, it decays into a photon of spin 1 and a neutrino, the latter carrying 1/3d of the energy of the decaying photon. The ensuing periodic drop in apparent brightness of QSO's causes the periodicity in their ln(1+z) histogram. The power spectrum of the histogram of seven complete samples of 440 QSO's displays this periodicity at a peak power of 29. The energy within a photon beam decreases as 1+z. The apparent magnitude of an object of M absolute magnitude is $m=5 \log 1/2 \sin 2\ln(1+z)+43+M+2.5 \log(1+z)$. This function implies that at high redshifts the slope of the Hubble plot of QSO's should reverse its sign, as this is indeed seen in four independent samples.

The histogram of lensed Seyfert galaxy nuclei (A.J. 1965) is $n=(\omega-1/2\sin 2\omega)\sin^2\omega\Delta\omega$; n becomes zero at the antipode, simulating a cutoff around z=3.81. The histogram matches that of the Osmer-Smith sample. Objects could be visible from two opposite directions.

Neutrinos which after several round trips have reached energies in excess of 10^9 eV and arrive focused upon their parent galaxy, supply, in the form of their relativistic mass, the invisible mass of galactic halos, the "missing mass." Transformation of radiant energy into rest mass, and vice versa, regulates R_o and insures the stability of the Einstein static universe.

For further details, see the 14 pages of "notes" distributed at the Symposium. Copies will be supplied upon request.

G. O. Abell and G. Chincarini (eds.), Early Evolution of the Universe and Its Present Structure, 55.

UNBIASED SEARCHES FOR QUASARS BEYOND A REDSHIFT OF 3.5

Ann Savage
Royal Observatory, Edinburgh.

Bruce A. Peterson
Mount Stromlo and Siding Spring Observatories
Australian National University

1. INTRODUCTION

Nearly a decade elapsed between the discovery by Wampler at al. (1973) that the QSO OQ172 has a redshift of 3.53 and the discovery by Peterson et al. (1982) that the QSO Pks 2000-330 has a redshift of 3.78. During that time, radio and optical searches were vigorously pursued to find QSO's with redshifts greater than 3.5, but none were found. In this paper, we discuss selection effects in optical and radio searches and show how these selection effects have limited the redshift range of previous surveys. We propose a combination of radio and optical techniques that may be used to find high redshift QSO's and provide us with an undistorted view of the Universe beyond a redshift of 3.5.

2. RADIO SELECTION EFFECTS

Fig. 1 shows a schematic radio spectrum of a source with an extended power law component and a compact component. The characteristic synchrotron self absorption peak is marked as ν_m. Below this frequency, the synchrotron source is opaque, and above this frequency it is optically thin. In Fig. 2, all the known radio sources with redshifts greater than 3 are shown. Each of these sources has a spectrum that shows a synchrotron self absorption peak ν_m and each source has a steep spectrum at frequencies greater than ν_m. If these sources were seen at higher redshifts, their observed spectra would be shifted to the left toward lower frequencies and down toward lower flux densities. They would then appear as weak radio sources with steep spectra between 2.7 and 5 GHz.

Fig. 3 gives the number magnitude distributions of complete samples of flat and steep spectrum QSO's from the survey of Fanti et al. (1979). The number of steep spectrum QSO's increases right up to the plate limit. The flat spectrum QSO's in this sample show an apparent magnitude cut-off and we interpret this to be caused by the

G. O. Abell and G. Chincarini (eds.), Early Evolution of the Universe and Its Present Structure, 57–64.

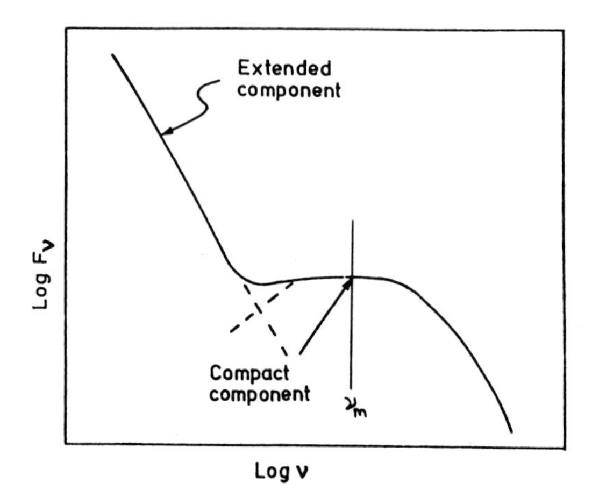

Figure 1 Schematic spectrum of a radio source with both extended and compact components.

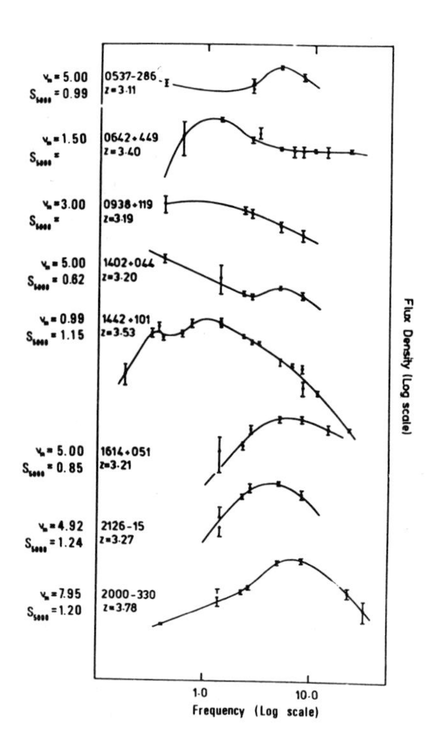

Figure 2 Radio spectra for all the known radio sources with Z > 3, showing pronounced synchrotron self absorption peaks.

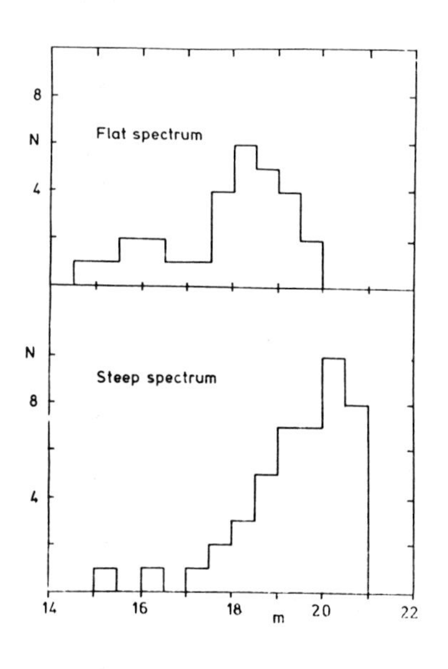

Figure 3 Number magnitude distributions for complete samples of flat and steep QSOs (Fanti et al. 1979), the N(m) for the steep spectrum QSOs continues to rise to the plate limit, while that for the flat spectrum QSOs peaks at $18^{m}.5$.

change in the form of the radio spectrum from flat to steep as the redshift increases. In our interpretation the missing faint, flat spectrum QSO's are at high redshifts and are observed as weak steep spectrum radio sources in the 2.7 to 5 GHz radio frequency window. The peak optical magnitude of the flat spectrum radio sources is found to be independent of the flux limit of the radio samples. Thus efforts to identify weak flat spectrum sources in deeper radio surveys will not find higher redshift QSO's, but will only sample the low luminosity portion of the luminosity function.

Spectral index distributions of complete samples of sources selected at 5 GHz (Condon and Ledden 1981) are shown in Fig. 4. The samples of strong sources ($S_5 > 600$mJy) show a double peaked spectral index distribution. The narrow peak at $\alpha = 0.75$ is characteristic of the electron energy distribution in optically thin synchrotron sources. The broad peak at $\alpha = 0.00$ is due to the "flat" synchrotron self absorbed sources. The samples of weak sources ($S_5 > 15$mJy) will be a mixture of sources at all redshifts, but among them may be some sources at large redshift. At high redshift the peak in the synchrotron self absorption spectrum is shifted to lower frequencies and the source is observed as an optically thin source with $\alpha = 0.75$. Thus the lack of weak flat spectrum sources in high frequency samples can be attributed to the radio spectrum shape at high redshifts, however spectral index distributions between 408 MHz - 5 GHz also show decreasing numbers of flat spectrum sources at low flux densities (Wall and Benn 1981) and the selection effect hypothesized here may only have a small effect.

3. OPTICAL SELECTION EFFECTS

Fig. 5 shows the (redshift independent) velocity widths (FWHM) of Ly α emission lines as a function of redshift for optically selected QSO's (●) (MacAlpine and Feldman 1982) and for 5 of the high redshift radio QSO's in Fig. 2 (▼). It can be seen that the Ly α emission line is becomming increasingly narrower at high redshift. We interpret this as due to increasing numbers of Ly α absorption lines cutting into the short wavelength side of the Ly α emission line. Our examination of the observed equivalent widths of the Ly α emission lines of the optically selected QSO's in Fig. 2 showed that in the redshift interval 1.8 to 2.8, the observed equivalent widths increased nearly as 1+Z, corresponding to an almost constant emitted Ly α flux for the QSO's found optically in this redshift range. However in the redshift range 2.8 to 3.2, the emitted Ly α flux dropped, consistent with Ly α absorption by intervening clouds which increases rapidly with increasing redshift (Peterson 1978, 1983). If the unabsorbed Ly α flux follows the relation shown by the line in Fig. 5, then the unabsorbed Ly α flux emitted by QSO's seen at a redshift of 4 would be 0.65 mag less than the unabsorbed Ly α flux emitted by QSO's seen at a redshift of 3, and the higher redshift QSO's would be systematically harder to identify

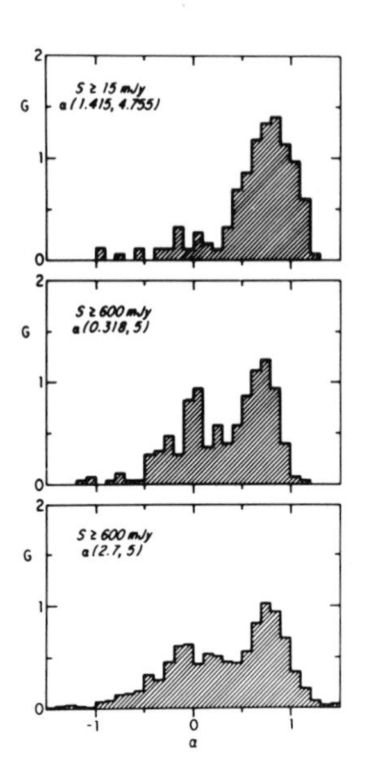

Figure 4 Spectral Index distributions of complete samples of sources at 5 GHz (Condon and Ledden 1981) showing the decrease in flat spectrum sources as the survey flux limit is decreased. These effects (Fig 3 and Fig 4) could be due to the shape of radio spectra of the high Z radio QSOs.

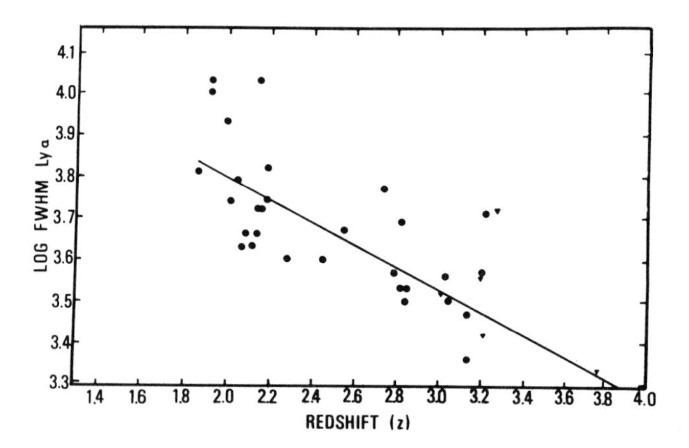

Figure 5 Redshift independent velocity widths (FWHM) of Ly α as a function of Z for some optical (●) (MacAlpine and Feldman 1982) and radio (▼) QSOs. The velocity widths of Ly α decreases by a factor of 2 for a redshift change from Z = 3 to Z = 4.

on prism plates.

Figures 6a and 6b show the optical spectra of two QSO's; (a) is a QSO with a redshift of 2.44, (b) is Pks 2000-330 with a redshift of 3.78. The prism spectrum insert on both of these figures is reproduced from objective prism spectra obtained with the UK Schmidt Telescope. Fig 6a was taken with the low dispersion prism (2200 A/mm) and has a resolution of 60 A. The IIIa-J emulsion covered a wavelength range of 3200 to 5400 A. Fig 6b was taken with the medium dispersion prism (2000 A/mm) and has a resolution of 50 A. The IIIa-F emulsion and filter combination covered the wavelength range 5700 to 6900 A. The emulsion sensitivity with wavelength is given for each emulsion. It can be seen that the sensitivity response of the IIIa-F emulsion is not smooth, and that the response drops by 0.5 mag between 6000 A and 6500 A. In addition the contrast of the IIIa-F emulsion drops by 30% from a peak at 6050 A to a low at 6900 A (Emerson, private communication).

The Ly α emission feature of the high redshift QSO (Fig 6b) is nowhere near as prominent as that for the low redshift QSO (Fig 6a). The observing conditions are the same for both QSO's except for emulsion type.

In Fig 6b, the spots on the IIIa-F emulsion sensitivity curve correspond to the wavelengths of the C IV 1549 emission lines used to identify the high redshift QSO's found in the CTIO 4m grism survey (Osmer 1982). The wavelengths of these lines are near the sensitivity peaks of the IIIa-F emulsion, and no C IV lines were detected in the range where the emulsion response drops by 0.5 mag. The failure of the CTIO 4m grism survey to identify any high redshift QSO's by their Ly α emission line may be due to a combination of the restricted wavelength region of peak emulsion sensitivity (and hence the smaller volume of space that was searched) and the systematic weakening of the flux in the Ly α line with increasing redshift.

4. CONCLUSIONS

In order to properly explore the space distribution of QSO's at redshifts greater than 3.5, we need to

(i) Investigate the redshift distributions of radio QSO's drawn from a 100 mJy sample at 2.7 GHz comprising both "flat" and "steep" spectrum radio QSO's that have been identified from accurate radio interferometer positions without regard to their optical morphology.

(ii) Conduct optical searches for high redshift QSO's with a sensitive wide field detector that has uniform wavelength sensitivity in combination with a grism that has 50 A resolution.

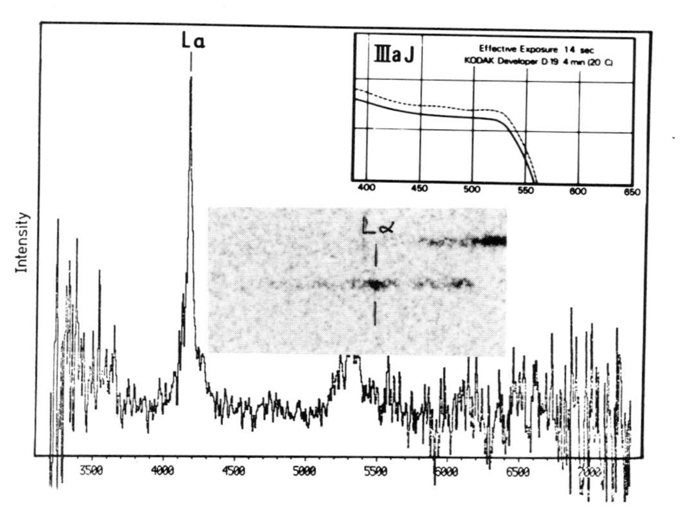

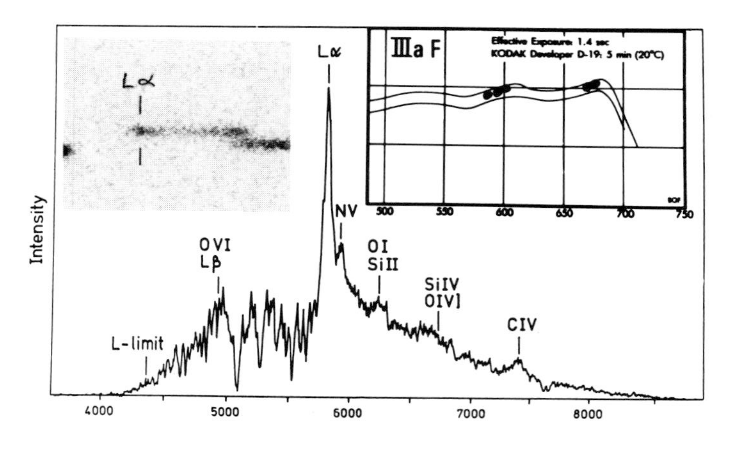

Figures 6(a) and (b) Optical spectra of two QSOs (a) SGP 0048-298 Z = 2.44 (b) PKS 2000-330 Z = 3.78. The prism spectra from the UKST objective prism plates are inserted. Ly α is not easily seen in the prism spectrum of PKS 2000-330 due to the variation in sensitivity response of the IIIa-F emulsion with wavelength and the loss of Ly α emission line flux due to increasing numbers of Ly α absorption lines.

. REFERENCES

ondon, J.J. and Ledden, J.E.: 1981, Astron.J., 86, 643.
anti, R., Feretti, L., Giovannini, G., and Padrielli, L.: 1979, Astron. Astrophys., 73, 40.
lacAlpine, G.M. and Feldman, F.R.: 1982, (preprint).
)smer, P.S.: 1982, Astrophys.J., 253, 28.
eterson, B.A.: 1978, in IAU Symp. No. 79, 389.
eterson, B.A.: 1983, in IAU symposium No. 104, (this issue), p. 349.
eterson, B.A., Savage, A., Jauncey, D.L., Wright, A.E.: 1982, Astrophys.J. (Letters) 260, L27.
all, J.V. and Benn, C.R.: 1981, in IAU Symp. No. 97, 441.
ampler, E.J., Robinson, L.B., Baldwin, J.A. and Burbidge, E.M.: 1973, Nature, 243, 336.

DISCUSSION

hklovsky: Part of the form of the spectra of QSOs can be explained by means of the relativistic Doppler effect (Scheuer and eadhead). In this case your argument must be changed.

avage: Since we find that half of the high redshift quasars have peaked spectra, at least for these sources, relativistic eaming does not smooth out the spectra. VLBI observations of peaked pectrum sources indicate that they generally are double sources with early equal component flux densities and no evidence for superluminal otion. Perhaps flat spectra sources consist of components with peaked pectra which are spread out by Doppler shifts.

hompson: Automatic measuring machines such as COSMOS have the ability to remove wavelength sensitivity from photographic objective rism surveys. Since the optical selection effects which you discuss an be partially removed by correcting for the wavelength sensitivity, o you plan to employ such automated techniques to analyze any of your urvey data?

avage: There is "automatic quasar detection" software written for COSMOS by Clowes, Cooke and Beard (see their poster paper t this symposium for recent results) which has already been used to elect quasars from the UK Schmidt telescope low dispersion objective rism plates taken on the IIIa-J emulsion. Parameters are input to llow for the variation of emulsion sensitivity variation with wavelength. Thus, this program can easily be used to search the UK Schmidt edium dispersion objective prism plates which are taken on IIIa-F mulsion. We hope to start such a search this October.

Baldwin: The radio selection effect you describe certainly affects a number of sources, but I think the number is very small at low flux densities. In a 20-MJy sample at 5 GHz of 100 sources, for which spectra extend down to 150 MHz, the number which show this effect is perhaps only one.

Savage: The survey you refer to is over a small area of 25 square degrees. The recent VLA mapping at 1415 MHz and 5 GHz of the PKS 2700 MHz deep survey regions (200 square degrees, 600 sources t 100 MJy) referred to by Peacock should give us a clearer picture of the percentage of "peaked" spectra sources in such a survey.

HE PROBLEM OF THE REDSHIFTS

Geoffrey Burbidge
Kitt Peak National Observatory*

It is my intention in this brief paper to present once again some f the evidence which appears to show that a major contribution to the edshifts of QSOs and perhaps some galaxies is not associated with the xpansion of the universe. The evidence continues to grow and it is of our kinds:

(a) Evidence concerning galaxies with discrepant redshifts.

(b) Connections between QSOs and galaxies.

(c) Statistical information concerning the association of QSOs with galaxies.

(d) Individual and very striking examples of pairings between QSOs and bright galaxies.

If any part of this evidence is accepted it means that we cannot rust the redshifts of QSOs and perhaps some galaxies as distance ndicators. It would suggest that the majority of the QSOs at least are o further away than ~ 100 Mpc. In turn this means that:

(1) Evidence for the strong evolution of QSOs as a function of epoch evaporates.

(2) Absorption in QSOs cannot be due to intervening galaxies and intergalactic clouds.

(3) Gravitational lenses as explanations of pairs of QSOs with identical redshifts cannot be correct.

Operated by the Association of Universities for Research in Astronomy, nc., under contract with the National Science Foundation.

. O. Abell and G. Chincarini (eds.), Early Evolution of the Universe and Its Present Structure, 65–71.

(4) Superluminal motions and certainly those requiring highly relativistic velocities don't occur.

Three years ago at the IAU Symposium No. 92 in Los Angeles (1) I discussed much of the evidence that I shall describe today. More evidence is available, but as far as I am aware none of the evidence given there and elsewhere (2,3) has been refuted. Consequently while I shall describe some of it again I shall be brief.*

GALAXIES WITH DISCREPANT REDSHIFTS

In this category of evidence we have small groups of galaxies such as Stefan's Quintet and VV172 where one object has a redshift very different from the others. There is extensive literature on this subject going back at least a decade. The statistical arguments underlying such evidence have often been discussed. Each case must be accidental if non-cosmological interpretations of the redshift differences are to be avoided.

More recently a number of galaxies with very different redshifts and luminous bridges between them have been shown (4,5). As far as I am aware no one has provided any evidence which disproves the existence of these bridges.

CONNECTION BETWEEN QSOs AND GALAXIES

The pre-eminent case of this kind is NGC 4319 and Markarian 205 (6). When Arp first claimed to show that a luminous connection existed, his paper was followed by a series of papers in which authors either argued that the connection did not exist, or was due to isophotal overlap between the two images (7,8,9). More recently Stockton, Wyckoff and Wehinger (10) claim to have detected a background galaxy which has MK 205 as its nucleus. Most recently Sulentic (11) has carried out a detailed study using digital image processing on the best photographic plates available of NGC 4319 and MK 205, and has concluded that there is indeed a connection with a continuous spectrum between the two systems which can be traced into the nuclear region of NGC 4319.

While this investigation will undoubtedly be subjected to more scrutiny, I conclude that it provides prima facia evidence that the connection is real.

---

*Illustrations of many of the more striking objects appear in (3) and will not be repeated here.

A second case of a similar kind is that of the radio galaxy 3C 303 where an N system with a redshift z = 0.141 gives rise to a radio structure which contains three faint optical objects one of which is a QSO with a redshift z = 1.57 (12).

STATISTICAL INFORMATION CONCERNING THE ASSOCIATION OF QSOs WITH GALAXIES

The statistical evidence is of several kinds. It is of:

(a) QSOs with $z \leqslant 0.45$ and galaxies at the same redshift.

(b) QSOs at all redshifts with bright galaxies which all have very small redshifts.

(c) QSOs and lists of both bright and faint galaxies.

We discuss these in turn.

(a) Stockton (13) chose all known QSOs with redshifts $z \lesssim 0.45$ and looked for galaxies lying within 45 arcsec of them. He reported that 13 galaxies in 8 fields (out of 27 fields) have redshifts within 1,000 km/sec of the QSO redshift. He estimated that the probability that this could occur by chance is $<1.5 \times 10^{-6}$. This is the primary evidence that some QSOs at least have redshifts of cosmological origin.

(b) Many QSOs have been found either by accident, or by searching close to bright galaxies with redshifts less than a few thousand kilometers a second. Many but not all of these were found by Arp. Approximately 100 of the QSOs in the catalogue of Hewitt and Burbidge (14) are in this category. I have carried out a statistical study from this sample of the QSOs lying within 10' of bright galaxies (3) and a table giving the results is reproduced below.

TABLE 1

A COMPARISON BETWEEN THE NUMBERS OF QSOs FOUND NEAR BRIGHT GALAXIES (N_0) AND THE NUMBER EXPECTED BY CHANCE (<n>)

Apparent Magnitude	$\theta \lesssim 180''$		$\theta \lesssim 600''$	
	N_0	$\langle n \rangle$	N_0	$\langle n \rangle$
$\lesssim 17$	5	0.69	11	7.8
$\lesssim 18$	14	2.38	28	25.9
$\lesssim 19$	30	6.93	58	77.9
$\lesssim 20$	45	23.2	83	259

The surface density of QSOs used in this study is based on surveys which are not in dispute. As can be seen from Table 1 the result is highly significant for $\Theta \leqslant 3'$. The fact that far fewer QSOs are seen beyond 3' from the galaxies than are expected by chance, either means that in making the calculations the surface density in the field was overestimated (i.e., there are far fewer QSOs in the field in general than the surveys would suggest), in which case the significance of the results for QSOs close to the galaxies is increased; or that the fields far out around galaxies have not been well searched. It is possible that there is no such thing as a field density of QSOs, but that they all originate in the galaxies, and fade before they move far away from them.

Another statistical survey of QSOs close to bright galaxies was that made in 1971 of the 3C QSOs and galaxies in the Shapley-Ames catalogue (15). This also showed a strong positive effect, but some later studies using other samples of radio QSOs failed to confirm it.

(c) More recently Seldner and Peebles (16) carried out a statistical analysis applying the cross correlation technique to the QSOs in the Burbidge, Crowne, and Smith catalogue (17), and galaxies in the Lick Catalogue (18). They have found statistically significant evidence of a correlation between the QSOs and the Lick counts of galaxies. Nieto and Seldner (19) have done a similar analysis using the bright galaxy catalogue of de Vaucouleurs et al (20) and have concluded that the only effect which may be real is due to a sub-class of QSOs.

Most recently Chu, Zhu, Burbidge and Hewitt (21) have done an extensive analysis using the cross-correlation technique on the QSO sample in (14) and galaxies in (20). They have found a strong positive result.

From all of these investigations I believe that the fairest statement which can be made is that there is strong statistical evidence that some QSOs are physically associated with galaxies at the same redshifts, but there is equally good evidence that many are physically associated with QSOs at very different redshifts.

INDIVIDUAL AND VERY STRIKING EXAMPLES OF PAIRINGS AND ALIGNMENTS BETWEEN QSOs AND GALAXIES

While statistical evidence is very important the primary evidence which often persuades scientists to rethink their ideas is evidence of individual events (in particle physics) or pictures (in astronomy). A good example with parallels for today is Wegener's original thinking about continental drift, which stemmed from simply noticing that the ontinents of the earth looked as though they could be fitted together nd thus were probably once physically associated.

Particularly striking examples which bear on the redshift arguments re:

(1) It has been pointed out by Arp that 3 QSOs lie within 2' of the center of the bright spiral galaxy NGC 1073 (3). The fact that they each have redshifts at the peaks of the redshift distribution discovered long ago (22) i.e., z = 0.6, 1.4, and 1.95 is even more remarkable.

(2) The discovery by Hazard, et al. (23) of a very close (within 5") pair of QSOs (1548 + 114 a,b) with very different redshifts (1.901 and 0.436) close to a group of galaxies with redshift 0.432.

(3) Two triple systems of QSOs each exactly aligned, with the lines nearly parallel (24). The redshifts in each triplet are very different namely 2.1, 0.51 and 1.7 in the northern triplet, and 2.1, 0.54 and 1.6 in the southern triplet. "Pairs" in the sense that they have approximately the same redshifts, lie in approximately the same positions relative to the other members of the triplet.

As I have stated before it does seem that the large body of evidence for the existence of non-cosmological redshifts requires us to rethink many of our basic concepts in extragalactic astronomy.

REFERENCES

1. Burbidge, G. 1980, IAU Symposium No. 92, Objects of High Redshifts, eds. G. O. Abell and P. J. E. Peebles, (Reidel: Dordrecht) pp. 99-105.
2. Burbidge, G. 1979, Nature **282**, pp. 451-455.
3. Burbidge, G. 1981, Annals N.Y. Acad. of Sciences, pp. 123-156.
4. Arp, H. C. 1971, Ap. Letts., **7**, pp. 221-224.
5. Arp, H. C. 1980, Ap. J. **239**, pp. 469-474.
6. Arp, H. C. 1971, Ap. Letts. **9**, pp. 1-4.
7. Ford, H. C. and Epps, H. W. 1972, Ap. Letts, **12**, pp. 139-141.
8. Adams, T. F. and Weymann, R. J. 1972, Ap. Letts **12**, pp. 143-146.
9. Lynds, C. R. and Millikan, A. G. 1972, Ap. J. (Letters) **176**, pp. L5-L8.
10. Stockton, A. N., Wyckoff, S. and Wehinger P. 1979, Ap. J. **231**, pp. 673-679.
11. Sulentic, J. W. 1983, preprint.
12. Kronberg, P. P., Burbidge, E. M., Smith, H. E., and Strom, R. G. 1977, Ap. J. **218**, pp. 8-19.
13. Stockton, A. N. 1978, Ap. J. **223**, pp. 747-757.
14. Hewitt, A. and Burbidge, G. 1980, Ap. J. Suppl. **43**, pp. 57-158.

15. Burbidge, E. M., Burbidge, G. R., Solomon, P. M. and Strittmatter, P. A. 1971, Ap. J. 170, pp. 233-240.
16. Seldner, M. and Peebles, P. J. E. 1979, Ap. J. 227, pp. 30-36.
17. Burbidge, G. R., Crowne, A. H. and Smith, H. E. 1977, Ap. J. Suppl. 33, pp. 113-188.
18. Shane, C. D. and Wirtanen, C. A. 1967, Publ. of the Lick Obs., 22, Part 1, (SW).
19. Nieto, J-L. and Seldner, M. 1982, Astron. Astrophys. 112, pp. 321-329.
20. de Vaucouleurs, G. et al. 1976, Second Reference Catalogue of Bright Galaxies (Univ. of Texas Press).
21. Chu, Y., Zhu, X., Burbidge, G. and Hewitt, A. 1982, submitted to Ap. J.
22. Burbidge, G. R. 1978, Physica Scripta 17, pp. 237-241.
23. Hazard, C., Jauncey, D. C., Sargent, W. L. W., Baldwin, J. A. and Wampler, E. J. 1973, Nature 246 pp. 205-208.
24. Arp, H. C. and Hazard, C. 1980, Ap. J. 240, pp. 726-736.

DISCUSSION

Chincarini: Is there any possibility that the correlation between angular separation and the distance from the galaxy is due to bias in the evaluation of the selected areas? (Their envelope runs parallel to the correlation line.)

G. Burbidge: Selection effects are present, as Browne and others have pointed out. However, I do not believe that they can account for the effect.

Nieto: About the specific question of QSO-galaxy associations, I would like to mention two contributions in collaboration with M. Seldner (Astronomy and Astrophysics, to be published) and with R. Bacon (submitted preprint) that are presented in this symposium as posters. The second one suggests a possible interpretation in terms of gravitational lenses due to individual stars in galactic haloes.

About "abnormal" associations between galaxies, the use of conventional distance criteria from a surface photometry study in collaboration with L. Tiennot of NGC 4156 (claimed to be associated by Arp with NGC 4151, whose redshift is much smaller) suggests that NGC 4156 is at its cosmological distance, and is in fact physically associated with a faint compact galaxy having the same redshift.

G. Burbidge: Your results are interesting. However, I feel that alignments of QSOs across bright galaxies make gravitational lens explanations very doubtful.

R. Ellis: Regarding Chincarini's comment, such a controlled experiment to test the significance of the angular separation-redshift effect you present has been performed. Several years ago Ian Browne (M.N.R.A.S.) showed that using a technique similar to that employed by Arp, one could derive a similar relationship between normal galactic stars and galaxies. The effect is entirely due to the selection procedure as he has repeatedly stressed to you.

G. Burbidge: Browne has certainly argued that the relation between θ and the distances of the bright galaxies is due to observational selection. However, I think the precision of the selection and its recent extension by Arp make this very unlikely.

Abell: I am unaware of any complete survey about a randomly selected sample of galaxies chosen, before knowledge of quasars near them in direction, and in which quasars have been searched for to a given limiting magnitude and to a given radial distance from the galaxies, both specified in advance. Until this is done, statistics of quasars near galaxies, I think, should be regarded with caution.

G. Burbidge: You are correct. However, Arp, A. Hewitt, and M.H. Ulrich are now doing such a study with us. Also, I believe that the analysis that I carried out, comparing all known QSOs within 10' of galaxies with those expected by chance, is a conservative study, as was the original study of 3 CR QSOs and bright galaxies carried out by Burbidge et al. in 1972.

I might add that Arp seems to have a knack of finding QSOs near galaxies. Since he finds far more than are present in the field according to all other workers, this, in effect, is evidence that the effect is real.

THE EVOLUTION OF THE RADIO GALAXY POPULATION AS DETERMINED FROM DEEP RADIO-OPTICAL SURVEYS

Harry van der LAAN, Peter KATGERT, Rogier WINDHORST and Marc OORT
Leiden Observatory, University of Leiden, The Netherlands

1. INTRODUCTION

A first step in the study of the evolution of the radio galaxy population is the determination of the radio luminosity function (RLF), i.e. $\rho(\log P, z)$, which results from (and must finally be interpreted in terms of) 'light curves' of individual objects (i.e. P(t)) and the 'birth rate' function $\dot{n}$ (log P, t).

Information on the RLF is obtained from deep optical identifications and spectroscopy of complete radio samples. At high radio flux densities identifications are virtually complete, as are spectroscopic redshift determinations. At lower flux densities (say below the "bump" in the normalized source counts) only about 50% of the radio sources can be identified on 4m plates, and as yet only a small, but quickly growing, fraction has spectroscopic redshifts.

In the absence of complete spectroscopic z information, photometric redshifts have been used. It is then assumed that the radio galaxies are ellipticals, for which a 'standard candle' hypothesis (SCH) is justified. As shown by Auriemma et al. (1977) the absolute magnitude of elliptical radio galaxies does not depend on radio luminosity when $\log P \gtrsim \log P^*$ ($\sim$ 25.0 at 1.4 GHz, for H_o = 50 km sec^{-1} Mpc^{-1}). When z_{photom} determined with the SCH yields $\log P > \log P^*$ one has at least consistently interpreted the data in terms of an elliptical radio galaxy. However, only when the ratio of radio to optical luminosity (or flux, with K-corrections taken into account) is larger than those found locally for Seyfert and spiral galaxies, is this interpretation (and therefore the redshift) very likely to be correct. On the other hand, when either the implied log P $< \log P^*$, or the ratio of radio to optical flux is in the range also populated by Seyferts and spirals, the interpretation is uncertain and z_{photom} is probably an overestimate.

G. O. Abell and G. Chincarini (eds.), Early Evolution of the Universe and Its Present Structure, 73–79.

Table 1. The radio surveys with deep optical data.

observer(s) +reference	freq. MHz	$S_{1.4}^{lim}$	pass band	m_{lim}	N_{gal}	$N(z_{spec})$
Peacock & Wall (MNRAS 194, 331)	2700	3.5 Jy	V	23^m	61	49
Laing et. al. (MNRAS 184, 149)	178	1.3 Jy	V	23	33	21
Katgert-Merkelijn et. al.	178	0.9 Jy	V	20.5	28	12
(A&A suppl. 40, 91)	408	0.9 Jy	V	20.5	41	19
Allington-Smith (preprint)	408	0.4 Jy	r	23	29	0
Grueff & Vigotti (A&A suppl. 20, 57)	408	0.4 Jy	V	22	21	0
Katgert et. al. (A&A 38, 87)	1412	8 mJy	B/R	22.5/21	39	0
de Ruiter et. al. 48" (A&A suppl. 28, 211)	1412	3 mJy	B/R	22.5/21	118	0
de Ruiter et. al. 4m (A&A suppl. 28, 211)	1412	3 mJy	B	23.5	13	0
Windhorst, Kron, Koo (preprint)	1412	0.2 mJy	U/J	24/24	175	46
			F/N	23/22		
Total					558	147

2. RADIO-OPTICAL SURVEYS OF LARGE DYNAMIC RANGE.

As described by Van der Laan and Windhorst (1982) combined deep radio and optical surveys, primarily at Westerbork and KPNO respectively have led to large identified samples. In Table 1 the surveys used in our studies are listed. $S_{1.4}^{lim}$ is the limiting flux of a complete 1.4 GHz sample taken from the sample defined at the original finding frequency. N_{gal} is the number of identified radio galaxies in the 1.4 GHz sample. The last column gives the present number of spectroscopically determined redshifts in each sample.

For $S_{1.4} \gtrsim 10$ mJy the elliptical radio galaxy SCH is quite likely to be correct for the majority of the radio galaxy identifications. On that basis we have determined $\rho(\log P, z)$ out to $z_{photom} \sim 0.85$ for log $P \gtrsim 24.6$ (see Van der Laan and Windhorst). This determination is based on more than half (the identified portion) of the radio galaxy populatio with $S_{1.4} \gtrsim 10$ mJy. We have therefore asked whether one can account for the remaining, non-identified fraction of the source count by very simpl extrapolation of the $\rho(\log P, z)$ determination.

As is well-known, the comoving density of radio galaxies increases strongly from $z \sim 0.3$ to ~ 0.85. In one model, we could reproduce the unidentified portion of the source count (and hence the total source count) for $S > 10$ mJy to within a few percent, by conservatively assumin $\rho(\log P, z)$ beyond $z \sim 0.85$ to be identical to $\rho(\log P, z \sim 0.85)$ except for a slight increase of the average log P towards higher redshifts. In this very simple model, $\rho(\log P, z)$ for radio galaxies must be set to zero beyond $z \sim 1.5$ in order not to exceed the observed count. This interesting, but clearly not unique model suggests that beyond $z \sim 1.5$ practically all radio galaxies appear as radio quasars (which we did not include in our RLF determinations, and which were separately accounted for in the source count, see Fig. 1).

3. VERY DEEP 1.4 GHz SURVEYS.

Choosing a field in the Lynx area ($8^h42^m, 44^o47'$, radius 0^o43) which lacks strong (S > 10 mJy) radio sources, a Westerbork survey (σ_W = 120 μJy) was deepened in a VLA exposure of 2 x 6h to σ_{VLA} = 45 μJy (Windhorst, Miley, Owen, Kron and Koo, 1983). From this a complete sample was distilled consisting of a total of 94 sources, 60 above the WSRT 5 σ limit and an additional 34 between the VLA 5 σ limit and the Westerbork limit. These counts are shown in Figure 1 by open circles and crosses. Also shown are VLA counts by Condon and Mitchell (1982) which are quite consistent with our result.

The interesting feature of the counts below 10 mJy would seem to be

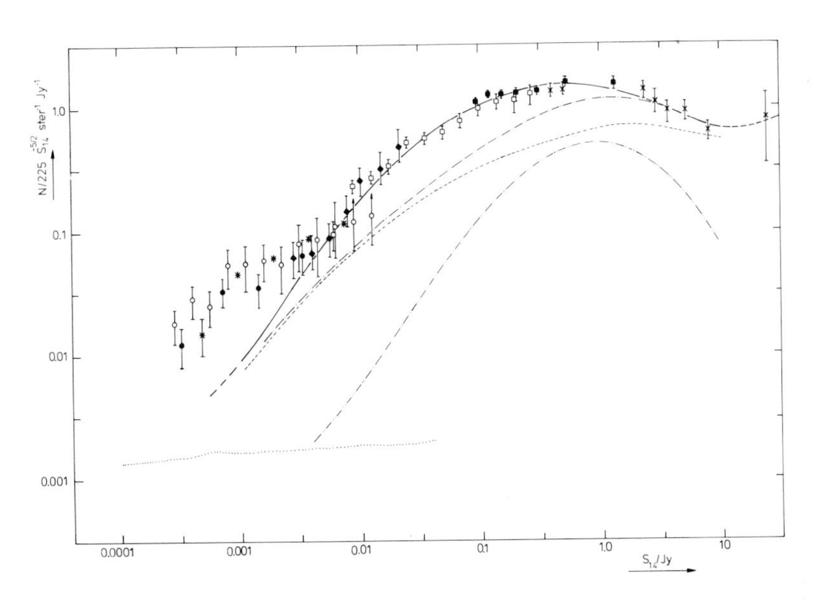

Figure 1. The 1.4 GHz differential source count normalized to a differential Euclidean count (225 $S^{-5/2}Sr^{-1}$ Jy^{-1}). References to older data (S ≳ 10 mJy) can be found in Willis et al. (1977). Filled circles: Condon and Mitchell (1982), open circles: Windhorst et al. (1983 5σ sample, asterisks: Windhorst et al. 6.5σ sample, - .. -: observed count of radio-loud quasars, ----: count of powerful radio ellipticals out to $z \sim 0.85$, calculated from observed RLF, and assuming non-evolving RLF below log P ∿ 25.0, —.—: sum of two above, ———: total count of quasars and powerful ellipticals, with elliptical RLF extrapolated out to $z \sim 1.5$ (as described in text), and again assuming no evolution below log P ∿ 25.0.

the lack of continued convergence, or rather the steepening, similarly visible in all datasets.In our opinion there is also consistency on this point with recent very deep 6 cm source count data (Fomalont, this volume

It is obviously of great interest to determine the nature of the faint source population responsible for the steepening of the count. One important step to this end is the optical identification of the sample, using the deep KPNO 4m plates available for this (and other) deep Westerbork field(s).

For the Lynx sample the identification status is given in Table 2. A first conclusion is that local spiral galaxies contribute to the source count in this flux range at the expected level. In Figure 1 that level, computed from the radio luminosity function data recently reviewed by Meurs (1982), is shown as a dotted line. It falls short of the actual counts by a factor 20, consistent with the bright spirals identification of this deep sample.

4. AN EVOLVING,WEAK SOURCE POPULATION.

Current data are insufficient to determine the nature of the weak sources' parent population. The fuzzy blue objects require both high angular resolution photometry, ultimately with Space Telescope, and modes resolution spectroscopy for the recognition of their character. The 4m multi-colour photometry is proceeding and will be reported in Windhorst, Miley, Owen, Kron and Koo (1983).

Spectroscopic work on blue galaxy identifications in the two to thirty mJy range has started, at McDonald for optically bright and with the KPNO 4m cryogenic camera for the fainter galaxies in the sample. First results (see also Koo, this volume) indicate a mixture, some galaxies close to the 'standard candle' Hubble curve, and others two to three magnitudes fainter. These results confirm that below 10 mJy the 'standard candle' hypothesis is no longer applicable. While at higher flux densities the well-known powerful radio galaxy population associated with luminous ellipticals dominates, below $\sim$ 10 mJy intrinsically fainter galaxies (possibly both ellipticals and spirals) are important.

Table 2. Lynx identifications.

600 μJy < $S_{1.4}$ < 10 mJy	225 μJy < $S_{1.4}$ < 600 μJy
60 sources	34 sources
2 bright spirals	3 bright spirals
1 bright elliptical	
1 fainter galaxy	1 fainter galaxy
20 faint (mainly fuzzy) objects	15 faint (mainly fuzzy) objects
4 stellar objects (2 galactic)	

In Figure 2 the hatched area roughly indicates the location of these sources, if they have redshifts in the range 0.2 - 0.6 as implied by the Lynx sample identifications for absolute magnitudes 2 to 3^m fainter than the elliptical standard candles. Whether this population is associated with weak ellipticals, or with spirals/Seyferts cannot be decided on present evidence. Recent near-IR observations (Windhorst, this volume) of these faint radio galaxies show both 'thermal' and non-thermal IR spectra.

Given the continuity of quasars, Seyferts and active galaxy nuclei (cf. Maccacaro and also Braccesi in this volume) one may speculate that this population consists of active spirals, relatively quiescent at z = 0 but evolving in the manner of the giant ellipticals, with a greatly enhanced comoving density in the range $10^{22.5} < P_{1.4} < 10^{24.5}$ at $z \sim 0.4$.

In addition to more optical spectrophotometry, high resolution radio maps may help to distinguish this class of sources from the strong source population. The latter are generally symmetric and large, the former are expected to be rather more amorphous and smaller.

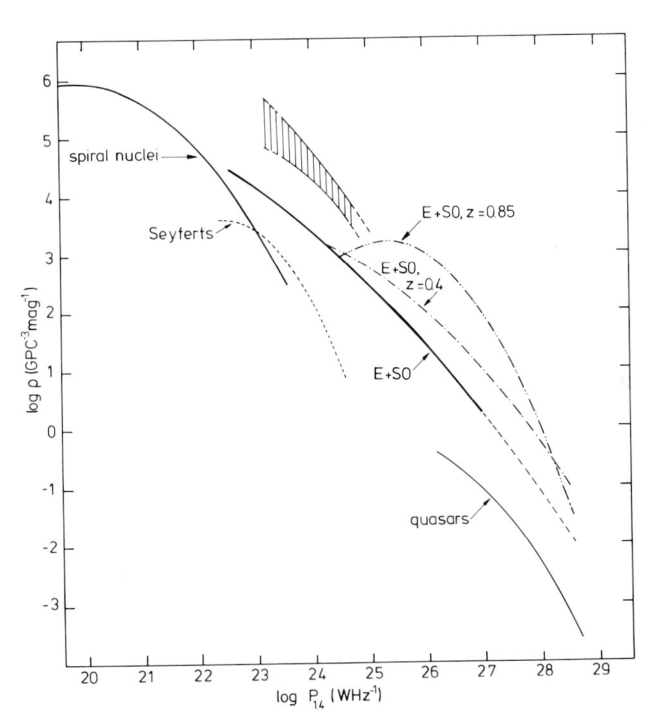

Figure 2. Schematic representation of the observed local RLF for: spiral galaxy nuclei (Hummel, 1980, thesis). Seyfert galaxies (Meurs, 1982,thesis), elliptical galaxies (Windhorst, 1983, thesis) and radio quasars (Fanti and Perola, 1977). Also shown are the elliptical RLF's at $z \sim 0.4$ and ~ 0.85. The hatched area indicates the possible location of mJy objects responsible for the steepening of the count, for an assumed redshift of 0.4 ± 0.2.

REFERENCES

Auriemma, C., Perola, G.C., Ekers, R., Fanti, R., Lari, C., Jaffe, W.J., Ulrich, M.H., 1977, Astron. & Astrophys. 57, 41.
Condon, J.J., Mitchell, K.J., 1982, preprint.
Fanti, R., Perola, G.C., 1977, Luminosity Functions for Extragalactic Radio Sources in "Radio Astronomy and Cosmology", Ed. D.L. Jauncey (Reidel, Dordrecht), p. 171.
Van der Laan, H., Windhorst, R.A., 1982, in "Astrophysical Cosmology", Eds. Brück, Coyne, Longair, p. 349.
Willis, A.G., Oosterbaan, C.E., Le Poole, R.S., de Ruiter, H.R., Strom, R.G., Valentijn, E.A., Katgert, P., Katgert-Merkelijn, J.K., 1977, Westerbork Surveys of Radio Sources at 610 and 1415 MHz in "Radio Astronomy and Cosmology", Ed. D.L. Jauncey (Reidel, Dordrecht), p. 39.
Windhorst, R.A., Miley, G.K., Owen, F.N., Kron, R.G., Koo, D.C., 1983, preprint.

DISCUSSION

Segal: In the analyses of radio luminosity functions that you quote, did you allow for the (probably severe) flux cutoff bias, and if not, isn't that a weak link in your argument?

van der Laan: We are fully aware of the pitfalls in converting the luminosity distribution of flux-limited complete samples to the luminosity function of comoving populations.

Ekers: Your use of the constancy of optical luminosity of radio galaxies introduces a dependency on the small sample of radio galaxies with known redshift (the mainly 3 CR sample of radio galaxies in your Hubble diagram). The bivariate luminosity function, for both elliptical galaxies (Auriemna et al. 1977) and spiral galaxies, shows that the probability of radio emission is a function of the optical luminosity. This will affect the predictions of your models and may also explain the difference in $\langle M_p \rangle$ which you presented for your weak radio galaxy sample.

van der Laan: There was no time to discuss the refinement of the bivariate luminosity function here. We have in fact used it in our analyses as published thus far. Whether the range of the bivariate luminosity function can be extended to the optical and radio luminosities at issue in this faint sample remains to be established. We can speculate about, but not yet make, the choice between spirals/ Seyferts or ellipticals as parent population for these weak sources. In both instances a strong epoch dependence is inescapable.

Menon: In making statements about luminosity evolution, it is usually assumed that it is a continuous function of time only. Is it physically more reasonable to assume that luminosity is a discontinuous function of time, the periods of discontinuity themselves depending on luminosity?

van der Laan: I have no quarrel with your statement as it applies to individual sources. After convolution over a whole source population, however, I expect the luminosity function to vary smoothly with epoch.

Melnick: Gas-rich dwarf galaxies (H II galaxies) appear to be very abundant in the universe. Their luminosity function is not known but there is some evidence that it may be significantly different from that of normal galaxies. H II galaxies are thermal radio sources just as normal H II regions, but significantly brighter (the luminosities reach 10^{42} ergs/sec). Have you estimated the contribution of these galaxies to your deep (VLA) radio source counts?

van der Laan: Although I am not aware of a systematic study of dwarf galaxies' radio luminosity function, enough is known to exclude dwarf galaxies as the main contributors. I say this on the basis of the identifications and the spectroscopic redshifts obtained thus far.

A 6-CM DEEP SKY SURVEY

E. B. Fomalont and K. I. Kellermann
National Radio Astronomy Observatory*

J. V. Wall
Royal Greenwich Observatory

In order to extend radio source counts to lower flux density, we have used the VLA to survey a small region of sky at 4.885 GHz (6 cm) to a limiting flux density of 50 μJy. Details of this deep survey are given in the paper by Kellermann et al. (these proceedings). In addition, we have observed 10 other nearby fields to a limiting flux density of 350 μJy in order to provide better statistics on sources of intermediate flux density.

A total of 13 sources from the Deep Survey and 26 sources from the Intermediate Survey were used to construct the source count. The new VLA data includes the weakest radio sources yet observed and extends the observable range of flux density by nearly two orders to magnitude to reach a source density of 600,000 sources $ster^{-1}$ (Fig. 1). For the first time, the observed range of flux density exceeds the width of the luminosity function, and this will allow a better definition of the luminosity function and its spatial evolution.

Below 100 mJy, the 6 cm source count converges, although less rapidly than in the 75 cm Cambridge 5C Surveys. None of the 13 sources found in the "Deep" Survey show a visible counterpart on the PSS; they are probably faint distant galaxies beyond the plate limit.

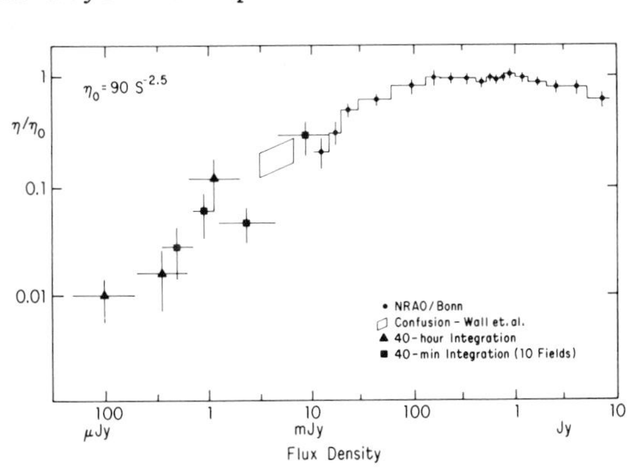

Figure 1. Differential source count normalized to a static Euclidean Universe. Data above 10 mJy are taken from the various NRAO/Bonn surveys.

* Operated by Associated Universities, Inc., under contract with the National Science Foundation.

G. O. Abell and G. Chincarini (eds.), Early Evolution of the Universe and Its Present Structure, 81.

NEAR INFRARED PHOTOMETRY OF FAINT RADIO GALAXIES

Rogier A. Windhorst, Sterrewacht Leiden
Jeffrey J. Puschell, Physics Department, U.Ca., San Diego
Trinh X. Thuan, Centre Etude Nucleaire, Saclay

This paper describes the first results of systematic near IR photometry of radio galaxies in the Leiden Berkeley Deep Survey. The LBDS contains ∿ 400 radio sources with $S_{1.4\ GHz} \geq 0.22$ mJy (for details see van der Laan et al., this symposium), 50% of which have reliable optical identifications on deep multicolor KPNO 4^m plates, being mainly faint blue radio galaxies (Windhorst et al., 1982). The aim of the present IR photometry program is to study their IR properties and to obtain clues as to the nature of these faint blue radio galaxies, which may be either ellipticals, possibly with recent star formation, or spirals with exceptionally high radio power. The additional optical-IR color baseline will help to solve these questions.

A complete subsample, with an m_V distribution representative for the total LBDS, was photometered in the passbands J(1.2μ), H(1.65μ) and K(2.2μ) with the $3\overset{m}{.}8$ UK and the $3\overset{m}{.}0$ NASA Infrared Telescopes (Puschell et al., 1983). Out of 48 objects, 45 were detected in K with errors typically between $0\overset{m}{.}05$ and $0\overset{m}{.}25$. The (J-H) vs (H-K) color-color diagram (fig. 1) shows the LBDS radio galaxies together with the 3CR radio galaxies of Lilly & Longair (1982). The IR color distribution of the mJy LBDS radio galaxies is similar to that of the 10^3x more powerful 3CR. The straight line is the locus of non-thermal spectra for different spectral indices. The IR colors of about half the LBDS sample are within the errors consistent with a non-thermal spectrum, as are all the 3CR N-galaxies. The other half of the LBDS sample differs significantly from the non-thermal locus and has IR colors that local E+Sp galaxies would have at redshifts in the range $0.2 \lesssim z \lesssim 0.6$, applying Bruzual's (1981) K-corrections.

The optical-IR color-magnitude diagram (V-K) vs K for the LBDS and the 3CR is shown in fig. 2. The errors in the still preliminary LBDS V-magnitudes may be uncertain by $\sim 0\overset{m}{.}75$, but precise photometry will be provided by Kron, Koo & Windhorst (1983). The hatched line shows the effective optical plate limit. The drawn lines are the predictions by Bruzual (1981) for a passively (C) and exponentially (μ=0.5) evolving elliptical and a non-evolving spiral (Sbc) galaxy, all three calculated for one single luminous $M_V=-23\overset{m}{.}9$ ($H_o=50$, $q_o=0$). (So galaxies with similar spectra, but with lower optical and IR

G. O. Abell and G. Chincarini (eds.), Early Evolution of the Universe and Its Present Structure, 83–84.

luminosities, will in first order appear to the right of the drawn lines). While most of the 3CR radio galaxies with K $\lesssim$ 16^m have (V-K) colors similar to these luminous ellipticals, this seems only true for a minority of the LBDS. Most mJy radio galaxies have (V-K) and also (R-K) colors up to 3^m redder - more than can be explained by the errors - than any of the models. The explanation could be, that they are overluminous in K. However, for a subsample spectroscopic redshifts have been measured (Koo, this symposium), yielding for only about half of the sample an M_K as luminous as the 3CR ellipticals and up to $\sim$2^m less luminous for the other half of the sample. In order to explain the very red (V-K) colors their M_V should be even more underluminous. The most likely explanation is that these objects have M_V's much fainter than the 3CR giant ellipticals, an UV upturn causing the blue optical colors and some IR excess causing the very red (V-K) and (R-K) colors. If this near IR excess is thermal, then the dust should be very hot ($\gtrsim$1000 K), or very dense, which is not very likely, if they were ellipticals. If the IR excess is non-thermal, the underlying galaxy is probably not elliptical either, because ellipticals with low radio power have generally not a strong non-thermal nucleus. It seems likely that a substantial fraction of the mJy LBDS radio galaxies with the very red optical-IR colors are spiral galaxies with some non-thermal IR emission like in Seyferts, or with thermally radiating hot dust.

References

Bruzual, G.A.: 1981, Ph.D. Thesis, University of California, Berkeley

Kron, R.G., Koo, D.C., Windhorst, R.A.: 1983, in preparation

Lilly, S.J., Longair, M.S.: 1982, M.N.R.A.S. 199, pp. 1053

Puschell, J.J., Thuan, T.X., Windhorst, R.A., Isaacman, R.: 1983, Ap.J., submitted

Windhorst, R.A., Kron, R.G., Koo, D.C., Katgert, P.: 1982, in Extragalactic Radio Sources, I.A.U. Symp. No. 97, pg 427, Ed. D.S. Heeschen & C.M. Wade (Reidel, Dordrecht).

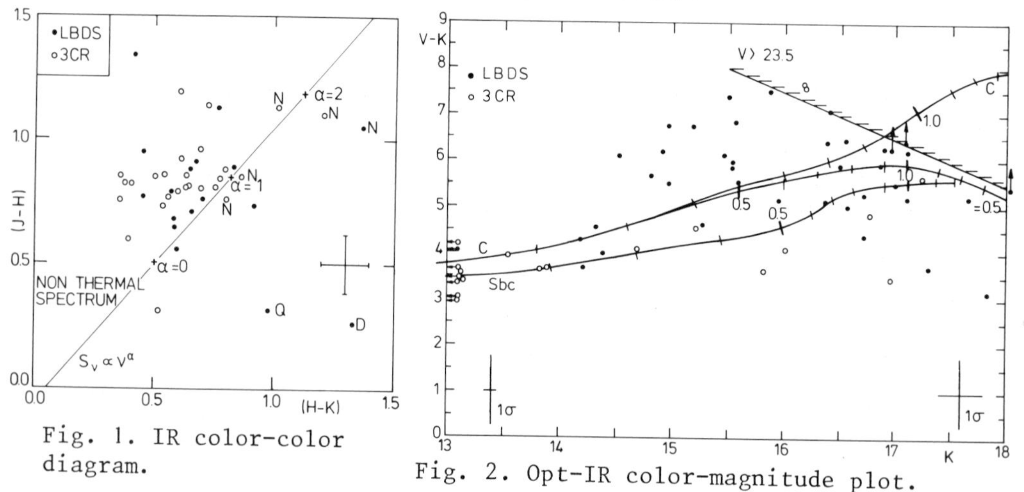

Fig. 1. IR color-color diagram.

Fig. 2. Opt-IR color-magnitude plot.

NEAR INFRARED AND RADIO OBSERVATIONS OF DISTANT GALAXIES

R. A. Laing, NRAO, Charlottesville, VA 22901, USA.;
F. N. Owen, NRAO, P.O. Box 0, Socorro, NM 87801, USA.; &
J. J. Puschell, CASS, University of California, San Diego,
La Jolla, CA 92093, USA.

This paper is concerned with the distant radio galaxies in a sample of bright sources selected at 178 MHz by Laing, Riley & Longair (1982). This sample is 96% complete for sources with $\theta < 10'$ and the bias of the 3CR catalogue against sources of large angular size has also been reduced. Deep optical searches have located many candidate identifications, but the probability of a chance coincidence with an unrelated object is appreciable, especially in the faintest cases, unless the area to be searched is small. We have therefore mapped the sources with candidate identifications having $V > 20$, using the VLA at a wavelength of 6 cm (Laing, Owen & Puschell, in preparation), in order to search for radio cores. We have so far located cores in 16/23 sources and set 5σ upper limits of 0.6 mJy for the remainder. None of the cores had been detected previously. In all cases, the cores coincide with optical objects, although one source (3C 340) had been misidentified. Several ambiguities have now been resolved.

The majority of these faint identifications are galaxies with $z > 0.4$ and there is consequently a substantial overlap in luminosity for the galaxies and quasars in the sample. Their space-density distributions have been compared directly by Laing et al. (1982), who conclude that the distributions of V/V_{max} for radio galaxies and quasars in the same range of luminosity are indistinguishable, and hence that the populations evolve in similar ways. The evolution is very strong, the mean values of V/V_{max} being 0.66 ± 0.03 for the powerful radio galaxies and 0.69 ± 0.04 for the quasars. If this result holds over a wide range of luminosity and redshift, then the factors which determine the rate of evolution are likely to be independent of the obvious differences in radio structure, nuclear emission, galactic morphology and perhaps environment between the two classes.

Deep identification surveys are extremely efficient at selecting very distant elliptical galaxies: most of those with known redshifts $z \simeq 1$ are in fact strong radio emitters. Whilst this allows us to study the evolution of the spectral energy distributions of elliptical

G. O. Abell and G. Chincarini (eds.), Early Evolution of the Universe and Its Present Structure, 85–86.

galaxies over appreciable look-back times, there is always the worry that radio galaxies are in some way unrepresentative. Lebofsky (1981), Lilly & Longair (1982) and Puschell, Owen & Laing (1982) have recently published near-infrared photometry of an appreciable number of galaxies from the sample of Laing et al. (1982). We refer to these papers for a detailed discussion, but wish to emphasize an unexpected result found by Puschell et al. The calculations of Bruzual (1981) suggest that the spectral shapes of elliptical galaxies in the near infrared should be independent of epoch out to $z \simeq 2$. Our results indicate, however, that powerful radio galaxies at $z > 0.4$ are significantly too blue in the wavelength range 1 - 3 μm, i.e., J - K is less than the predicted value. It is entirely possible that these galaxies are anomalous in some way, perhaps because of non-thermal contributions to their infrared emission, and we therefore plan to study optically-selected and weak radio galaxies at $z \sim 1$ in the near infrared.

References

Bruzual, G., 1981. Thesis, University of California, Berkeley.
Laing, R. A., Riley, J. M. & Longair, M. S., 1982. MNRAS, in press.
Lebofsky, M. J., 1981. Ap. J., 245, L59.
Lilly, S. J. & Longair, M. S., 1982. MNRAS, 199, 1053.
Puschell, J. J., Owen, F. N. & Laing, R. A., 1982. Ap. J., 257, L57.

STUDIES OF FAINT FIELD GALAXIES

Richard S. Ellis
Durham University
England

Although claims are often made that photometric surveys of faint field galaxies reveal evidence for evolution over recent epochs ($z<0.6$),it has not yet been possible to select a single evolutionary model from comparisons with the data. Magnitude counts are sensitive to evolution but the data is well-mixed in distance because of the width of the luminosity function (LF). Colours can narrow the possibilities but the effects of redshift and morphology can only be separated using many passbands.

The conclusions of Koo's(1981) UJFN analysis do not differ significantly from that of Kron (1978) who used counts in J and F. Moreover the data can be fitted without recourse to evolution if the LF is strongly dependent upon colour/morphology. The adjustment necessary to eliminate evolution in this way is rather extreme but it highlights two ways in which we can make further progress in this important subject.

Firstly, uncertainties in intepreting faint data reflect those in local galaxy parameters (Ellis 1980). One remedy is to derive statistical information on nearby galaxies from local redshift surveys.Here I discuss results based on the AAT redshift survey (Peterson et al 1982) which comprises 5 Schmidt fields to J = 16.7 i.e. well beyond local inhomogeneities. Secondly, the difficulties in resolving the many possibilities encountered with faint photometry could be resolved with redshifts. To obtain redshift distributions for faint samples is now feasible via multi-object spectroscopy. At intermediate magnitudes ($J\sim20$) such distributions test the faint end of the galaxy LF; at faint magnitudes ($J\sim22$) they offer a direct evolutionary test.

The AAT redshift survey reveals two important points. Firstly, the LF normalisation is lower than previous estimates - as might be expected if local effects were important in other surveys. Figure 1 shows that the AAT survey counts together with deeper Schmidt counts are well-fitted by a model incorporating no evolution,K-terms reviewed by Ellis(1982) and a morphological variation of the LF defined by the AAT redshift data. The Kirschner et al(1978,KOS) counts have been transformed by photometering their SP3 field in our system; their counts show a northern normalisation significantly higher than the AAT value. The effect of the low value (60% lower also than that adopted by Koo and co-workers) is to strengthen the

G. O. Abell and G. Chincarini (eds.), Early Evolution of the Universe and Its Present Structure, 87–91.

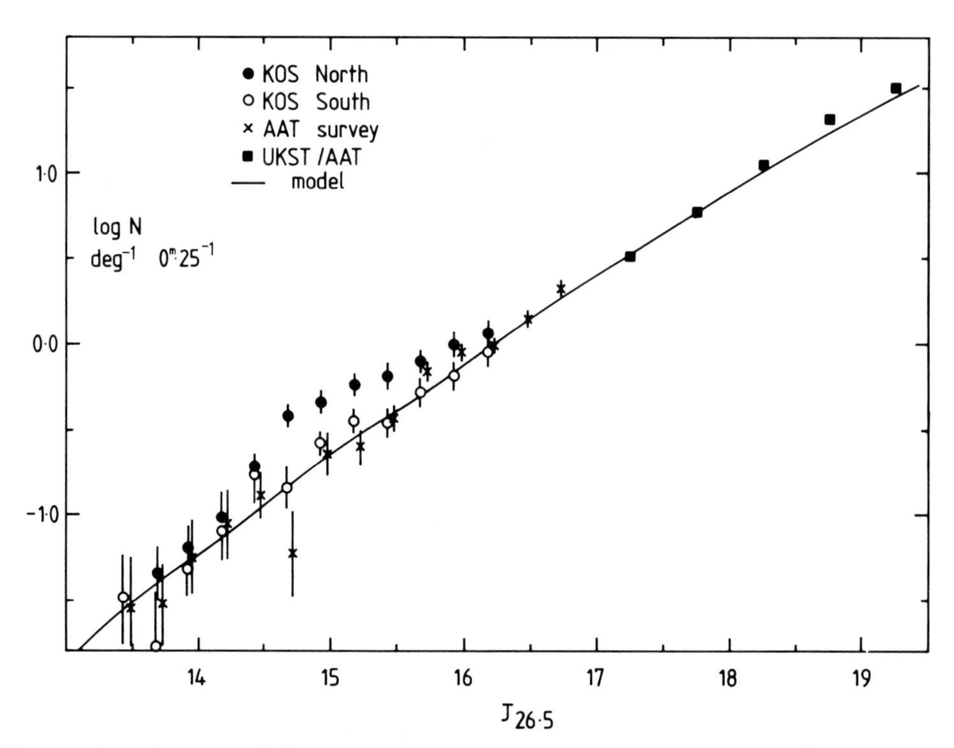

Figure 1: Differential galaxy counts transformed to the J (IIIa-J+Schott GG385) system at an isophote of 26.5 mag/sq arcsec. The model assumes the AAT survey LF, K-terms from Ellis(1982) and no evolution. Note the high normalisation for the northern KOS da

evolution needed at faint magnitudes. Whilst it is possible that the area covered by the AAT survey are anomalous, this seems unlikely given the volumes sampled, the uniformity of the counts from field to field and the agreement between the count slope and that expected for a homogeneously distributed sample. The results also show the importance of obtaining wel calibrated photometry in the deeper non-evolving region $17< J <19$.

The AAT survey also shows (Figure 2) that the LF over all types does not vary significantly from one sample to another <u>provided</u> each sample is analysed in the same way. The parameters reported by Huchra (this symposi for the CfA survey LF are also close to those in Figure 2. It appears tha M^* is known to within $\pm$ 0.1m and α to within $\pm$ 0.15; the effect of these uncertainties on faint galaxy predictions is small. The remaining problem here is the morphological/colour variation in the LF, specifically for late types whose K-terms are small and for which M^* may be very faint. Kron (1982) mentions 13 galaxies with $B<18.9$ and $U-B<-0.35$ for which $\langle M\rangle = -17.5$ ($H_0=100$) whereas the latest class in the AAT survey, Sdm/Irr with $U-B\sim-0.15$, gives $\langle M\rangle=-18.9$. Larger or deeper surveys should resolve this question.

Simple considerations show that the redshift distribution of 2-300 galaxies at $J\sim22$ should test evolution of the form discussed by Bruzual (1980) and Tinsley(1978). The long integrations necessary to determine

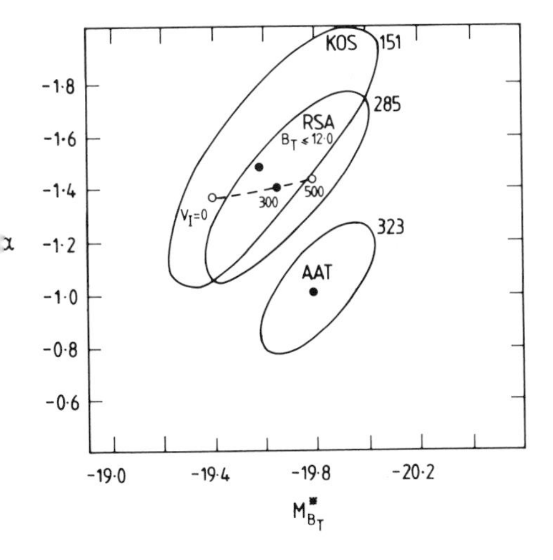

Figure 2: 1σ error contours for Schechter LF parameters determined for various redshift catalogues (H_0=100) after Efstathiou et al 1982. The Revised Shapley Ames (RSA) results are sensitive to the Virgo infall velocity as indicated (km/sec). The sample size is indicated alongside each contour.

the redshift of a faint galaxy has led to searches for quick approximate alternatives to direct spectroscopy. At present there seems to be no generally accepted alternative. Colours have been considered by various groups; infrared colours are somewhat better than visible ones because, for modest redshifts they should not depend on morphology. However the current i-r/redshift relation for optically-selected galaxies shows much scatter for z>0.3 although the average is close to that expected for no evolution (Ellis and Allen 1982). The situation may be better than current data imply because, of necessity, the relationship is derived using cluster members where visible and i-r anomalies are often found. The optical colour-z relation for clusters(Kristian et al 1978) is not easy to reconcile with the field colour-z tests performed by Koo. Since the i-r method would be very efficient if imaging were possible, it is important to continue studying the i-r colour/redshift relation using field galaxies. The combination of i-r and optical colours allows, in principle, the separation of morphology and redshift. With 47 optically-selected galaxies Ellis and Allen were unable to find any high z objects with rest-frame s.e.ds like elliptical galaxies.

Projects using genuine spectroscopy of faint field galaxies presently involve only small (∿50) samples but should provide an impressive impact on the subject over the next 2 years. Turner(1980) has obtained redshifts for 58 galaxies with J∿19-21 and Gunn(1982) has claimed that the redshift distribution does not agree at all well with model predictions. However, as Koo(1981) remarks, the sample is not a random subset of a complete photometrically-defined sample and the magnitude distribution peaks near J=20 but extends beyond J=21. The selection criteria assumed by Gunn, viz <J>=20, may well be inappropriate for the sample; as an example, if we assume the sample is complete to J=21 then the redshift distribution is in fair agreement with a no evolution model based on the AAT survey parameters.

I have begun a survey using the AAT multi-object spectrograph (Ellis et al 1982) which currently reaches J=21.2; spectra of 50 galaxies have been secured and provisional redshifts for 20 do not indicate any serious discrepancy with the AAT survey-based no evolution prediction. The only results on fainter galaxies where evolution might be observed in small samples are those of Koo(this symposium); at J∿22.5 the redshifts of 54 galaxies indicate mild evolution although, at present, the sample is biased to those galaxies with strong, easily detectable, features includ emission lines.

This raises the general difficulty of determining the redshift of a galaxy whose morphology and redshift is not known a priori. The waveleng range over which familiar features are recognisable is limited by the detector and sky and is crucially important in the red. At faint magnitu where z∿1 galaxies might occasionally be expected, it must be demonstrat convincingly that such a redshift could have been measured successfully before claiming the absence of high z objects is significant. This is an important point because, with multi-object facilities, there may a reluctance to go back and clean up individual failures.

REFERENCES

Bruzual, G. 1980. Ph.D. thesis, University of California, Berkeley.
Efstathiou, G.,Shanks, T.,Ellis,R.S.,Bean,J.,Peterson,B.A. and Zou,Z. 1982. in preparation.
Ellis,R.S. 1980. IAU Symposium 92, eds. G.O. Abell and P.J.E. Peebles (Dordrecht: Reidel), pp. 23-30.
Ellis,R.S. 1982. Lectures at VIIth Course of Cosmology and Gravitation, Erice, (Dordrecht: Reidel) in press.
Ellis,R.S. and Allen, D.A. 1982. Mon. Not. R. astr. Soc., in press.
Ellis,R.S., Carter,D. and Gray, P.M. 1982. in preparation.
Gunn, J.E. 1982. Astrophysical Cosmology, Proceedings of the Vatican Stu Week on Cosmology and Fundamental Physics, eds. H.A. Brueck, G.V.Co and M.S. Longair (Specola Vaticana).
Kirschner, R.P., Oemler, A. and Schechter, P.L. 1978. Astron.J.,83,pp.15
Koo, D.C. 1981. Ph.D. thesis, University of California, Berkeley.
Kristian, J.,Sandage, A.,Westphal,J.A. 1978. Astrophys. J.,221,pp.383-39
Kron, R. 1978. Ph.D. thesis, University of California, Berkeley.
Kron, R. 1982. Vistas, in press.
Peterson, B.A.,Ellis,R.S.,Efstathiou,G.,Shanks,T.,Bean,J. and Zou,Z. 1982. in preparation.
Tinsley, B.M. 1978. Astrophys. J., 220, pp816-821.
Turner, E. 1980. IAU Symposium 92, eds. G.O. Abell and P.J.E.Peebles (Dordrecht; Reidel), pp71-72.

DISCUSSION

Segal: Non-parametric analyses by an optimal statistical technique or large low-redshift galaxy samples due to de Vaucouleurs (1979), Visvanathan (1979), and the revised Shapley-Ames catalog show a very goo fit by the Lundmark quadratic law and corresponding discrepancies from the Hubble law. Such discrepancies have a systematic redshift dependenc convolved with the Lundmark law luminosity function, and so appear as evolution within the framework of the Friedmann cosmology. Is there any special reason to doubt that the evolution you describe may originate in this way and that physically there is no evolution?

Ellis: It is certainly true that the discrepancy between the Friedmann predictions and the observations is conventionally explained as luminosity evolution with little reason other than the expected changes in the stellar populations from considerations of the H-R diagram in our own galaxy. I will say, however, that the accumulation of large well-defined redshift samples at various magnitude limits will help clarify the non-Friedmann possibilities. Those data will be available to you at the earliest opportunity.

Xiang: Does your result of no evidence for evolution not support the high value of $q_o(+1)$, consistently found by Sandage and co-workers from the Hubble diagram of cluster galaxies, assuming no evolution?

Ellis: The no-evolution model implied by the AAT survey fits only the faint data to J $\sim$ 21.5, or equivalently z $\lesssim$ 0.3; thereafter, mild evolution is required. Since the Hubble diagram data covers a wider redshift range, evolutionary corrections would be necessary before deriving q_o. In a recent review (Ellis 1982), I was unable to reconcile the evolution implied by the field counts with that indicated on the Hubble diagram, assuming q_o is small (as local dynamical arguments imply). However, this was partly because the published Hubble diagrams are not in agreement with one another. Furthermore, one might question whether the giant ellipticals in rich clusters evolve in the way expected for normal early-type galaxies.

F PHOTOMETRY OF FOURTEEN DISTANT RICH CLUSTERS OF GALAXIES

W.J. Couch*, E.B. Newell
Mount Stromlo and Siding Spring Observatories, Research School of Physical Sciences, Autralian National University

ABSTRACT

We have recently completed a photometric study of fourteen rich clusters of galaxies in the redshift range $0.18 \leqslant z \leqslant 0.39$. The data are based on JF photographic photometry of each field. We report on the analysis of the cluster galaxy colour distributions; in particular we find that all the clusters in our sample with $z \gtrsim 0.26$ contain an excess number of blue galaxies (i.e., show the Butcher-Oemler effect). The blue excess, which was measured in terms of the ratio of the fraction of blue galaxies observed to that expected on the basis of Dressler's (1980) [morphological mix, local projected galaxy density]-correlations, ranges from 2 to ∿5. The highest value of 4.8 found in the cluster Cl0024+1654 (z=0.39), confirms Butcher and Oemler's (1978) observations of this cluster.

*Present Address: Physics Department, Durham University, England.

G. O. Abell and G. Chincarini (eds.), Early Evolution of the Universe and Its Present Structure, 93.

OLOR DISTRIBUTION OF FAINT GALAXIES AND QUASI-STELLAR OBJECTS

Richard G. Kron
Yerkes Observatory
The University of Chicago

The apparent colors in a flux-complete sample of gal-xies depend on the redshifts and the spectral types in the ample. These in turn depend on both luminosity evolution and volution of the shapes of the spectra. Thus in principle a reat deal about the evolution of galaxies can be learned rom complete multicolor surveys, especially now that large amples can be routinely generated by automatic machine easurement. The following will review what has been learned rom recent work.

An example is a new survey by workers at Bell Lab-ratories (Tyson, Valdes, and Jarvis, private communication). welve 4-m fields in the two bands J^+ and F are under analy-is with new image-classification software by F. Valdes 1982). Similar studies have indicated a distribution with a lue tail or with blue median colors (e.g., Karachentsev 980; Harris and Smith 1981). The trend to the blue seems to ave an amplitude of only a few tenths of a magnitude, and to et in fainter than B = 21. This phenomenon has also been laimed from Schmidt data (Phillipps, Fong, and Shanks 1981), ut is not always evident even in fainter samples (cf. color magnitude diagrams for field regions by Couch 1981). The lue trend has often been taken to be evidence for galaxy volution. However, Pence (1976) predicted a shift to bluer edian colors between B = 22 and B = 24, without evolution; his arises from the differential k-correction between those alaxies with hot spectra and those without.

Koo (1981) has studied complete samples in the four ands UJ^+FN. The random errors and incompleteness factors ere evaluated in detail, and are included in the models. hese models adopt Bruzual's (1981) evolving energy dis-ributions, and predict what the distribution of galaxies ould look like in each cell of color and magnitude. In color - color diagrams, the data show the expected change in

. O. Abell and G. Chincarini (eds.), Early Evolution of the Universe and Its Present Structure, 95–99.

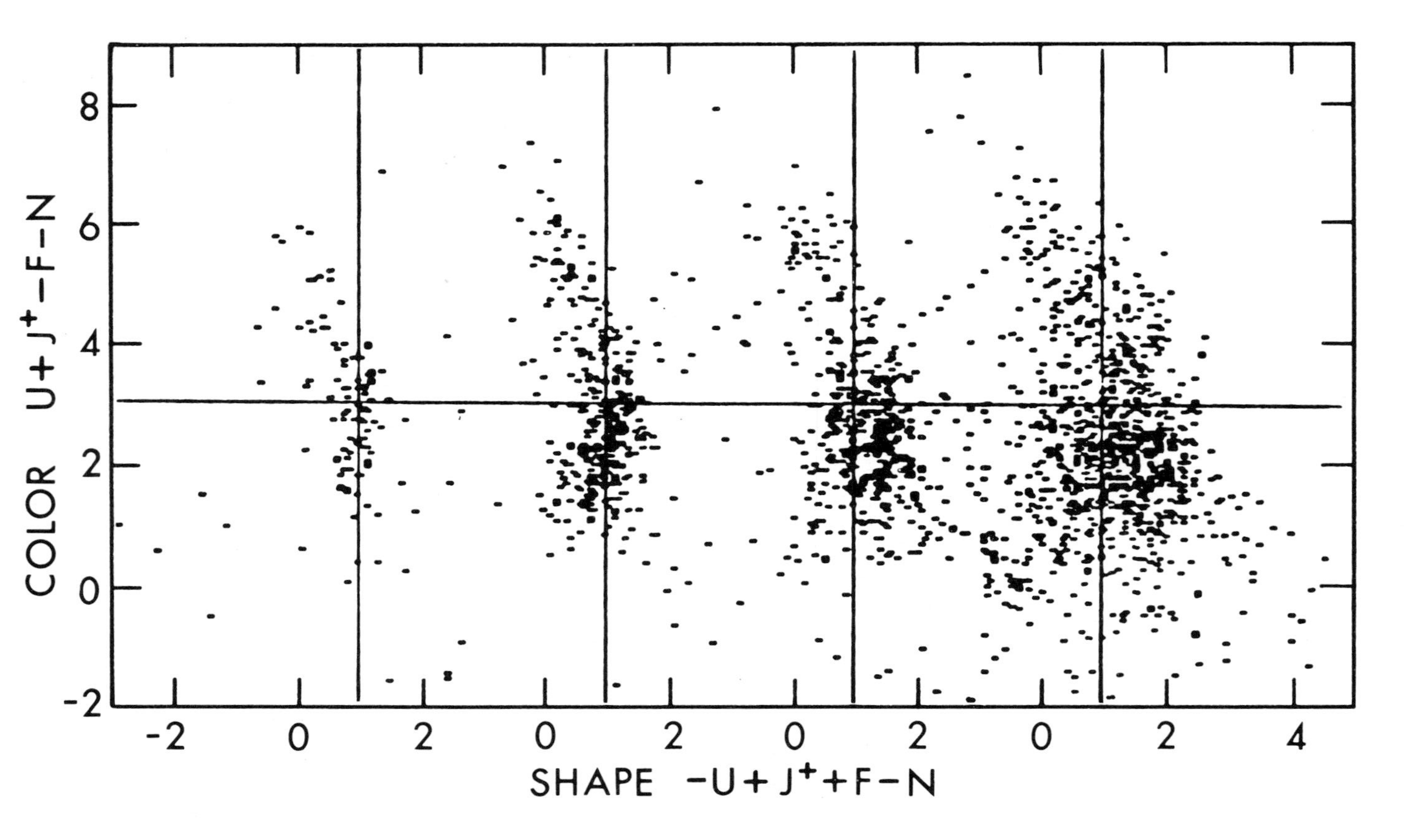

Figure 1. Koo's color-shape plots for galaxies in SA 68. The four segments are bins of J^+ magnitude: 20-21; 21-22; 22-22.5; 22.5-23. Lines of constant redshift are roughly vertical, and are separated by $\sim$0.25 mag in the shape parameter for each $\Delta z = 0.1$

the position of the centroid with increasing magnitude (Figure 1). In general Koo finds encouraging agreement between data and theory, but there are limitations. For example, the models do not reproduce the curvature in the distribution of points in Figure 1. Whether or not the apparently faint blue galaxies have high redshifts turns out to be a good test for the existence of large evolutionary effects, but the only unambiguous way to tell would be faint spectroscopy (Koo 1983).

Progress in infrared detector technology in the past few years enables measurement of many galaxies near optical detections limits, and in fact some radio source positions were first identified in the near infrared (Grasdalen 1980). Because of the small k-corrections, such observations are especially valuable for work on distant galaxies with cool spectra, like many radio sources. The study of the (K, z) Hubble diagram for radio galaxies is thus a natural first step (Grasdalen 1980; Lebofsky 1981; Puschell, Owen, and Laing 1982; Lilly and Longair 1982). (The evolutionary corrections which apply to the near infrared band may not necessarily be simpler nor smaller than evolutionary corrections in the optical.) High-z radio galaxies which have been measured for infrared colors, like 3C 265, 3C 6.1, 3C 184, 3C 34, and 3C 280, often have very strong narrow lines in the optical band [Spinrad (1982)] and could perhaps be peculiar in continuum shape. Lilly and Longair (1982) pointed out that if the redshift is high, infrared colors cannot distinguish between galaxies with and without a nonstellar power law contribution to the light. For some applications it may turn out that the apparent magnitude is a better redshift estimator than the colors, which if demonstrable would reflect on the true information content of colors.

Ellis and Allen (1982) have measured 47 faint field galaxies in the J band and, of these, 26 galaxies in the K band. The optical color-infrared color diagram can then be used to argue for a particular redshift distribution and morphological type distribution. The idea is thus similar to purely optical techniques discussed by Koo (1981) and Butchins (1981; 1982), except that Ellis and Allen use the fact that the J-K colors at z = 0 for most galaxies display only weak dependence on morphological type (Aaronson 1977) — therefore, a universal J-K _vs._ z relation may exist, at least for sufficiently low z. (For $z \gtrsim 0.25$, the dependence of J-K on z unfortunately is weak.) Six out of the 26 galaxies were found to have unusually red colors, with $\langle J\text{-}K \rangle = 1.89$. This would indicate high redshifts, but then one of the calibrating galaxies has J-K = 1.77 ± 0.10 and a spectroscopic redshift of only 0.30. [Some radio-quiet galaxies were found by Lebofsky (1981) to be very _blue_ compared with expectations.]

A variety of problems are in some way connected with counts of quasi-stellar objects, such as statistics relating gravitational lenses, the evolution of nonstellar light in galaxies, the integrated X-ray background, and the incidence radio-quiet BL Lac objects. Many techniques to isolate samp of QSO's have been developed, but the classical UV-excess criterion continues to be competitive (e.g., Formiggini _et a_ 1980; Notni 1980; Richer and Olson 1980; Usher and Mitchell 1982; and Arp and Surdej 1982). Brighter surveys pick up sta like white dwarfs and sdB's, but it can be argued that this contamination problem should be less severe at faint limits. the other hand, a critical problem is that _galaxies_ may have colors like QSO's; faint blue galaxies greatly outnumber QSO

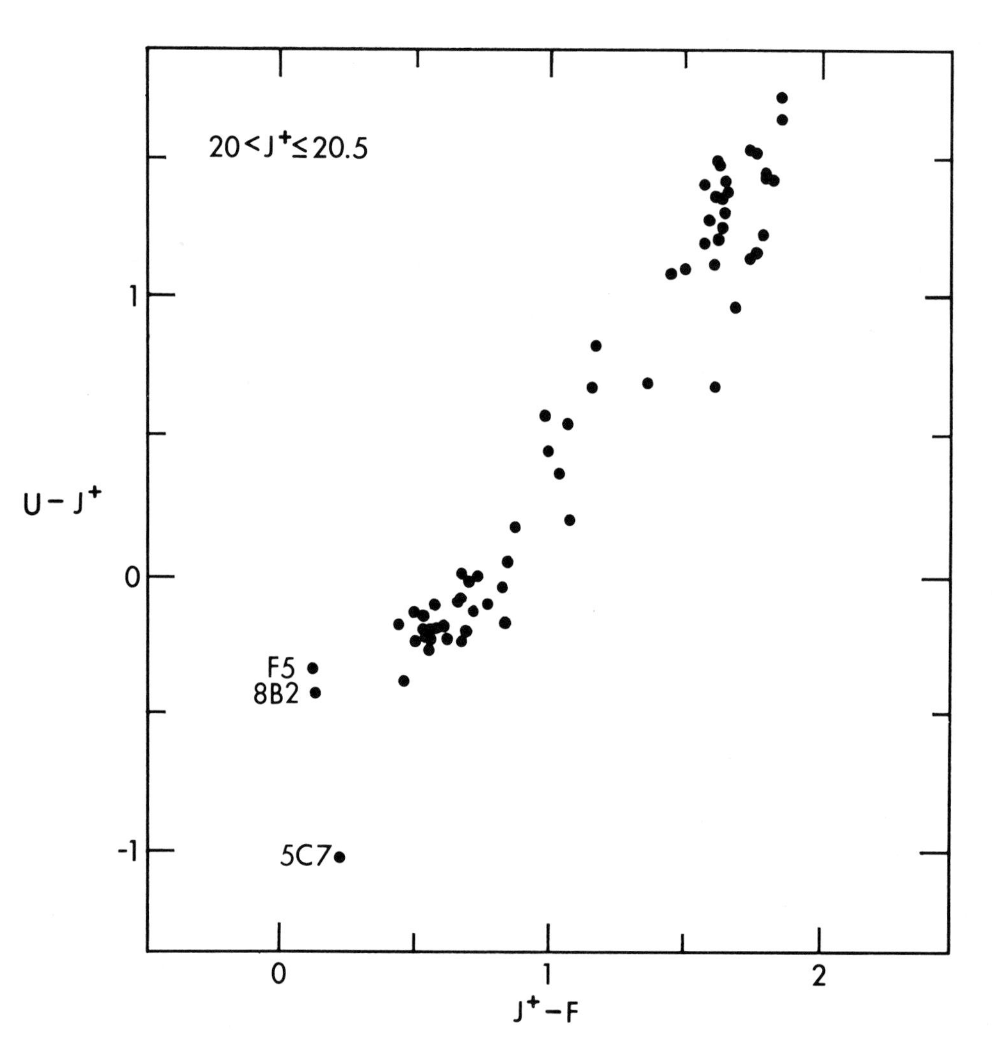

Figure 2. $U\text{-}J^+$ _vs_. $J^+\text{-}F$ for stellar objects in SA57.

and may be only marginally larger than the seeing disk. Koo and Kron (1982) used eight 4-m plates to obtain accurate colors and image sizes for all objects in the field of SA 68. They found that the total optical light from QSO's is converging for $B \gtrsim 21$; also, almost all of the objects which did not have the colors of ordinary stars had UV excess (see also Kron and Chiu 1981). An important qualification of Koo and Kron's work was the lack of spectroscopic verification. In the meantime, Koo and Kron (unpublished) have repeated the survey for the field of SA 57, for which some spectroscopic checks are available (Kron and Chiu 1981). Figure 2 shows a color - color diagram for all stellar objects within a half-magnitude interval, with three known QSO's identified. The other magnitude intervals confirm the accuracy of the color classification, so the contamination from hot stars should indeed be small. Still, this demonstration does not test the nature of the B = 22.5 candidates (see Koo 1982).

REFERENCES

Aaronson, M.: 1977, thesis, Harvard University.
Arp, H., and Surdej, J.: 1982, A.&A. 109, pp. 101-106.
Bruzual-A., G.: 1981, thesis, University of Calif., Berkeley.
Butchins, S.A.: 1981, A.&A. 97, pp. 407-409.
Butchins, S.A.: 1982, preprint.
Couch, W.J.: 1981, thesis, Australian National University.
Ellis, R.S., and Allen, D.A.: 1982, AAO preprint No. 167.
Formiggini, L., Zitelli, V., Bonoli, F., and Braccesi, A.: 1980, A.&A Suppl. 39, pp. 129-132.
Grasdalen, G.L.: 1980, in IAU Symp. No. 92, eds. G.O. Abell and P.J.E. Peebles (Dordrecht: Reidel), pp. 269-276.
Harris, W.E., and Smith, M.G.: 1981, A.J. 86, pp. 90-97.
Karachentsev, I.D.: 1980, Sov. Astron. Lett. 6, pp. 1-2.
Koo, D.C.: 1981, thesis, University of Calif., Berkeley.
Koo, D.C.: 1983, this volume, p. 105.
Koo, D.C., and Kron, R.G.: 1982, A.&A. 105, pp. 107-119.
Kron, R.G., and Chiu, L.-T.G: 1981, P.A.S.P. 93, pp. 397-404.
Lebofsky, M.J.: 1981, Astrophys. J. (Lett.) 245, pp. L59-L62.
Lilly, S.J., and Longair, M.S.: 1982, Mon. Not. Roy. Ast. Soc. 199, pp. 1053-1068.
Notni, P.: 1980, Astron. Nachr. 301, pp. 51-67.
Pence, W.: 1976, Astrophys. J. 203, pp. 39-51.
Phillips, S., Fong, R., and Shanks, T.: 1981, Mon. Not. Roy. Ast. Soc. 194, pp. 49-62.
Puschell, J.J., Owen, F.N., and Laing, R.: 1982, Astrophys. J. (Lett.) 257, pp. L57-L61.
Richer, H.B., and Olsen, B.I.: 1980, P.A.S.P. 92, pp. 573-575.
Spinrad, H.: 1982, P.A.S.P. 94, pp. 397-403.
Usher, P.D., and Mitchell, K.J.: 1982, Astrophys. J. Suppl. 49, pp. 27-52.
Valdes, F.: 1982, in S.P.I.E. Instrumentation in Astronomy IV, 331, March 1982, Tucson, ed. D.L. Crawford.

PECTROSCOPY OF DISTANT GALAXIES IN CLUSTERS

Alan Dressler
Mount Wilson and Las Campanas Observatories of the
Carnegie Institution of Washington

Butcher and Oemler have reported that many clusters of galaxies t high redshift have enhanced populations of blue galaxies when ompared to nearby clusters of similar type. Since this claim is ased on broad-band colors alone, it is essential to obtain spectra f these blue objects to determine if they are actually cluster members. These spectra can also be used to provide a rough morphological lassification of these objects that are too distant for detailed maging with ground-based telescopes.

Gunn and I are conducting such a study using a low-noise CCD etector and transmission grating spectrograph (the PFUEI) at the ale 5-m. Our first results are contained in a paper in the Dec. 15, 982, Astrophysical Journal. Here I report on new data that has ecently been obtained for one of the original Butcher-Oemler clusters, he one containing the radio galaxy 3C 295. We now have 26 good pectra of 6 red (V-R > 1.3) and 20 blue objects in the field studied y Butcher and Oemler. Our three principal conclusions are as follows:

(1) All six red galaxies we have studied are cluster members (Z $\approx$ 0.46) and have spectra similar to present-day ellipticals or S0 galaxies.

(2) Of the 20 blue objects, 11 are not members of the cluster (foreground and background), 6 are confirmed members and 3 have no determined redshifts. Thus, instead of a $\sim$40% population of blue galaxies, the true membership appears to be $\sim$ 20%. This is not an excessive fraction of blue galaxies compared to present-day clusters so the Butcher-Oemler effect is not confirmed in this first well-studied case.

(3) On the other hand, this small population of blue objects do not have the spectra of normal spiral galaxies (low-excitation emission) as is typical of the blue galaxies in nearby clusters, which are usually spirals. Instead, three of the six blue cluster members have the spectra of active galactic nuclei (high-excitation emission) in-

G. O. Abell and G. Chincarini (eds.), Early Evolution of the Universe and Its Present Structure, 101–103.

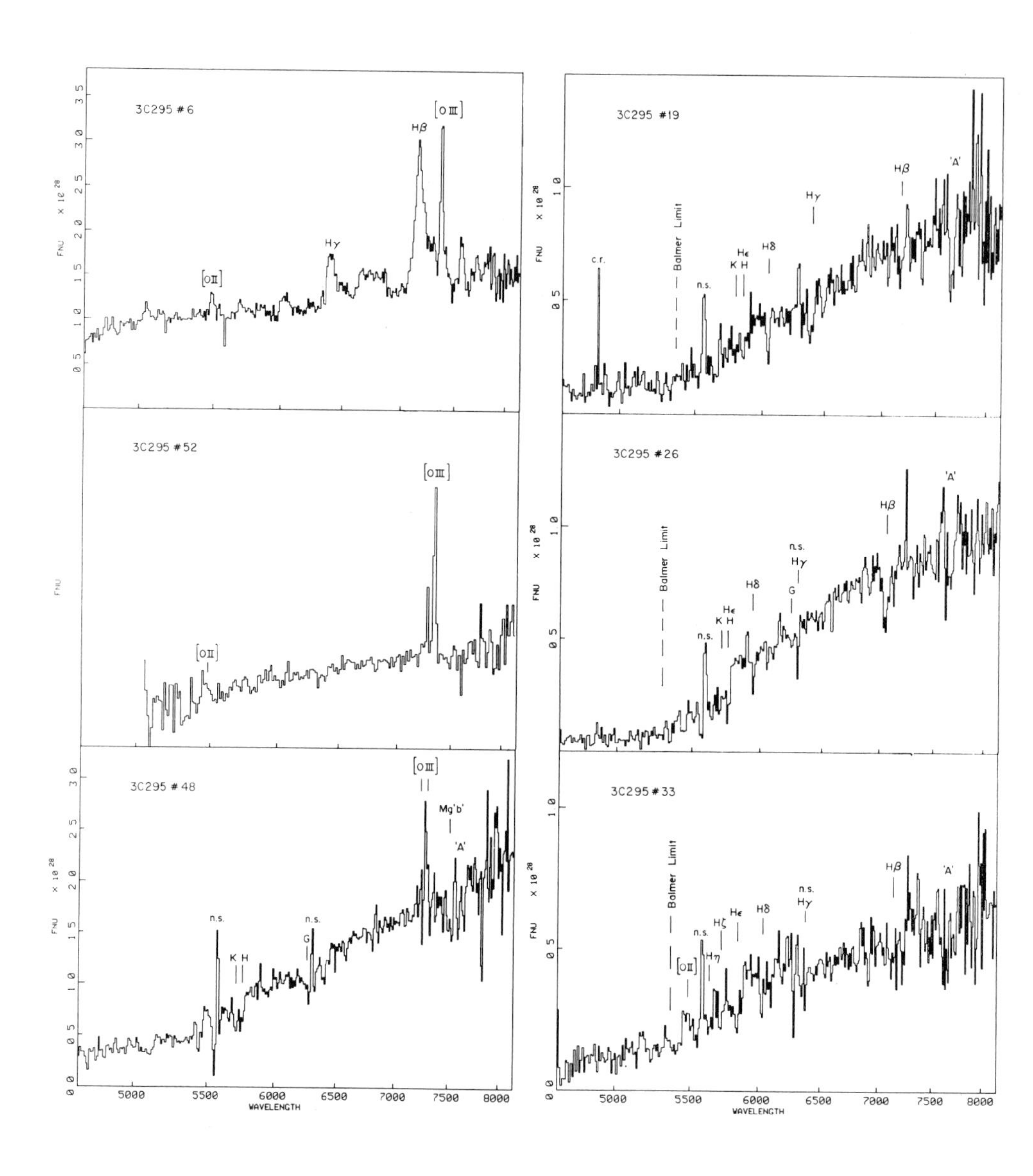

Fig. 1. The spectra of the six blue galaxies that have been found to be members of the 3C 295 cluster. The left panel shows the spectra of three galaxies with active nuclei. The right panel shows the spectra of three galaxies whose strongest features are Balmer absorption lines, probably resulting from a burst of star formation. Key: n.s. - night sky feature; 'A' - atmospheric A band.

cluding a Seyfert 1 and a Seyfert 2. The other 3 have very strong Balmer absorption lines with no emission features, which is indicative of a strong burst of star formation $\sim 10^9$ years old. The spectra of these six active galaxies are shown in Figure 1. Both of these types of galaxies are present today but they are rare, comprising only a few percent of the population. Therefore, if the much larger fraction of active galaxies in 3C 295 is typical of other high-redshift clusters, this would indicate a strong evolution of these types.

A REDSHIFT SURVEY OF VERY FAINT ($B \leq 22.5$) FIELD GALAXIES, RADIO SOURCES, AND QUASARS

David C. Koo

Department of Terrestrial Magnetism
Carnegie Institution of Washington
Washington, D. C. USA 20015

As part of a three year program to study the evolution of galaxies, quasars, and radio sources (in collaborations with R. Kron and R. Windhorst), we will use over 30 nights on the KPNO 4-m CCD Cryogenic Camera System, in the multi-aperture mode, to measure the redshifts of several hundred very faint objects. These will be selected from four fields (SA68 00h+15; Lynx 08h+45; SA57 13h+30; Herc 17h+50); for each we possess 4-m plates in four bandpasses (limits: $U \simeq 23$, $B \simeq 24$, $V \simeq 23$, $I \simeq 21$) as well as Westerbork interferometer maps to 1 mJy at 21cm (a 50cm survey is in progress). Galaxies are selected randomly by their apparent red magnitudes as measured with a PDS; quasar candidates to $B \approx 22.5$ by stellar objects which lie apart from Galactic stars in the multi-color diagram ; and radio sources by their positional coincidence with optical images ($\sim$50% of the 300 radio sources are identified). To date , 10 nights have produced spectra of $\sim$250 field galaxies, $\sim$20 radio sources, and $\sim$40 quasar candidates. Reductions are still underway, but we do have a few preliminary results:

1) We find that the magnitude-redshift distribution of a subsample of 54 field galaxies (Fig. 1) favors mild evolution, with the caveat that our current sample is obviously biased, e.g., by including only those galaxies with two or more strong spectral features, often emission lines. We have sacrificed completeness to assure reliability of our redshifts. This subsample still contains very faint galaxies ($B \approx m(6100)+1.5$). Yet not one galaxy has an unusually high redshift $z > 0.6$; evolution, if present, has been mild. Evidence for evolution can be seen, however, by noting that 24 galaxies in Fig. 1 lie <u>above</u> the no-evolution 75% line. About 14 are predicted, so this deviation is significant at the $(24-14)/\sqrt{14}$ or 2.7 sigma level. In contrast, the dashed 75% line of the evolutionary model does indeed separate a sample of $\sim$14. Both models are detailed by Koo (1981) and use Bruzual's (1981) evolutionary synthesis of galaxy spectra.

2) In the radio sample, a few apparently blue, very faint galaxies ($V \geq 21$) yielded low redshifts $z \sim 0.3$; the implied <u>intrinsic</u> blueness ($B-V \sim 0.5$) and normal luminosities ($M_V \sim -20.5$) both serve as warnings that distant radio galaxies do not all belong to one class of luminous

G. O. Abell and G. Chincarini (eds.), Early Evolution of the Universe and Its Present Structure, 105–106.

early-type galaxies.

3) We have counted 65 quasar candidates in a 1050 arc min^2 field in SA6 to B$\approx$22.5 (Koo and Kron 1982); we also counted 58 in SA57 (unpublished These results support a universe that is homogeneous and isotropic to within 10% at quasar redshifts and favor a luminosity evolution (LE) ove a density evolution (DE) model of quasars. We now have spectra of 26 candidates with B$\sim$22 ±0.5 for an area$\sim$600 arcmin^2: 9 show broad emission lines typical of quasars (5 have z$\sim$2.3); 6 have strong narrow emission in Hβ and [OIII] 5007,4959 (all with z$<$0.6 and UV excess; see Fig. 2 for one example); 3 possess a single narrow emission line (probably [OII] 3727; if so, z$<$0.65); and 8 display no strong feature so their identification is uncertain. The nature of the narrow-lined objects is unknown. Perhaps they are a compact subset of Markarian or Haro galaxies (at z$\approx$0.4, 7kpc equals 1" for H_o = 50). The log10 of the number density per Gpc^3 per magnitude interval at $M_B\sim$-20 (H_o = 50, q_o = 0, and no K corrections assumed) is 4.4 ±0.2. As for the high-redshift broad-lined quasars, the 5 with z$\sim$2.3 imply a density, in the above units but at $M_B\sim$-25, of 3.0,a factor of 30 below the prediction o DE models. In contrast, even if the 8 unknown candidates were all at z$\sim$2.3, the higher density would be better fit by models with LE (e.g., see Table 9 of Braccesi et al. 1980); with LE the brightness increase with z (for z$<$2.5) is close to $(1+z)^n$, where n$\approx$4.2 ±0.2. This survey thus lends spectroscopic support to a scenario in which the number densi of quasar objects or events has been constant since z$\sim$2, but in which they are today fainter by a factor of 100.

References

Braccesi, A., Zitelli, V., Bónoli, F., Formagginni, L.: 1980, Astron. Astrophys. 85,pp. 80-92.
Bruzual, G.A.: 1981, Ph.D. Dissertation, Univ. of Calif., Berkeley
Koo, D.C.: 1981, Ph.D. Dissertation, Univ. of Calif., Berkeley
Koo, D.C., Kron, R.G.: 1982, Astron. Astrophys. 105,pp. 107-119.

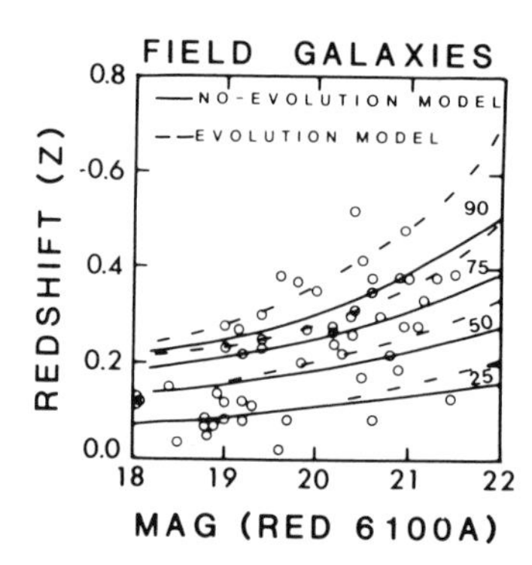

Fig. 1. Magnitude-redshift plot of 54 galaxies. Lines show model predictions of fraction (in percent) of total sample that should lie below.

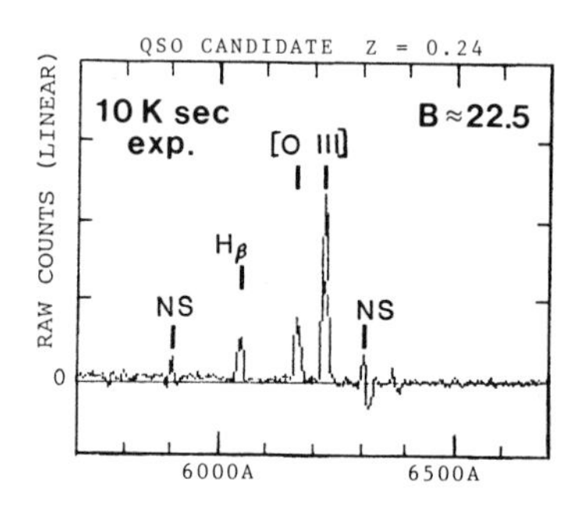

Fig. 2. Typical spectrum of a narrow-line quasar candidate. Night sky lines are labeled NS.

SPECTRAL ENERGY DISTRIBUTIONS OF GALAXIES IN MODERATE REDSHIFT CLUSTERS

W.J. Couch, R.S. Ellis
Physics Department, Durham University, England.
J. Godwin
Astrophysics Department, Oxford University, England
D. Carter
Mt. Stromlo and Siding Spring Observatories, Research School of Physical Sciences, Australian National University

ABSTRACT

We have observed 3 moderate redshift clusters using a combination of 7 intermediate band filters and 2 CCDs in order to derive photometric information for cluster galaxies from 400 nm to 900 nm. Preliminary results are presented for 2 clusters: Abell 1942 (z=0.224) and Abell 1525 (z=0.259) from 580 nm to 860 nm. The CCD photometry reaches a limit equivalent to R_F=21 mag with a precision of better than 0.1 mag. The galaxy colours derived from the intermediate band measurements are generally consistent with those expected at the appropriate redshift. However, in Abell 1525, and to a lesser extent in Abell 1942, a large proportion of cluster members have far red (720-860 nm) colours redder than expected. Many of these galaxies have blue photographic B_J-R_F colours. A possible explanation for the anomalous CCD colours is that these galaxies possess a strong emission line component which enters the far red filter at z=0.25.

G. O. Abell and G. Chincarini (eds.), Early Evolution of the Universe and Its Present Structure, 107.

DISTORTION OF THE MICROWAVE BACKGROUND BY DUST FROM POPULATION III

Michael Rowan-Robinson
Queen Mary College, Mile End Road, London, E.1.

Abstract:
Population III has been invoked to explain the missing mass in the haloes of galaxies, the first heavy elements in our Galaxy and even to explain the whole microwave background. However there are alternative explanations for each of these phenomena. The most compelling evidence for the existence of a pregalactic generation of objects is the observed distortion of the microwave background in the millimetre range, which can be explained as radiation from Population III objects absorbed and re-emitted by dust grains. If the distortion is confirmed, we can probably conclude that the density fluctuations in the early universe were isothermal and that no neutrino can have a mass in the astrophysically interesting range 1-100 eV.

1. DISTORTION OF THE MICROWAVE BACKGROUND

The idea of a pregalactic generation of objects, Population III, was revived by White & Rees (1978) in order to explain the dark haloes of galaxies, which they suggested could be made up of 10^6 $M_\odot$ black hole remnants of Population III objects. Truran & Cameron (1974) had earlier extensively discussed the possibility that the first metals in our Galaxy were made in a pregalactic generation of massive stars. Neither of these roles for Population III is essential, though. Massive neutrinos or hypothetical particles like gravitinos (which are a natural consequence of Grand Unified Theories) can account for the missing mass. The first metals in galaxies may equally well have been made in a transient generation of massive stars early in the life of the galaxies.

The idea that the microwave background radiation might in fact be background light from pregalactic objects rather than the relic of the hot Big Bang (Rees 1978 and references therein) has always come up against the difficulty of thermalising the radiation. To do so with normal dust grains requires excessively high primordial heavy element abundances (Rowan-Robinson et al 1979). The suggestion of Narlikar et al (1976) that long graphite whiskers could provide the thermalisation

G. O. Abell and G. Chincarini (eds.), Early Evolution of the Universe and Its Present Structure, 109–112.

has been developed further by Rana (1981) and Wright (1982), who find that very large length-to-radius ratios are required. Carr (1981) suggested that free-free absorption at early epochs ($z \sim 1000$) could provide the thermalisation provided the gas was highly clumpy, but this claim is disputed by Wright (1982). Unless a glaring inconsistency in the Big Bang predictions of the primordial abundances of ^2H, ^3He, ^4He, and ^7Li develops, these ideas do not seem very attractive.

The one role for Population III which may turn out to be essential, and which would have far-reaching implications for our ideas on the history of the early universe, is that light from Population III objects absorbed and re-emitted by dust grains can explain the distortion of the microwave background spectrum from a blackbody observed by the Berkeley group (Rowan-Robinson et al 1979, Negroponte et al 1981, Rowan-Robinson 1982). Initially Woody & Richards (1979) announced that their measurements were inconsistent with a blackbody spectrum at the 5σ level. A subsequent, more detailed analysis reduced the significance of this to 2.7σ (Woody & Richards 1981), though the distortion still appears significant when compared to the ground-based radio and molecular-line measurements. In their review Danese & de Zotta (1977) estimated that the latter implied a mean blackbody temperature

$$T = 2.73 \pm 0.05 \text{ K}$$

whereas Woody and Richards found the best blackbody fit to their data to be 2.96 K. Fig. 1 shows a comparison of two of the models of Negoponte et al (1981) with the Berkeley data and with the earlier ground-based measurements. Also included are the recent rocket measurements by Gush (1981) which show an even more drastic distortion from a blackbody form in the opposite sense to, and inconsistent with the Berkeley data. This experiment was unfortunately affected by the rocket motor moving into the field of view of the telescope: the data plotted have been corrected for the effect of this. Clearly this remarkable distortion needs to be confirmed and the inconsistency with the Berkeley data resolved. At present there seem to be four options:

(1) To believe the Berkeley data and assume that the Gush data are invalidated by calibration problems or the rocket motor incident. In this case the obsidian grain model of Negroponte et al provides an excellent fit to the observations.

(2) To believe the Gush data and assume that the Berkeley data are, for some reason, invalid. A Population III type of model would still probably be the simplest interpretation of the distortion, though the redshift at which the grains are assumed to form would need to be reduced to ~ 30. A Compton distortion (see eg Danese & de Zotta 1977) is also a possibility, though it would be expected to produce a more gradual distortion than that claimed by Gush.

(3) To suppose that each experiment is broadly correct over its range of maximum sensitivity, i.e. $\gtrsim 1$ mm for the Berkeley data and $\lesssim 1$ mm for the Gush experiment. In this case the dirty silicate model of Negroponte et al gives results broadly consistent with the observations though predicting a somewhat lower amplitude distortion than observed.

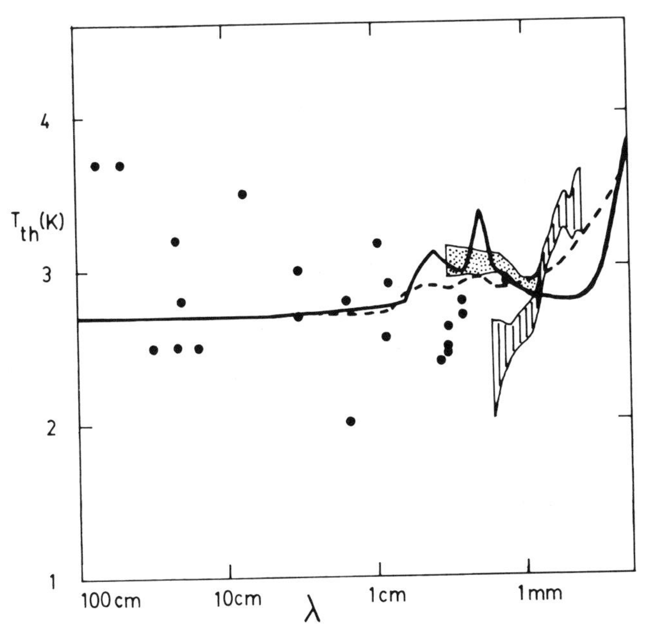

Fig.1: Microwave background spectrum, in form of plot of thermodynamic temperature against wavelength, for 2 models given by Negroponte et al (1981): solid curve, obsidian grains (z_f = 225, Z = 5 x 10^{-6}, Ω = 1), broken curve dirty silicate grains (z_f = 225, Z = 2.5 x 10^{-6}, Ω = 1). Dotted shaded area: Woody & Richards data (± 1 σ limits). Filled circles: ground-based measurements, taken from the review by Danese & de Zotta.

(4) To believe neither distortion and suppose that the microwave background spectrum is, within present observational limits, a blackbody with 2.7 K ≲ T ≲ 3.0K.

2. IMPLICATIONS OF POPULATION III

The implications of options (1) to (3) above, which seem to require the existence of Population III, are interesting because they run counter to several contemporary bandwagons.

(a) For Population III to have existed we need isothermal density fluctuations to have been present prior to the decoupling era. The natural prediction of the GUTs, however, is adiabatic fluctuations (e.g. Nanopoulos 1982), though scenarios in which isothermal fluctuations are produced can be invented (e.g. Barrow & Turner 1981).

(b) · No neutrinos can have a mass in the astrophysically interesting range 1 to 100 eV, since the Jeans mass at decoupling would have been far too high for 10^6 $M_\odot$ objects to have formed (Bond et al (1980). However GUTs can supply other particles, e.g. gravitinos, to solve the missing mass problem, and these could have a high enough rest mass for the relevant Jeans mass at decoupling to be reduced to the range needed for Population III to be able to form. Both these implications lend added interest to the experiments now in progress to measure the spectrum of the background accurately in the millimetre range.

Peter Tarbet and I have been looking into the helium and heavy element yields expected from Population III stars (Rowan-Robinson & Tarbet 1982, Tarbet & Rowan-Robinson 1982). Making plausible assumptions about the mass function of the stars, mass loss and the helium and heavy element yields from the stars (which we suppose to extend from 10 to 10^6 $M_\odot$), we find that the light required to make the distortion postulated by Negroponte et al and the heavy elements needed to form the thermalising dust grains can be generated without violating observational constraints on primordial helium and heavy elements in the Galaxy, provided the mass function extends up to 10^6 $M_\odot$ accreting black holes. Models can also be found which generate the whole microwave background and the observed primordial helium. However if very massive stars ($500 \leq M/M_\odot \leq 10^5$) give a small yield of heavy elements ($\sim$ 2%) as predicted by Klapp (1982), these models generate far too high a primordial heavy element abundance.

The interesting consequences of Population III make it worthwhile to examine in more detail the astrophysics of the pregalactic era, including the interaction of Population III objects with the pregalactic interstellar medium (reionisation, Compton distortion of the background spectrum) and the possibility that the more massive of the Population III black hole remnants may gravitate to galactic nuclei to form the power house for quasars and related phenomena.

REFERENCES

Barrow, J.D., & Turner, M.S., 1981, Nature 291, 469.
Bond, J.R., Efstathiou, G., & Silk, J., 1980, Phys.Rev.Lett. 45, 1980.
Carr, B.J., 1981, Mon. Not. R, Astr.Soc. 195, 669.
Danese, L., & de Zotta, G., 1977, Riv. Nuovo Cim. 7, 277.
Gush, H.P., 1981, Phys.Rev. Lett. 47, 745.
Klapp, J., 1982, Ph. D. thesis, Oxford Univ.
Nanopoulos, D.,1981, 'Cosmology & Particles', ed. J. Audouze et al, p.89.
Narlikar, J.V., Edmunds, M.G., & Wickramasinghe, N.C., 1976, 'Far Infrared Astronomy', ed. M. Rowan-Robinson (Pergamon, Oxford), p. 131.
Negroponte, J., Rowan-Robinson, M. & Silk, J., 1981, Astrophs.J. 248, 38
Rana, N.C., 1981 Mon.Not. R. Astr.Soc. 197, 1125.
Rees, M., 1978, Nature 275, 35.
Rowan-Robinson, M., 1982, Varenna Summer School on 'Gamov Cosmology'.
Rowan-Robinson, M., & Tarbet, P.J., 1982 'Progress in Cosmology', ed. A. Wolfendale (Reidel, Dordrecht) p.101.
Tarbet, P.J., & Rowan-Robinson, M. 1982, Nature (in press).
Truran, J.W., & Cameron, A.G.W., 1974, Astrophys. J. 190, 605.
White, S.D.M., & Rees, M.J., 1978, Mon.Not. R. Astro.Soc. 183, 347.
Woody, D.P., & Richards, P.L., 1979, Phys. Rev.Lett. 42, 925.
Woody, D.P., & Richards, P.L., Astrophys. J. 248, 18.
Wright, E.L., 1982 Astrophys. J. 255, 401.

WAS THE BIG BANG HOT?

Edward L. Wright
Department of Astronomy, UCLA

ABSTRACT: I consider experiments to confirm the substantial deviations from a Planck curve in the Woody and Richards spectrum of the microwave background, and search for conducting needles in our galaxy. Spectral deviations and needle-shaped grains are expected for a cold Big Bang, but are not required by a hot Big Bang.

I. INTRODUCTION

The temperature of the Big Bang is critical to an understanding of the early Universe. One normally assumes that the 3 K background measures this temperature, but the possibility that Population III stars could have produced the background after the Big Bang must be considered (Rees 1978). Woody and Richards (1981, hereafter WR) measured a substantial deviation from a Planck curve in the 3 K background, with a large excess flux near the peak, that cannot be explained by the simplest hot Big Bang models. Negroponte, Rowan-Robinson and Silk (1981) and Wright (1981) have shown that the WR excess at 1.5 mm can be explained by hot silicate dust at a redshift of 150. Such models require that 30-40% of the total energy in the 3 K background is added well after the Big Bang, but do not significantly alter the events before decoupling. Cold Big Bang models attempting to produce 99-100% of the background after the Big Bang were tried (Layzer and Hively 1973; Carr 1981) but a mechanism to thermalize the long wavelength tail was lacking. New work (Rana 1979; Wright 1982) has shown that very small abundances of needle-shaped conducting grains can provide the required opacity, so a cold or tepid Big Bang is possible. In this paper I will consider ways of verifying the WR spectrum, which is the experimental data behind dust-distorted models; and I will see whether needles can be seen in our galaxy.

II. INDIRECT METHODS OF VERIFYING THE WOODY-RICHARDS DISTORTION

Two techniques have been proposed for obtaining information about the absolute intensity of the background without doing an absolute radiometric experiment. One is to measure the frequency dependence of

G. O. Abell and G. Chincarini (eds.), Early Evolution of the Universe and Its Present Structure, 113–118.

the dipole anisotropy (Lubin and Smoot 1981; Danese and De Zotti 1981) of the background and the other is to measure the frequency dependence of the Sunyaev-Zeldovich (hereafter SZ) effect in clusters of galaxies (Rephaeli 1981). Both of these methods use the Doppler shift to convert a spectral gradient into a spatial inhomogeneity. Since a spatial variation can be measured using position switching, atmospheric effects and stray light are much easier to control. This advantage is so great that it is practical to look for 0.3 mK effects in the anisotropy in order to confirm a 300 mK distortion in the spectrum.

Since the Doppler shift produces a frequency shift proportional to the frequency, a logarithmic frequency variable is useful. Therefore I will define a variable s by $\nu = \nu_o \exp(s)$ where ν_o is an arbitrary frequency normalization. The corresponding intensity variable should be measured in photons per logarithmic frequency interval, which is proportional to I_ν. Thus I can write an unperturbed blackbody spectrum as

$$I(s) = B_\nu(T_o) \qquad \text{with } \nu = \nu_o e^s.$$

The change in I due to a Doppler shift giving a redshift z is

$$-(1+z)\partial I/\partial z = 3I - \partial I/\partial s$$

Note that $I \propto \nu^3$, a constant density in phase space, gives $\Delta I = 0$.

The SZ effect involves a random distribution of redshifts and blueshifts due to thermal motion of the scattering electrons. This leads to a diffusion in frequency plus a net increase due to an excess of blueshifts. The change in I for $h\nu << kT_e << mc^2$ is

$$\partial I/\partial y = \partial^2 I/\partial s^2 - 3\,\partial I/\partial s \qquad \text{with } y = \tau_e kT_e/mc^2 .$$

$I_\nu \propto \nu^3$ gives $\Delta I = 0$ as before, but now $I_\nu \propto \nu^0$ also gives $\Delta I = 0$. Thus the SZ effect changes sign near the peak of I_ν vs. ν.

In order to evaluate $\partial I/\partial s$ and $\partial^2 I/\partial s^2$ a smooth model flux is needed. It is not possible to numerically differentiate noisy experimental data and get reasonable results. Thus I have constructed an ad hoc model to fit the WR spectrum and the low frequency points. The form of this model is

$$I_\nu = [1 + a \exp(-b \ln\{\nu/\nu_1\}^2)]\, B_\nu(T_o)$$

with $a = 0.252$, $b = 4.50$, $\nu_1 = 6.158\ \text{cm}^{-1}$, and $T_o = 2.792$ K. This four parameter fit to the WR plus low frequency data gives $\chi^2 = 21.9$ with 22 degrees of freedom. Given this model fit to the WR spectrum I can compute ΔI_d (for Doppler or dipole) and ΔI_{sz} for both the WR spectrum and the null hypothesis, a blackbody (BB) spectrum with T = 2.734 K. In the following Table, all intensities have been expressed as Rayleigh-Jeans brightness temperature for the convenience of radio astronomers. The dipole anisotropy columns have been normalized to

3.78 mK in the low frequency limit (Boughn, Cheng and Wilkinson 1981) while the SZ effect is normalized to -1 mK. Note that the second derivative in the SZ effect emphasizes the sharply peaked excess flux in the WR spectrum, giving an effect at 150 GHz that is 100% higher than the BB spectrum. Unfortunately the SZ effect has never been observed with enough precision at two appropriate frequencies, so there is no data available now to confirm the WR distortion using the SZ effect.

TABLE 1: Blackbody (BB) vs. Woody-Richards (WR)

ν	Dipole			SZ		
(GHz)	BB	WR	Ratio	BB	WR	Ratio
30	3.693	3.696	1.001	-0.954	-0.956	1.002
60	3.449	3.446	0.999	-0.832	-0.815	0.980
90	3.083	2.933	0.951	-0.654	-0.544	0.832
120	2.646	2.495	0.943	-0.459	-0.544	1.185
150	2.187	2.386	1.091	-0.277	-0.581	2.097
180	1.748	2.220	1.270	-0.129	-0.356	2.760
210	1.356	1.842	1.358	-0.022	-0.071	3.227
240	1.024	1.386	1.354	+0.046	+0.101	2.197
270	0.757	0.985	1.266	+0.082	+0.159	1.939
300	0.548	0.681	1.243	+0.095	+0.157	1.653

The dipole anisotropy has recently been measured at 90 GHz and 184 GHz by two different, highly sensitive balloon experiments. The 90 GHz experiment (Lubin 1982) measured a dipole magnitude of 2.95 ± 0.1 mK, while a preliminary analysis of the MIT 184 GHz channel gives 1.6 ± 0.3 mK (Wright, Halpern and Weiss 1982). The ratio of 184 to 90 GHz dipoles is 0.54 ± 0.10, while the predicted ratio is 0.55 for BB and 0.67 for WR, so neither spectrum can be ruled out. Current experiments have adequate sensitivity for a definitive result, but careful cross-calibration will be essential.

III. DO CONDUCTING NEEDLE-SHAPED GRAINS EXIST?

One byproduct of measurements of the anisotropy of the microwave background is an estimate of the emission from our galaxy. The MIT experiment used 4 frequencies in order to determine the spectrum of the galactic emission. This offers a chance to look for emission from conducting needle-shaped grains that could thermalize the long wavelength tail of the microwave background.

An anisotropy experiment cannot measure the absolute intensity of the galactic emission, but only its spatial gradient. Thus, in order to determine T_p, the brightness temperature at the galactic pole, one has to compare the difference between the galactic plane and the pole with a model. For the MIT data with a 16° beam I have used a csc(b) model with a smooth cutoff at the galactic plane that approximates the FWHM of the beam. In addition I have included a galactic plane term with longitude variation. The high frequency channels have strong galactic emission but very little dipole signal, so I have used these signals to define the shape of the galactic model. Then I use a four parameter fit of dipole plus galaxy to find the emission spectrum of the galactic dust. The results are given in Table 2, along with lower frequency data from Lubin (1982) and Wilkinson (1982). Remember that these numbers are quoted at the pole, but are really measured close to the galactic plane. The average brightness of the galactic plane over the observed range of $75^o < \ell < 240^o$ is 14 times the value given for T_p in the four MIT channels.

TABLE 2: Galactic Emission

ν(GHz)	25	90	184	429	729	925
T_p(μK)	250 ± 70	45 ± 23	58 ± 19	69 ± 16	93 ± 20	116 ± 25

There is an excess emission in the 90-184 GHz data, but the excess is only slightly significant. Also, comparing different experiments is difficult unless they have identical sky coverage; but the MIT data is only from the outer galaxy while the low frequency experiments have better coverage.

If the excess galactic emission at 90-184 GHz is real, a fit to a sum of ordinary dust with emissivity $\propto \nu^2$ plus long needles with constant emissivity gives an optical depth of $(6 \pm 3) \times 10^{-5}$ due to needles with T = 3.8 K and an optical depth of $(4.5 \pm 0.8) \times 10^{-6}$ at 1 mm due to normal dust with T = 15 K. In this model the needles radiate 0.21% of the total galactic power. Since needles emitting just 0.1% of the total power could be cosmologically significant, it is very important that better 30-300 GHz spectra of the emission of cold galactic dust be obtained. Measurements of small, visually opaque dark clouds using large ground-based telescopes should give better data on dust emission than all-sky, large beam measurements of the entire galaxy.

This work was supported in part by NASA contract NAS 5-26994 to UCLA.

REFERENCES

Boughn, S. P., Cheng, E. S., and Wilkinson, D. T. 1981, Ap.J.(Letters), 243, L113.
Carr, B. J. 1981, M.N.R.A.S., 195, 669.
Danese, L. and De Zotti, G. 1981, Astr. and Ap., 94, L33.
Layzer, D. and Hively, R. 1973, Ap.J., 179, 361.
Lubin, P. M. 1982, private communication.
Lubin, P. M., and Smoot, G. F. 1981, Ap.J., 245, 1.
Negroponte, J., Rowan-Robinson, M., and Silk, J. 1981, Ap.J., 248, 38.
Rana, N. C. 1979, Ap.Space Sci., 66, 173.
Rees, M. J. 1978, Nature, 275, 35.
Rephaeli, Y. 1980, Ap.J., 241, 858.
Wilkinson, D. T. 1982, private communication.
Woody, D. P., and Richards, P. L. 1981, Ap.J., 248, 18.
Wright, E. L. 1981, Ap.J., 250, 1.
________. 1982, Ap.J., 255, 401.
Wright, E. L., Halpern, M., and Weiss, R. 1982, Bull.AAS, 14, 576.

DISCUSSION

(The discussion of the papers by Professors Rowan-Robinson and Wright was deferred until after Wright's paper. The following question, from Dr. Segal, was directed to Rowan-Robinson, but answered by Wright; Rowan-Robinson has waived the offer to submit a written response to Dr. Segal, thereby yielding to Wright's reply to Segal. Ed.*)*

Segal: The cosmic background radiation is, of course, not uniquely indicative of a Big Bang, but a Planck law for the background photons is implied by any temporally homogeneous theory in which the energy is modelled, as usual, by the infinitesimal time evolution generator. A very simply quasiphenomenological explanation of the Woody-Richards anomaly is a postulated non-vanishing isotropic angular momentum for the CBR in, for example, the vicinity of the Local Group. This provides a very good fit to their data, depends only on a single contemporary parameter rather than by hypothetical events at redshifts such as 200 or 1000, and automatically displaces the pure black-body law in the observed direction, rather than the opposite direction, as early discussions of perturbations of a Big Bang predicted. Therefore, isn't this scientifically more economical and in principle empirically accessible explanation for the Woody-Richards anomaly more natural than those presented that require a complete scenario hardly capable, in principle, of independent substantiation?

Wright: The Jakobsen, Kon, and Segal model (1979, Physical Review Letters, 42, 1788, hereafter JKS) of the Woody and Richards (WR) spectrum has two basic flaws. The first flaw is that it does not fit the data if the low frequency results are included. The Planck brightness temperature of the JKS model is a nonincreasing function of frequency, while the observed data rises from 2.7 K at low frequencies to 3.0 K at the peak, then falls to 2.8 K on the high frequency side of the peak. The JKS model matches the WR spectrum at the peak and higher frequencies, but predicts 3.4 K at low frequencies (see accompanying figure).

The second flaw in the JKS model is that the predicted background is inhomogeneous and anisotropic (Wright, 1980, Physical Review D, 22, 2361). The local perturbation just proposed by Segal is also manifestly inhomogeneous. An inhomogeneous background violates the cosmological principle, and is thus incompatible with all modern cosmological models, including the chronometric cosmology of Segal.

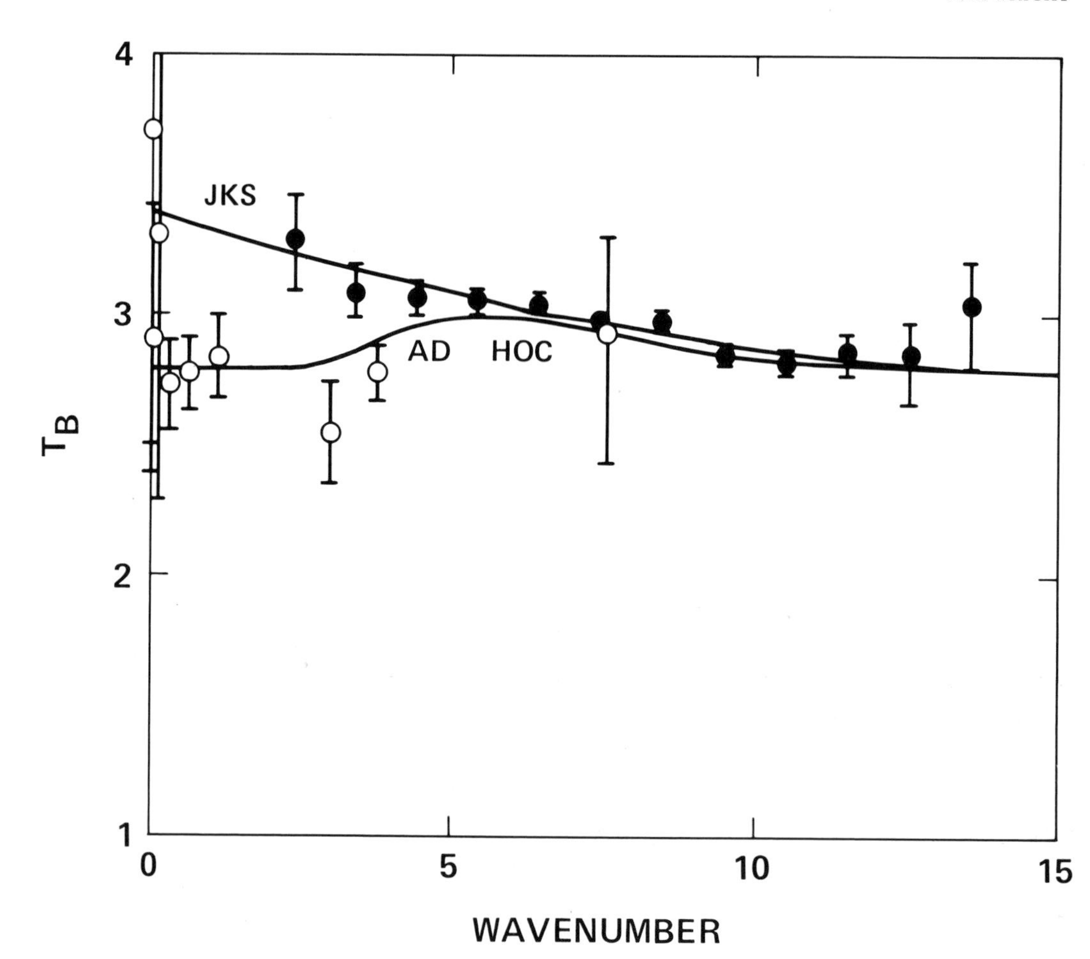

Comparison of the Jakobson, Kon and Segal model (dashed curve) and Wright's *ad hoc* fit (solid curve) to the Woody and Richards data (filled symbols) and the ground-based and CN data (open symbols. While the JKS model fits the W-R data, it is not consistent with the ground-based low-frequency data.

ON POPULATION III STAR FORMATION

A. Kashlinsky and M.J. Rees
Institute of Astronomy,
Madingley Road, Cambridge CB3 OHA.

If primordial fluctuations were isothermal their amplitude at recombination would be non-linear on scales $M_o \simeq 10^{6\div 9}\ M_\odot$. Since the Jeans mass after recombination is $M_{Jo} \simeq 8 \times 10^5\ \Omega^{-\frac{1}{2}}\ M_\odot$ the clouds of mass M_o would be able to form the first generation of compact objects, the so-called Population III. These clouds would acquire angular momentum via tidal interactions with their neighbours. The importance of rotation can be conveniently characterised by the spin parameter $\lambda = V_{rotation}/V_{free-fall}$ and tidal interactions lead to a spin $\lambda_o = 0.07 \pm 0.03$. As the cloud collapses λ increases as $r^{-\frac{1}{2}}$. Any fragment forming in a rotating cloud would have the same spin λ as the whole cloud. It could therefore collapse only by $\simeq \lambda_o{}^2$ in radius before centrifugal forces intervened, thus leaving a large geometrical cross-section for coalescence to be important. At radii $r \lesssim \lambda_o{}^{8/5}$ $(M_o/M_{Jo})^{2/15}\ r_o$ the coalescence time is shorter than the free-fall time and no fragmentation is possible below this radius. In the primordial clouds two major factors prevent fragmentation at larger radii. First, the background radiation is still 'hot' and the trapping of it would prevent fragmentation until the whole cloud has collapsed to a radius $10^{-2}\ x^{-2/3}\ r_o$. Here $x = 10^{-2}(M/10^7\ M_\odot)^{1/3}$ is the ionization fraction given by the balance between gravitational contraction and recombination cooling. Furthermore, any small density fluctuation would lead to fragmentation only after the paternal cloud had collapsed by a factor $(\delta/5)^{2/3}$ in radius. For these reasons fragmentation is unlikely until centrifugal forces halt the collapse and a disk forms. The disk will be initially at $T \simeq 10^4$K but after a small fraction of H_2 forms it will cool to $T_3 \simeq T/10^3\text{K} \simeq 1$ and the final fragments mass could be as low as $\simeq 0.2(\lambda_o/0.07)^4\ T_3{}^2(M_{Jo}/M_o)^{1/3}\ M_\odot$.

After the disk has fragmented the two-body interactions between stars will provide effective viscosity which would redistribute the angular momentum: the system will 'sphericalise' and evaporation of stars will begin. After some fraction of them have evaporated collisions between stars would become important and the likely outcome of it would be a formation of supermassive object (SMO). Thus, two different types of object would form: SMOs, and low-mass stars. We discuss the

G. O. Abell and G. Chincarini (eds.), Early Evolution of the Universe and Its Present Structure, 119–120.

proportions of these constitutents in terms of M_o and λ_o, and consider whether clusters of low-mass Population III stars should survive.

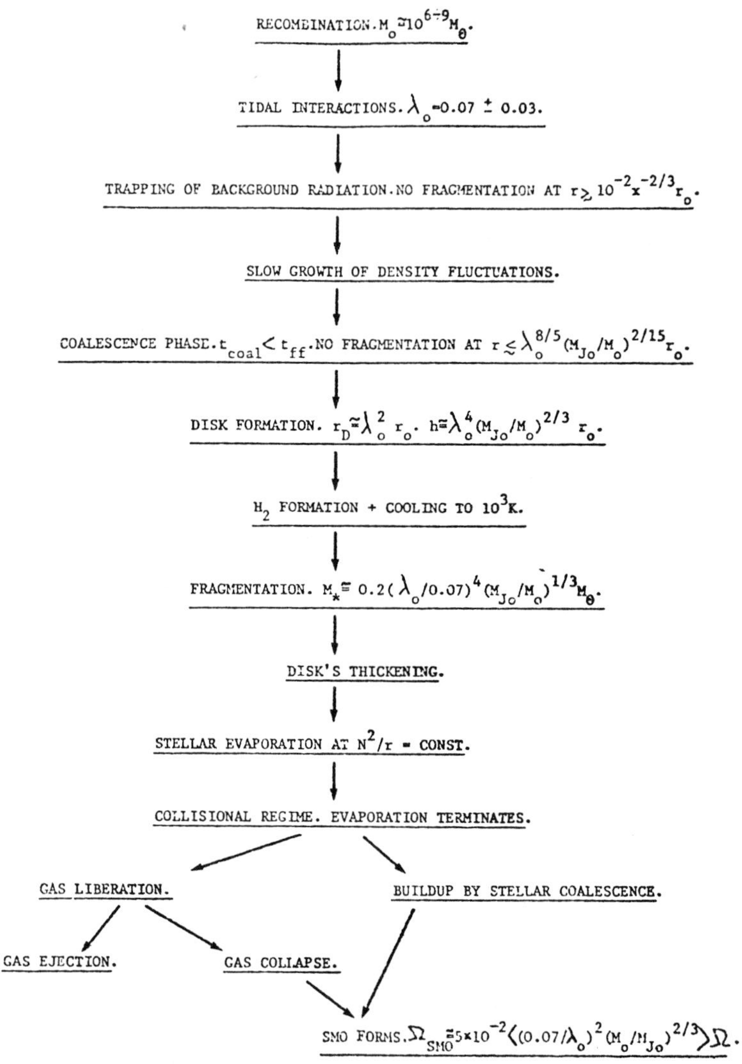

The Figure illustrates the evolution of the Population III systems.Co lisional regime will start once the velocity dispersion becomes comparabl to the escape velocity V_e from the surface of the star,when the total mas left in the system is $(c_o/\lambda_o V_e)^2 (M_o/M_{Jo})^{5/3} M_{Jo}$ (c_o is the speed of sound at 10^4K).Since this is also the mass of the SMO that forms,the fraction o mass in SMOs would be $\Omega_{SMO}=0.05\ (0.07/\lambda_o)^2 (M_o/M_{Jo})^{2/3}\ \Omega$.The gas liberate in stellar collisions will be ejected from the system if the collisional luminosity exceeds the Eddington limit;otherwise the gas will be retaine and will collapse to form SMO.All these processes along with the conditio that most of the Population III systems lose their individual identity by the present epoch definedifferent regions on (λ_o, M_o) plane and we discuss them in more detail elsewhere (Kashlinsky and Rees in preparation).

SEARCH FOR SMALL SCALE ANISOTROPY OF THE 3K EMISSION OF THE UNIVERSE

A.B. Berlin, E.V. Bulaenko, V.V. Vitkovsky, V.K. Kononov, Yu.N. Parijskij, and Z.E. Petrov.
Special Astrophysical Observatory, Nizhnij Arkhyz Stavropolskij Kraj, 357140, USSR

One of the greatest mysteries of nature is the absence of any trace of the present structure of the nearby universe in its relict 3K emission.

If we live in an evolving world (which evolves from extremely smooth to extremely structural), our radio telescopes should see observable temperature variations of 3K in practically any world model.

The second, and maybe even greater, mystery is the (observable by radio means) fact of thermodynamic equilibrium of different volumes of the primordial gas, which are separated by such distances that they are causatively independent in standard Big Bang theory.

These problems provide strong motivation for observers. It is not possible to review here all observations in the U.S., U.S.S.R., and Europe (see Boynton 1974; Partridge 1979). We realize that there has been an overinterpretation of Soviet results (many wrong corrections and statements, and even confusion about observing facilities). Our early (from 1968) results may be found in the references to Parijskij (1973), A short summary of the next attempt with RATAN-600 was published by Parijskij _et al_. (1977). Here we shall review our recent 1980-1981 results, again with RATAN-600.

The observations were made during March-May 1980 in meridian (transit mode) and in February-March 1981 in azimuth 30° (transit mode again) at wavelengths of 1.38 cm, 2.08 cm, 3.9 cm, 7.6 cm, 8.2 cm, and 31 cm with real sensitivities 70 mK, 30 mK, 15 mK 2 mK, 7 mK, and 50 mK (τ = 1 s). From about 120^d (2880^h) of the telescope time it was possible to collect 64^d (1536^h) of observations with our best 7.6 receiver (Berlin _et al_. 1981) and 2100^h observations with all other receivers. This time was spent on the narrow strip of the sky centered at a declination of SS 433 and about 10' wide at half power level at 7.6 cm, and about 45' at the level 0.1.

At 7.6 cm the resolution in right ascension is 0!9. At other wavelengths the beams are proportionally scaled. Thus, we have considerable information about the anisotropy of the sky on scales from 7" to 2π. The best results were obtained at 7.6 cm. We used 31 cm to allow for the galaxy noise, and 3.9 cm and 2.08 cm for atmospheric thermal emission correction. A special feed system and screens were used to reduce

G. O. Abell and G. Chincarini (eds.), Early Evolution of the Universe and Its Present Structure, 121–124.

spillover effects down to 2%. The feed system was optimized to a maximum brightness temperature sensitivity (with some losses in gain). After averaging over 64^d we expect to have "thermal receiver noise" on the main protocluster scale ($\sim$ 7'), a sensitivity of 1.2×10^{-5} in $\Delta T/T$ and 3000 independent points of the mean curve. Thus, to check the "null hypothesis," the expected 1σ accuracy in $\Delta T_A/T$ is $1.2 \times 10^{-5}/\sqrt{3000} = 2.2 \times 10^{-7}$, or 6×10^{-7} in $\Delta T_B/T$, when all the usual corrections were made.

Up to now only 5% of the data have been fully reduced, the main problem being not the thermal receiver noise but the confusion, the atmosphere and galaxy emission and man-made interference. Figure 1 shows r.m.s. fluctuations of antenna temperature as a function of scale. At the left the galactic and atmospheric noises dominate, at the right the confusion. We have realized that at the deepest point of the curve receiver noise dominates.

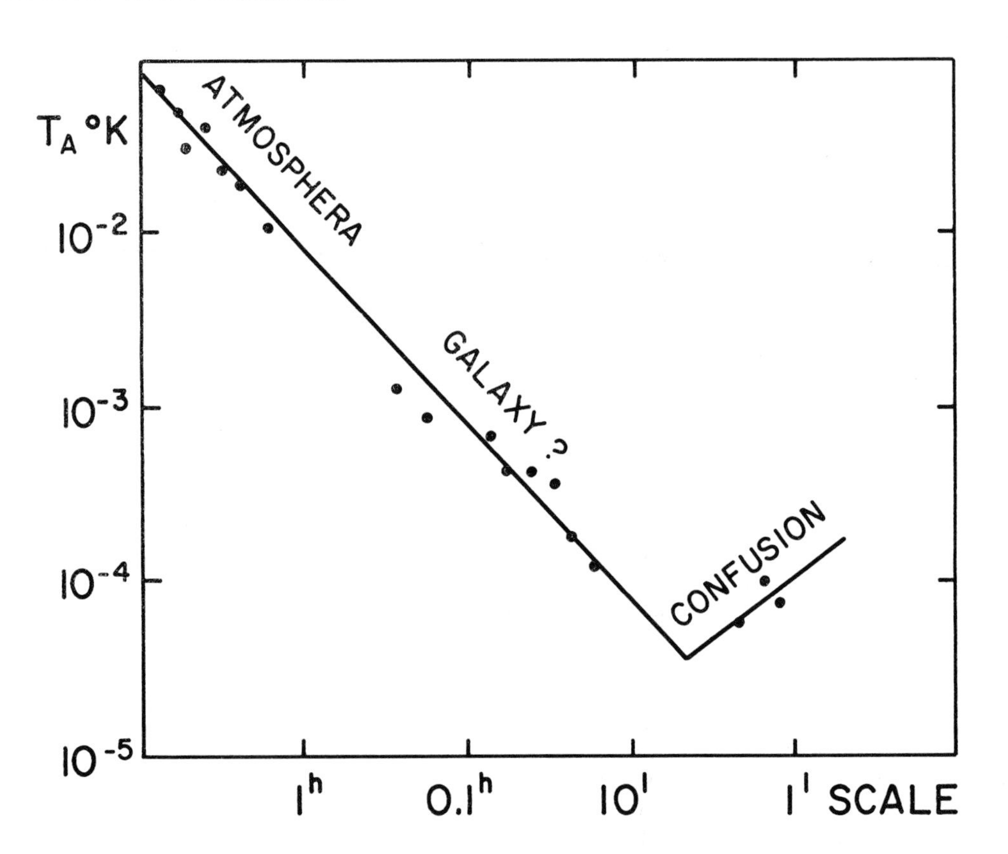

Figure 1. Antenna temperature as a function of scale at λ 7.6 cm. At least at some scales at the left galactic noise dominates. Confusion noise definitely dominates at the scales below 2!5. We expect to have receiver (radiotelescope) noise on scales smaller than 1' due to the small integration time.

"Null hypothesis" correlation control confirms that the 1σ level in less than 10^{-5} in $\Delta T/T$ on scales from 4!5 to 9'. Rough estimates of the upper limits for the anisotropy on other scales give the following results: 10^{-5}; 2×10^{-4}; 2×10^{-2}; 1 for mass scales of 10^{15} $M_\odot$, 10^{11} $M_\odot$, 10^{5} $M_\odot$, and 1 $M_\odot$, respectively, at the 1 σ level.

No trace on the supercluster scale is visible. In a scale of 1°, the upper limit we have reached is about 3×10^{-5} (1 σ level).

Our results confict with all the published predictions in 'fragmentation theories," inclusing massive neutrino variants. The disagreement is less for "Clustering Theories," with early developed galaxies (or globular clusters or stars). But our present philosophy is that it is practically impossible to reject absolutely any world model by our negative results alone.

We have a long-term program to reach the 10^{-6} level and we hope to achieve positive results which alone can tell about the real history of the universe. Black body isotropic emission gives us zero information on that subject, as a state of maximum entropy.

REFERENCES

Boynton, P.E.: 1977, in I.A.U. Symposium No. 79, ed. M.S. Longair and J. Einasto, Reidel, Co., Dordrecht.

Partridge, R.B.: 1979, in "The Large Scale Structure of the Universe," proceedings of a summer school at Jodlowy Dwor, Poland, M. Demianski, ed., Springer Verlag.

Parijskij, Yu.N.: 1973, Ap. J. Letters, 180, L47.

Parijskij, Yu.N., Petrov, Z.E., Cherkov, L.: 1977, Pisma v Ast. Zh., 3, 483.

DISCUSSION

de Zotti: The fluctuations due to discrete sources on angular scales of about 10' at λ ∿ 7 cm, as estimated from source counts, exceed by more than a factor of 10 the upper limit to $\Delta T/T$ you quoted; therefore, a very delicate subtraction of sources must be needed to derive your result. Can you comment on the uncertainties related to such subtraction?

Wilkinson: At a wavelength of 7 cm, confusion from radio sources is a major problem, as you have found. How accurately do you have to subtract radio sources in order to reach your upper limit of $(\Delta T/T)_{2.7} < 10^{-5}$ for 4.5' to 9' angular scales? In other words, what would your limit be if sources were not subtracted?

Parijskij (in answer to both *de Zotti* and *Wilkinson*): We have our own experience in counting radio sources down to S < 1 mJy at 7.6 cm and we can prove that many old nondirectly obtained lgN-lgS curves suffer from great overestimation of the number of very weak sources. Direct counting at Westerbork at 21 cm and the VLA at 6 cm and 21 cm confirms this, at least down to 1 mJy. Preliminary P(D) analysis which I have shown here gives us some indication that the flattening of lgN-lgS at 7.6 cm increases even at a lower level (below 1 mJy). At the present time we expect that there is no great problem in subtraction of the sources above 1 mJy. There is some indication that of the 0.1 - 0.3 mJy level about 30% of the energy is connected with more numerous, very weak sources. We hope to have a more definitive statement after reduction of all our data. To go deeper in $\Delta T/T$ estimates, we have to change frequency and/or use the aperture synthesis mode of observations at RATAN-600. We have just finished testing observations of this kind.

Baldwin: With what accuracy do you expect to be able to remove atmospheric fluctuations by observations at more than one frequency?

Parijskij: We expect to decrease the atmospheric noise by a factor of close to ten in 80% of the records.

Stecker: In view of the problems with theory, is it at all possible that in your analysis you have oversubtracted or missed fluctuations of cosmological significance, resulting in a background which is too smooth?

Parijskij: We really lose all information concerning scales comparable with beamwidth (1' - 3'); but in the most interesting region (4!5 - 9'), we hope there is no "overfiltration."

Wilkinson: I believe that your projected accuracy of $\Delta T/T)_{2.7\ K} < 10^{-6}$ is too optimistic. Foreground effects from: 1) ground radiation (if you beam switch); 2) atmospheric radiation (if you don't beam switch); 3) galactic emission; and 4) radio sources present exceedingly difficult problems at this level of accuracy. Using the NRAO maser on the 140' telescope (T_{SYS} = 50 K, λ = 1.5 cm, θ = 1.5 arcmin). Juan Uson and I only barely managed to overcome these difficulties at an accuracy of $\Delta T/T)_{2.7\ K} < 10^{-4}$. I don't believe that a two-order-of-magnitude improvement is possible from any telescope, and especially at λ = 7 cm.

Parijskij: The 10^{-6} goal is set by the limit imposed by thermal receiver noise with an integration time of about one year. With our multifrequency method, we hope to filtrate atmospheric noise and confusion noise close to that figure. We also expect to use shorter wavelengths (4 cm; 2.6 cm).

NEW LIMITS TO THE SMALL SCALE FLUCTUATIONS IN THE COSMIC BACKGROUND RADIATION

K. I. Kellermann and E. B. Fomalont
National Radio Astronomy Observatory*

J. V. Wall
Royal Greenwich Observatory

The VLA has been used at 4.9 GHz to observe a small region of sky in order to extend the radio source count to low flux density (Fomalont et al., these proceedings) and to look for small scale fluctuations in the 2.7 K cosmic microwave background radiation.

The new VLA observations were made in September 1981 in the D-configuration which synthesized a 700-m diameter aperture to give a resolution of 18 arcsec (FWHP) at 4.9 GHz. Forty hours, spaced over four different nights, were spent integrating on a field centered at $\alpha = 00^h15^m24^s$, $\delta = 15°33'00''$ (epoch 1950.0). The field of view was limited by the primary beam of the 25-m antennas with a diameter of 8.9 arcmin (FWHP). The system noise temperature was 60 K and we observed in both left and right circular polarizations (LCP and RCP) using a 50 MHz bandwidth for each polarization.

The instrumental and atmospheric gain and phase fluctuations were monitored by observing the nearby calibrator source 0007+171 for two minutes at 30 minute intervals. The flux density scale was obtained from an observation of 3C48 which we assumed has a flux density of 5.36 Jy at 4.9 GHz. Radio maps were made in the usual manner and the CLEAN algorithm was used to deconvolve the effects of the sidelobes in the synthesized beam pattern. The map size contains 512 x 512 pixels each with a separation of six arcsec. This corresponds to a field of view of 51.2 arcmin and thus the map extends well beyond the primary beam response pattern of the 25-m antennas. The fluctuations across the radio maps were analyzed by examining distribution of the observed deflections. Only the negative portion of the distribution was used in calculating the fluctuation characteristics since the faint discrete radio sources produce a positive tail in the histograms.

Fluctuations in the radio maps may arise from receiver noise,

* *Operated by Associated Universities, Inc., under contract with the National Science Foundation.*

G. O. Abell and G. Chincarini (eds.), Early Evolution of the Universe and Its Present Structure, 125–126.

instrumental instabilities, tropospheric path-length changes during the observations, and real sky brightness fluctuations. The difference map between the LCP and RCP observations is a good indication of the actual observational sensitivity since many instrumental effects and all tropospheric emission which is not circularly polarized are canceled. This distribution for the entire 512 x 512 map of the quantity (LCP-RCP)/$\sqrt{2}$ gives an rms of 10.3 μJy which is in reasonable agreement with the value of 8.5 estimated from the relevant system parameters. For the inner 6.4 arc minutes of this map alone (64 x 64 pixels), the distribution of noise fluctuations is essentially identical to that of the entire 512 x 512 map.

The analysis of the distribution of the average of the two polarizations (LCP + RCP)/2 gives a value of 11.2 mJy over the entire 512 x 512. The rms fluctuations in the inner part of the averaged map covering 6.4 arcmin (well within the FWHP primary beam) are 15.4 μJy, corresponding to an additional fluctuation of 10.3 μJy (the quadrature difference between 15.4 and 11.2).

We also have examined radio maps with 60 arcsec resolution by heavily weighting the data associated with the shorter spacings. A similar analysis of this data to that of the 18 arcsec resolution data gives the following results. The rms of the averaged map is 42 μJy over the 512 x 512 map; the rms of the averaged map in the inner 6.4 arcmin area is 54 μJy. This corresponds to an excess of 34 μJy in the center of the map with a resolution of 60 arcsec. This excess may be caused by real sky brightness fluctuations or by subtle instrumental effects, and we have taken the observed excess as the maximum possible true value. We thus place the following upper limits to possible small scale fluctuations in the cosmic background radiation, including a correction for the average primary beam attenuation of 20%.

TABLE 1. Limits to Background Fluctuations at 4.9 GHz

Angular scale	rms fluctuations	ΔT/T
18 arcsec	13 μJy	7.8×10^{-4}
60 arcsec	40 μJy	2.1×10^{-4}

The present results appear to be limited by receiver noise and instrumental effects at about the same level. It will be difficult to improve them significantly using the VLA without a large increase in observing time and bandwidth and a better understanding of the magnitude and effect of small correlator offsets and very low cross talk between the antennas.

The detection of measurable fluctuations in the microwave background on scales of the order of tens of arcsec would be of particular interest since this corresponds to the scale associated with mass fluctuations which might be associated with the formation of galaxies in the early universe.

) to 60 ARCMIN FLUCTUATIONS IN THE COSMIC MICROWAVE BACKGROUND

A.N. Lasenby and R.D. Davies
University of Manchester
Nuffield Radio Astronomy Laboratories
Jodrell Bank, Macclesfield, Cheshire, U.K.

Considerable astronomical interest attaches to sensitive measure-
ents of temperature fluctuations of the microwave background on scales
f a few arcmins to a few degrees. We describe here the first in a
eries of experiments being carried out at Jodrell Bank which are aimed
t providing reliable and repeatable information on these angular
cales. The first experiment used the MK II 25m dish at $\lambda 6$ cm and
overed scales from 10 arcmin (the beamwidth) to 60 arcmin (twice the
eamthrow). At this relatively long wavelength, the atmosphere has
egligible effect on the observations and day to day repeatability
onsistent with receiver noise alone was obtained. The observations
ere made in wagging mode near the North Celestial Pole (NCP) so that
he two beams alternately traced out a reference circle on the sky,
adius 30 arcmin, over the course of 24 hours. The NCP field was chosen
ince a high sensitivity discrete source survey (Pauliny-Toth et al.
978) was already available covering the area. This meant that antenna
emperatures around the reference circle could be corrected for any
ource induced effects $\gtrsim 0.4$ mK. In addition control observations were
ade at positions 30 arcmin East and West of the central field, so that
ystematic effects due to interaction of telescope sidelobes with
bjects in the immediate telescope environment, could be monitored and
emoved. Because of the high degree of repeatability of the data we
ere able to carry out this process in detail, and derive three tracks,
ach 188 arcmin long, fully corrected for known sources and systematic
ffects. The main statistical analysis was carried out on the track for
he central region (which had the fullest coverage) after deletion of
oints observed during daylight hours, which were affected by the Sun.
n the 10 arcmin scale this left 11 independent points around the
eference circle. An explicit estimator for the amount of intrinsic
ky variance compatible with the observed residual variations was
onstructed and an upper limit (at 95% confidence) of 0.86 mK (in
rightness temperature) was found, corresponding to a limit to $\delta T/T$ of
$.0 \times 10^{-4}$, The upper limits for the 30 and 60 arcmin scales, found by
ombining points in groups, were 4.6×10^{-4} and 2.3×10^{-3} respectively.
n addition the temperatures of the three observing centres relative to
he mean temperature around their reference circles could be found.

. O. Abell and G. Chincarini (eds.), Early Evolution of the Universe and Its Present Structure, 127–129.

These showed significant displacements away from zero and after allowing for possible effects of weak sources below the limit of the Pauliny-Toth et al. survey, a tentative indication of intrinsic anisotropy at a level of 9×10^{-5} between these points (30 arcmin apart) was found. Further observations both at Bonn (to detect weaker sources) and at Jodrell Bank (to include more observing centres) have been made to shed light on this possible detection and to refine further our upper limits on variations around the reference circles.

Our present limits on the 10 to 60 arcmin scales rule out existing theories of galaxy formation through adiabatic perturbations in the early universe, but are not yet sufficiently sensitive to rule out theories based on isothermal perturbations. We are examining ways of analysing the data via an angular autocovariance function, which as stressed by Davis & Boynton (1980) is the natural way of expressing the predictions of such theories.

As part of this overall study we have used the same observing methods to investigate the possible microwave decrement in the clusters A576 and A2218. We find clear evidence for emission at λ6 cm at the centres of both clusters.

REFERENCES

Davis, M. & Boynton, P., 1980. Astrophys.J., 237, pp365-370.
Pauliny-Toth, I.I.K., Witzel, A., Preuss, E., Baldwin, J.E. & Hills, R.E., 1978. Astr.Astrophys.Suppl., 34, pp253-258.

DISCUSSION

Jones: Rosie Wyse and I recalculated the damping of adiabatic perturbations through recombination and the resultant temperature fluctuations in the microwave background. Generally speaking, we find that previous estimates of $\delta T/T$ are too large and our numbers are easily consistent with your observations.

Bonometto: I have recently recomputed small scale fluctuations due to adiabatic fluctuations. What we find is a dependence on the spectral index if it is sufficiently large. Just choosing $|\delta(k)|^2 \propto k^u$ with u = 0 before recombination, adiabatic fluctuations give results that are not in contradiction with the microwave background limits.

Kaiser: I have calculated the small angle fluctuations of the CBR in the adiabatic scenario and I have found it possible to construct models in which nonlinear structure forms by z = 3 and the temperature fluctuations do not exceed 10^{-4} even with Ω as low as 0.1. These fluctuations occur predominantly on an angular scale $\simeq 3 \times \Omega^{1/2}$ arcmin

It seems plausible that in the isothermal picture reioniza-may occur, in which case the last scattering shell is broad and small-scale fluctuations would be "washed out."

asenby: I have recomputed the predictions for RMS(δT/T) in the minimal isothermal case (that is, just Doppler scattering), nd I have found results that are more optimistic from an observational oint of view than those given by Davis and Boynton. Even so, the current upper limits will have to be reduced by factors of at least 3 to 4 efore they come into conflict with the revised predictions, so invoking econdary ionization as a way out is not necessary yet.

EEP RADIO SOURCE COUNTS AND SMALL SCALE FLUCTUATIONS OF THE MICROWAVE ACKGROUND RADIATION.

L. Danese and G. De Zotti
Istituto di Astronomia, Padova (Italy)

N. Mandolesi
Istituto TESRE/CNR, Bologna (Italy)

Source counts which now extend to surface densities of $\sim 10^5$ ources/sr make possible a direct evaluation of the radio source contri-ution to the small-scale fluctuations in the microwave background on cales larger than $\sim 10'$, at wavelengths $\gtrsim 6$ cm. Comprehensive radio spec-ral data permit a straightforward and largely model-independent extrapo-ation of the N(S) relation to shorter wavelengths. On the other hand, eacock and Gull (1981, hereafter PG) have constructed a set of models hich incorporate a wealth of additional data, such as local luminosity unctions, luminosity/redshift distributions, luminosity-spectral index orrelations; they can therefore be exploited to optimize the extrapola-ions both to higher frequencies and to fainter flux densities. Only ne of these models, however, namely No. 4, is consistent with the recent (D) results (Wall *et al.* 1982; Ledden *et al.* 1980) which provide infor-ation on the areal density of sources at $s \sim 1$mJy; therefore, in the ollowing we shall focus on it. (It is interesting to note, in passing, hat P(D) counts do not reflect the faint end of counts at 408 MHz as it

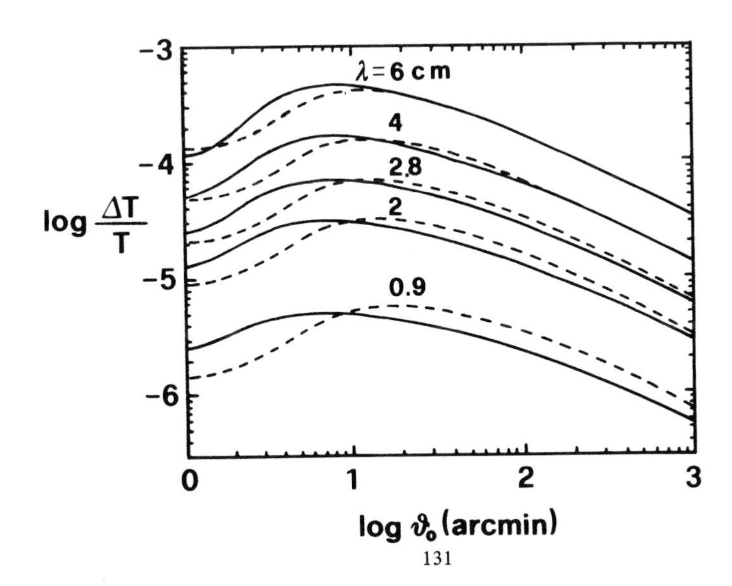

G. O. Abell and G. Chincarini (eds.), Early Evolution of the Universe and Its Present Structure, 131–133.

would be expected in view of the rapid convergence of the flat spectrum source counts below 500 mJy. In fact, PG model 4 predicts that 5 GHz counts at S<1mJy are dominated by flat spectrum sources. The figure shows the temperature fluctuations as a function of the angular scale as predicted, at several wavelengths, by PG model 4 (continuous lines) and by Kulkarni's (1978) model A (dashed lines). The latter is used to illustrate the fact the results are essentially independent of the adopted radio luminosity function. The small discrepancies between the two models at angular scales <10' simply reflect the uncertainties in the counts at mJy levels; the widening of differences with increasing frequency is the effect of different assumptions about the spectral index distributions. The largest uncertainty in our calculations is associated with the assumption that all sources have a simple power law spectrum with constant spectral index up to very high frequencies; in fact, it is known that the spectra of many compact sources steepen above some frequency $\nu_s \simeq 20 - 30$ GHz. At $\nu > \nu_s$ an increase $\Delta\alpha$ of the mean spectral index results, for a given flux density, in a decrease of the counts by a factor $(\nu(1+z_e)/\nu_s)^{\Delta\alpha(1-\beta)}$, where z_e is the effective redshift of counted sources and β is the slope of differential counts. For $\nu < \nu_s$, the effect of steepening is restricted to those flux densities where counts are dominated by sources at $z > (\nu_s/\nu-1)$. It is easly seen that, for reasonable values of ν_s, z_e and $\Delta\alpha$, the effect is small, at least for $\lambda > 3$ cm and even at $\lambda = 0.9$ cm the amplitude of expected fluctuations can hardly decrease by more than a factor of 3 to 5.

The amplitude of fluctuations $\Delta T/T$ peaks at angular scales $\theta_o \sim 10'$ where it reaches values $> 10^{-4}$ for $\lambda > 3$ cm. This means that the most sensitive searches for primordial anisotropies at cm wavelengths have already come close to the source confusion limit. Primordial anisotropies of amplitude $\Delta T/T \sim 10^{-5}$ on angular scales in the range from a few arcminutes to $\sim 1°$ can possibly be detected only if observations are made at $\lambda \lesssim 1.5$ cm. However, inhomogeneous atmospheric emission may be a problem at such wavelengths.

References

Kulkarni, V.K.: 1978, Monthly Notices Roy. Astron. Soc., 185, 123-135.

Ledden, J.E., Broderick, J.J., Conden, J.J., Brown, R.L.: 1980, Astron. J., 85, 780-788.

Peacock, J.A., Gull, S.F.: 1981, Monthly Notices Roy. Astron. Soc., 196, 611-633.

Wall, J.V., Scheuer, P.A.G., Pauliny-Toth, I.I.K., Witzel, A.: 1982, Monthly Notices Roy. Astron. Soc., 198, 221-237.

DISCUSSION

Ekers: It is possible to correct for the effect of discrete sources by obtaining higher angular resolution observations to determine the contribution from discrete sources and subtracting these from the lower resolution observation.

le Zotti: The results I presented were obtained by assuming that sources are subtracted when they provide a deflection larger han the r.m.s. value by a factor $\gtrsim 3$. The technique you mention may ertainly decrease the discrete source contribution to fluctuations; I uspect, however, that the uncertainties inherent in such subtraction vill not allow a decrease by a large factor.

'ilkinson: I would like to make two points which contradict statements made by several speakers: 1) for small-scale anisotropy ieasurements at wavelengths of 1 cm or less, radio sources are not a roblem. System noise will provide a higher limit; 2) on angular scales f a few arcminutes, atmospheric fluctuations do not limit accuracy at . = 1.5 cm.

These comments are based on a recent small-scale measurement lone by Juan Uson and me, using the NRAO maser and 140-foot telescope. 'rom a beam size of $1\overset{'}{.}5$ and a beam throw of $4\overset{'}{.}5$, the result is $\delta T/T <$ $.1 \times 10^{-4}$ with 95% confidence; 24 spots along $\delta = 87°$ were observed. The result is only two times larger than the ideal limit for a 50 K sys-.em noise. About one-third of the total observing time was useful.

le Zotti: 1) The figure clearly shows that at $\lambda \simeq 1.5$ cm, fluctuations due to discrete sources are at least one order of magnitude)elow your upper limit to $\Delta T/T$; therefore, I can only agree with your comment.

2) It is also not a surprise that atmospheric fluctuations at that wavelength and on an arcminute angular scale are well below $\delta T/T \sim 10^{-4}$; on the other hand, we expect that they may become a problem for substantially more sensitive future experiments ($\delta T/T \sim 10^{-5}$), especially on larger angular scales and for still shorter wavelengths.

Kaiser: The CBR fluctuations in the adiabatic picture are predicted to display 20% linear polarization, so sensitive polarimetry may enable us to detect or constrain cosmological fluctuations at a level below the amplitude of the discrete source fluctuations.

Windhorst: Do the latest 21 and 6 cm VLA deep source counts below 1 mJy enable you to make a more definite choice between the various models?

de Zotti: For our purpose, the VLA counts do more than allow a better choice between models. They permit us to extend the direct, model-independent estimates of fluctuations at $\lambda \gtrsim 6$ cm to essentially the full range of angular scales we have considered. Preliminary checks show that the models we have adopted are in reasonable agreement with the VLA data and lead to the estimate that our extrapolations below 10', at $\lambda \gtrsim 6$ cm, cannot be in error by more than a factor of 1.5.

MEASUREMENT OF THE 3 K COSMIC BACKGROUND NOISE IN THE FAR INFRARED

G. Dall'Oglio[1], P. de Bernardis[2], S. Masi[2] and F. Melchiorri[2]

[1]Institute for Atmospheric Physics, CNR, Rome
[2]Institute of Physics, Rome

1. ABSTRACT

Quantum fluctuations of the cosmic background have been measured in the 900 to 2000 micron range (H.P.B.W.) by means of a balloon-borne correlator operating between 5 and 150 Hz. Preliminary results indicate an upper limit $\sqrt{\langle dP^2 \rangle} \lesssim 2.1 \times 10^{-17}$ watt/(cm^2 srad Hz)$^{1/2}$, corresponding to the noise of a blackbody at a temperature T $\lesssim$ 3.1 K at 1σ.

2. THE CORRELATOR

The instrument consisted of two Germanium Composite Bolometers operating at 0.3 K, a mylar beamsplitter and Cassegrain optics at 1.8 K. The electrical NEP was 8×10^{-16} watt/Hz$^{1/2}$ and the throughput was 0.05 cm^2 srad with a field of view of 2 deg. Anisotropy measurements were alternated to noise measurements by wobbling the secondary mirror for half the observing time. The intrinsic, uncorrelated noise of the two detectors has been subtracted by using an analog multiplier followed by an integrator. The expected noise is shown in Figure 1.

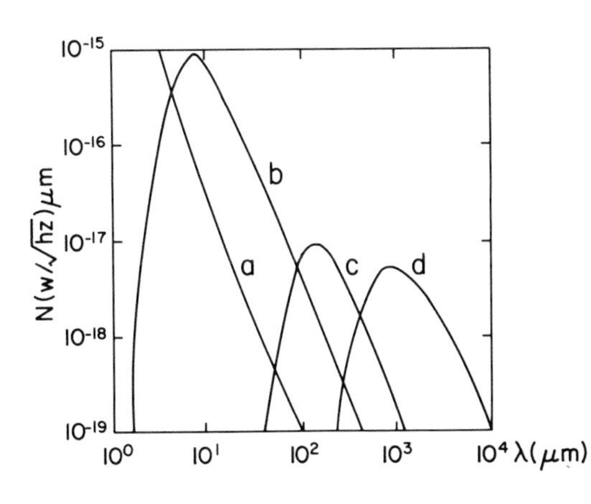

Figure 1.
Expected noise scenario in the 1 to 10^4 μm band. Lines a, b, c and d are:

a) Zodiacal light (scattered)
b) Zodiacal light (thermal)
c) Galactic dust (12 K)
d) C.B.R.

Power is computed for unitary throughput.

G. O. Abell and G. Chincarini (eds.), Early Evolution of the Universe and Its Present Structure, 135–137.

The balloon was launched August 8 from Milo, Sicily. The correlated noise is plotted versus elevation in Figure 2. The secant law fit has been done by disregarding the point at highest elevation. We used the relation:

$$\sqrt{\langle dP^2 \rangle} = 2.77 \times 10^{-18} \left[T^5 \tau \int_{x(2000\ \mu m)}^{x(900\ \mu m)} \frac{(e^x - 1 + \tau)\ x^4\ dx}{(e^x - 1)^2} \right]^{1/2} \ \mathrm{W/cm^2\ srad\ Hz)^1}$$

with $x = h\nu/kT$ and τ = efficiency of the system estimated to be about 0.2. We get an upper limit of 3.1 K for the temperature of the Cosmic Background Radiation at 1σ.

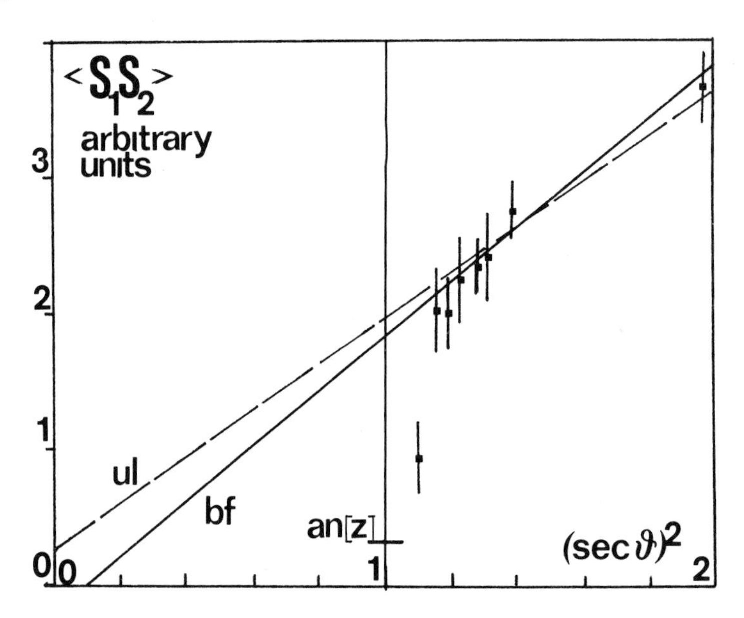

Figure 2. Signals at different tipping angles are plotted. The right side of the figure shows the extrapolation to sec $\theta = 0$; bf is the best fit secant law; ul is the upper limit at 1σ; an(z) is the expected atmospheric noise at the zenith (T = 250 K, emissivity = 5×10^{-4}).

ACKNOWLEDGMENTS

This work has been made possible with the help of the Italian and French teams at the balloon base of Milo. Our warm thanks are due to Dr. Cosentino and Dr. A. Soubrier. The work has been supported by Servizio Attività Spaziali del CNR, Roma.

DISCUSSION

Wright: Do you measure the cross correlation function versus lag to get the cross power spectrum, or do you measure only the correlation at zero lag? If the latter, how do you know you are not seeing dewar temperature fluctuations?

de Bernardis: Up to now we measured only the mean square value of the correlated signal (zero lag). Since the temperature of the dewar is lower than 3 K and the fluctuations in the dewar add quadratically to fluctuations in incoming radiation, the result is a little offset to the measure. We measured it by closing the liquid helium section of the dewar and fixing the "zero level" in this way.

Bishop Irineos, Metropolitan of Kissamos and Selinos. *(Courtesy, K. Brecher)*

ON THE LARGE-SCALE ANISOTROPY OF THE COSMIC BACKGROUND RADIATION IN THE FAR INFRARED

P. Boynton[1], C. Ceccarelli[2], P. de Bernardis[2], S. Masi[2], B. Melchiorri[2], F. Melchiorri[2], G. Moreno[2], V. Natale[3]
[1]Dept. of Physics, University of Washington, Seattle
[2]Istituto di Fisica Università di Roma
[3]Istituto Onde Elettromagnetiche, C.N.R., Firenze

1. ABSTRACT

We report preliminary results relative to a balloon-borne search for the large-scale anisotropy carried out in 1980 by means of two far infrared photometers centered at 400 and 1100 microns. While these results are consistent with those obtained in an earlier flight, the second, shorter wavelength channel included in the 1980 work provides interesting insights into the influence of galactic dust on such far infrared observations.

2. INTRODUCTION

Fabbri *et al.* (1980) have reported the existence of a distortion in the dipole anisotropy observed at millimeter wavelength (500 to 1500 microns). Subsequent analysis by Melchiorri *et al.* (1981) and by Ceccarelli *et al.* (1982) have shown that galactic emission could be responsible for at least part of the observed signal. Furthermore, the analysis of the celestial intensity distribution in terms of spherical harmonics is less than satisfying due to the limited sky coverage. Ceccarelli *et al.* (1982) found that Q_2 and perhaps Q_5 are above the noise, while Fabbri *et al.* (1982) found that Q_3 and Q_4 are marginally significant. In any case, the agreement with the data of Cheng *et al.* (1981), who claim to have detected Q_5 at 5 sigma, could well be fortuitous. Due to limited sky coverage, we believe that the best representation of our 1978 data is provided by the isophotes of Fig. 1. We note several sources (which are believed to be galactic, since they concentrate in the galactic plane), as well as a large-scale "hot spot" on the right of the figure. It corresponds to a gradient of about 0.1 mK over 6 degrees of wobbling amplitude and it turns out to be of the order of 0.3 to 0.9 mK when interpreted as part of a general quadrupole distribution. The dipole anisotropy is well evident in the figure with the maximum concentrated toward the lower left corner around 8 to 9 hours of Right Ascension.

3. RESULTS OF 1980 FLIGHT

In order to clarify the nature of the distortion, we carried out another set of observations using two detectors operating at 400 and

G. O. Abell and G. Chincarini (eds.), Early Evolution of the Universe and Its Present Structure, 139–141.

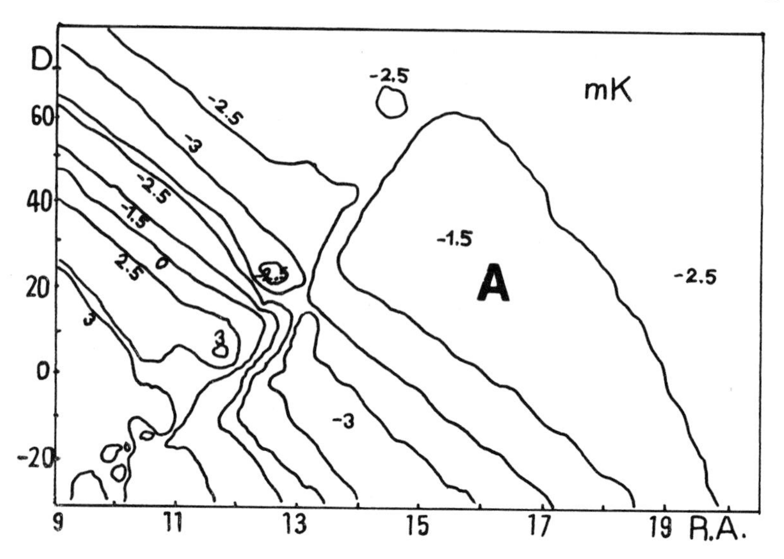

Figure 1. Isophotes of 1978 flight. The "hot spot" A is discussed in the text.

1100 microns, respectively, with roughly unit fractional bandwidth. As in the case of the 1978 flight, the dipole anisotropy was observable in real time in the far infrared channel. (See Figure 2.)

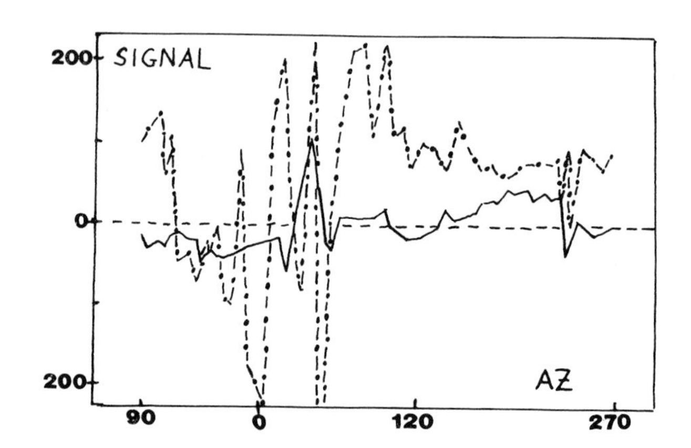

Figure 2. Far infrared (continuous line) and middle infrared channel (dotted line) during one revolution of the gondola in the 1980 flight.

Galactic sources present in both channels have been used in order to normalize the relative amplitude of the two channels. Subsequent subtraction of the 400 micron from the 1100 micron channel should eliminate the local galactic contribution, if all sources (extended as well as unresolved at 5 degrees of resolution) have the same spectrum. It is clear, however, from Fig. 2 that several 400-micron sources are not present in the far infrared channel. Therefore, application of this procedure tends to transfer the galactic contribution from the 400-micron into the 1100-micron channel. We decided on filtering out the unresolved sources in both channels and then we subtracted the 400-micron channel from the far infrared channel. The new isophotes are plotted in Fig. 3.

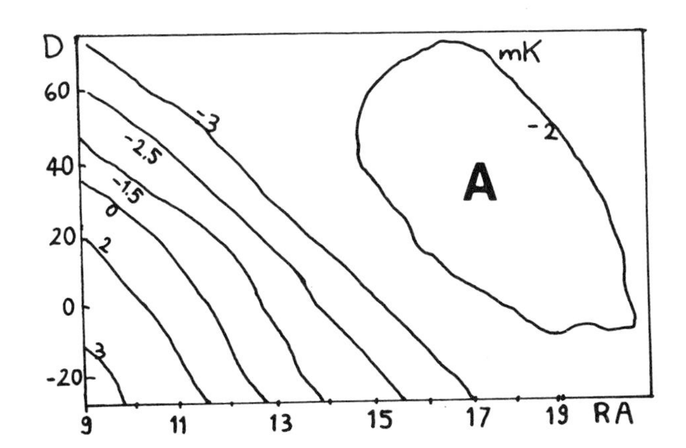

Figure 3. Isophotes of the 1980 flight after subtraction of 400-micron sources from the 1100-micron channel. Note the presence of the "hot spot" A.

Neither the amplitude of the inferred dipole anisotropy nor that of the extended "hot spot" distortion are significantly affected by this source correction technique. This insensitivity of the properties of large scale features to the 400-micron contribution suggests that these features could be in fact cosmological. That is, neither the dipole anisotropy nor the "hot spot" are evident in the 400-micron channel as would be consistent with features having a radiation temperature around 3 K. However, the difficulty we found in correcting for point-like sources is troublesome. Our result implies that correction for galactic emission is a rather difficult task. Therefore, the existence of a large-scale cosmological distortion could be proved beyond any reasonable doubt only after a detailed survey of dust emission at different wavelengths becomes available.

4. REFERENCES

Ceccarelli, C., Dall'Oglio, G., Melchiorri, F., Pietranera, P. 1982, Ap. J., 260, 484.

Fabbri, R., Guidi, I., Melchiorri, F., Natale, V. 1980, Phys. Rev. Lett., 44, 1563.

Boughn, S.P., Cheng, E.S., Wilkinson, D.T. 1981, Ap. J. Lett., 243, L113.

Melchiorri, F., Melchiorri, B., Ceccarelli, C., Pietranera, P. 1981, Ap. J. Lett., 250, L1.

Fabbri, R., Guidi, I., Natale, V. 1982, preprint.

LARGE-SCALE ANISOTROPY AT CENTIMETER WAVELENGTHS

DAVID T. WILKINSON
Joseph Henry Laboratories
Physics Department
Princeton University
Princeton, N. J. 08544
U.S.A.

Scientific interest in large-scale anisotropy measurements is now focused on intrinsic effects, which could tell us much about the early Universe. Current experimental precision of better than 10^{-4} K begins to probe for interesting physical processes. However, at these levels of precision systematic effects and foreground sources present serious difficulties. Some recent results from balloon flights of a maser radiometer (λ 1.2 cm) and a cooled mixer (λ 3 mm) are discussed and interpreted. The dipole effect gives a velocity for the Local Group in the general direction of the Virgo cluster. The Earth's motion is clearly seen. There is no quadrupole detected at a level of $\Delta T/T \sim 5 \times 10^{-5}$.

1. INTRODUCTION

Precision observations of the large-scale anisotropy of the 2.7K cosmic microwave background radiation have demonstrated the extragalactic origin[1] of the radiation, and have measured the Earth's velocity[2,3] with respect to the radiation - presumably the co-moving reference frame of the Universe. As experimental accuracy improves, the scientific focus for this work has become intrinsic effects that might have originated in the very early Universe, or been caused by interactions with anisotropic matter at the time of decoupling of radiation and matter. Several possible effects with $\Delta T/T \leq$ few x 10^{-4} have been theoretically discussed[4], and two groups[3] have reported tentative detections of a quadrupole effect in the 2.7K radiation. However, recent results from two measurements show no quadrupole distribution larger than $\Delta T/T \sim 5 \times 10^{-5}$. Apparently, even better measurements are needed to see the imprints of early processes on the 2.7K radiation.

This paper discusses the recent experimental work at centimeter wavelengths, and some implications of the latest results. But first, to put the results into perspective, some techniques and experimental difficulties are discussed.

G. O. Abell and G. Chincarini (eds.), Early Evolution of the Universe and Its Present Structure, 143–148.

2. TECHNIQUES AND PROBLEMS FOR CENTIMETER WAVELENGTHS

The basic problem is to measure the difference in the radiation temperature from two directions in the sky with an accuracy of better than 10^{-4} K. Carefully designed room-temperature Dicke radiometers integrate their random noise down to this level in about a day; the best cooled radiometers require about 10^3 sec of integration per pair of points. Radiometer front-ends for large-scale anisotropy measurements have evolved from tube-type, room-temperature, mixers to maser amplifiers and mixers[6] cooled to liquid helium temperature. However, system noise is not the main problem with these instruments. Instabilities of 10^{-5} in reference sources, Dicke switch balance, and antenna wall radiation cause serious systematic errors at sensitivity levels of 10^{-4} K. Differential (two horn) techniques, mechanical and electrical symmetry, temperature control, and magnetic shielding are all essential. In addition, rotation of the instrument[7] (secondary Dicke switching) is necessary because such a high degree of instrument symmetry cannot be maintained for longer than a few minutes.

Local radiation sources present another difficult problem at centimeter wavelengths. Ground radiation ($\sim$ 300 K) must be excluded to an extraordinary level (10^{-6}) by special antennas and shielding. Nonuniform atmospheric emission has forced all recent experiments onto balloons or high-flying aircraft. Plans are being made for satellite experiments to fly in this decade.

It now appears that foreground Galactic radiation (synchrotron and bremsstrahlung) will limit the accuracy of large-scale anisotropy measurements at wavelengths longer than 5 mm. Synchrotron emission has been mapped at longer wavelengths[8], but must be extrapolated over more than a factor of 50 in wavelength to reach the 1 cm to 3 mm wavelength. Spectral index variations with wavelength or position on the sky cannot be modeled. Bremsstrahlung radiation presents even more serious problems. It is not mapped on large angular scales, because at long wavelengths Galactic synchrotron emission dominates the bremsstrahlung emission. H II regions and intense emission near the Galactic plane have been mapped at centimeter wavelengths, but weak diffuse emission off the plane can be a problem for anisotropy measurements, and little is known about this. Sensitive maps of Hα emission[9] give lower limits on emission measure along some lines of sight, but unknown extinction corrections prevent accurate predictions of bremsstrahlung radiation, even if the plasma temperature is known.

At shorter wavelengths ($\sim$ 3 mm) the foreground radiation problem becomes dust emission. Even less is known about this than about synchrotron and bremsstrahlung radiation. The distribution of dust is poorly known, and the spectral index is uncertain by at least a factor of 2. However, the radiation intensity is much lower than at $\lambda \sim 1$ cm.

In principle, the foreground radiation problem can be overcome by anisotropy measurements at several frequencies, using spectral signatures to separate foreground from background components. If foreground sources can be understood, the experimental accuracy will be limited by systematic errors.[7] The most worrisome systematic effects are those that are synchronous with instrument rotation such as, coupling to the Earth's magnetic field, asymmetric ground radiation entering antenna side lobes, radio interference, and instrument heating from asymmetric infrared sources.

3. RECENT RESULTS

3.1 Dipole Distribution

The largest anisotropy in the 2.7K background radiation is a dipole (cos θ) distribution, most of which is due to the Sun's velocity with respect to the radiation frame. This effect was anticipated from the beginning[1], but ten years of instrument and technique development were needed before the dipole was clearly seen. A number of dipole measurements have now been reported[2,3] at wavelengths between 1.2 cm and 1 mm. All are in reasonable agreement on the magnitude and direction of the effect. Furthermore, the spectrum of the dipole is approximately blackbody, as expected for the velocity effect. The spectrum rules out Galactic radio sources for the dipole.

Two very recent results also show good agreement with older dipole results. The Berkeley group has flown, in balloons, a cooled 3 mm mixer radiometer with system noise 125K. The Princeton group has flown a 1.2 cm maser radiometer with a system noise of 33K. Data from two northern hemisphere flights of each instrument cover over 60% of the sky. At 3 mm, corrections for Galactic radiation are small, but the dust contribution is unknown. At 1.2 cm, Galactic radiation is important, and uncertainties in the model limit the accuracy of the results, especially with the very low instrumental noise of the maser. The preliminary dipole results of these two experiments are shown in the following table.

λ (cm)	T_x (mK)	T_y (mK)	T_z (mK)	Group
0.33	-3.48 ± 0.08	0.61 ± 0.08	-0.32 ± 0.08	Berkeley[6]
1.2	-3.02 ± 0.04	0.85 ± 0.04	-0.41 ± 0.03	Princeton[5]

The temperatures given in the table are thermodynamic temperatures, assuming that the cosmic background is a 2.7K blackbody; errors are statistical only. The reference frame is the Sun.

The velocities of the Sun and the Local group are easily derived from the dipole anisotropy by assuming that all of the effect is due to motion through the radiation, and that the radiation temperature is 2.7K. Averaging the two results above, and estimating the error from the difference, gives

$$v_{SUN} = (372 \pm 25)\,{}^{km}/sec, \text{ towards } \alpha = 11^h\!.2 \pm 0^h\!.2,\ \delta = -6^\circ \pm 3^\circ.$$

Taking the canonical value for solar motion with respect to the local group (300 km/sec towards $\ell = 90^\circ$, $b = 0^\circ$) gives,

$$v_{LG} = (610 \pm 50)\,{}^{km}/sec, \text{ towards } \alpha = 10^h\!.5 \pm 0^h\!.4,\ \delta = -26^\circ \pm 5^\circ.$$

These should be regarded as preliminary results, as the data are still being analyzed for systematic errors and sensitivity to Galactic radiation corrections. However, it is particularly interesting to note that this direction of $\vec{v}_{LG}$ is only 17° from the direction of $\vec{v}_{LG}$ found by Hart and Davies[10] from HI observations of Sbc galaxies with redshifts between 1000 km/sec and 5500 km/sec. The magnitude of the velocity from that work is (436 ± 55) km/sec, substantially smaller than the cosmic background result. The $\vec{v}_{LG}$ direction is 49° away from the Virgo cluster.

The high sensitivity of the maser instrument makes possible a clear detection of the Earth's velocity with respect to the 2.7K radiation. The instrument was flown on Dec. 10-11, 1980 and again on June 29-30, 1981 - an interval of about 6 months. The difference of the Earth's velocity on those dates predicts a dipole difference of

$$\Delta T_x = -0.528 \text{ mK},\ \Delta T_y = 0.016 \text{ mK},\ \Delta T_z = 0.007 \text{ mK}.$$

The dipole results for the individual flights are

Dec. 80: $T_x = -3.37 \pm 0.06$ mK, $T_y = 0.80 \pm 0.05$ mK, $T_z = -0.19 \pm 0.04$ mK,

June 81: $T_x = -2.71 \pm 0.07$ mK, $T_y = 0.92 \pm 0.06$ mK, $T_z = -0.33 \pm 0.05$ mK.

The measured dipole difference, due to the Earth's velocity is

$$\Delta T_x = -0.66 \pm 0.09 \text{ mK},\ \Delta T_y = -0.02 \pm 0.08 \text{ mK},\ \Delta T_z = 0.14 \pm 0.06 \text{ mK}.$$

Comparing the predicted and measured dipole differences shows that the Earth's velocity has been detected, and is about a 7σ effect in the maser data.

3.2 Quadrupole Distribution

The recent Berkeley and Princeton results have also been searched for evidence of a quadrupole distribution in the 2.7K radiation. No significant quadrupole effects are seen; the results are,

λ (cm)	Q_2[11] (mK)	Q_3 (mK)	Q_4 (mK)	Q_5 (mK)	Group
0.33	0.23 ± 0.12	0.26 ± 0.10	-0.10 ± 0.08	0.06 ± 0.07	Berkeley[6]
1.2	0.06 ± 0.06	0.05 ± 0.04	0.16 ± 0.04	0.00 ± 0.03	Princeton[5]

Again, the quoted errors are only statistical and do not include calibration errors, uncertainties in the Galactic radiation model, or correlations introduced by incomplete sky coverage. The apparently large effects, such as Q_4 in the Princeton results, are not believable, because of the extreme sensitivity to systematic effects at this level. For example, if the Galactic radiaion model is perturbed by adding the term, (60 μK)cosec b, then Q_4 becomes 0.07 mK, and Q_2 becomes 0.19 mK. Since there is no independent way of detecting a Galactic component as small as this perturbation, we must assume that Q's less than $\sim$ 0.20 mK are due to inaccuracy of the Galactic radiation model.

The Princeton results can, however, be interpreted as an upper limit on possible quadrupole effects in the 2.7K background. Taking an rms of the Q values, integrated over the sky, gives an upper limit on Q_{rms} of 0.13 mK, with 90% confidence.[5] Clearly, the reported[4] quadrupole effects (e.g., $Q_5 = -0.54\pm0.14$ mK) are not in the 2.7K background. The centimeter work was done with radiometers having 10 times more noise than the maser; systematic errors were very hard to detect. Also, a much improved model for Galactic foreground radiation has been used for analyzing the maser data. The older data are being reanalyzed with the new model, and the instruments are being further tested for systematic effects. The millimeter observation of quadrupole-like effects could have been caused by large-scale dust emission. Little is known about the dust distribution at high Galactic latitudes, and no spectral information was obtained to look for dust emission.

4. FUTURE PROSPECTS

The preliminary results discussed above represent a substantial improvement over earlier results.[2,3] But the measurements are now reaching fundamental limits due to foreground sources, systematic errors, and instrument noise. Further improvements will require multiple frequency observations, with cooled radiometers, good sky coverage, and long integration times. More balloon work is planned at λ 0.65 cm and λ 0.33 cm, however, satellite[12] experiments planned for late in this decade hold the most promise for reaching the best possible accuracy.

REFERENCES

1. Partridge, R.B., and Wilkinson, D.T.: 1967, Phys. Rev. Letters, 18, p.557.
2. Smoot, G.F., and Lubin, P.M.: 1979, Ap. J., 234, pL83. Weiss, R.: 1980, Ann. Rev. Astron. Astrophys., 18, p.489.

3. Fabbri, R., Guidi, I., Melchiorri, F., and Natale, V.: 1980, Phys. Rev. Letters, 44, p1563; erratum: 1980, Phys. Rev. Letters, 45, p401. Boughn, S.P., Cheng, E.S., and Wilkinson, D.T.: 1981, Ap. J., 243, pL113.
4. Peebles, P.J.E.: 1981, Ap. J., 243, pL119. Wilson, M.L., and Silk, J.: 1981, Ap. J., 243, p14.
5. Fixsen, D.J.: 1982, "A Balloon Borne Maser Measurement of the Anisotropy of the Cosmic Background Radiation", Ph.D. Thesis, Princeton University. Fixsen, D.J., Cheng, E.S., and Wilkinson, D.T.: 1982, in preparation.
6. Lubin, P.M.: 1982, Varenna Conf. Course LXXXVI, preprint. Lubin, P.M., Epstein, G., and Smoot, G.F.: 1982, in preparation.
7. Wilkinson, D.T.: 1982, Phil. Trans. Roy. Soc., Oct. 15, 1982.
8. Haslam, C.G.T., Salter, C.J., Stoffel, H., and Wilson, W.E.: 1982, Astron. and Astrophys. Suppl., 47, p1.
9. Reynolds, R.J.: 1980, Ap. J., 236, p153. Sivan, J.P.: 1974, Astron. and Astrophys. Suppl., 16, p163.
10. Hart, L., and Davies, R.D.: 1982, Nature, 297, p191.
11. See Boughn, et al., ref. 3, for a definition of the Q's.
12. Mather, J.: 1980, Physica Scripta, 21, p671.

I would like to acknowledge support from the National Science Foundation and NASA.

DISCUSSION

F. Melchiorri: Is our "hot spot" at a level of 0.1 mK, extending about 20 by 20 degrees, detectable in your data?

Wilkinson: Yes. An object of this size would be detectable at the 0.1 mK level in the maser data. A fit could easily be made, if you give us the coordinates of your cloud.

Davies: The impressive new results on the dipole anisotropy by the Berkeley and Princeton groups show a formal difference in amplitude of about 15 percent. This difference translates into a difference in the derived solar motion and is important in comparisons with other determinations of the solar motion. I note that the two estimates of the solar motion differ by ~ 4 times the sum of the quoted errors. Would the two authors care to comment on this discrepancy?

Wilkinson: The errors given in the tables include only statistical errors; they will probably double when systematic errors are included. All three pairs of three measurements by the Berkeley-Princeton groups have shown discrepancies not quite covered by the error bars. We believe that calibration errors have been underestimated. Calibration in flight is difficult, and only recently have the techniques improved. Both groups are aware of this problem and are working on it. For now, I suggest that the results be averaged and that errors include both results, as has been done in this paper.

LARGE-SCALE ANISOTROPY OF THE COSMIC RELIC RADIATION IN SPATIALLY OPEN COSMOLOGICAL MODELS

V.N. Lukash
Space Research Institute, Moscow

The observed microwave background radiation is a sensitive tool for studying the fundamental features of the universe. A puzzling constancy on the celestial sphere of the temperature, T, of the equilibrium relic radiation coming to us from causally nonrelated regions of space-time points to the global spatial homogeneity and isotropy of the cosmological expansion. On the other hand, a small anisotropy of the relic background can tell a lot about the physics of the beginning of the universal expansion, where primordial cosmological perturbations, which later affect the relic isotropy, formed (see, e.g., [1,2] and other reviews on the early universe). We would like to emphasize another factor that forms mainly the large-scale structure of relic anisotropy: the spatial curvature of the background Friedmann Universe. In the light of the discovery of the large-scale anisotropy of the cosmic radiation [3-5], this problem becomes very important.

Let us consider the observable structure of the largest-scale perturbations in the spatially flat, closed and open cosmological models. Quantitatively, the space curvature within the present cosmological horizon is characterized by Ω, the ratio of the mean-to-critical density.

It is well known that any small perturbations of infinite scale in the flat background model do not disturb spatial curvature and the space deformation they cause results in a quadrupole anisotropy of the relic temperature ($\Delta T/T$). These infinite-scale perturbations are spatially homogeneous (they belong to the Bianchi Type I cosmological models) and any of their linear superpositions has the same properties [6-8]. The largest possible scale perturbations of the closed Friedmann model disturb the cosmic radiation in a similar manner: the characteristic angular scale of the temperature variations is of the order of unity and, in this sense, the T/T structure is qualitatively the same for Ω - 1 and $\Omega > 1$. (Subtler differences may be revealed after measurements of the following low-order moments of the relic background and taking into account the discrete spectrum of perturbations in the closed space.)

A qualitatively new effect appears in the open background model with the Lobachevski hyperbolic 3-space [9]. The simplest infinite-scale perturbation mode does not disturb spatial curvature, it is spatially homogeneous (Biachi Type V) and deforms the space along a bundle of

G. O. Abell and G. Chincarini (eds.), Early Evolution of the Universe and Its Present Structure, 149–152.

parallel lines [7]. Figure 1 demonstrates the principal directions of the shear deformation in the 2-plane to which the line going through the observer (in the center) belongs. Relic quanta propagate to the observer along the radii, the circle is the locus of the last scattering of relic photons (its inverse radius $\simeq \Omega$ for $\Omega << 1$). Figure 2 shows the relic temperature anisotropy caused by this homogeneous mode. The radius is the temperature in the given direction (dashed and solid curves correspond to $\Omega \simeq 1$ and 1/3, respectively). Thus, with Ω decreasing, the quadrupole $\Delta T/T$ transforms to the ring (spot) with angular dimension $\Delta\Theta \simeq 2\Omega$ [9,10]. Arbitrary infinite-scale perturbations are linear superpositons of the homogeneous modes considered, so large-scale relic anisotropies in the open universe have characteristic angular variations of temperature $\sim \Omega$ [7,8]. The relic spot-anisotropy caused by large-scale perturbations is as natural a property of the Lobachevski space as is the quadrupole for the Euclidean one

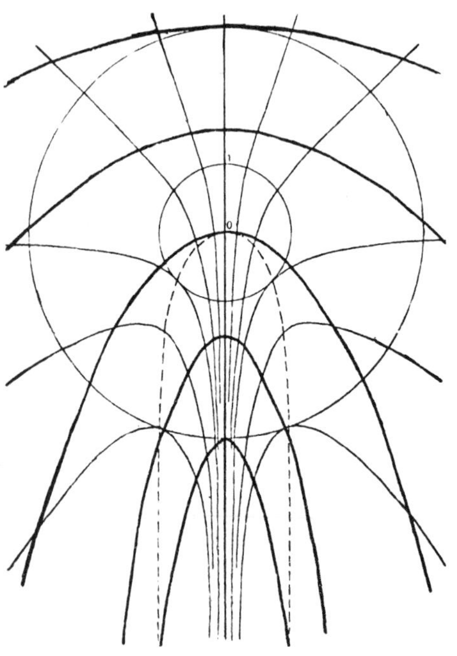

Figure 1

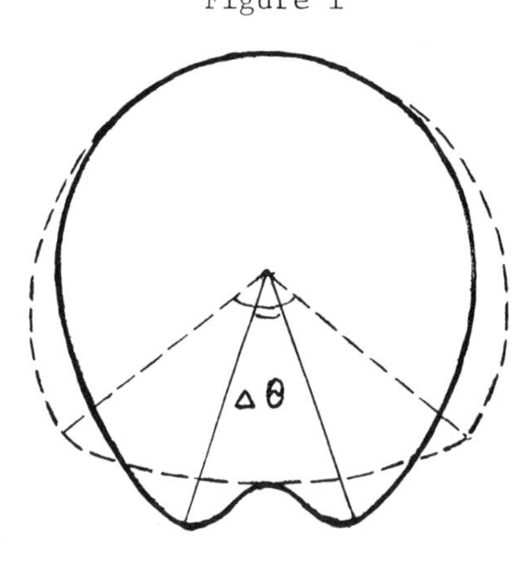

Figure 2

Next, we consider the technique of detecting this effect on the celestial sphere. The Lifshitz formalism of spherical harmonics [11] is good for flat and closed spaces (see, e.g., [12,13], etc.), but not for the hyperbolic one. The point is that the infinite-scale homogeneous perturbation, which is constant over the Lobachevski space and gives the $\Delta T/T$-spot, has a wide spectrum of all-order moments while being expanded on spherical harmonics. (Note that the harmonics with a characteristic variation of $\Delta\Theta \sim \Omega$ will be slightly singled our [14]; in the Euclidean case only a quadrupole moment does not vanish.) That is why, in order to single out the pure spot-effect, it is necessary to use the formalism of parabolic waves, which form a full set of eigenfunctions in the Fourier-integral expansion and are analogs of plane waves in the Lobachevski space [7,8]. Each of these waves gives the $\Delta T/T$ spot with its microstructure depending on the wave-number K. For example, the scalar convergent and divergent parabolic waves have the form:

$$Q^{\pm} = (\sqrt{1 + h^2\vec{x}^2} \mp h\vec{e}\vec{x})^{\mp ik/h}, \quad k \geq 0 ,$$

in the space coordinates $d\ell^2 = dx^2 + dy^2 + dz^2 - (h\vec{x}d\vec{x})^2/(1 + h^2\vec{x}^2)$, where $\vec{e}$ is the unit vector in the direction of the wave propagation. Evidently, $Q^{\pm} = e^{i\vec{k}\vec{x}}$ for h = 0 (Euclidean space) and the spot scale $\Delta\Theta \simeq 2|\vec{x}_{last\ sct}|^{-1} \simeq 2\Omega << 1$ for h = 1 (Lobachevski space); $\vec{k} = k\vec{e}$, $\vec{e}\vec{x} = |\vec{x}| \cos \Theta$.

References

1. Lukash, V.N. Novikov, I.D. Proceedings of I.A.U. Symposium 104, (1983), p. 457. Aug. 30 to Sept. 2, 1982.
2. Lazarides, G. Proceedings of I.A.U. Symposium 104, Crete, Greece, (1983), p. 469.
3. Smoot, G.F., Lubin, P.M. Ap. J., 234, L83, 1979.
4. Fabbri, R., Guidi, I., Melchiori, F., Natal, V. Preprint, 1979.
5. Boughn, S.P., Cheng, E.S., Wilkinson, D.T. Ap. J., 243, L113, 1981.
6. Lukash, V.N. Nuovo Cimento, 35 B, 268, 1976.
7. Lukash, V.N. Proceedings of the Eighth International Conference on Gravitation, 237, Waterloo, Canada, 1977.
8. Bisnovatyi-Kogan, G.S., Lukash, V.N., Novikov, I.D. Proceedings of Fifth Regional Meeting (IAU/EPS), Liège, Belgium, 29 to 31 July 1980.
9. Novikov, I.D. A. Zh., 45, 538, 1968.
10. Doroshkevich, A.G., Lukash, V.N., Novikov, I.D. A. Zh., 51, 940, 1974.
11. Lifshitz, E.M. Zh.E.T.F., 16, 587, 1946.
12. Grishchuk, L.P., Zeldovich, Ya.B. A. Zh., 55, 209, 1978.
13. Silk, J., Wilson, M.L. Ap. J., 244, L37, 1981.
14. Wilson, M.L. Ap. J., 253, L53, 1982.

Discussion

Smoot: Novikov visited Berkeley about four years ago and pointed out that a quadrupole anisotropy in expansion would give a large-scale anisotropy of the spot shape in the cosmic background radiation. A graduate student, Chris Wilebsky, and I read his paper and used a small dish antenna to search for such an effect near the maximum in the dipole anisotropy measured by our U2 experiment. Since that time the bulk of the dipole signal is better explored and is accepted as due mainly to the motion of the galaxy. Thus, you make a very valid point that fitting by using spherical harmonics does not efficiently search for these spot size anisotropies. Although we make maps now that the receivers are so improved, using your parabolic wave expansion will provide a much better qualitative limit. For $\Omega \lesssim 0.05$ the antenna beam size will smear the spot significantly from its characteristic shape which Chris and I referred to as the "Navel of the Universe" because of the shape like a navel orange and its signature as a relic of the birth of the universe.

Lukash: I would like to note that the search for the spot anisotropy must differ from that for the dipole one. The point is that you can, in principle, detect the dipole while investigating any region on the celestial sphere of an angular scale ~ 1. But if the actual anisotropy is of the spot type (even with the characteristic angular scale ~ 1 for $\Omega \sim 1$!), then, for example, you find nothing in the southern sky and detect the spot only in the northern sky.

OMMENTS AND SUMMARY ON THE COSMIC BACKGROUND RADIATION

George F. Smoot
Space Sciences Laboratory and Lawrence Berkeley Laboratory
University of California, Berkeley, California 94720

his session on the cosmic background radiation is remarkable for the uality of the interesting new results presented and the controversy ver substantive issues. This paper reviews the status of measurements f the spectrum and anisotropy of the cosmic background and summarizes hat is known of the properties of the cosmic background radiation: ts spectrum, anisotropy, polarization, and statistics.

PECTRUM

I would like to comment on the spectrum of the cosmic back-round radiation, the one area where no direct new experimental data as contributed. There are two experiments with results pending: 1) a repeat measurement with new detectors planned by Richards' group nd (2) the low frequency measurement whose first phase we have just oncluded. Figure 1 shows the currently published spectrum measure-ents with arrows indicating the five low frequencies at which we have ust made measurements and the five high frequencies Richards' group lans to measure.

These low frequency measurements were a collaborative effort ith G. Sironi from Milano responsible for the 12 cm radiometer, N. andolesi of Bologna responsible for the 6 cm radiometer, and B. artridge of Haverford College responsible for the 3.2 cm atmospheric onitor. My group from UC Berkeley was responsible for the cold-load alibrator, the support system, and the three higher-frequency radi-meters. S. Friedman, G. De Amici, and C. Witebsky were responsible or the 3, 0.9, and 0.33 cm wavelength radiometers respectively.

After a full test at Berkeley the experiment was conducted at he UC White Mountain Research Station at a dry (2-3 mm H_2O), high-ltitude (3800 m or 12,500 ft) site. The cold-load calibrator was set n the ground, suspended from a railroad. Each spectrum-measuring adiometer was mounted on a cart allowing it to be rolled along the ailroad and positioned above the cold-load calibrator. The radiometers easured the emitted power from the cold-load calibrator and then from he sky, each relative to the power entering the reference antenna. By ifferencing the power measurements and by knowing the temeprature of he cold-load calibrator and the conversion from power to antenna tem-erature, we determined the temperature of the sky.

G. O. Abell and G. Chincarini (eds.), Early Evolution of the Universe and Its Present Structure, 153–158.

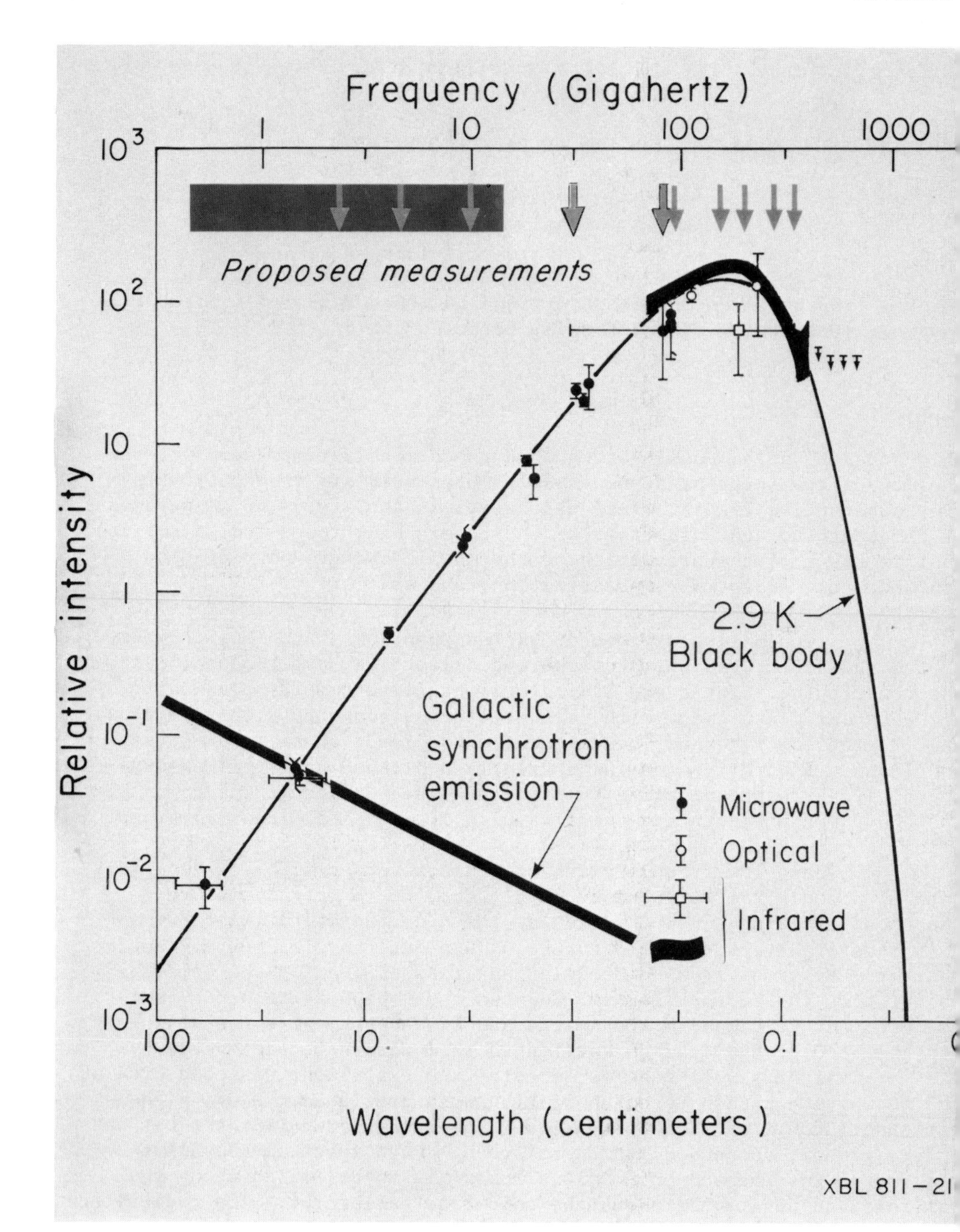

Figure 1: Current and Proposed Measurements of Cosmic Background Radiation Spectrum.

The antenna temperature of the sky is the sum of the cosmic ackground radiation, galactic emission, atmospheric emission, and mission from the earth and surrounding objects. The emission from the tmosphere was determined through zenith scans. Low-sidelobe antennas nd ground shields reduced pickup of emission from the earth to neg-igible levels. At the lowest frequency the minimum galactic emission ives 0.2 to 0.3 K of antenna temperature and decreases rapidly with ncreasing frequency. We arranged the observation schedules so that the owest frequency radiometers measured the spectrum during the times of inimum galactic emissions. The three lowest frequencies also measured he galactic emission by scanning in right ascension.

The cold-load calibrator was a large open-mouth dewar con-aining a very good microwave target. Measurements indicate that the alibrator was a blackbody with an emissivity greater than 0.999. The old-load calibrator was operated first with LN as the cryogen, then ith LHe to make our best measurements. At the three lowest frequencies e expected the temperature difference between the cold-load calibrator approximately 3.8 K) and the sky (2.8 K cosmic background radiation lus 1 K atmosphere) to be small. As we operated the 3 cm radiometer e could see that the temperature difference was less than 0.1 K--enerally on the order of 0.05 K. At the higher frequencies the atmos-heric emission was higher, about 4.5 and 12 K at 0.9 and 0.33 cm res-ectively. These values were about what we expected from our atmos-heric model and were in good agreement with previous measurements by urselves and others. We expect to have preliminary results out by the nd of the year.

We plan to return to White Mountain next summer (1983) to epeat the measurements with improved systems and a new radiometer. The ew radiometer will be tunable from 1.7 cm to 15 cm to provide contigu-us measurements of the spectrum at many wavelengths. A single tunable adiometer has a number of potentially important advantages over several ixed-frequency radiometers. Although the absolute accuracy will be omparable to the fixed-frequency radiometers, the relative accuracy f one measurement to the next should be improved. This is a great enefit in looking for spectral deviations. Traditionally systematic rrors have limited the accuracy measurements. Use of a single receiver ith one measurement procedure, rather than multiple receivers will elp reduce this problem. A tunable system is more likely to find arrow features than fixed-frequency systems and can also detect and void isolated points of RF interference more readily.

On the other hand, good broad-bandwidth components are more ifficult to design and manufacture than those designed for a single ixed frequency. A particularly important example is the antennas. ixed-frequency, narrow-bandwidth antennas can be made with very low idelobes and good impedance matching (low reflection). Thus we cannot se the tunable radiometer to determine the atmospheric emission ac-urately through zenith scans, since it is too sensitive to the earth's mission. Instead, we will make atmospheric zenith scans with the ixed-frequency radiometers. The repeated and improved fixed-frequency easurements serve also as a cross check of the tuned measurements.

The use of anisotropies and possibly the Sunyaev-Zeldovitch effect to look for spectral distortions has entered a new stage. There is now a sufficient number of quality measurements of the first-order anisotropy over enough frequencies to do comparison fitting to a Planckian spectrum versus a Woody-Richards spectrum. Wright pointed out that the data slightly favor the Planckian spectrum. The near future holds the promise of improved data both from existing data not yet analyzed and from upcoming measurements.

ANISOTROPIES

This system has unveiled a new generation of experiments on both the large and small angular scale. The new large-angular-scale data from the Princeton, Berkeley, and MIT group confirm and more accurately measure the first-order anisotropy, and set upper limits on quadrupole and higher-order anisotropies. New Princeton and Ratan 600 measurements have reduced the upper limits on small-angular-scale anistoropy.

What characterizes this new generation is: (1) significantly improved system performance, (2) substantial sky coverage (and its corollary pointing), (3) improved understanding of the galactic foreground signal, and (4) good calibration. Previous experiments typically had sensitivities of 50 mK for one second of integration; this new generation has sensitivities of about 10 mK for a second of integration. Parijskij reported the best sensitivity, 2 mK for a second of integration at 7.6 cm and sensitivities of 7 and 15 mK at 8.2 and 3.9 cm respectively. These improved sensitivities are the result of new technologies being adapted to anisotropy experiments. Currently the maser systems are providing sensitivity at about the 5 mK level for a one second integration while cryogenically-cooled receivers and bolometers are worse by about a factor of two. All of these represent an astounding improvement in system performance, evidenced by the fact that these systems show the first-order anisotropy in real time during the experiment, while five years ago the effort was to discover and detect that anisotropy.

The improved performance coupled with substantial sky coverage is important for two reasons. First, balanced sky coverage uncorrelates the various moments of a spherical harmonic expansion fit to the data. This allows better limits, simplifies understanding of the stated errors, and reduces the chance that one spherical harmonic might feed power to another. Second, it is now possible to generate sky maps where each resolution element has enough sensitivity to see galactic sources if they are present and with sufficient coverage to see the galactic pattern on the sky. These maps are key in improving understanding of the galactic foreground signal and thus ultimately in understanding the data on cosmic background anisotropy. The structure of the map points out to the experimenter the necessity of understanding the galactic signal and provides a vehicle for testing galactic models. Most of the old and new data for understanding the galactic emission does not come from the anisotropy experiment but from radioastronomical and infrared measurements. These data must be synthesized into a galactic model and tested against the sky map for completeness and correctness. As Wilkinson pointed out there is reason o believe that an incorrect galactic emission model probably accounts or his previously claimed low-frequency quadrupole signal. Possibly alactic dust accounts for the large hot region reported by Melchiorri, abbri, and collaborators.

Several groups are soon to be in a position to present ad-itional large-scale anisotropy data. Wilkinson's group not only has early complete coverage of the northern sky but also has data taken rom a balloon flight in Brazil which they have not yet analyzed. Ber-eley (Epstein, Lubin, myself) plans to get southern hemisphere data ate this fall or early this winter with our 3 mm (90 GHz) system. eiss at MIT is trying to arrange for another flight to improve his orthern sky coverage and to use his new, more sensitive bolometers. hus, not only is there new data from the 24 GHz Princeton maser, the 0 GHz Berkeley helium-cooled radiometer, and the 250 GHz MIT pumped-elium-cooled bolometer reported here, but also the expectation of ad-itional data soon to come.

Calibration is the fourth significant improvement that I isted. Calibration is important in two ways: one, the expected and he other, more subtle. With improved sensitivity the first-order nisotropy is measured with smaller statistical errors. In order to ompare the results of different measurements, either to look for spec-ral deviations or to determine the velocity of the galaxy, one needs ood calibration. With the improved sensitivity and the attendant esire and need for improved sky coverage, experiments are having to ombine data taken from two balloon flights occurring six months or more apart. Flights roughly six months apart are necessary to cover the part of the sky obscured by the sun in the first flight, since the quality data can only be taken at night. If the calibration changes from one flight to the next it introduces a spurious quadrupole signal into the data. Our 3-mm-wavelength anisotropy experiment has a small pop-up calibrator to check the spurious relative gain during the flight and from flight to flight. We are now converting this small pop-up calibrator to a full-beam calibrator in the expectation that it will provide an absolute inflight calibration.

The small scale anisotropy measurements reported by Wilkinson and Parijskij are spectacular in the scale of the apparatus used and the technology of the receivers. They are equally impressive in the scope of their procedures and data analysis. Wilkinson reported a 95% confidence level upper limit of $\Delta T/T$ of 1.1×10^{-4} at a frequency of 19.5 GHz and a beamwidth of 1.5 arcminutes. Parijskij reported a large number of upper limits at different angular scales and frequencies. His most stringent limits were 10^{-5} and 5×10^{-5} for 5 arcminutes and one degree at 7.6 cm wavelength. These very low limits generated spirited discussion about the data processing and source confusion, particularly about how the sources are subtracted. Whatever one's view on the sub-ject, it is now clear that the small-scale anisotropy experiments are also in the business of understanding the foreground galactic emission in order to interpret the anisotropy data properly.

The results of new experiments at both the large- and small-angular scales are very impressive and are a substantial improvement over past measurements. The field of anisotropy measurements has matured greatly as is evident in the sophistication of techniques and understanding of experimental pitfalls.

SUMMARY OF PROPERTIES OF COSMIC BACKGROUND RADIATION

The spectrum is presently well-fitted by a 2.7 K Planckian spectrum in the Rayleigh-Jeans region, although Woody and Richards report a 15% distortion of the best-fitted Planckian spectrum at and above the peak.

The first-order (dipole) anisotropy is now confirmed and well measured by several groups to be at the 10^{-3} level. The reports of quadrupole anisotropy at the 0.5 to 1 mK level are disputed. There is no evidence for any anisotropies except the diple at greater than the 10^{-4} level.

The upper limit for fractional linear polarization is less than 10^{-4} or 0.3 mK at the 95% confidence level. At Berkeley we have set an upper limit of 10^{-3} on the variation of circular polarization from position to position on the sky. (We have not reported these results yet because there is no easy way to separate a circular polarization signal internal to the apparatus from one in the cosmic background radiation. Thus we can only do difference measurements from position to position and can only fully trust those along fixed declinations for which the polarimeter is not moved or changed.)

The only measurement of the time dependence of the cosmic background radiation that anyone has is a measurement of the photon statistics. If the cosmic background radiation has a thermal origin, then the photons should have a Planckian or at least Bose-Einstein distribution, with the corresponding statistics. Several different groups have begun experiments to check this hypothesis. Bruce Allen, a student working with R. Weiss at MIT, was the first to set up such an experiment. As a result other groups have become interested. De Barnadis, Dall Oglio, et al. reported here the first preliminary results, indicating that in the infrared (near the peak and shorter wavelengths) the background radiation appeared to have the proper (i.e. thermal) photon statistics.

In my opinion one must be prepared for the possibility that at the limits set by astrophysical foregrounds, the cosmic background radiation is described as having a Planckian distribution and is isotropic except for a Doppler-shift induced dipole anisotropy. This possibility provides a very simple description of a very fundamental attribute of the universe.

EDSHIFTS AND LARGE SCALE STRUCTURES

G. Chincarini
University of Oklahoma and European Southern Observatory

The effort to measure the geometry of space by experiment, hat is, the determination of the Hubble Constant, H_O, and of the eceleration parameter, q_O, led toward the end of the first half of the entury, to the classical paper by Humason, Mayall, and Sandage (1956). heir catalogue contains 920 redshifts collected over a twenty-year eriod (1935-1955). Further redshifts of galaxies were measured to efine such determinations and to study the dynamics of clusters (Zwicky 933).

With the advent of fast optics, it became possible to systemtically observe faint objects. Page (1960) and Burbidge (1975) attacked he problem of the determination of the masses of galaxies. Minkowski 1958), among others, obtained spectra of radio sources. Schmidt (1963) bserved and interpreted the spectra of quasars; Mayall (1960) and wicky (1957), among others, called further attention to the dynamics of lusters of galaxies, etc. It was, however, only after the advent of anoramic intensified detectors, following the pioneering work of allemand (1962) that redshift surveys of large samples of galaxies ecame feasible. Some of these surveys, which had been planned to invesigate the size of clusters of galaxies, naturally led to the problem of he large scale space distribution of galaxies.

As discussed by Professor Oort (1983) at this symposium, such urveys revealed a segregation in the distribution of redshifts, the resence of very large structures, $\gtrsim$ 50 h^{-1} Mpc, and the presence of egions which, down to the limiting magnitude of the sample, are void of alaxies. Voids seem not to be in contradiction with gravitational lustering models (Aarseth and Saslaw 1982) and well accounted for in the ork of Doroshkevich, Shandarin and Zeldovich (1982). Their dynamical ffects as negative density fluctuations have been studied by Peebles 1982) and Salpeter (1983). They tend to form over density ridges and to ause small fluctuations in the velocity field of the order of 100 to 50 km/sec.

The presence of empty regions of space contradicts the concept f a uniform background density of galaxies as conceived by Hubble (1934) nd, in a somewhat different form, by Zwicky (see Chincarini 1982). If

G. O. Abell and G. Chincarini (eds.), Early Evolution of the Universe and Its Present Structure, 159–165.

such backgrounds exist, voids can be used to set an upper limit to the background density.

On purely observational grounds, we cannot exclude that such voids may be populated by faint galaxies (in this case, however, we would have to explain variations of the luminosity function on a scale size of the voids) or by other forms of matter.

In addition to surveys in selected regions of the sky, we need to understand: 1) the structure and dynamics of the local supercluster (see early maps by Shapley and Ames, 1932, and the work by de Vaucouleurs over the last 30 years, 1978, and references therein); 2) the distribution of galaxies over the whole sky; and 3) the discrepancies in the measurements of the density parameter, Ω, as derived from different samples and methods. Efforts in these directions have produced: 1) the 21-cm survey by Fisher and Tully (1981); 2) the survey by Sandage (see the revised Shapley-Ames catalogue of bright galaxies by Sandage and Tammann 1981); and 3) the Harvard Survey by Davis, Huchra, Latham and Tonry (1982). Other surveys, for instance the one by Rubin, Thonnard, Ford and Roberts (1976), have been very significant in supplying a useful data base, in measuring the anisotropy of the Hubble flow and in stimulating research in this direction. The Rubin-Ford effect is not yet fully understood.

All this work contributed to form a redshift data base that, according to the catalogue by Palumbo, Tanzella-Nitti and Vettolani (1982), updated to 1980, amounts to 13,672 redshift measurements of 8250 galaxies with cz $\lesssim$ 100,000 km/sec. Huchra is presently compiling a catalogue with entries for about 9000 galaxies, so that we can estimate we have at the present 12,000 to 14,000 published and unpublished redshifts of galaxies. A large contribution to this has been given by recent surveys, yet to be published, carried out in 21-cm at the Arecibo Observatory. Finally, according to M.P. Veron (1982), there are at present redshifts for about 1800 quasars.

Clusters as Markers of the Large-Scale Structure

Abell (1958) noticed that the distribution of clusters is not random and measured a cell size of about 40 h^{-1}. His catalogue and subsequent redshift measurements (Sarazin *et al*. 1982 list redshifts for about 329 clusters) have been the objects of various statistical analysis; among them are Kiang (1967), Kiang and Saslaw (1969), Hauser and Peebles (1973) and Rood (1976). Einasto and his collaborators (1980) used clusters to map large-scale structures, in particular the Perseus/Pisces region. Most recently, Bahcall and Soneira (1982) obtained the important result that clusters are correlated to separations of about 150 h^{-1} Mpc.

Clusters present, however, some limitations in the study of the large structures (superclusters) and in their use as markers of the distribution of matter. Clusters, in fact, define the density peaks of the large structures so that the information is limited to the location of the large density enhancements. Furthermore, as pointed out by Peebles (1980), the contrast in density for clusters is larger than the contrast in density for galaxies so that both clusters and galaxies cannot be good tracers of the large-scale mass distribution. The galaxies seem to be more reliable.

Galaxies allow one to map regions of low density, to study the effects of environmental conditions, and in a more reliable way to study the geometry of the large structures and to measure the deviations from the Hubble flow.

Geometry and Peculiar Motions

It is quite clear from the redshift-defined structures that their shape is highly asymmetric. In most cases, no limit of the structure has been found within the region of the sky sampled so that there may be continuity from one structure to the other. While the shape may partly depend on "how" we define the structure as mapped by the redshift surveys, filaments (often spaghetti-like?) seem to be quite common and I do not see much evidence of predominant pancake structures or of a defined cell size as suggested by Corwin (1981). On the other hand, within such structures a scale size can be defined by the two-point angular correlation function. This is in agreement with the slope of the cosmic correlation function derived by Peebles.

Evidence of the above statements is observed in the Her/A2197-99 supercluster which extends in the Serpens-Virgo region and may then continue toward Coma. Even more impressive is the sample in Perseus-Pisces and I refer to its discussion by Giovanelli (1983).

I find it quite interesting that some of these structures are very narrow, at least in one dimension. The velocity dispersion along the line of sight for Perseus-Pisces, Coma A1367 and Horologium is $\sigma \leq$ 500 km/sec. Selected features have velocity dispersions which are even smaller. The velocity dispersion may be partly contaminated by peculiar motions (however, away from concentration of matter - clusters - perturbations of the velocity field are expected to be small). Structures in Perseus-Pisces are thin, also, as seen projected on the celestial sphere. The Lynx-Ursa major supercluster observed by Giovanelli and Haynes (1982) has a projected width of only a few megaparsecs. It cannot be that in all cases we are observing projection effects in disk-like structures. Neither do I see evidence, at the present, that the geometry could be determined by the observational definition of a structure or by the limitation of the samples. The dispersed supercluster component (non-cluster galaxies) is not very luminous and is of about 5×10^{10} $L_\odot/Mpc^2$. Determination of distances by the Tully-Fisher effect will allow deeper understanding of the geometry and possibly the measurement of non-Hubble velocities near noncluster density peaks. Such a program is now in progress.

Except for the local supercluster, little is known about peculiar motions. As is known, infall velocities in a local and fair (large enough) sample of the universe and measurements of the two-point angular correlation function in the direction parallel and perpendicular to the line of sight $\xi(\sigma,\pi)$ allow, following the method outlined by Peebles (1980), a measure of the density parameter Ω. Such determinations are unaffected by virialization (clusters) or evolutionary (q_o) problems and, assuming a fair sample, should give a realistic value of Ω. The value determined by the above methods oscillates, at present, in the range of 0.1 to 0.3 (Davis and Peebles 1982). It is such values that point to an open universe and yet they are too large to be consistent with the primordial helium abundance if the matter density is dominated by baryons ($\Omega b\ h_o^2 \simeq 0.01$). To conclude, I see some of the future observational work aimed at: 1) better defining the geometry of the large structures; and 2) tackling the difficult problem of non-Hubble flow in selected structures. Advances and new experiments in high-energy physics will tell us whether the helium problem can be solved by massive neutrinos.

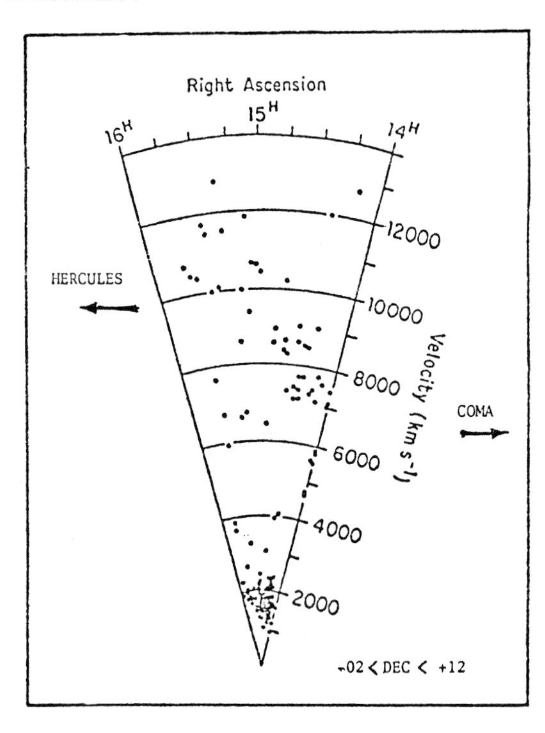

Fig. 1. Cone diagram of a sample of galaxies in the southern extension of the Hercules supercluster. It may be indicative of a connection to the Coma supercluster. (from Giovanelli Haynes, 1982)

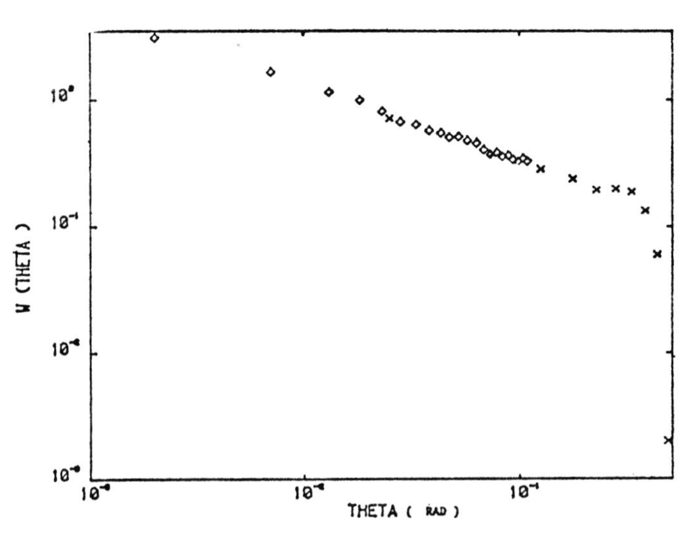

Fig. 2. Two-point angular correlation function for the Perseus-Pisces supercluster.

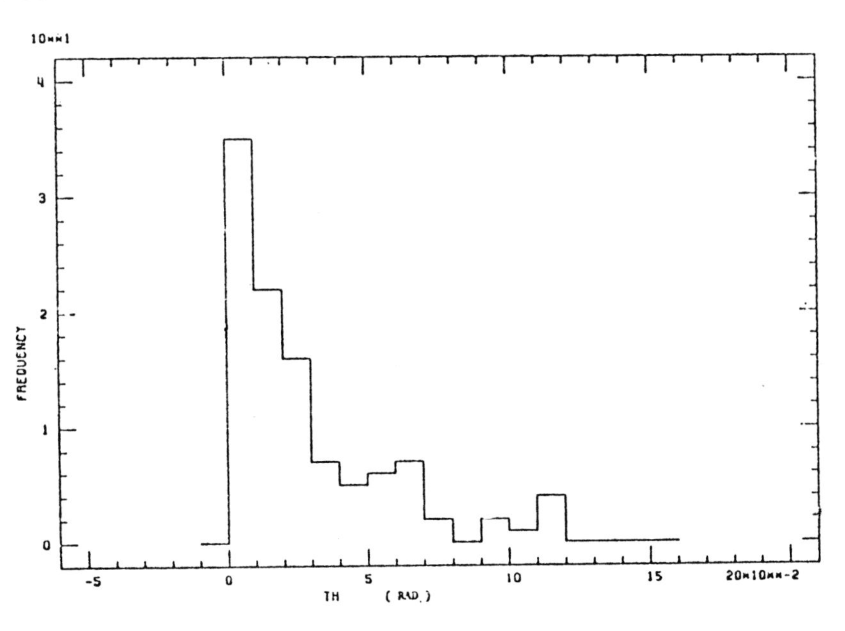

Fig. 3. Integrated width on the celestial sphere of the Perseus-Pisces supercluster (0.01 rad $\simeq$ 1 Mpc, H_o = 50 km/sec/Mpc).

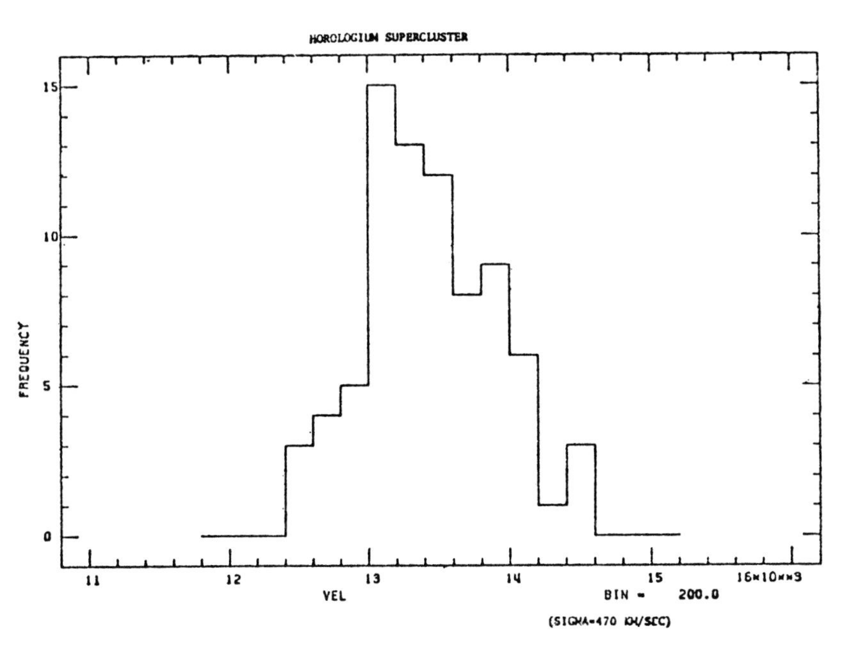

Fig. 4. Redshift distribution of the clump at V = 13436 in the Horologium supercluster.

REFERENCES

Aarseth, S.J., and Saslaw, W.C. 1982, Ap. J., 258, L7.
Abell, G.O. 1958, Ap. J. Suppl., 3, 211.
Bahcall, N.A., and Soneira, R.M. 1982, Princeton Observatory Preprint.
Burbidge, E.M., and Burbidge, G.R. 1975, Galaxies and the Universe, ed. A. Sandage, M. Sandage and J. Kristian, p. 81 (Chicago: University of Chicago Press).
Chincarini, G. 1982, Proceedings, III Escola de Cosmologia e Gravitacac Rio de Janeiro, February 1982, and University of Oklahoma Preprint (and references therein).
Corwin, H. 1981, Ph.D. Thesus, University of Edinburgh.
Davis, M., and Peebles, P.J.E. 1982, preprint.
Davis, M., Huchra, J.P., Latham, D.W., and Tonry, J. 1982, Ap. J., 253, 423.
de Vaucouleurs, G. 1978, Proceedings, I.A.U. Symposium 79, ed. M. Longair and J. Einasto, p. 205 (Dordrecht: Reidel).
Doroshkevich, A.G., Shandarin, S.F., and Zeldovich, Ya.B. 1982, Comment on Astrophysics (and references therein), Vol. IX, 265.
Einasto, J., Jôeveer, M., and Saar, E. 1980, M.N.R.A.S., 193, 353.
Fisher, J.R., and Tully, R.B. 1981, Ap. J. Suppl., 47, 139.
Giovanelli, R., and Haynes, M.P. 1982, private communication.
Hauser, M.G., and Peebles, P.J.E. 1973, Ap. J., 185, 757.
Hubble, E.P. 1934, Ap. J., 79, 8.
Humason, M.L., Mayall, N.U., and Sandage, A.R. 1956, A. J., 61, 97.
Jôeveer, M., and Einasto, J. 1978, Proceedings, I.A.U. Symposium 79, ed. M. Longair and J. Einasto, p. 241 (Dordrecht: Reidel).
Kiang, T. 1967, M.N.R.A.S., 135, 1.
Kiang, T., and Saslaw, W.C. 1969, M.N.R.A.S., 143, 129.
Lallemand, A. 1962, in Advances in Electronics and Electron Physics, XVI, 1, Academic Press.
Mayall, N.V. 1960, Ann. Astrophys., 23, 344.
Minkowski, R. 1958, Paris Symposium on Radio Astronomy, ed. R.N. Bracewell, p. 315 (Stanford, Calif.: Stanford University Press).
Oort, J.H. 1982, These Proceedings (and references therein).
Page, T. 1960, Ap. J., 132, 910.
Palumbo, G., Tanzella-Nitti, G., Vettolani, G. 1982, private communication.
Peebles, P.J.E. 1982, Ap. J., 257, 438.
Rood, H.J. 1976, Ap. J., 207, 16.
Rubin, V.C., Thonnard, N., Ford, W.K., and Roberts, M.S. 1976, A. J., 81, 719.
Salpeter, E.E. 1982, these proceedings, p. 211.
Sandage, A., and Tammann, G.A. 1981, A Revised Shapley-Ames Catalog of Galaxies, Carnegie Institution of Washington, Publication 633.
Sarazin, C.L., Rood, H.J., and Struble, M.F. 1982, Astron. Astrophys., 108, L7.
Shapley, H., and Ames, A. 1932, Harvard Obs. Ann., 88, No. 2.
Veron, M.P. 1982, private communication.
Zwicky, F. Helv. Phys. Acta, 6, 110.
Zwicky, F. 1957, Morphological Astronomy (Berlin: Springer-Verlag).

DISCUSSION

Inagaki: My question is similar to that of Prof. Dick Miller to Prof. Oort. Are there aggregations of galaxies ranging continuously from small groups to superclusters or are they hierarchical?

Chincarini: A cluster size can be defined dynamically or by using density criteria. Since clusters are embedded in superclusters with no discontinuity in the number density distribution of galaxies (see, for instance, Coma-A1367), clusters fade into superclusters, and a clear separation in the density distribution is not possible. In the redshift-defined large structures (for example, Coma-A1367 and Perseus-Pisces), supercluster clusters and groups are embedded in the structure and appear as density enhancements, which are probably bound. (For some groups the observations are equivocal, however.) I see no evidence of a purely hierarchical structure, groups of groups or clusters of clusters, independent of their supercluster environment.

Ellis: Would you comment on the existence of large structures as general features of the galaxy distribution? Redshift surveys in "interesting areas" may reveal such structures, but their significance as general features can be assessed only by performing deep surveys in randomly-chosen directions. The AAT survey (Bean et al., this symposium) do not show statistically convincing evidence for these large-scale features.

Chincarini: Certainly we need to have a fair sample of the universe, as you suggest. I believe that such samples are available. The observations I made with Martins (1975) refer to a noncluster region (Seyfert sextet). Structure is seen in the horologium sample (poster by Chincarini et al.), and especially in Perseus-Pisces, the Arecibo sample presented by Giovanelli (this covers a region of 30° in declination and 6h in right ascension), Hercules, and Coma-A1367. Finally, I refer to the work of Kirshner et al. and Davis et al. (Harvard Survey). If a sample does not show a similar structure, I am not surprised, since it may reflect the distribution of objects in that region and depend on various inclination effects of the structure. I am not familiar with your sample; it may reflect, however, a region of small density fluctuations. To use the correlation function analysis (your poster). You may possibly need a smooth distribution in depth. Finally, there seems to be no contradiction with some models in what is observed (see Shandarin, these proceedings). I hope deeper surveys will soon be available so that we can better understand the proposed geometry.

THE CENTER FOR ASTROPHYSICS REDSHIFT SURVEY

M. Davis, J. Huchra and D. Latham
Harvard-Smithsonian Center for Astrophysics

We have completed the survey of radial velocities for all 2400 galaxies brighter than 14.5 at high galactic latitude in the northern hemisphere. This data set has already been used to derive a good measure of the local mean mass density, describe the overdensity and the dynamics of the local supercluster and to analyse the dynamics of groups and clusters of galaxies within the sample volume.

I. INTRODUCTION

Since the discovery by Hubble of a linear relation between radial velocity or redshift and distance and the development of a theoretical basis for cosmology provided by the universe models of Einstein, De Sitter, Lemaitre, and Friedmann, we observational cosmologists have been mapping the universe with surveys of galaxy redshifts. The goals of most such surveys have been to find the mean local luminosity density of galaxies, measure masses for larger systems of galaxies by applying the virial theorem to "bound" clusters of galaxies, and study large scale cosmological effects by measuring deviations from the redshift-magnitude law of Hubble.

The CfA redshift survey has its origins in the theoretical work of Jim Peebles and his students who described the application of "statistical" virial theorems to galaxy catalogs (eg. Geller and Peebles 1973; Peebles 1976; and Davis, Geller, and Huchra 1978). They found that a local mass density estimate could be derived from accurate measurements of the velocity dispersion for galaxies in the field and the potential energy in galaxy clustering measured by the correlation function. Additional impetus was provided by the work of Turner and Gott who used the Zwicky catalog of galaxies (Zwicky et al. 1961-1968) and very limited velocity information to derive local cosmological parameters (eg. Gott and Turner 1976; Turner and Gott 1975; Turner and Gott 1976). By the mid-1970's, a large number of people realized that a careful, complete and relatively deep survey of galaxy redshifts could be used to derive

G. O. Abell and G. Chincarini (eds.), Early Evolution of the Universe and Its Present Structure, 167–173.

considerably more reliable information about the three-dimensional distribution and relative motions of galaxies in space. Also about this time, a small number of observers realized that such a survey was practicable.

Almost all earlier redshift survey work produced velocities with accuracies worse than 100 km/s, not good enough for studies of the "cosmic" virial theorem or of small galaxy systems with small velocity dispersions. Two major advances in the art of redshift measurements, however, improved the obtainable accuracy to better than 30 km/s. The first is 21-cm spectroscopy, unfortunately applicable only to gas rich galaxies. The second and most important for this work is optical spectroscopy with photon-counting detectors and data reduction via cross-correlation techniques. We can now obtain a redshift for almost any galaxy brighter than 15th magnitude on a 60-inch telescope in 60 minutes or less. With instrumental help from Steve Shectman, the NSF and the Smithsonian Institution plus a guarantee of sufficient telescope time on the 1.5 m on Mt. Hopkins, we started observing in the spring of 1978.

II. THE DATA

The original goal of the survey was to obtain accurate radial velocities for all the galaxies in the Zwicky or Nilson (1973, Zwicky missed a few) catalogs brighter than m_{pg} 14.5, above galactic latitude 40°, and north of declination 0°. We extended this goal to include redshifts for those galaxies below galactic latitude -30° and above declination -2.5^{o}. There are 2400 galaxies in this sample which covers 2.7 steradians (1.83 in the north and 0.83 in the south). This survey was basically completed in April of 1981 - we have spent the last year remeasuring poor or discrepant velocities and starting major extensions of this survey which will be described later.

The catalog of redshifts for the above sample has been accepted for publication of the Ap. J. Supplements (Huchra, et al. 1982). We have measured approximately 62% of the velocities in the catalog, with the remainder coming primarily from 21-cm measurements at Green Bank and Arecibo (19%) and the work of Sandage and Tammann (1981, 5%) and de Vaucouleurs, et al. (1976, 5%). The median quoted error in this catalog is 27 km/s, with only 35 velocities (1.5%) with errors larger than 100 km/s.

At present, a total of approximately 5500 spectra have been taken of approximately 4000 galaxies with both the 1.5 m and the Multiple Mirror Telescope. In addition, we have assembled a computer readable catalog containing these plus most other published galaxy radial velocities. This catalog contains the culled velocities for approximately 9000 galaxies, and covers two "complete" samples - the one described above (hereafter called NZ40) and a whole sky sample of approximately 1300 galaxies to m_p 13.2 very similar to the Revised Shapley-Ames sample.

II. RESULTS

Analyses of the CfA survey have already produced several nteresting results. These can be grouped into four catagories: 1) Studies of the local supercluster and its dynamics, (2) the arge scale space distribution of galaxies, (3) the identification nd dynamics of virialized systems, and (4) the luminosity density f normal and "active" galaxies, I will briefly try to summarize ach in the following paragraphs.

With the discovery and "accurate" measurement of the infall of he local group into the local supercluster (see Mould, et al. in his symposium or Aaronson, et al. 1982 for references and a more etailed discussion) we can derive an estimate of the cosmological arameter Ω if we determine the magnitude of the relative overdensity nside our position in the supercluster. The exact determination of he overdensity requires a model for the infall (to place galaxies n their proper positions). With an assumed infall of 250 km/s, he mean internal overdensity is 2.1 and the computed Ω is only 0.2 Davis and Huchra 1982).

The maps made using the quasi-three-dimensionality of the edshift data show some old friends and some surprises. Figures a-c show three cuts in velocity in the northern galactic hemisphere. n fig. 1a we plot all galaxies with velocities between 0 and 000 km/s. The local supercluster dominates the distribution and irgo dominates that. The classical Ursa Major and Leo groups are een as relatively diffuse extensions to Virgo. In fig. 1b we see alaxies between 3000 and 6000 km/s. There is a hint of a long ilamentary structure between Abell 1367 and Coma, but extending uch further east than Coma itself. There is almost nothing behind irgo! In fig. 1c, which shows galaxies between 6000 and 10000 km/s, oma and A1367 stand out, as well as a diffuse group near 14^h and $+10^o$ dentified with a Zwicky cluster. There are several structures like his, as well as several empty regions, with scales in excess of 0 Mpc. A comparison of these maps to similar "maps" made from -body simulations which can reproduce the observed correlation unction show <u>no</u> similarity of structure (Davis, et al. 1982).

We have developed several competing algorithms for finding roups of galaxies and clusters in redshift catalogs (Huchra and eller 1982; Press and Davis 1982). Even with a much deeper survey e can reproduce the classical de Vaucouleurs groups without running nto foreground/background confusion problems. All techniques ield roughly the same answer for the mass-to-light ratio for groups - pproximately 200 in solar units. With the mean mass density erived below, this also gives a value of Ω of $\simeq 0.1$. If selection ffects are properly taken into account, we do not find any evidence or a correlation between M/L and size for groups and clusters. he measured M/L for Virgo is between 190 and 320. In producing

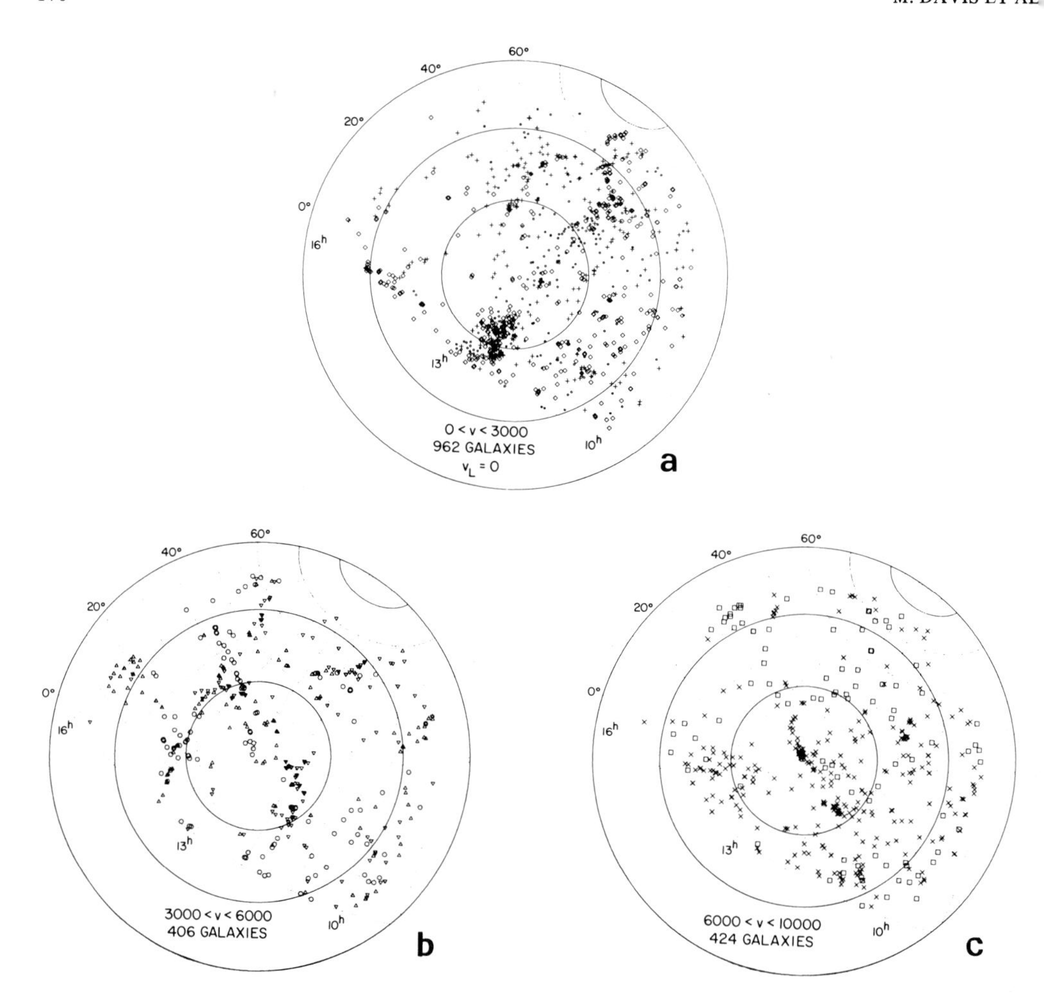

Figure 1. The surface distribution of galaxies with velocities between 0 and 3000 km/s (a), 3000 and 6000 km/s (b), and 6000 and 10000 km/s (c).

density enhancement contours to search for groups, we see both heirarchical and concentric (self similar?) structures (Fig. 2).

Because this sample is deeper and/or contains more galaxies than all previous samples used to find the luminosity density, we can do quite a reasonable job. Parametric and non-parametric techniques give very similar results for the shape of the function - which is relatively poorly fit by a Schechter form. The bright end does <u>not</u> turn over as fast as an exponential, and the faint end has too much curvature. The best Schechter function parameters (assuming H_o = 100 km/s/Mpc and an infall velocity of 250 km/s) are $\alpha = -1.3$, $M^* = -19.4$ and $\phi^* = 12.5 \times 10^{-3}$ Gal/Mpc^3/mag. The luminosity function is still very poorly determined below $M_p = -15.5$ because of the sample statistics and clustering.

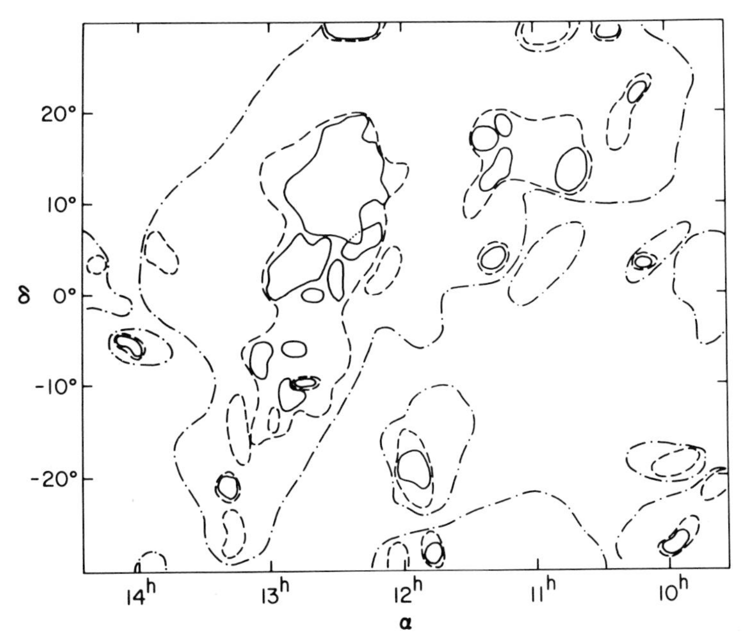

Figure 2. Three dimensional contour plot for groups of galaxies in the Virgo region.

One interesting byproduct of this survey was the discovery of a large number of new Seyfert galaxies (Huchra, Wyatt and Davis 1982). The spectroscopic coverage includes both Hβ - [OIII] and Hα - [NIII] - [SII]. We have doubled the number of bright Seyfert 1 and 2 galaxies, thus almost doubling their estimated space density. This includes finding objects that show broad Hα but no visible broad Hβ - objects that could not have been classified without coverage in the red. Curiously, almost all the absolutely bright galaxies in our sample of 2400 are active. Even more curiously, we found no active nucleus in any galaxy with an absolute magnitude less than -18.5.

IV. FUTURE GOALS

In order to accomplish the original goals of the CfA redshift survey, we have established considerable machinery in terms of computers, detectors, software and staff. We can efficiently and effectively obtain accurate radial velocities for galaxies. As usual, while analysing the CfA survey we have come across a number of problems that can only be answered with more information. As seen in the last section, we see structures on scales almost as large as our sample depth. We are blind to 80% of the sky. Well known nearby clusters show correlated velocity and spatial structure. We still don't know the shape of the luminosity function at the faint end, nor have we accurately discerned the density of "active" galaxies.

We are thus continuing the CfA redshift survey. Our basic goals are to:

(1) Extend the survey to uncovered areas of the sky.
(2) Increase the sample depth in NZ40 to magnitude 15.5.
(3) Perform wide field velocity surveys of nearby clusters.
(4) Complete the sample of high quality digital spectra of the bright galaxies in NZ40.

Our present projects include extending the survey to the southern hemisphere, wide field surveys of Virgo, Pegasus and a few other nearby Abell clusters, and completing the spectroscopic surveys of bright galaxies and remeasuring poor and discrepant redshifts. We are extending the survey into the southern hemisphere using the Vorontsov-Velyaminov catalog and the ESO catalog to select objects. This work is being done in collaboration with a group from Brazil using both Mt. Hopkins and the Brazilian National Observatory 1.5 m with a photon-counting reticon built at CfA. We expect to produce in excess of 1000 high quality redshifts per year. In addition. we already have velocities for 4000 of the approximately 8000 galaxies in NZ40 to magnitude 15.5.

The CfA Redshift Survey and its analyses have been funded by NSF Grant 80-00876, NASA Grant NAGW-201 and by continuous and generous support from the Smithsonian Institution. Special thanks go to S. Dillon Ripley and the Secretary's Fluid Research Fund to enable the presentation of this work.

REFERENCES

Aaronson, M., Huchra, J., Mould, J., Schechter, P., and Tully, R. B.: 1982, Ap. J. 258, 64.
Davis, M., Geller, M., and Huchra, J.: 1978, Ap. J. 221, 1.
Davis, M., Huchra, J., Latham, D., and Tonry, J.: 1982, Ap. J. 253, 423.
Davis, M., and Huchra, J.: 1982, Ap. J. 254, 437.
de Vaucouleurs, G., de Vaucouleurs, A., and Corwin, H.: 1976, Second Reference Catalog of Bright Galaxies, Austin, University of Texas Press.
Geller, M., and Peebles, P. J. E.: 1973, Ap. J. 184, 329.
Gott, J. R. II, and Turner, E.: 1976, Ap. J. 209, 1.
Huchra, J., and Geller, M.: 1982, Ap. J. 257, 423.
Huchra, J., Wyatt, W., and Davis, M.: 1982, A. J. in press.
Nilson, P.: 1973, Uppsala General Catalog of Galaxies, Uppsala Astron. Obs. Annuals V, Vol. 1.
Peebles, P. J. E.: 1976, Ap. J. 205, L109.
Press, W., and Davis, M.: 1982, Ap. J. in press.
Sandage, A., and Tammann, G.: 1981, Revised Shapley-Ames Catalog, Washington, Carnegie Institution of Washington.
Turner, E., and Gott, J. R. III: 1975, Ap. J. 197, L89.
Turner, E., and Gott, J. R. III: 1976, Ap. J. 209, 6.
Zwicky, F., Herzog, E., Wild, P., Karpowicz, M., and Kowal, C.: 1961-1968, Catalog of Galaxies and of Clusters of Galaxies, Vol. 1-6, Pasadena, California Institute of Technology.

DISCUSSION

Wampler: Is the velocity dispersion in the filaments that connect clusters and superclusters lower than the dispersion found in the clusters and superclusters?

Huchra: Yes, it appears to be lower by perhaps a factor of two, which is similar to what is expected in cluster collapse.

I.E. Segal: Your proposed motion toward Virgo, when added to the redshifts of various samples due to Visvanathan, de Vaucouleurs, etc., and represented as complete, generally impairs the (m,z) relation, independently of cosmology, unlike, for example, the CBR-determined motion. That is, for example, the standard deviation of the residuals from the theoretical relation become larger. Can you explain this?

Huchra: You really want to ask this question of Jeremy Martel later today, but a quick answer is that Yahil, Sandage and Tammann find a result substantially in agreement with ours for the Virgo infall by using a technique similar to the one you describe. Note both their result and ours assumes the direction to be radial towards Virgo. Also, the infrared Tully-Fisher technique is much more powerful than the magnitude-redshift relation.

THE ANGLO-AUSTRALIAN REDSHIFT SURVEY

J. Bean[1] G. Efstathiou[2] R.S. Ellis[1] B.A. Peterson[3] T. Shanks[1] and Z-L. Zou[4].

1 Department of Physics, South Road, Durham, England.
2 Institute of Astronomy, Madingley Road, Cambridge, England.
3 Mount Stromlo and Siding Spring Observatory, Woden, Canberra, Australia.
4 Peking Observatory, Peking, China.

The aim of the survey is to sample a relatively large, randomly chosen volume of the Universe in order to study the large-scale distribution of galaxies using the two-point correlation function, the peculiar velocities between galaxy pairs and to provide an estimate of the galaxian luminosity function that is unaffected by density inhomogeneities and Virgo infall.

We have measured magnitudes and redshifts for a sample of 340 galaxies located in 5 small (3.8^{o} x 3.8^{o}) fields complete to a magnitude limit $J \sim 16.75$. The radial velocities were measured using the Image Photon Counting System on the Anglo-Australian telescope and the spectra were analysed using cross-correlation techniques. For most of the sample the radial velocities have been measured with an accuracy better than 50 km sec^{-1}. The characteristic depth of the survey is $D^{*} = 200\ h^{-1}$ Mpc (h is Hubble's constant in units of 100 km sec^{-1} Mpc^{-1}) hence the sample is not biased by the Local Supercluster.

The results for the two-point correlation function show that the galaxy pattern is consistent with a Poisson distribution on scales larger than 10 h^{-1} Mpc. This result does not necessarily conflict with observations of large superclusters and voids but indicates that such structures must be rare. We find that the r.m.s. peculiar velocity between galaxy pairs is ~ 250 km sec^{-1} and applying the cosmic virial theorem we infer a low value for the cosmological density parameter, $\Omega = 0.1 \times 2^{\pm 1}$. We have determined the luminosity function for our sample and find good agreement with luminosity functions obtained using data from published surveys.

G. O. Abell and G. Chincarini (eds.), Early Evolution of the Universe and Its Present Structure, 175.

CATALOGUE OF RADIAL VELOCITIES OF GALAXIES

G. Palumbo[1], G. Tanzella-Nitti[2], G. Vettolani[2]

[1]Istituto TESRE, CNR, Bologna, Italy
[2]Istituto di Radioastronomia, CNR, Bologna, Italy

A catalogue of radial velocities of galaxies has been completed which lists all radial velocities published until the end of December, 1980, for galaxies in both the northern and the southern hemispheres.

For each galaxy the catalogue provides R.A. and Declination, name(s) of the galaxy (NGC, IC, 3 CR, etc.), each known heliocentric radial velocity with its reference, and the type of observation (whether optical or radio).

The comments on dubious identifications, wrong measurements, misprints in the original papers, and similar misleading items are given in the note section.

Statistics of the data contained in the catalogue is given in Table I. The Catalogue of Radial Velocities of Galaxies is being published by Gordon and Breach Science Publishers

Table I

Total number of galaxies with redshift	8250
Total number of redshifts	13,672
Total number of optical measurements	11,132
Total number of quoted references	2540

G. O. Abell and G. Chincarini (eds.), Early Evolution of the Universe and Its Present Structure, 177.

THE SOUTHERN CLUSTER SURVEY

G.O. Abell
University of California

Harold G. Corwin, Jr.
University of Texas, Austin

We are searching for clusters of galaxies on photographs taken for the Southern Sky Survey with the U.K. Schmidt telescope at Siding Spring, and we are preparing a comprehensive catalogue of the southern clusters. The purpose is to extend the early cluster survey made from the Palomar Sky Survey (Abell 1958) to full-sky coverage. We are taking special pains to delineate those clusters in the new southern survey that as nearly as possible match those criteria set for the northern survey clusters to be included in a homogeneous statistical sample. It is our expectation, therefore, that there will be a homogeneous all-sky sample of clusters for studying the large-scale distribution of matter in the universe.

The new southern cluster survey is now approximately half complete. We have also revised and updated the northern cluster catalogue, and we plan to publish both together.

THE REVISED NORTHERN CATALOGUE

The revised cluster catalogue, assembled from the Palomar Survey, contains many corrections that we have found or that have been called to our attention (the vast majority being slight adjustments to the equatorial coordinates), and some additional information. The following are tabulated for each cluster:

1. Catalogue number.
2. Equatorial coordinates to ± 1' for the equinox of 1950.
3. Precession in right ascension and declination.
4. Galactic coordinates on the modern system.
5. The photored magnitude, R_{10}, of the tenth brightest galaxy.
6. Distance class.
7. Richness class.
8. Raw counts of galaxies within the Abell radius, corrected for the field.
9. Redshifts, when available.

The catalogue numbers of the clusters will remain the same, inasmuch as these numbers are in rather wide use. As a consequence, because the coordinates have been precessed, the clusters will no longer be numbered precisely in order of increasing right ascension.

G. O. Abell and G. Chincarini (eds.), Early Evolution of the Universe and Its Present Structure, 179–184.

The identification of clusters that meet the criteria for inclusion in the homogeneous sample remains as before, and clusters that do not meet the criteria are indicated with an asterisk. The criteria are: 1) the cluster must be in a region clear enough of the Milky Way to assure reasonably complete identification. The galactic latitude limits vary somewhat with longitude (see Abell 1958); 2) the redshift, estimated from R_{10}, must lie in the range $0.02 \leq z \leq 0.2$; 3) there must be 50 or more galaxies within the Abell radius that are brighter than $R_3 + 2$, after correction for the background.

THE SOUTHERN CLUSTER CATALOGUE

For each cluster in the southern catalogue the following data will be given:

1. Catalogue number, in order of increasing right ascension, beginning with 2713.
2. Equatorial coordinates (to ± 1'), for the equinox 1950.
3. Precession in right ascension and declination.
4. Galactic coordinates on the modern system.
5. Plate coordinates, in millimeters, measured from the southeast corner.
6. Bautz-Morgan class.
7. Abell class (regular or irregular).
8. Visual magnitudes, V_1, V_3, and V_{10}, of the first, third, and tenth galaxies.
9. Raw counts of galaxies within the Abell radius, corrected for the field.
10. Distance class.
11. Richness class.
12. Redshift, when available.

As in the northern catalogue, many clusters not meeting the criteria for the homogeneous sample will be included. Because of the far higher quality of the U.K. Schmidt survey plates, many more of such clusters are being catalogued than in the northern survey; most of these are clusters of estimated $z > 0.2$. Also listed in the southern catalogue but marked as not in the homogeneous sample, are clusters in the fields that overlap the northern survey but which were missed by Abell (1958). A few of these do meet the criteria for the sample (see below), but are nevertheless marked with an asterisk in order not to contaminate any statistical differences that may exist between the completeness levels of the northern and southern catalogues.

PROCEDURES FOR THE SOUTHERN CLUSTER SURVEY

Experience since the northern catalogue was completed has taught us to include some additional data, for example, the morphological classes of clusters and the estimated magnitudes of the first and third brightest galaxies, in addition to the tenth. There are, however, four significant differences between the procedures used in the northern and southern catalogues: 1) the color sensitivity of the plates used for the survey; 2) the use of both glass originals and film copies; 3) the use of visual rather than photored magnitudes; and 4) the manner of correction for field galaxies.

BLUE VERSUS RED PLATES

In the northern survey all clusters were identified and measured on the red (103a-E) plates of the Palomar Sky Survey. Because of their redshifts, even clusters of moderate distance ($z \sim 0.2$) are far more conspicuous on the red (E) than on the blue (O) emulsion. In the southern survey, however, all data are being taken from the blue (IIIa-J) Siding Spring plates rather than from the red plates taken with the ESO Schmidt in Chile. There are three reasons: 1) the U.K. blue survey is now nearly complete, and its use saves us an estimated two years; 2) the southern survey fields are only 5 degrees across, because the field of view of the ESO Schmidt can cover only that area with a modest overlap. The U.K. Schmidt, however, like the Palomar Schmidt, has a field of 6.6 degrees, which provides a very large overlap between adjacent fields, giving us a splendid opportunity to check thoroughly our completeness and consistency; and 3) despite the redshifts of remote clusters, the U.K. J plates reveal more faint clusters and with higher resolution than do the ESO red plates. The ESO exposures are not as full as would be desirable, we understand because of telescope flexures.

In fact, the U.K Schmidt J plates show more distant and fainter clusters, and with better resolution, than do the early 103a-E (red) plates of the Palomar survey, despite the color advantage of the latter. The U.K. Schmidt is an improved version of the Palomar instrument, with less flexure, it has automatic guiding (not available in the 1950s, when the Palomar survey was completed), and the J emulsion has substantially higher resolution than the 103a-E emulsion. Thus, although we are using blue rather than red plates, we are obtaining far better results than was possible with the earlier Palomar survey.

PLATES VERSUS FILM COPIES

All data for the northern cluster were taken from original glass plates of the Palomar Sky Survey. On the other hand, we have utilized both original U.K. Schmidt plates and film copies in our southern survey. The plates are either those taken for the Southern Survey that had been rejected for the purpose of reproduction, but for reasons that do not affect their value for our work, or plates centered on Southern Survey fields, taken under identical conditions of exposure and processing, but for non-survey purposes. Most of these plates are on file at the Royal Observatory Edinburgh, and all of those available (nearly half the total southern sky) have been searched for clusters (the overwhelming majority by HGC).

The survey of the remaining fields is being done on the film copies, mostly by GOA. Such would not have been possible with the Palomar Survey; the paper print copies were prepared to maintain the limiting magnitude of the originals, but at the expense of greatly increased contrast, making recognition of many distant clusters, clearly visible on the original plates, impossible on the prints. Even the glass negative copies of the Palomar Survey have contrasts too great to be optimum for a cluster survey.

On the other hand, the film copies of the U.K. Schmidt Survey plates are generally of excellent quality. Our experience has shown that for our purposes the copies are nearly indistinguishable in quality from the original plates. Fortunately, the large overlap between adjacent fields makes it possible for us to verify that no loss in completeness or precision results from use of the film copies.

MAGNITUDES

All our magnitude estimates are made with step scales consisting of galaxy images assembled from cut-outs of images on film copies of U.K. Schmidt photographs. The step-scale images are calibrated by comparing them with images of galaxies of known magnitude on the survey plates, by exactly the same procedure with which we compare the step-scale images with galaxies in our identified clusters. The magnitudes of galaxies serving as standards come from a variety of sources, including photoelectric measures by one of us (HGC) in South Africa. Because we record our step-scale values, we can refine our calibration at any time.

What needs to be explained is how we reconcile photored magnitudes in the northern survey and visual magnitudes in the south, estimated from blue plates. The brightest galaxies in most rich clusters are ellipticals and early spirals, which have B-V colors in the range 0.7 to 1.0 -- roughly all the same to the precision of our step-scale estimates. Consequently, we assume that visual magnitudes can be estimated on blue plates, and that we can give appropriate transformation equations to relate blue, visual, or red magnitudes. Fortunately, we can check our assumptions with the substantial overlap data in fields also included in the northern survey. We have taken care to account for the differences in the K-corrections in different magnitude systems in our calibration.

FIELD CORRECTIONS

When the northern survey was carried out there was no objective knowledge of the apparent luminosity function (distribution of galaxies with apparent magnitude). Consequently, corrections for the background and foreground field were made by counting galaxies both in the clusters and in several regions of the same area and in the same magnitude range in the surrounding field; the latter were averaged and subtracted from the former.

Because of superclustering, field counts so obtained are quite unreliable, and we suspect this to be a serious source of error in the homogeneity of the northern cluster catalogue. Now we can avoid the difficulty; there exist extensive counts of galaxies as a function of apparent magnitude. Those counts in the magnitude range relevant to our work are by Brown (1979) and by Rainey (1977), and they are in excellent agreement with each other (Abell 1978). Here we use Rainey's counts, which are complete to $V = 19.5$ in three widely separated directions in the sky. We simply subtract from our count of galaxies within the Abell radius and prescribed magnitude interval in a cluster the number of galaxies Rainey's observations indicate should be due to the average field.

CONSISTENCY OF NORTH AND SOUTH SURVEYS

There are 135 clusters in common with the northern survey and partially completed southern survey. For these, the right ascensions measured on the northern and southern survey plates, α_N and α_S, respectively, and the declinations δ_N and δ_S have the following mean differences:

$$<\alpha_N - \alpha_S> = 0\overset{m}{.}04; \quad \sigma = 0\overset{m}{.}16;$$
$$<\delta_N - \delta_S> = 0'.54; \quad \sigma = 1'.92.$$

The mean ratio of raw counts of galaxies in the southern to the northern clusters, after correction for the field, is:

$$<N\ [\leq(V_3 + 2)]/N\ [\leq(R_3 + 2)]> = 1.17; \quad \sigma = 0.49.$$

We expected a large difference in the counts of galaxies in clusters, not only because of the sensitivity of the counts to small errors in the step-scale magnitude estimates, but because of the effect of superclustering on the background correction mentioned above. We are gratified that the agreement between the richnesses of the clusters as observed in the north and south is as good as it is.

The mean magnitude difference between V_{10} in the south and R_{10} in the north is:

$$<V_{10} - R_{10}> = 0.63 \text{ mag}; \quad \sigma = 0.83.$$

The scatter is large, as expected for step-scale magnitude estimates. Nevertheless, the mean difference between the visual and red magnitudes is almost precisely the expected value for galaxies with B-V colors of 1.0. We believe that most of the large scatter is due to the estimates of magnitudes in the north survey. Twenty-five years more experience, far superior plate quality, better calibration, and a much higher altitude above the horizon lead us to expect the magnitude estimates in the south to be more reliable. (The zone of overlap between the two surveys is low in the Palomar sky, where atmospheric extinction and seeing are at their worst.)

Of the 135 clusters in common:
17 were classed richness "0" in both north and south;
19 were classed richness "0" in the north but not the south;
16 were classed richness "0" in the south but not the north.

The mean difference between the richness classes r(N) and r(S) in the north and south, respectively, is:

$$<r(N) - r(S)> = -0.19 \text{ class}; \quad \sigma = 0.87 \text{ class}.$$

Abell (1958) had originally intended the richness classes to be two standard deviations wide. He was evidently slightly too optimistic, at least for the southernmost clusters in the northern survey.

COMPLETENESS

Among the clusters in the overlap region, in the south nine were missed in the first screening (before consulting the Abell catalogue); only three would make the statistical sample. However, 66 clusters were picked up in the south that are richer than $N[\leq (V_3 + 2)] = 50$. Of these, 17 have $V_{10} \leq 18.5$ (the limit corresponds to $R_{10} = 17.9$ after correction for extinction, the red magnitude corresponding to $z = 0.2$).

Thus, the 135 clusters found in the north survey should have been 152, suggesting an incompleteness of 11% in the northern catalogue at declinations south of -18 degrees.

REFERENCES

Abell, G.O. 1958, Ap. J. Suppl., 3, 211.
Abell, G.O. 1978, The Large-Scale Structure of the Universe, ed. M. Longair and J. Einasto (Dordrecht: Reidel), p. 253.
Brown, G.S. 1979, A. J., 84, 1647.
Rainey, G.W. 1977, Ph.D. Thesis, University of California, Los Angeles.

DISCUSSION

I.E. Segal: Isn't it the case that your concept of cluster is inherently distance-dependent (and hence presumably redshift-dependent) That is, requirement associated with a fixed magnitude difference implies different physical diameters for clusters at different distances, indeed a rather systematic trend with distance for physical diameters, does it not?

Abell: The identification of clusters depends on their recognizability; we have no a priori information on redshift, magnitudes or distances. The data given for each cluster are operationally defined and are independent of cosmological assumptions. (Although the counting radius, defined in terms of redshift estimated from V_{10}, is a constant linear distance only for conventional cosmologies.) Of course, the interpretation of the catalogued data is up to the cosmologist who uses the catalogue.

Thompson: In the revised version of the northern cluster survey, will the Coma cluster be changed to richness class 3? Such a change is important because too many astronomers presume that Coma is an average rich cluster, while, in fact, it is uncommonly rich.

Abell: You are right. I am inclined, for consistency with original counts, to put this sort of correction under "comments."

CATALOGUE OF CLUSTERS THAT ARE MEMBERS OF SUPERCLUSTERS

M. Kalinkov, K. Stavrev, I. Kuneva
Department of Astronomy
Bulgarian Academy of Sciences, Sofia

An attempt is made to establish the membership of Abell clusters in superclusters of galaxies. The relation

$$\log z = -3.637 + (0.135 \pm 0.014)\, m_{10} = (0.179 \pm 0.030) \log P \qquad (1)$$

is used to calibrate the distances to the clusters of galaxies with two redshift estimates. One is m_{10}, the magnitude of the ten-ranked galaxy, and the other is the "mean population," P, defined by:

$$P = p/2\pi^2 \ , \qquad (2)$$

where p = 40, 65, 105 . . . galaxies for richness groups 0, 1, 2 . . . , and r is the apparent radius in degrees given by:

$$r = 0.0286\ (1 + z_1)^2/z_1 \ . \qquad (3)$$

The first iteration for redshift, z_1, is obtained from m_{10} alone:

$$z_1 = -4.568 + 0.216\ m_{10} \ .$$

The standard deviation for Eq. (1) is 0.105, the number of clusters with known velocities is 342 and the correlation coefficient between observed and fitted values is 0.921. With z_i from Eq. (1), we define Cartesian galactic coordinates $X_i = R_i h^{-1} \cos B_i \cos L_i$, $Y_i = R_i h^{-1} \cos B_i \sin L_i$, $Z_i = R_i h^{-1} \sin B_i$ for each Abell cluster, i = 1,...,2712, where R_i is the distance to the cluster (Mpc), and $H_o = 100$ h km s^{-1} Mpc^{-1}.

The mean density of clusters from Abell's sample for $B > 50°$ is $D_s = 5.5 \times 10^{-6}\ h^3\ Mpc^{-3}$ and for all Abell clusters is $D_a = 7.65 \times 10^{-6}\ h^3\ Mpc^{-3}$. These values have been adopted for the entire investigated space and a limiting density D_1 has been introduced.

Our procedure for supercluster searches consists of the following: Let us take an arbitrary cluster, C_1, from the catalogue, let C_2 be its nearest neighbor, and O_2 the mean point between the two clusters. Let the sphere defined by the radius $R_2 = O_2C_1 = O_2C_2$ have volume V_2; the mean density is D_2 is $2/V_2$. If $D_2 < D_1$, we conclude the

G. O. Abell and G. Chincarini (eds.), Early Evolution of the Universe and Its Present Structure, 185–186.

examination of cluster C_1 and pass to another cluster. If $D_2 > D_1$, we look for the third nearest nearest neighbor of O_2. Let it be Cluster C_3. We define O_3 as a barycenter of the three clusters, supposing equal masses. We determine V_3 with radius $R_3 = \max(O_3C_1, O_3C_2, O_3C_3)$ and the mean density is $D_3 = 3/V_3$. If $D_3 < D_1$, we should say that clusters C_1 and C_2 form a double cluster. If $D_3 > D_1$, we seek the fourth nearest neighbor of O_3, and so on, until we reach $D_{q+1} < D_\ell$. In that case, q clusters satisfy the inequality $D_q > D_\ell$. For a high value of D_ℓ, we arrive at the same configuration of q clusters, regardless of which cluster is used to start the execution of the described procedure. In this manner, we select triple, quadruple,...clusters.

Among all clusters in Abell's sample, 290 clusters have been identified as participating in multiple clusters ($q \geq 3$), satisfying the condition $D_\ell = 50\ D_s$. Some examples: i) A2600 is situated near the center of a supercluster, including A2608, 2606, 2605, 2579, 2580, 2599 and 2583. The diameter is 26 h^{-1} Mpc and the mean density is 160 D_s; ii) A2540 is near the center of a supercluster with A2531, 2550, 2547, 2521, 2542 and 2509. At a distance of 19 h^{-1} Mpc from the center of this supercluster is located A2579 from supercluster i); iii) A1183 is almost in the center of a supercluster with A1201, 1209, 1159 and 1157, with a diameter about 20 h^{-1} Mpc and $\bar{D} = 200\ D_s$; iv) A998 is at the center of a supercluster with A1005, 968, 1046 and 1006, with a diameter 16 h^{-1} Mpc and $\bar{D} = 400\ D_s$.

Setting $D_\ell = 10\ D_s$, the number of clusters and members of configurations of high multiplicity increase sharply. The largest configuration then includes 36 clusters and it is around A2550, enclosing i) and ii).

It is interesting to note that the nearest neighbor of some clusters is at a great distance. The mean distance between clusters for the density D_s is 31 h^{-1} Mpc. From the distribution function for the nearest neighbor, we have $F_1(70) = 0.9996$, which means that for all clusters in Abell's sample we would expect only one case with a distance greater than 70 h^{-1} Mpc. Our processing shows that spheres with radii greater than 70 h^{-1} Mpc can be drawn around 21 clusters from Abell's sample. Examples are A380, 565, 583, 732, 1451, 1577, 1780, and 2155.

Another result concerning all clusters from Abell's catalogue is that almost one-half of the clusters are contained in regions with $\bar{D} > 10\ D_a$.

CHARACTERISTIC SIZE OF SUPERCLUSTERS

M. Kalinkov
Department of Astronomy
Bulgarian Academy of Sciences, Sofia

A new processing of the surface distribution of clusters from the catalogues of Abell and Zwicky has been carried out for an area of 2000 square degrees around the NGP. Four statistical tests (M. Kalinkov, [illegible]. Kuneva, 1980, C.R. Acad. Bulg. Sci., 33, 1305) are applied to search apparent characteristic sizes. Assuming a mean distance R = 440 h^{-1} Mpc (H_0 = 100 h km s^{-1} Mpc^{-1}) to Abell clusters, a characteristic size Q = 52 Mpc for superclusters has been obtained. For Distance Group 5 (R = 400 Mpc) and for Distrance Group 6 (R = 530 Mpc), we have Q = 48 Mpc and Q = 40 Mpc, respectively. For Zwicky D clusters (R = 390 Mpc, assuming a uniform distribution in the volume from 300 to 450 Mpc, Q = 55 Mpc and for ED clusters R ≈ 650 Mpc), Q = 46 Mpc. Therefore, the mean characteristic size of superclusters is 50 h^{-1} Mpc. There is an exception, however, for VD clusters (R = 540 Mpc) for which Q ≈ 100 h^{-1} Mpc.

G. O. Abell and G. Chincarini (eds.), Early Evolution of the Universe and Its Present Structure, 187.

QUASARS AND SUPERCLUSTERS

Patrick S. Osmer
Cerro Tololo Inter-American Observatory
Casilla 603
La Serena, Chile.

The topic of quasars and superclusters is only a few years old. Although the first pairs of quasars with small angular separations on the sky were found ten years ago (Stockton 1972, Wampler et al. 1973), the pair members had very different redshifts. The surface density of samples with available redshifts at that time was far too low for cases of quasars with both small angular separations and small redshift differences to turn up. Setti and Woltjer (1977) pointed out that if quasars occur in the nuclei of giant elliptical galaxies, then clustering should be apparent at 20th magnitude and fainter. In 1979 Walsh, Carswell and Weymann found a very close pair with identical redshifts; that, of course, was the first discovery of a gravitational lens. Also in 1979 Arp, Sulentic, and di Tullio showed that some of the quasars near NGC 3389 had similar redshifts, although at that time they did not discuss the hypothesis of the quasars being associated with superclusters. Subsequently Burbidge et al. (1980) confirmed that a compact (5 minutes of arc) group of 3 quasars found by Hoag on a 4m grating prism plate of the M82 field had very similar redshifts. Indeed, the group had the dimensions of a galaxy cluster, not a supercluster. Oort, Arp, and de Ruiter (1981) then specifically called attention to the fact that enough pairs of quasars with similar redshifts were known to suggest that quasar associations on the scale of galaxy superclusters do exist.

However, independent analyses of the Cerro Tololo surveys for large redshift quasars (Osmer and Smith 1980, Hoag and Smith 1977, Osmer 1980) did not find evidence for clustering of the quasar population as a whole. Osmer (1981) noted that many groups and pairs with similar redshifts could be found that had separations of supercluster size but demonstrated that the eye was a poor judge of their statistical significance. After he allowed for the various selection effects in the catalogs and applied binning analysis, the nearest neighbor test, and the correlation function method to the data, he concluded that there was no evidence for a departure from randomness in the quasar distribution. Webster (1982), in a separate study of the Curtis Schmidt quasars, came to the same conclusion by using power spectrum analysis, with the exception that he showed that one low redshift ($z \sim 0.37$) group is a significant enhancement.

G. O. Abell and G. Chincarini (eds.), Early Evolution of the Universe and Its Present Structure, 189–194.

The present state of the topic is that there is agreement that quasars with separations comparable to or less than the size of superclusters are known, but there is no agreement that quasars in general tend to cluster. It is my opinion that the available data are not adequate to demonstrate the latter point, although surveys currently under way are likely to provide much improved limits on the clustering of quasars.

The potential importance of any connection between quasars and superclusters is clear. At large redshifts ($z \gtrsim 1$), quasars are still the only available indicators of the large scale structure of the univer[se]. Now that large samples with high surface density are becoming available, it is possibly to study the quasar distribution on much smaller scales than could be done previously. For example, in the Hoag-Smith (1977) 4m survey, the average nearest neighbor distance at $z \sim 2$ is 64 h^{-1} ($h = H_o/100$) Mpc (Osmer 1981) in present epoch coordinates or 21 h^{-1} Mpc at a time corresponding to $z = 2$. The expectations of finding significant structures on such scales has increased recently as a result of work on a region of low galaxy density (Kirshner et al. 1981) which appears to be surrounded by a region of galaxy overdensity (Bahcall and Soneira 1982a). Therefore the strength of galaxy clustering on supercluster scales of 50 h^{-1} Mpc appears to be much higher than previously thought.

The theoretical aspects of clustering and their relation to cosmology are an extensive field in themselves, which is amply discussed elsewhere in this symposium. Here I believe it is important to concentrate on the observational side of the problem. A sound understanding of the data base and its limitations must be attained before theoretical conclusions about the observations can be drawn. The recent literature already contains several articles having entirely opposite conclusions about the significance of clustering just in the Cerro Tololo samples (cf. Arp 1980, Sulentic 1981 in addition to the references mentioned already).

In my opinion the most dangerous pitfall in analyzing data for clustering is to assume uniformity in the selection process. While the assumption is a natural one to make and allows simple analytic estimates of expected probabilities, it is not sufficiently appreciated that <u>any deviation from uniformity in the selection process will produce apparent clustering</u> in analyses that assume uniformity. A discussion of this problem is given by Osmer (1981). For example, large scale trends in the data can produce apparent clustering. Such a trend in right ascension in the Curtis Schmidt survey has much to do with the effects noted by Arp (1980) and Sulentic (1981).

The steepness of the apparent quasar luminosity function is another difficulty in achieving adequate uniformity in surveys. If the number of quasars increases by a factor of 8 per magnitude, then a 0.2 mag variation in limiting sensitivity in a survey will lead to a 50% difference in the expected number of objects. Such a variation could occur between different observations or be caused in part by variable interstellar absorption. Clearly such effects must be taken into account in any analysis of the data.

A related problem is the uneven sampling in redshift that occurs in any magnitude-limited survey. At larger redshifts only the most luminous objects are detected, and their space density is lower than that of less luminous objects. Of course, any variation in the overall space density with redshift must be considered as well. Finally, all observational techniques for finding quasars have redshift biases of their own that have to be considered.

Specifically, the main difference between the work of Osmer (1981) and Webster (1982), who concluded that quasars on the whole do not show evidence for clustering, and that of Oort et al. (1981) as well as Sulentic (1981), who claimed evidence for associations of quasars with each other or with bright galaxies, is that the latter papers based their expected values on the assumption of uniform sampling while the former two made allowance for the non-uniformities in the data. Again, there is general agreement that some close pairs and groups of quasars with similar redshifts exist and that they occur on the scales of superclusters or smaller, but there is definitely not agreement that quasars in general show evidence for clustering.

It is clearly important to continue work on this purely observational side of the problem. New surveys in previously unstudied areas of the sky are needed for at least two reasons. First, many of the surveys done to date were not set up to study clustering, and often were centered on already known quasars or other previously studied objects. Thus the results are not statistically independent from previous ones. Second, Wills' (1978) precept continues to be very timely -- the best check of an unusual result or configuration found after the fact in a given survey is to look again for the <u>same thing</u> in another field. Finding something else from what was sought in the new survey does not answer the question. For example, Arp has first claimed that quasars are associated with bright galaxies and subsequently found associations of quasars with companions to bright galaxies.

It should be possible to improve significantly the sensitivity of new surveys to the presence of weak clustering with respect to the previous ones, which in general were not designed to study clustering. For example, numerical simulations (Osmer 1981) show that the arrangement of 4m grism fields into a long thin rectangular strip will give much improved information on the quasar distribution on the scales of superclusters. In addition, by overlapping adjacent fields it will be possible to establish limiting magnitude differences independently of the quasar results themselves, which will also increase the detectability of weak clustering. It is important to note that large surface densities are required to investigate the distribution on supercluster scales. Consequently the 4m grism and UK Schmidt objective prism data are the best suited for work at high redshift, as they have high surface densities and favor a limited redshift range, which produces a high <u>space</u> density. The UK Schmidt does not yield as high a surface density, as the 4m but the large surface area covered on a single plate makes it very attractive, as there is no need to tie together several plates. The ultraviolet excess technique should not be overlooked, although to use it for the three dimensional problem will require follow up spectroscopy of a larger number of faint quasar candidates than may be required with the objective-prism approach.

At CTIO I have obtained good 4m grism plates of 21 fields near 12^h, $-11°$ to investigate just this problem. The region of the sky is previously unstudied for optically selected quasars, in contrast to much previous 4m work that has been done on regions near known quasars. An overall improvement of at least a factor of two in the detection limit for clustering may be expected for scales of 15-50 h^{-1} Mpc (present epoch). Such a limit would allow correlations of the type mentioned by Bahcall and Soneira (1982b) for superclusters to be detected if they exist in quasars at $z \sim 2$. Bahcall and Soneira estimate that the spatial correlation function for clusters of galaxies drops to unit value at 25 h^{-1} Mpc, while the new survey should be able to detect a value of 0.6 at 25 h^{-1} Mpc at the 2σ level of confidence.

As the results of this survey and those being carried out by other workers become available in the next few years, we can expect a considerable advance in our knowledge of the space distribution of large redshift quasars.

REFERENCES

Arp, H., Sulentic, J.W., and di Tullio, G 1979, Ap. J. 229, 489.
Arp, H. 1980, Ap. J. 239, 463.
Bahcall, N.A., and Soneira, R.M. 1982a, Ap. J. (Letters) 258, L17.
Bahcall, N.A., and Soneira, R.M. 1982b, preprint.
Burbidge, E.M., Junkkarinen, V.T., Koski, A.T., Smith, H.E. and Hoag, A. A. 1980, Astrophys. J. (Letters) 242, L55.
Hoag, A.A., and Smith, M.G. 1977, Ap. J. 217, 362.
Kirshner, R.P., Oemler, A., Jr., Schechter, P.L., and Shectman, S.A. 1981, Ap. J. (Letters) 248, L57.
Oort, J.H., Arp, H., and de Ruiter, H. 1981, Astron. Astrophys. 95, 7.
Osmer, P.S., and Smith, M.G. 1980, Ap. J. Suppl. 42, 333.
Osmer, P.S. 1980, Ap. J. Suppl. 42, 523.
_____. 1981, Ap. J. 247, 762.
Setti, G., and Woltjer, L. 1977, Ap. J. (Letters) 218, L33.
Stockton, A.N. 1972, Nature Phys. Sci. 238, 37.
Sulentic, J.W. 1981, Ap. J. (Letters) 244, L53.
Walsh, D., Carswell, R.F., and Weymann, R.J. 1979, Nature 279, 381.
Wampler, E.J., Baldwin, J.A., Burke, W.L., Robinson, L.B., and Hazard, C. 1973, Nature 246, 203.
Webster, A. 1982, M.N.R.A.S. 199, 683.
Wills, D. 1978, Physica Scripta 17, 333.

Discussion

Wampler: 1) Did you include the redshift information in your Monte Carlo surveys, or did you check for clustering on the surface distribution without regard for redshift?

2) Doesn't your explanation for the observed increase in surface density of quasars found on 4-meter plates as compared to Curtis Schmidt plates require surface clustering?

Osmer: 1) Most of the work made use of the redshifts to examine the three-dimensional distribution. Although the question of surface clustering has not been thoroughly investigated, there is no obvious evidence for it.

2) You raise a very good question. The approaches I did use cannot answer it very well. Rather, an independent estimate of the surface density to be expected for the 4-m survey is needed. But I fully agree that if the surface density turns out to be high for the 4-m fields, it will be very strong evidence for surface clustering.

Inagaki: You showed the diagram of the distances versus the number of pairs. What is the pair correlation function [$\xi(r)$] of QSOs? I think that it is possible to convert the diagram to $\xi(r)$. Are there enough data to calculate the covariance function from the observed distribution of QSOs?

Osmer: Yes, I have used the correlation function to investigate if quasars cluster. Since the function gave no indication of clustering, there is as yet no information on what the covariance function actually is for quasars.

Oort: I want to stress the utmost importance of homogeneity in a survey aimed at discovering density fluctuations at large redshift (which is a most important undertaking). It should also be stressed that, in view of the huge dispersion in absolute magnitude of quasars, it is essential to obtain redshifts.

Osmer: I couldn't agree more.

Miller: The Einstein serendipitous sources already give evidence for QSO correlation on the sky. Additional X-ray sources in the field of X-ray QSO's are often QSO's (usually intrinsically faint), where such sources seldom show up in control fields. Admittedly, this is not a nice survey with control on magnitude, redshift, and so on, but it already says there are pretty strong QSO correlations on the sky.

Osmer: You raise a good point, although as you say, it is important to determine just what is expected from the control fields. It is also worth noting that the X-ray-selected quasars have lower redshifts than the ones in the optical samples I discussed.

M. Burbidge: Whereas redshifts determined from grism and objective prism plates are usually approximately correct, there is a certain proportion in which the two lines detected are not the identifications first assumed. Will you follow up any groups of QSOs found by the survey with slit spectrograms for redshift verification?

Osmer: Although I hope some progress can be made from the grism redshifts themselves, you are quite correct that followup spectroscopy will be essential.

Tyson: With an average nearest neighbor separation of 64 h^{-1} Mpc, your present survey probably would not detect the presence of clustering on scales 130 h^{-1} Mpc or less, due to undersampling.

Osmer: I do not agree that clustering could not be detected on scales smaller than 130 h^{-1} Mpc. After all, if the mean nearest neighbor distance for a sample turned out to be significantly smaller than expected, then we would conclude that clustering was present. However, it may be that little information on the nature of the clustering could be derived.

URTHER INVESTIGATIONS ON POSSIBLE CORRELATIONS BETWEEN QSOs AND THE ICK CATALOGUE OF GALAXIES.

J.-L. Nieto[1] and M. Seldner[2]
[1]Observatoires du Pic-du-Midi et de Toulouse, Laboratoire Associé CNRS 285, F-65200 Bagnères-de-Bigorre, France
[2]Princeton University, Department of Physics, Princeton, NJ 08544, USA

e analyze the different contributing factors in a previous study by eldner and Peebles (1979) who found statistically significant evidence or a correlation between a list of 382 QSOs at $|b| > 40°$, $\delta > -23°$ and he Lick counts of galaxies ($m < 19.0$), namely, that there are on the verage 1.45 ± 0.39 more galaxies within 15' of a QSO than expected if SOs were placed at random across the sky. Taking into account these ifferent factors and using a larger sample of QSOs whose detection is ot expected to bias our statistical results, we conclude that: i) There s no longer statistically significant evidence for an excess of Lick alaxies close to QSOs from our list considered as a whole. After removng the different possible biases, if there was any excess left in that revious study, it was rather due to radio QSOs and not to optical QSOs. his effect is marginal if we adopt a mean density of galaxies in a ring etween 2° and 5°, and null if the mean density is taken between 1° and ° from QSOs; ii) The whole correlation function $w(\theta)$ shows in general hump between 1°5 and 2°. This is either an off-set value or a maximum alue preceding a slow decrease in $w(\theta)$ between 1°5 - 2° and 5°. This eature seems to pertain more to high redshift QSOs ($z > 0.4$) and to adio QSOs than to optical QSOs. It seems slightly more pronounced at he polar galactic caps and at very low galactic latitudes; iii) Numerial simulations seem to indicate that the error bars determined by ssuming that cell numbers are independent are slightly underestimated. hey reproduce fairly well some features within 1° from the QSOs and uggest that the hump at 1°5 - 2° would not be as unlikely as indicated y the error bars.

Reference

Seldner, M., and Peebles, P.J.E. 1979, Ap. J., 227, 30.

G. O. Abell and G. Chincarini (eds.), Early Evolution of the Universe and Its Present Structure, 195.

URVEY OF THE BOOTES VOID

R. P. KIRSHNER
University of Michigan

A. OEMLER
Yale University

P. L. SCHECHTER
Kitt Peak National Observatory

S. A. SHECTMAN
Mount Wilson and Las Campanas Observatories of the
Carnegie Institution of Washington

ABSTRACT. Radial velocities have been measured for 231 galaxies chosen by apparent magnitude from 282 small fields spanning the area on the sky thought to contain the Bootes void. The galaxy distribution exhibits a spherical volume 6000 km/s in diameter in which no objects are found. The rms velocity difference of close pairs in our sample is less than 180 km/s.

INTRODUCTION

Based on an apparent-magnitude limited radial velocity sample of galaxies in just three fields, separated from each other by typically 30° on the sky, we supposed that a large volume of space in the direction $(\alpha, \delta) = (14h40m, +50°)$ might be empty of galaxies in the velocity range from 12,000 to 18,000 km/s (Kirshner, Oemler, Schechter and Shectman, 1981). In this case, much of the area of the sky between our three fields should exhibit the same gap in the distribution of galaxy radial velocities.

OBSERVATIONS

In order to effectively sample the galaxy distribution in such a large volume, we have conducted a new survey in 282 small fields distributed on the sky between the original three. Each new field has an area of .0625 sq. deg.

G. O. Abell and G. Chincarini (eds.), Early Evolution of the Universe and Its Present Structure, 197–201.

Photographic enlargements of the 282 fields, made at Kitt Peak from glass copies of the red Palomar Sky Survey plates, were distributed to the 4 collaborators. The fields were chosen in two batches, 122 for the 1981 observing season and 160 for 1982. For each batch, each of us compiled a list of galaxies in order of estimated apparent magnitude. An averaging procedure was used to produce a master list of galaxies ranked in order of brightness. Because of the increased contrast of the photographic enlargements, there is a tendency to overestimate the luminosity of galaxies of lower surface-brightness. Variations in photographic sensitivity, both on the original plates and in the copying process, also contribute scatter to the final rankings.

Radial velocities for galaxies on the lists were measured from optical spectra obtained with the cassegrain spectrographs at the Palomar 5-meter telescope, with the Reticon photon-counting detector, and at the Kitt Peak 2.1-meter telescope, with the IIDS. The velocity sample is roughly complete above the limiting rank. The overall distribution of velocities is not very different from more rigorously chosen photometric samples, with a moderate enhancement of low velocity, low surface-brightness galaxies.

Figure 1a shows the positions on the sky of the 282 fields. The 1981 fields were evenly distributed on a grid in the sampled area. Based on the results from the first 102 velocities, the 1982 sample was chosen to double the field density of the original grid and quadruple the density of fields in the central region where the hole in the galaxy distribution appeared to be most conspicuous.

RESULTS

The distribution on the sky of galaxies with measured velocities is shown in Figure 1b, c, and d for three ranges in radial velocity which contain in all 156 galaxies. The remaining 32 percent of the sample lies closer than 6,000 km/s or more distant than 24,000 km/s. The absence of galaxies from the central part of Figure 1c, between 12,000 and 18,000 km/s, is most conspicuous just where the density of survey fields is highest.

Note that two galaxies in Figure 1c at (14h10m, +48°) and (14h10m, +42°) have velocities of 17,910 and 12,060 km/s respectively. A sphere 6,180 km/s in diameter, centered at 15,000 km/s and (14h48m, +47°), contains no measured velocities.

The empty sphere intersects the line of sight from 164 of the 282 survey fields. Judging from the smoothed velocity distribution of the sample, the expected number of galaxies in the spherical volume is 23. This estimate is conservative in the sense that we have not yet tried to delineate a larger but irregularly shaped empty volume, or one of reduced, but not zero, density. Such a larger volume is indicated by the lines of sight to our original three fields, which lie entirely

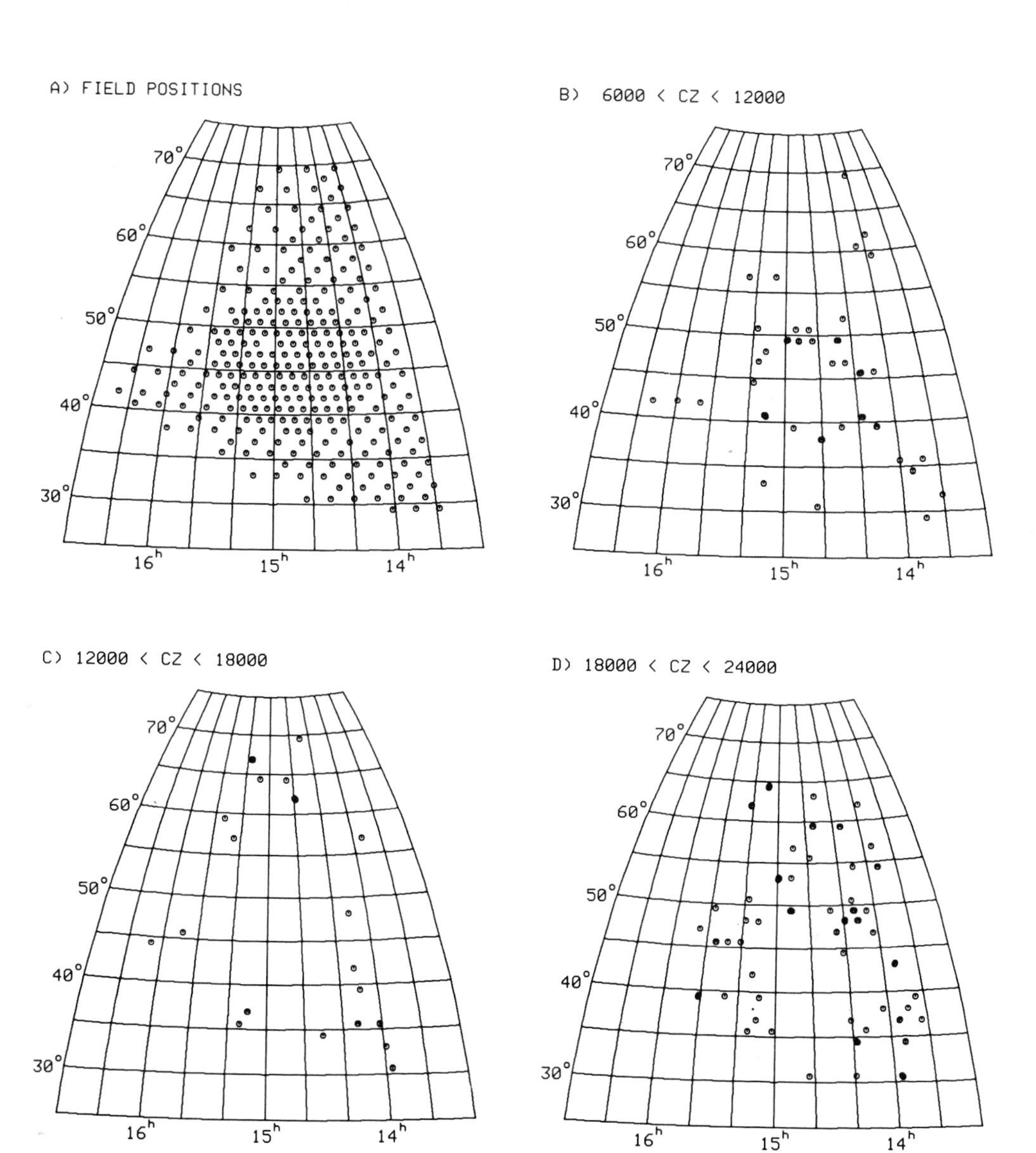

Figure 1-(a) Positions on the sky of 282 survey fields; (b), (c), (d) Positions of galaxies with measured velocities in three ranges from 6,000 to 24,000 km/s.

outside the boundary of the present empty sphere, yet exhibit a strong galaxy deficit in the same velocity interval.

One advantage of a survey conducted in fields of such small size is that the correlation of galaxy velocities is much lower. In our previous survey, velocities occur in groups of typically three or four. In the present survey, groups or small clusters tend to appear in single fields. A list of all groups of velocities which occur in single fields where the redshift difference is less than 1000 km/s, contains 144 singl objects, 29 doubles, 7 triples, and 2 quadruples. The average number of velocities in each group is only 1.3. The 23 velocities missing from the empty sphere would be expected to occur in 18 such groups.

The probability of observing no events when 18 are expected is only 1.5×10^{-8}. Some allowance must be made for additional correlation of velocities between neighboring fields, and also for choosing the empty sphere after the fact. Since the entire sample contains 231 galaxies, there are only of order 10 independent ways to find an empty volume of this size. Thus, the void is very unlikely to be a statistical artifact There is a probability of at least a few percent, however, that the density of galaxies in the empty sphere is as high as 0.25 times the average density.

The distribution of velocity differences within single fields is itself remarkable. If the galaxies in our velocity sample were distributed at random, 15 such pairs would be expected with velocity differences less than 1000 km/s. In fact, there are 62 such pairs in our sample (1 in each double, 3 in each triple, 6 in each quadruple). However, only 9 of the pairs have velocity differences between 400 and 1000 km/s--just the number expected by chance. The rms velocity difference of the remaining pairs, correcting for the 6 expected at random, is only 180 km/s. This value can be taken as an upper limit since there is no reason to believe that the measurement error is much lower.

REFERENCE

Kirshner, R. P., Oemler, A., Schechter, P. L., and Shectman, S. A.: 1981, Ap. J. (Letters), 248, L57.

Discussion

Inagaki: How does the size of the void depend on the luminosity of galaxies?

Shectman: A population of fainter galaxies could exist in the empty volume. Such galaxies would not show up in our apparent magnitude limited sample.

luchra: Are any of the galaxies in the sample of Weedman and Balzano (Markarian galaxies) or Sanduleak and Pesch in your empty sphere?

Shectman: No, Balzano and Weedman, in particular, consider an even larger area on the sky than that occupied by the empty sphere.

ON THE ORIGIN OF THE VOIDS

Yehuda Hoffman
Dept. of Physics & Astronomy, Tel-Aviv University, Israel
Jacob Shaham
Racah Institute of Physics, The Hebrew University,
Jerusalem, Israel

In the past few years the large scale structure and dynamics of the galaxy distributions have been studied intensively, mainly by means of red-shift surveys. From these surveys it seems that the distribution of galaxies is characterised by large, empty voids of (20 - 40) h_{100}^{-1} Mpc in diameter with a typical density contrast of $\delta \sim -(0.7 - 0.8)$.

The mere existence of the voids poses a severe challenge to all theories of the formation of galactic systems. In particular, it seems to be difficult to reconcile the observed structure with the gravitational instability theory in a large Ω_o(i.e. $\Omega_o \lesssim 1.0$) Universe. The point is that if a void of radius d_o is to be formed via random fluctuations, galaxies should have large peculiar velocities, $v_P \sim d_o H_o^{-1} = v_H(d_o)$ for $\Omega_o = 1$. (For $\Omega_o < 1$ linear theory predicts $v_P \sim d_o H_o^{-1} \Omega_o^{0.6}$.) These large peculiar velocities are not observed. Now, suppose that the progenitors of the voids are negative density perturbations in which δ is smoothly varying, such that the gravitational force field acts, mainly, radially outwards from some center. Under these conditions the average distance travelled is only $d_o/4$, about one third of the random motion value, 7/8 d_o. In this case the peculiar velocities are expected to be much less than in the random case.

Our basic claim is that such a structure is a natural consequence of a combined spectrum of adiabatic and isothermal perturbations. The adiabatic component is truncated on scales less than $M_D \sim (10^{13} - 10^{15}) M_o$ where M_D is the damping mass, while the isothermal perturbations may be the "seeds" from which structure on scales less than M_D has evolved. In the following discussion we concentrate only on the adiabatic component. A truncated density perturbation power spectrum implies a wiggly covariance function, $\xi(r)$. As the density excess $\delta(r)$, around sufficiently high density maxima (minima) is proportional to $\xi(r)$, out to some coherence radius, it should be nearly spherically symmetric and manifests a wiggly structure.

G. O. Abell and G. Chincarini (eds.), Early Evolution of the Universe and Its Present Structure, 203–206.

The main point is that while high positive density peaks are the progenitors of rich clusters, negative amplitude density troughs are the progenitors of the voids. These high negative density peaks are surrounded by shells of positive δ. As the density contrast within the voids decreases the density excess of the shell increases. There is a coherent flow of galaxies, formed within the void, to its boundaries, i.e. to the shells.

The dynamical evolution of the voids has been investigated by means of a spherical shells model. The power spectrum of the primordial adiabatic component is assumed to be some power law truncated at k_D. At the initial epoch the spherical shells system is assumed to have a density excess profile given by: $\delta(r) = \delta_o\ \xi(r)$, ($\xi(r)$ is normalized so as to make $\xi(0) = 1$) and an initial Hubble-like expansion velocity field. Thus, specifying Ω_o, H_o, the initial epoch, δ_o and the perturbations power spectrum index, n, the perculiar velocity and density field at the present epoch is calculated.

We note, that we are interested in such negative (density peak) perturbations that, were they positive, would have formed at present superclusters and rich clusters whose central density contrast is $\delta_{RC} \sim (10^3 - 10^4)$. The initial epoch was taken to be $Z_i = 10^3$, and the present density profiles and peculiar velocity fields were calculated for $0.1 < \Omega_o < 1.0$ and $-1.5 < n < 1.5$ as a function of r (in units of $(1+Z_i)k_D^{-1}$). A void ($\delta < 0$) is formed surrounded by a shell of positive δ only for $n > -1$. For positive (density peak) perturbations that reach $\delta > \delta_{RC}$ at the present epoch, the negative counterparts reach $\delta < -0.85$ ($\Omega_o = 1.0$), $\delta < -0.75$ ($\Omega_o = 0.45$) and $\delta < -0.65$($\Omega_o = 0.1$) in their centers. A typical void has a central density contrast of order $\delta \sim -(0.8 - 0.9)$ and is surrounded by a shell with a positive density excess of $\delta \sim (1.0 - 10)$. The thickness of the shell is $\sim (10 - 20)$% of the radius of the void. Negative (density peak) perturbations form voids with a radius 10 times larger than that of clusters that would have formed from their positive counterpart. The density field is, relatively insensitive to Ω_o, compared to the velocity field which is a rather sensitive function of Ω_o. In a void of $\delta \sim -(0.8 - 0.9)$ the maximal relative peculiar velocities galaxies would have is of the order of $(v_p/v_H) \sim (0.4 - 0.5)$ for $\Omega_o = 1$, $\sim (0.2 - 0.25)$ for $\Omega_o = 0.45$ and $\lesssim 0.09$ for $\Omega_o = 0.1$. These values should be compared with the values of 1, 0.62 and 0.25, respectively, for the random motion scenario.

Ω_o can be inferred from studying the dynamics of galaxies on scales larger than $\sim 10\ h^{-1}$ Mpc in two independent ways. One way is from the velocity field of galaxies in superclusters e.g. the Virgo supercluster, the other is from the dynamics of the voids in the galaxy distribution as described above. If these two independent estimates of Ω_o are found to be consistent it will give a major support to the support to the dissipationless gravitational instability theory.

Discussion

Bonometto: I) Is the value of n you quote the one after recombination?
II) Do you take into account the spike around the Jeans mass (prior to recombination) and the change of slope above it?

Hoffman: I) Yes.
II) No, as the scales considered here are smaller than the Jeans scale prior to the recombination epoch.

Inagaki: I) In your calculation, how does the void evolve? Does it evolve self-similarly? Recently Bill Saslaw and his research student developed a theory concerning voids based on equilibrium statistical mechanics. Their theory agrees excellently with N-body simulations. What do you think about it?
II) In a spherical model the gravitational force is always towards the center. There is no mutual attraction between shells. I am afraid that the result is quite different if you remove the assumption of spherical symmetry. What do you think about it? Have you compared your results with those of N-body simulations?

Hoffman: I) The evolution is self-similar only until $\delta\rho/\rho$ becomes appreciable. After that it is strongly nonlinear, as the formation of the ridges shows.
II) Yes, N-body calculations can show the effect of small-scale fluctuations on the development of the mean large-scale structure. In particular, the ridges might thicken in some places and not in others; this possibility is omitted in spherical calculations.

Thompson: I would like to point out that your assumption that the M/L ratio increases with increasing scale is an uncertain one. It is, for the most part, only an assumption, since dynamical data on large scales (2 to 10 Mpc) is very sparse. In a paper to be presented later in this session, I will show evidence for a low M/L ratio over large scales.

Hoffman: The basic prediction of the M/L model (Hoffman et al., 1982, Ap. J., 262, 413) is that M/L is a nondecreasing function of scale. This seems to be confirmed by most of the observational data (see also Dr. Harms' paper in these Proceedings). The scale on which M/L reaches its asymptotic value depends on Ω_0.

Huchra: I'd like to add to Laird Thompson's comment. In the groups and clusters we've analyzed in the CfA survey, there is no evidence for M/L increasing on the large scale -- binaries, groups and clusters all give M/L $\sim$ 100 to 200.

Hoffman: The main feature of the Hoffman et al. (1982) model is that M/L is a nondecreasing function of the linear scale. The results quoted by Dr. Huchra are consistent with the model for a $\Omega_0 \sim 0.1$ universe. For any reasonable values of Ω_0, M/L is constant over scales larger than a few Mpc and hence the voids are devoid of matter. The v_P/v_H field depends strongly on the limiting value of M/L, i.e., Ω_o.

PROPERTIES OF GALAXIES IN LOW DENSITY REGIONS

N. Brosch
Laboratory Astrophysics, Huygens Laboratorium, Postbus 9504
2300 RA Leiden, The Netherlands

Isolated galaxies seem to have more nuclear activity and less disk activity than nonisolated objects. This is interpreted as an expression of the mass-profile in their halos.

The discovery of voids in the three-dimensional distribution of galaxies has rapidly been followed by claims that in at least one such void some objects, like Markarian and emission-line galaxies, are present. Why are only active galaxies found in these voids? Should we also expect the opposite to be true, namely that active galaxies ought to be found more often in voids, or in regions of the Universe with lower-than-average density? There are indications that this seems to be the case, as Gisler (1978) found that emission-line galaxies, Markarians and possibly Seyferts avoid environments of dense clusters. Differences between "field" and "cluster" populations are probably due more to the enhanced local density of galaxies than to the nature of the overall environment.

We have studied a sample of objects identified by Huchra and Thuan (1977, hereafter HT) as isolated, to check how different are the observed properties of galaxies at the low end of the neighborhood-density distribution. We (Brosch and Shaviv, 1982) found that the 12 HT galaxies show nuclear ultraviolet excesses when compared to the average properties of objects from de Vaucouleurs (1961). These excesses could not be modeled by reasonable mixtures of stellar populations. Brosch and Isaacman (1982) found that at least for half the HT sample, the spectral distribution of the nuclei between 3500 Å and 3.5 μm resembles that of late K stars. Brosch and Krumm (1982) detected compact nuclear 5 GHz sources in two HT objects, but their control sample (referred to as BK) showed no radio sources among nuclei of 17 galaxy members of sparse groups, "open clusters" in the nomenclature of Gisler (1978).

Balkowski and Chamaraux (1981) found that HT galaxies have more neutral hydrogen than group galaxies. The full-synthesis maps produced at Westerbork (Krumm and Shane 1982) confirm their values for the total hydrogen masses. The data available in the literature permit comparisons to be made between the HT and BK samples. We find that the BK objects show a tight correlation (r = 0.94) between the logarithm of the total hydrogen mass, M(H), and the logarithm of the optical physical diameter,

G. O. Abell and G. Chincarini (eds.), Early Evolution of the Universe and Its Present Structure, 207–209.

D. A similar correlation has been reported by Giovanelli (1981) from the Arecibo survey of "isolated" galaxies drawn from the list of Karachentseva (1973, hereafter KI). Giovanelli finds $M(H) \propto D^{1.85}$, while we find $M(H) \propto D^{2.6}$. The HT objects show consistently higher values of M(H) for all values of D. The excess of hydrogen is more pronounced at the small-diameter end of the regression.

The HT galaxies have lower average blue surface brightness than the BK sample (90% confidence level). This is consistent with what Arakelyan and Magtesyan (1981) found from a comparison between galaxies in pairs and KI objects. The percentage of reported line emission in the optical spectra is also different; only 6/12 HT galaxies show emission, while the fraction among BK objects is 11/17. Adopting Gisler (1978) statistics, we should have observed 9/12 HT and 10/17 BK galaxies with emission in their spectra.

Finally, the HT galaxies show some properties characteristic of later morphological types than those assigned from their optical images. There are the colors U-B and B-V as measured through the larges aperture by Brosch and Shaviv (1982), the distribution of neutral hydrogen in one object and the rotation curve derived from HI in the second galaxy studied by Krumm and Shane (1982). It appears therefore that there are several real differences between isolated and nonisolated, but noncluster, galaxies. Most of these differences indicate a trend toward less disk activity in HT objects, while the nuclear UV excesses and the discovery of compact nuclear radio sources (despite the small statistica sample) imply enhanced central activity relative to nonisolated objects.

This can be understood in the light of results from N-body simulations of slow tidal encounters between galaxies (Dekel et al. 1980 where it was shown that interactions modify the inner regions of galaxie by flattening their mass-profiles relative to those of unperturbed galaxies. Thus, HT objects probably have nearly unperturbed halos, whil BK and other nonisolated galaxies have had their halos partially strippe and their inner regions "puffed up." A sharper mass-profile eases inflo of matter to the central regions and also delays star formation in the infalling gas. This implies that any activity would preferentially be confined to the nuclei. Similar arguments might apply to galaxies in voids; this would explain their being active as well as predict their being gas-rich.

References

Arakelyan, M.A., and Magtesyan, A.P. 1981, Astrophysics, 17, 28.
Balkowski, Ch., and Chamaraux, P. 1981, Astron. Astrophys., 97, 223.
Brosch, N., and Isaacman, R. 1982, Astron. Astrophys., in press.
Brosch, N., and Krumm, N. 1982, in preparation.
Brosch, N., and Shaviv, G. 1982, Ap. J., 253, 526.
Dekel, A., et al. 1980, Ap. J., 241, 946.
de Vaucouleurs, G. 1961, Ap. J. Suppl., 5, 233.
Giovanelli, R. 1981, in "Extragalactic Molecules," NRAO, 141.
Gisler, G.R. 1978, M.N.R.A.S., 183, 633.
Huchra, J., and Thuan, T.X. 1977, Ap. J., 216, 694.
Karachentseva, V.E. 1972, Comm. Spec. Astr. Obs. USSR, 8, 3.
Krumm, N., and Shane, W.W. 1982, preprint.

Discussion

Giovanelli: One comment and one question: The comment: I would advise caution in the inference of integral properties from such a small sample. Intrinsic scatter is very large. The question: Did I read correctly in one of your graphs that the difference in the hydrogen content between isolated galaxies and those in groups is one order of magnitude? If so, it sounds far too high.

Brosch: I) I agree with the need for extreme caution when dealing with small samples, but the HT galaxies are the only sample of bright "isolated" galaxies. The KI sample, as well as that of isolated pairs (KP; Karachentsev 1979, Comm. Spec. Astr. Obs. USSR, 7, 1), are heavily contaminated by members of systems. I found that among those KI and KP systems detected at 6 cm by Stocke and his collaborators and having published redshifts, those nearby (within $\sim$ 2000 km s^{-1}) have a significant number of companions within 0.5 Mpc (H_o = 100 km s^{-1} Mpc^{-1}), and within 150 km s^{-1} of the primary system. The HT galaxies have effectively no companions. Thus, studies based on the Soviet samples of isolated systems should be approached with extreme caution because the samples are not clean.

II) I must emphasize that the M(HI) values for the BK and HT samples come from different authors. If you believe the Balkowski and Chamaraux HI masses, then the discrepancy between the two samples is indeed one order of magnitude at the low end of the M(HI)-D relation. It would be extremely useful to have both samples observed by one single person or group, with the same system.

van Woerden: Your claim that isolated galaxies tend to have strong nuclear radio-continuum sources appears at variance with the results of a large survey of galaxies by Hummell at Westerbork. Hummel (Astron. Astrophys., 89, L1) finds that in interacting pairs nuclear sources are more frequent than in single galaxies.

Brosch: I) We have observed at 5 GHz while Hummel (1981, Astron. Astrophys., 93, 93) looked at 1.4 GHz. Thus, we are looking at different things.

II) Our nonisolated galaxies have been selected not to be interacting or peculiar objects, because we wanted to check only the effects of slightly increased density of galaxies. This sample has been compared to the "cleanest" sample of isolated galaxies. Hummel (1981) had, on the other hand, used as the "isolated" class a quite badly defined sample.

DYNAMICS OF VOIDS AND CLUSTERS AND FLUCTUATIONS IN THE COSMIC BACKGROUND RADIATION

E. E. Salpeter
Cornell University, Ithaca, NY, USA

1. INTRODUCTION

I want to "make propaganda" for some calculations carried out at Cornell; a detailed paper is available on this work (Hoffman, Salpeter and Wasserman 1982, hereafter HSW), so I only summarize it briefly (Sect. 2) and concentrate mainly on implications. Like previous work by Peebles (1982) and by Occhionero et al (1982), these calculations use spherically symmetric models without dissipation for the dynamical development of large voids and galaxy clusters (and superclusters) from small underdensities and overdensities, respectively, at the recombination era. We now know how this development depends on various parameters and on the asymmetries between over- and under-densities. I discuss conjectures for more complex geometries in Sect. 3.

With a detailed parameter study on the development of voids and clusters one can invert the process to infer the density fluctuations which must have been present just after the recombination era to produce some present-day configuration. Fluctuations in the present-day cosmic background radiation (on scales of $\sim$10 arcmin) are related to this and their inferred amplitude depends very strongly on the present-day value of the cosmological density parameter Ω. The relation to observed upper limits on these fluctuations are discussed in Sect. 4.

2. DYNAMICAL MODELS

We start with a small spherically symmetric density perturbation (but no perturbation in the Hubble velocity field) just after the proton-electron recombination era. We use a simple shape for the density profile (a "rounded-off stepfunction") so that the initial conditions are characterized by the (sign and size) density enhancement $\delta\rho_{rec}$, the radius of the enhanced (or depressed) region and the value of the dimensionless cosmological density parameter Ω at the initial time. We only consider regions small compared with the Hubble radius, so the size (or mass M_1) of the perturbed region only enters as a scalefactor; thus we have only two dimensionless parameters for a single, isolated, spherical perturba-

G. O. Abell and G. Chincarini (eds.), Early Evolution of the Universe and Its Present Structure, 211–216.

tion. These two parameters can be expressed as the present-day value of Ω plus the ratio t/t_1, where t is the present epoch and t_1 the time (defined more quantitatively) when the perturbation "separated out from the rest of the universe."

Numerical results for the present-day density profile for models with different values of t/t_1 and of Ω are now available (see HSW); I only summarize the qualitative features, first for cases with (present-day) Ω appreciably less than unity. For $t/t_1 < 1$ the perturbations grow without change of shape (symmetrically for overdensities and underdensities) but the nonlinear growth for $t/t_1 >> 1$ is rather different: Once an overdensity has developed into a cluster it grows in mass only a little (from mass outside M_1 "falling in") and after a while the mass and radius of the cluster stays almost constant (in fixed coordinates, not in comoving coordinates). Since the density of the uniform outside background decreases continuously, the density contrast (expressed as the ratio of densities) continues to increase steadily. Two other features for dissipationless models for clusters are (a) the internal density decreases appreciably from the center of the cluster outward and (b) no ridges of underdensity develop outside the cluster. For an initial spherical underdensity, on the other hand, it is the density contrast (the ratio) which gets "frozen in" at the constant value reached as soon as t/t_1 exceeds unity and Ω decreases below ~ 0.4. Thus, a large value of t/t_1 is no guarantee for a large density contrast if $\Omega << 1$. Two other features for voids are (a) the internal density varies little and (b) dense ridges (spherical shells of high density) form outside of the void.

The formation of dense ridges is particularly important for applications to more complex geometries (Sect. 3). It is a rather general phenomenon, mathematically similar to the formation of a shock-front, where a positive density gradient outwards means that inner material is decelerated less and catches up with the slower matter further out. The time (in units of t_1) and the mass-shell (in units of the mass M_1 of the initial "density step") for the occurrence of the singularity depend on the details of the rounding off of the density step but qualitatively some dense ridge always forms after a few times t_1.

The redshift of the recombination era is known uniquely ($z_{rec} \sim 10^3$) which fixes the relationship between the cosmological density parameters then and now, Ω_{rec} and Ω respectively, but not their actual values. The numerical calculations for density ρ show that the ratio of $(\delta\rho/\rho)_{now}$ to $(\delta\rho/\rho)_{rec}$, increases quite considerably with increasing Ω. For an underdensity this trend is seen most easily by noting that the density contrast "freezes in" soon after Ω drops below ~ 0.4; for an overdensity by noting that $(1 - \Omega_{rec})$ increases with decreasing Ω and that the value of $(\delta\rho/\rho)_{rec}$ required for a bound cluster increases with increasing $(1 - \Omega_{rec})$. Although described differently, the trend is actually similar for clusters and voids with $(\delta\rho/\rho)_{rec} \sim 3\times10^{-3}$ and 3×10^{-2}, respectively, needed to give an "appreciable" contrast now if $\Omega = 1$ and 0.05, respectively.

I discuss next another hypothetical construct, a regular cellular structure (resembling a honeycomb) with a void at the center of each cell. In such a regular lattice the mean density inside each cell equals the mean cosmological density ρ_∞ and the cell boundaries expand with pure Hubble flow. Somewhat as in solid state physics, each polygonal cell can be approximated by a "Wigner-Seitz sphere" containing the same mass M_2 and having the same volume as the real cell. In the real lattice equal mounts of matter flow across each boundary-face in opposite directions; this is mimicked in a spherical calculation by using a single Wigner-Seitz sphere but with a perfectly reflecting spherical wall which expands precisely with Hubble flow (and internal mean density precisely ρ_∞). For such models there is an additional parameter, the ratio of the cell mass M_2 to the mass M_1 of the initial, central underdense region. For $M_2/M_1 \to \infty$ such models reduce to those for a single, isolated void. For $M_2/M_1 \sim 2$ to 10 the development of the central underdensity into a void is essentially unchanged by the reflection boundary condition, a dense spherical shell (a ridge) forms as before but in these models the dense matter remains "piled up" in a dense narrow shell just inside the boundary wall. Note that the permanence of these thin dense shells does not require any dissipation in the models.

3. MORE COMPLEX GEOMETRIES

Cellular honeycomb structures have been discussed before by Einasto et al (1980) and Doroshkevich et al (1980); the formation of "pancakes" with dissipation by Sunyaev and Zel'dovich (1970). As anticipated by Dekel (1982) and Peebles (1982), we have found dense spherical shells ("curved pancakes") which remain thin (but massive) even without any dissipation. We do not understand enough about the formation of individual galaxies to judge the importance of dissipation on theoretical grounds: (i) if galaxies can form easily from a modest density increase, they will have formed <u>before</u> any gaseous shock forms and there is little dissipation; (ii) if <u>it is</u> difficult to form galaxies, shock dissipation and radiative cooling will happen first. Our model results show that observations of "thin" pancakes (without a quantitative measure of "thin") do not necessarily prove the presence of dissipation.

A general discussion of density singularities has been given by Arnold, Shandarin and Zel'dovich (1982); but I only make some conjectures from generalizations of our highly symmetric models. Consider first a "slightly irregular lattice" of perturbations just after the recombination era, with some cells having overdensities at their centers and others underdensities. We saw that dense, thin ridges form easily surrounding an underdensity but deep, narrow holes do <u>not</u> form around overdensities. There is a further asymmetry: A <u>slightly</u> irregular underdensity develops into a void of similar shape, but (as previous work had shown) a slightly irregular overdensity develops into a dense, thin pancake (possibly with an even denser central core). Thus, if there were as many overdense as underdense centers initially, the final appearance <u>may</u> be closer to one of a hole-centered honeycomb with dense sheets (<u>with c</u>luster cores giving the sheet a mottled appearance) for boundaries. Ironically, it is because sheets of low density do not form readily that one "sees" cells dominated by (centered on) underdensities.

The distribution of galaxies today depends not only on Ω and on the initial density perturbations at $z \sim 10^3$ (including the amplitude, which controls t/t_1), but also on the "microphysics" of galaxy formation: If galaxies form easily and if t/t_1 is not very large, galaxies formed without dissipation and occupy a complex set of moving galaxy sheets, isolate superclusters, etc.; if a large compression ratio is required to form a galaxy and if t/t_1 is large, then the dense sheets, protoclusters and ridges merged (with dissipation of bulk motion and with radiative cooling before galaxy formation and galaxies would now be seen only along well-defined, continuous, orderly cell-walls. The observational picture emerging at this Symposium does not seem to be well-ordered, which may be indirect evidence against the importance of dissipation.

With redshifts known for several thousand galaxies we can now map (at least crudely) the nearest few superclusters as well as our own Local Supercluster (centered on the Virgo cluster). These superclusters are by no means arranged in a regular lattice network, but one can nevertheless attempt to fit models for a cluster-centered lattice to the data (I refer now to clusters closer than the nearest large void (Kirshner et al 1981, Davis et al 1982) and not to large hole-centered cells with many clusters in their walls). One interesting feature of such fits is that M_2, the mass of a Wigner-Seitz sphere (total mass per supercluster but also including galaxies outside the supercluster) is very much larger than M_1, the mass of the initial overdense region. The total superclust region which has a noticeable density enhancement (e.g. a Virgo-centered region extending out to our location for the Local Supercluster) is more massive than M_1 (closer to the traditional Virgo cluster itself), yet even this region has a mass of only about 0.1 M_2. At the moment it is not clear if this feature is due to (a) a property of the initial density perturbations; (b) a tendency for rare, strong fluctuations to "swallow up" more common, weaker fluctuations during the dynamic development; or merely due to (c) the greater ease of identifying a rare, strong feature from an imperfect observational data base. Whichever the cause, we know fairly reliably that of order 10% of the mass of the universe is made up of regions like that extending from the Virgo cluster to our location, with a mean density about three times ρ_∞. We shall use this knowledge in the next section.

4. ANISOTROPIES IN THE COSMIC MICROWAVE BACKGROUND

Disregarding smearing by radiative diffusion, an adiabatic perturbation of baryonic density contrast $(\delta\rho/\rho)_{rec}$ at the end of the recombination era gives rise to temperature perturbation of $\delta T/T \approx (1/3)(\delta\rho/\rho)_{rec}$ in the present-day cosmic background radiation (the situation is qualitatively similar for isothermal perturbations). The models relate the present-day density perturbation $(\delta\rho/\rho)_{now}$ to $(\delta\rho/\rho)_{rec}$ and predict the temperature perturbations from an observed $(\delta\rho/\rho)_{now}$. Previous work (see Silk 1968, Peebles 1981) emphasized sinusoidal perturbations, whereas I will concentrate on the consequences of presently observed clusters, superclusters and voids. Present-day regions of density $\sim 3\rho_\infty$ (containing $\sim$10% of all mass) lead to predictions for $\delta T/T$ slightly larger than the observational upper limit (Partridge 1980) if $\Omega = 1$, but much larger if Ω is small (multicored superclusters and large voids might add appreciably to $\delta T/T$).

If most of the mass of the universe is in the form of massive neutrinos, the predicted $\delta T/T$ would be smaller because (a) $\Omega_{tot} \approx \Omega_\nu$ is large, (b) the baryon density is small and (c) "neutrino clusters" have larger radii (less dissipation, more thermal motion) than inferred from galaxies and thus a smaller density contrast. I hope we shall soon have detections (instead of upper limits) of δT. We predict a non-Gaussian distribution, with "hot-spots" of $\sim$(5 to 10) arcmin over $\sim$10% of the sky mimicking "discrete, extended sources", but with a blackbody spectrum. This work was supported in part by NSF grant AST81-16370.

REFERENCES

Arnold, V.I., Shandarin, S.F., and Zel'dovich, Ya.B.: 1982, Geophys. Astrophys. Fluid Dyn. 20, p. 111.
Davis, M., Huchra, J., Latham, D., and Tonry, J.: 1982, Ap.J. 253, p. 423.
Dekel, A.: 1982, Ap.J. 262 (in press).
Doroshkevich, A.G., Kotok, E.V., Novikov, I.D., Polyudov, A.N., Shandarin, S.F. and Sigov, Yu.S.: 1980, M.N.R.A.S. 192, p. 321.
Einasto, J., Joeveer, M., and Saar, E.: 1980, M.N.R.A.S. 193, p. 353.
Hoffman, G.L., Salpeter, E.E., and Wasserman, I.: 1982, Cornell Univ. Report CRSR 793.
Kirshner, R., Oemler, A., Schechter, P. and Schechtman, S.: 1981, Ap.J. (Letters) 248, p. L57.
Occhionero, F. et al.: 1981, Astr. Ap. 99, p. L12.
Partridge, R.B.: 1980, Ap.J. 235, p. 681.
Peebles, P.J.E.: 1981, Ap.J. 243, p. 119.
Peebles, P.J.E.: 1982, Ap.J. 258, p. 415.
Silk, J.: 1968, Ap.J. 151, p. 459.
Sunyaev, R.A., and Zel'dovich, Ya.B.: 1980, Ann. Rev. AA 18, p. 537.

Discussion

Shandarin: When you considered the evolution of overdensities and underdensities, did you assume that the scale of perturbation is much larger than Jeans mass?

Salpeter: Yes. The models are for zero-temperature matter.

Hoffman: What fixes the scale of the holes and why don't we see holes on scales smaller than $\sim$ 10 Mpc?

Salpeter: The model calculations themselves are dimensionless for zero temperature and consequently they have no scale. For real matter with a finite temperature, the models do not apply for scales less than the Jeans mass.

Occhionero: In a poster paper shown here, we outline a simple algorithm by which the density profiles and velocity fields for holes can be evaluated straightforwardly. In particular, we call attention to two parameters which define: 1) the depth of the hole, and 2) the shape of the surrounding mass ridge.

Giovanelli: From the picture of evolution of inhomogeneities that you presented, could you comment on how filaments arise from the cell structure and would structures preferentially form that resemble filaments rather than cells?

Salpeter: Our spherical models are only an approximation to models for a lattice cell. If filaments are the intersections of two cell walls, they CANNOT be approximated by our models.

B. Jones: Do the ridges formed around the holes meet before or after they have had time to attain a sufficient density contrast to fragment into galaxies? This is important if you want galaxies to lie on thin sheets around the voids.

Salpeter: This depends mainly on the "microphysics" of galaxy formation: If only a moderate density increase is required, the ridges may already contain galaxies before they merge into "cell walls."

Vignato: I do not understand why it is not possible to obtain void structure by simply supposing the presence of a number of elliptical perturbations.

Salpeter: No comment.

CONDENSATIONS AND CAVITIES

F.Occhionero, P.Santangelo, and N.Vittorio

Istituto di Astrofisica Spaziale, C.N.R., Frascati
and
Istituto Astronomico, Università di Roma, ITALY

We present a unified algorithm which describes the non-linear growth 1) of condensations surrounded by cavities or 2) of cavities surrounded by condensations (i.e. ridges of higher density) in the Hubble flow. The main idealization is that of pressureless spherical symmetry (Tolman-Bondi solution); overall algebraic details and results for problem 1) are given in previous work (Occhionero, *et al.*, 1981 a and b); results for problem 2) will be given elsewhere (Occhionero, *et al.*, 1982).

Each perturbation is embedded in a Friedmann model; for the simplest Einstein-de Sitter case results are expressed in analytic form; the generalization to any other cosmological model is straightforward. Our models have two parameters which relate to the amplitude and the shape of the initial perturbation. The first of these defines either 1) the height of the density excess in the condensation or 2) the depth of the hole; the second defines the shape either 1) of the surrounding cavity or 2) of the surrounding mass ridge.

Recent tridimensional studies of the distribution of cosmic matter suggest the existence of large scale voids surrounded by superclusters of galaxies. In the scenario of problem 2), we may infer that galaxies form where the expanding ridges originating from nearby holes collide (and possibly coalesce in presence of dissipative phenomena).

REFERENCES

Occhionero, F., Veccia-Scavalli, L., and Vittorio, N., 1981 a, Astr.Ap., 97, 169
--, 1981 b, Astr.Ap., 99, L12
Occhionero, F., Santangelo, P., and Vittorio, N., 1982, Astr.Ap., in press

G. O. Abell and G. Chincarini (eds.), Early Evolution of the Universe and Its Present Structure, 217.

ORRELATION FUNCTIONS, MICROWAVE BACKGROUND, AND PANCAKES

S.A. Bonometto*°, F. Lucchin*
*Istituto di Fisica "G. Galilei," Via Marzolo, 8, Padova, Italy
°I.N.F.N., Sez. di Padova

The isothermal theories of galaxy formation encounter serious difficulties in accounting for the existence of large voids (see, e.g., Aarseth and Saslow 1982). The alternative picture of adiabatic fluctuations and pancake collapse (see, e.g., Doroshkevich et al. 1978 and references therein) faces difficulties due to: i) observed correlation functions, and ii) the absence of small-scale fluctuations in the microwave background.

As far as i) is concerned, a puzzling aspect is the absence of any trace of the damping scale, M_D (this scale should have already reached a sufficient density to fragment into galaxies). All published results on M_D based on reliable recombination physics agree, within 60%, with the relation:

$$M_D/M_\odot = 10^{12.25} (\Omega h^2)^{-1.4}.$$

Here we discuss some results concerning ii). Let us suppose that, before recombination, $|\delta_m(k)|^2 = |\delta\rho_m/\rho_m|^2 = Ak^n$ from a scale k_J up to a scale k_D. We shall assume $\lambda = 2\pi/k$, and $M = (\pi/6)\rho_m\lambda^3$.

Below λ_D fluctuations are damped because of photon diffusion (see, e.g., Bonometto and Lucchin 1979). From λ_D up to λ_J (Jeans length), fluctuations oscillate from their entry in the horizon (at $z_H(\lambda)$) until recombination. Thence, above λ_J, the amplitude of fluctuations is greater by a factor $z_H(\lambda)/z_{rec}$, depending upon λ.

For $k < k_J$, therefore, A is increased by a factor $(z_H(\lambda_J)/z_{rec})^2$, while n becomes n + 4. Moreover, let the mass variance, after recombination, be $\delta M/M \propto M^\alpha$. If n is unchanged (for $\lambda_D < \lambda < \lambda_J$) during recombination, it turns out that $\alpha = -n/6-1/2$. Therefore, for $n > -3$, $\delta M/M$ has its maximum just above M_D. We then normalize $\delta_m(k)$ by requiring that $\delta_m(k_D)$ becomes nonlinear at $z \sim 3$. For $n \lesssim -3$ the situation is different and the most sensible normalization procedure refers to correlation functions. This case was treated by Silk and Wilson, who considered n = -3, -4, -5. They found fluctuations in the microwave radiation nearly independent of n and exceeding radiation observational limits.

The results we present here concern the values n = 1, 0, -1.

G. O. Abell and G. Chincarini (eds.), Early Evolution of the Universe and Its Present Structure, 219–221.

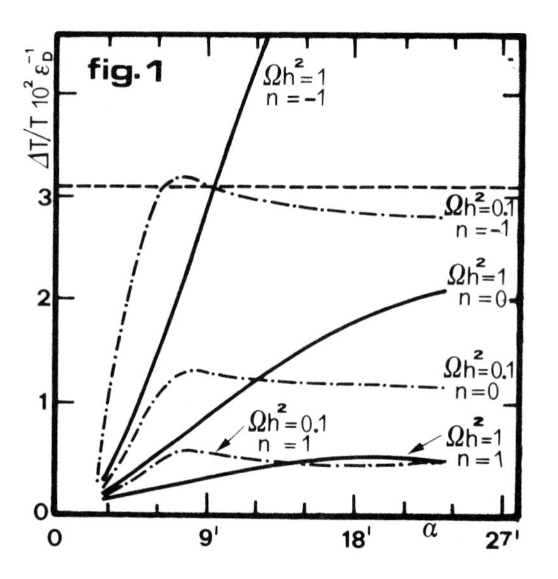

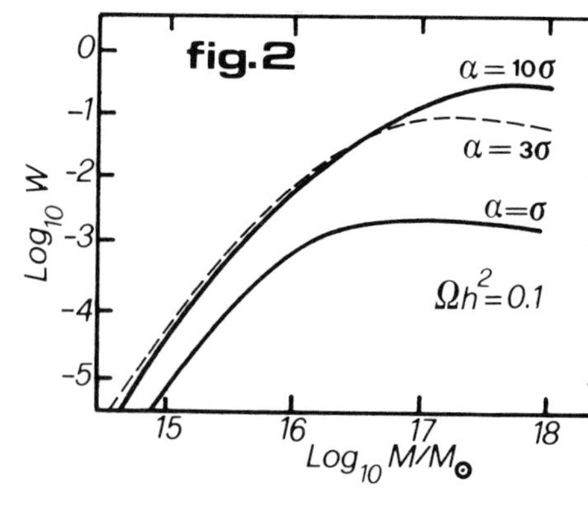

Figure 2. The weight, W, of th different mass scale in determining $\Delta T/T$ is plotted for σ = 1.8' and several values of α.

Figure 1. The square average $\Delta T/T$ of the relative temperature difference between two directions at angle α is plotted for a beam width σ = 1.8'; here h = 0.75; A is normalized in order that the initia mass variance over the scale M_D is ε_D. Values $\varepsilon_D \gtrsim 10^{-2.5}$ are required The dashed horizontal line yields the level corresponding to $\Delta T/T = 10^{-}$

This result is obtained by using a full kinetic treatment of radiation; we made an expansion of the phase space distribution into spherical harmonics and checked that considering 50 of them is sufficie down to a redshift where γ-e scattering can be neglected and the evolution of photon density perturbations can be treated exactly (see Bonometto et al. 1982). In Figure 1 we plot the amplitude of the expected small-scale fluctuations of the microwave radiation which are compatible with observational limits and dependent on n. In fact, the scales from which microwave fluctuations are mostly generated widely exceed the normalization scale M_D (see Figure 2); a larger n (steeper spectrum) causes smaller microwave fluctuations.

It has been suggested that a nonzero neutrino mass could result in microwave fluctuation limits that are consistent with observations. Our point here is that this is not a necessary requirement.

References

Aarseth, S.J., and Saslow, W.C. 1982, Ap. J., 258, L7.
Bonometto, S.A., and Lucchin, F. 1979, M.N.R.A.S., 187, 611.
Bonometto, S.A., Caldara, A., and Lucchin, F. 1982, submitted to Astror and Astrophys.
Doroshkevich, A.G., Zeldovich, Ya.B., and Sunyaev, R.A. 1978, Sov. Astron., 22, 523.
Peebles, P.J.E., and Yu, J.T. 1970, Ap. J., 162, 815.
Silk, J., and Wilson, M.L. 1981, Ap. J., 243, 14.

Discussion

zalay: The formula for the damping mass assumes that the universe is fully dominated by baryons. If the universe contains a arge amount of dark matter (e.g., neutrinos), the expression for the amping mass should be:

$$M_D = 3 \times 10^{13} \Omega_B^{-3/2} \Omega_T^{1/4} h^{-5/2} M_\odot ,$$

s given by Bond, Efstathiou and Silk (1980, Phys. Rev. Lett., 45, 1980-984). If $\Omega_B < 0.1$, then $M_D \gtrsim 10^{15} M_\odot$, and the problem with the absence f a feature in the correlation function disappears.

onometto: I feel that it is fairly important that all computations of M_D for a baryon-dominated universe, based on different but cceptable approximations, arrive at results in good agreement. This is ot directly evident from the literature, where different relations mong k, λ, M, and different values of h are used; furthermore, the volution of fluctuations before decoupling is sometimes mixed up with ransmission through decoupling.

Certainly, the "use" of massive neutrinos helps to solve uite a number of problems. My main point here was that at least one of hese problems (microwave radiation small scale fluctuations) can be olved without the need for massive neutrinos.

PHOTOMETRIC AND MORPHOLOGICAL INVESTIGATION OF VERY REMOTE CLUSTERS

A. Kruszewski and R.M. West
EUROPEAN SOUTHERN OBSERVATORY, Garching bei München, FRG

bstract

About 500 remote, mainly compact clusters of galaxies have een identified on ESO (R) Schmidt plates; around 40 of these have been bserved with the ESO CCD camera on the Danish 1.5-m telescope on a Silla. We describe the method of identifying the clusters and give ome preliminary photometric results.

A search is being conducted on ESO (R) Schmidt plates IIIa-F + RG 630, 120 min) obtained with the ESO 1-m Schmidt telescope or the (R) half of the joint ESO/SRC Survey of the Southern Sky. Twelve ields have been searched by a visual/automatic method, as described by 'est and Kruszewski (Irish Astr. Bull., 15, 25, 1982). Cluster candi-ates are identified visually and the corresponding 5 × 5 arc min^2 areas re scanned. Individual objects are identified automatically by recently eveloped VAX software and main image parameters are determined. The mages are subjected to simplified, but efficient, automatic cleaning rocedures, removing contributions from blended objects and are separated nto stars, galaxies and defects.

In order to calibrate photometrically the Schmidt plates, CCD bservations have been made of selected clusters found during this earch. Standard Gunn filters (g, r, i, z) were placed in front of the SO CCD camera attached to the Danish 1.5-m telescope at La Silla, and 5- to 45-min exposures were obtained of about 40 clusters, not all of hem, however, in all colours. Observations were made of standard stars Hoessel, private communication) in order to transform into the standard ystem. In Fig. 1 we show an example r versus (g-r) for the cluster 630-566 (z = 0.35) in the instrumental system, but with calibrated zero-oint. The limiting magnitude in this sample is defined by the need to eparate stars and galaxies with high confidence (about r = $21^{m}.5$ for 15 in exposures). Also shown are histograms for the (g-r), (r-i) and (i-z) olour indexes. The hatched areas and the filled circles refer to bjects inside a circle with 40 arc sec diameter, centered on the clus-er. The colours can be seen to cluster near certain values, which can pparently be quite accurately determined from observations like these. 'e also note a dependence of the (g-r) colour on the r magnitude in the

. O. Abell and G. Chincarini (eds.), Early Evolution of the Universe and Its Present Structure, 223–225.

sense that fainter objects are bluer.

In Fig. 2 we show the cluster 0630-566 as seen on the (R) Schmidt plate and on a 30-min (r) CCD frame. The limiting magnitudes are approximately $22^{m}_{.}0$ and $23^{m}_{.}0$ for the Schmidt and CCD, respectively (A = 19.18; B = 22.26).

Figure 3 compares results of the automatic object identification and magnitude determination program applied to both images presented in Fig. 2. Objects classified as galaxies on both frames are denoted with filled circles. Objects classified as galaxies on the CCD frame and as stars on the Schmidt plate are shown with open circles. Small arrows indicate magnitudes of galaxies which have not been detected on the Schmidt plate.

During the continuation of this project, calibrated magnitudes of individual galaxies in cluster candidates will be used to establish a homogeneous sample of distant, rich clusters in selected sky regions. It is the intention to measure redshifts of selected objects with a CCD spectrograph at the ESO 3.6-m telescope and to use the same as a well-defined starting point for the study of cosmological and evolutionary effects.

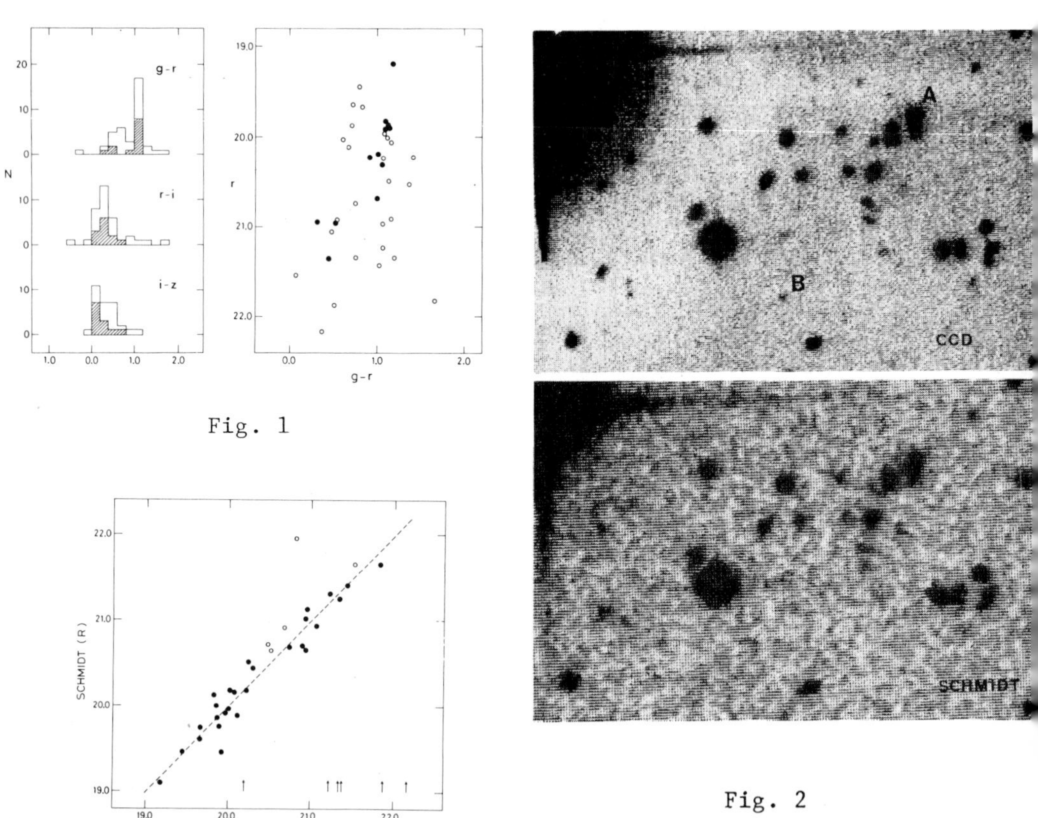

Fig. 1

Fig. 2

Fig. 3

his paper was read for the authors by Dr. M. Tarenghi. Eds.)

Discussion

llis: How many galaxies per cluster (on the average) were photometered with the CCD, and how were they selected?

renghi: The cluster 0630-566 is a good example of the rich clusters selected for the CCD observations.

ell: Our experience, based on film copies of ESO Schmidt plates in the first shipment from ESO of the Southern Survey, is that ey definitely show fewer faint galaxies than the blue (J) Siding Spring hmidt plates. Were these plates (used by West and Kruszewski) taken pecially for their program, with longer exposure?

(Of course, everything else being equal, the red plates should ow faint galaxies better; this is true, for example, for full-exposure d plates taken with the Siding Spring Schmidt with the new red correctg lens, compared with the Siding Spring J-plates.

(Perhaps the difference between our experience and that of est and Kruszewski is in the J plates used for comparison; is this ssible? Where were the comparison J-plates taken?)

renghi: West and Kruszewski did their search on a usual plan copy of the ESO/SRC atlas. More technical information, including a mparison with the blue plates, is given in their Irish Astr. Bull. aper.

PERNOVAE AS A COSMOLOGICAL TOOL

Virginia Trimble
Astronomy Program, Univ. of Maryland, College Park, MD 20742
Dept. of Physics, Univ. of California, Irvine, CA 92717 USA

Cosmology can mean many different things to different people. ındage (1970) once described it as "the search for two numbers" (H_o ıd q_o). At the other end of the spectrum, it may comprise almost all ıe interesting bits of astronomy and physics that bear on how the ıiverse got to be the way it is. Supernovae can probe many of these its because they are bright, have been going on for a long time, and ontribute directly to the chemical and, perhaps, dynamical evolution of tructure in the universe.

The most obvious application is the use of supernovae as dis- ance indicators. The underlying philosophy is that it should be easier o understand the physics of one star than of a whole galaxy. (I'm not ure this is true; compare, for instance, psychiatry and sociology.) hree approaches have been tried. First, as originally suggested by ilson (1939) and Zwicky (1939), one can regard SNe, or some subset, as tandard candles, calibrating them on nearby galaxies and applying the esult to more distant ones. A recent application of this method Sandage and Tammann 1982) yielded $M_B^{max} = -19.65 \pm 0.19$ for Type I's and $_o$ = 52±5 km/sec/Mpc, in good accord with that found by the same authors sing other methods.

Second, one can modify the Baade-Wesselink method for deter- ining distances to variable stars and get an effective temperature for supernova from its colors and its radius from integration of the adial velocity curve. Recent applications (e.g., Branch _et al_. 1981) ave also yielded H_o near 50, at least toward the Virgo cluster. Con- ersion of colors to temperatures and line profiles to velocities equires, in practice, a fairly detailed model atmosphere. All analyses o date have neglected opacity due to scattering (Wagoner 1981) and the ossibility of a shallow density profile in the SN envelope, both of hich cause luminosities to be overestimated, and the observed UV defi- iency of Type I's, which goes the other way (Arnett 1982).

Finally, one can start from a set of hydrodynamical models for he explosions, pick the right one out of the set using distance- ndependent properties like time scale, and use the model luminosity to et distance. Arnett's (1982) use of this method gave H_o = 70±5 km/sec/ Ipc for Type I's in both Virgo and field ellipticals and a one-sigma

. O. Abell and G. Chincarini (eds.), Early Evolution of the Universe and Its Present Structure, 227–229.

dispersion in peak brightness of $0^m.4$. Sufficiently detailed models of the explosions will eventually predict line profiles and colors as well as light curves and radii, so that methods two and three will coalesce around a physical understanding of the events. Twenty percent accuracy in H_0 should be possible from the ground and 50% in q_0 from the Space Telescope (Colgate 1979).

Some participants in this meeting may live long enough to see light curves and spectra for supernovae at $z \gtrsim 1$. Heaven forbid that these should by then be needed to resolve the traditional cosmological problems or to measure distances to the parent galaxies, thus giving L(z), etc. But they will enable us to probe early galactic evolution and nucleosynthesis directly. Interesting data would include: a) supe nova rates and types, b) distributions of SN positions in parent galaxies, and c) masses and compositions of parent stars (found by model fitting to light curves and spectra) all as a function of cosmological epoch and galaxy type. In addition, we could expect to see (or not see at significant levels) supernovae contributing to nucleosynthesis in sites other than galaxies of recognizable types -- intergalactic and intercluster space; protogalactic star clusters or gas clouds; quasi-stellar objects and whatever else. Events recorded so far include one in a Seyfert galaxy (the Type I 1968a in NGC 1275) and one at least 50 kpc from the center of any galaxy (the Type I 1980i centered in a triangle formed by the ellipticals NGC 4375, 4387, and 4496; Smith 1981

To make full use of such information, we clearly need the sam kinds of statistical data for supernovae in galaxies (etc.?) here and now. These can be acquired with existing technology but require more systematic identification and study of extragalactic SNe than has so fa been carried out. This is a plug for the supernova search projects currently underway or contemplated in Berkeley (Kare *et al*. 1982), Cambridge (Cawson and Kibblewhite 1982), New Mexico (Colgate 1982) and elsewhere.

References

Arnett, W.D. 1982, *Ap. J.*, 254, 1.
Branch, D., S. Falk, M. McCall, P. Rybski, A. Uomoto, and B. Wills. 1981, *Ap. J.*, 244, 780.
Cawson, M., and E. Kibblewhite 1982, in Reese and Stoneham (1982), p. 341.
Colgate, S.A. 1979, *Ap. J.*, 232, 404.
Colgate, S.A. 1982, in Rees and Stoneham (1982), p. 319.
Kare, J., C. Pennypacker, R. Muller, T. Mast, F. Crawford, and M. Burns 1982, in Rees and Stoneham (1982), p. 325.
Rees, M.J., and R.J. Stoneham (Eds.) 1982, Supernovae: A Survey of Current Research (Dordrecht: Reidel).
Sandage, A.R. 1970, Physics Today (February), p. 34.
Sandage, A.R., and G. Tammann 1982, *Ap. J.*, 256, 339.
Smith, H.A. 1981, *A. J.*, 86, 998.
Wagoner, R.V. 1981, *Ap. J.*, 250, L65.
Wilson, O.C. 1939, *Ap. J.*, 90, 634.
Zwicky, F. 1939, *Phys. Rev.*, 55, 726.

Discussion

Tuchra: When people compute H_0 from the supernovae distances to Virgo, do they include the effect of our infall velocity?

Trimble: Everybody allows something for infall, but as they allow the same amount they obtained from other arguments, each author tends to report the same value for H_0 from supernovae that he gets from other methods.

M. Burbidge: In a Commission 28 (Galaxies and Cosmology) meeting in Patras, it was unanimously voted that supernovae should be reinstated in Commission 28 (where they belong), whether or not they are also included in the Commission on Variable Stars. It was remarked that what is needed is volunteers to form a Working Group, especially for the supernova search.

Trimble: An excellent idea! I would be happy to join the Working Group in its endeavors (except that I don't belong to Commission 28).

SUBSTRUCTURE IN CLUSTERS OF GALAXIES

M. J. Geller and T. C. Beers
Harvard-Smithsonian Center for Astrophysics

Rich galaxy clusters containing multiple condensations are common. This subclustering affects many cluster properties and provides insight into cluster evolution.

"Clumpy" clusters of galaxies are interesting, in part, because N-body simulations indicate that multi-component substructure occurs before complete collapse and virialization (White 1976). Interesting systems can be identified both from x-ray surface brightness maps (Forman et al. 1981) and from contour maps based on galaxy counts (Geller and Beers 1982).

Here we present optical contour maps of the four clusters identified by Forman et al. (1981): A98, A115, A1750, and SC0627-54 (Figure 1).

The correspondence between the x-ray surface brightness map and the optical contour map for A98 is excellent (Henry et al. 1981). A cosmological timing argument applied to this "two-body" system suggests that the two clumps were at maximum expansion 3.5 billion years ago and that they will coalesce in another 3 billion years (Beers, Geller, and Huchra 1982).

The cluster A115, at a redshift of 0.2, has three clumps of galaxies, only two of which (marked by arrows in Figure 1) have associated extended x-ray emission (Beers, Huchra, and Geller 1983).

The optical map of A1750 is almost exactly like the x-ray image; the cluster is triple in both cases. The southernmost clump has a very small velocity dispersion ($\sigma_{los} \simeq 300$ km s^{-1}) and is quite faint in the x-ray. The two large clumps of galaxies are separated in velocity by ~ 1000 km s^{-1}.

The northeastern clump in SC0627-54 is again the same in both the x-ray emission and the galaxy distribution. In the southwest,

G. O. Abell and G. Chincarini (eds.), Early Evolution of the Universe and Its Present Structure, 231–233.

the apparent deficiency of galaxies relative to extended x-ray emission is probably due to contamination of the x-ray image by emission from the radio source PKS 0625-545. In fact, there are also strong radio sources in A98 and A115 (3C 28) with associated high-energy radiation which presumably contaminates the x-ray observations. Upon correction for this contamination, the agreement between the x-ray surface brightness and galaxy distributions is excellent on scales $\gtrsim$ 0.12 Mpc. This agreement can be used to place constraints on the properties of the "missing mass" (Gioia et al. 1982).

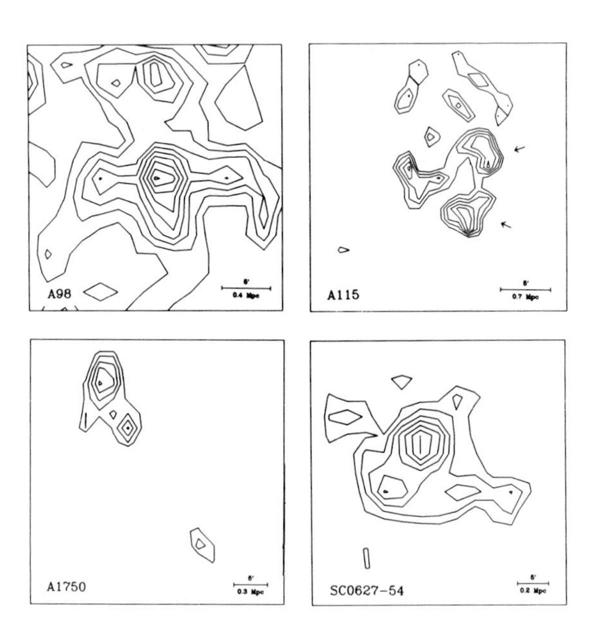

Figure 1: Galaxy surface number density maps of the original x-ray double clusters. The contour levels are linearly spaced. The binning scale is 0.24 Mpc. The lowest contour corresponds to 2 (A98), 29 (A115), 9 (A1750) and 16 (SC0627-54) galaxies/bin and is 3σ above the local background; the highest levels are 22, 36, 23, and 44 galaxies/bin respectively. The original galaxy positions are from Dressler (A98), KPNO 4-m plates (A115 and A1750), and the ESO quick blue survey (SC0627-54). See the review by T. Maccacaro and I. M. Gioia in this volume for the x-ray contours.

REFERENCES

Beers, T. C., Geller, M. J., and Huchra, J. P.: 1982, Ap. J. 257, 23.
Beers, T. C., Huchra, J. P., and Geller, M. J.: 1983, Ap. J. (in press).
Geller, M. J. and Beers, T. C.: 1982, Publ. Astron. Soc. Pac. 94, 421.
Forman, W., Bechtold, J., Blair, W., Giocconi, R., Van Speybroeck, L., and Jones, C.: 1981, Ap. J. 243, L113.
Gioia, I. M., Geller, M. J., Huchra, J. P., Maccacaro, T., Steiner, J. E. and Stocke, J.: 1982, Ap. J. (Letters) 255, L17.

Henry, J. P., Henriksen, M. J., Charles, P. A., and Thorstensen, J. R.: 1981, Ap. J. 243, L137.
White, S. D. M.: 1976, Mon. Not. R. Astron. Soc. 177, 717.

Discussion

Dekel: Is there a correlation between having a substructure and a cD galaxy, perhaps with a double nucleus?

Geller: We do not yet have a sufficiently homogeneous sample to be able to tell.

FILAMENTS

J. Richard Gott, III
Department of Astrophysical Science
Princeton University
Princeton, NJ 08540 USA

I would like to make a brief comment on some work that John Moody, Edwin Turner and I have done on filamentary structure in the Shane-Wirtanen galaxy counts. The picture of the Shane-Wirtanen galaxy counts to the 19th magnitude appears quite filamentary to the eye. The existence of such filaments is quite important theoretically, since filaments could easily be Zeldovich pancakes seen edge-on, and filamentary structure is expected in general when the perturbation spectrum is cut off at small scales. So if the filaments are real, they are telling us something quite important about the universe. The experience of the Martian "canals" reminds us, however, that the eye is very good at picking linear features in random data, and here we are dealing with a clumpy distribution where such effects may be enhanced.

To investigate this problem, we devised a computer algorithm which identified filament pixels as those which lay on ridge lines. This algorithm is quite successful at identifying just those filaments which one's eye detects. We then applied the algorithm to the Soniera-Peebles simulation which contains a hierarchical pattern of clustering, reproduces the two- and three-point correlation functions, but which contains no intrinsic filamentary structure. Of course, random alignments of clumps in this model can produce filaments and the simulation does appear somewhat filamentary to the eye, but not as dramatically so as the real sky. When our filament algorithm is applied to the simulation, it produces a pattern of filaments which is astonishingly similar to that in the real sky: the fraction of pixels that are filament pixels and the lengths of the filaments are statistically indistinguishable in the real sky and the simulation.

There are some differences, however. The filaments in the real sky are somewhat brighter than those in the simulation, making them more prominent. Whether this difference could be eliminated by a hierarchical simulation with somewhat different parameters is not clear. So while there are differences between the real sky and the simulations, their similarities are striking. Thus, a good deal of the filamentary appearance of the Shane-Wirtanen counts may simply be due to hierarchical clustering.

G. O. Abell and G. Chincarini (eds.), Early Evolution of the Universe and Its Present Structure, 235–237.

Discussion

Scott: Some years ago, Scott, Shane and Swanson (1954, Ap. J., 119, 91) constructed a synthetic plate in order to compare to the Shane and Wirtanen Lick survey. The synthetic plate was constructed by Monte Carlo clustering. That is, the cluster centers were assigned in space by a three-dimensional Poisson distribution, using a table of random numbers. The number of galaxies assigned to each cluster followed a geometric distribution as in Neyman, Scott and Shane (1953, Ap. J., 117, 92); galaxy coordinates with respect to centers were trivariate normal; luminosity function, etc., also, as in Neyman, Scott and Shane (1953), using random numbers. The resulting galaxies were projected onto two dimensions; galaxies falling on a hypothetical 6° × 6° plate were retained. They were next retained if brighter than varying limiting apparent magnitude. They were next corrected for random errors in counting. The final synthetic plate, entirely based on random numbers under the hypothesis of complete clustering, was compared to actual plates in the Lick survey. Among other comparisons, we looked for "chains," "crescents," etc., that would now be called filaments. We found good agreement between the synthetic and actual plates, but filaments on the synthetic plates were clearly optical, since the distribution was composed of galaxies from different clusters at different distances. Due to construction by hand, we kept track of which cluster each galaxy comes from. It was convenient to do by distance shells. Therefore, apparent filaments will be found when none really exist -- when we have just simple clustering.

Shandarin: Filamentary structure exists in any distribution of particles; for example, in a pure Poisson one. Thus, it is not a qualitative question but a quantative one.

Gott: We have run our filament algorithm on Poisson data, and the results are highly statistically significantly different from the sky or the hierarchical simulation. But the sky and the hierarchical simulation were statistically indistinguishable, except for the brightness effect discussed.

B. Jones: The Shane-Wirtanen survey is not very uniform and the filamentary structures seen in their survey are not always as evident on modern surveys.

Osmer: Will your approach allow you to distinguish among different types of clustering, e.g., hierarchical or power law?

Gott: So far we have just tested it on the sky and on the hierarchical clustering simulation, but it would be interesting to try it on a variety of types of simulations.

Peacock: Are you then saying that the real sky shows a deficiency of faint filaments?

Gott: Yes. The number and lengths of the filaments in the simulations and the real sky are the same and the filaments in the sky are on average slightly brighter. So the sky contains, therefore, fewer faint filaments.

MORPHOLOGY OF THE LOCAL SUPERCLUSTER*

R. Brent Tully
Institute for Astronomy, University of Hawaii

The intent of this brief note is to summarize some of the fundamental properties of the region, rich in galaxies, in which we live. A more complete account can be found in The Astrophysical Journal, 257, p. 389, 1982.

1. The Local Supercluster contains three components: the Virgo Cluster core (containing 20% of the luminous galaxies), a flat disk (containing 40% of the luminous galaxies), and a "halo" consisting of a small number of discrete clouds (containing 40% of the luminous galaxies).

2. The disk component is irregular in shape and can be separated into two principal clouds of galaxies. Overall, this component has the axial ratios 6:3:1. The global rms scale height along the short axis is $\pm 1.1\ h_{100}^{-1}$ Mpc.

3. The thinness of the disk suggests that either the supercluster is just collapsing today or random motions perpendicular to the disk are less than 100 km s^{-1}.

4. Line-of-sight random motions for galaxies within 4 h_{100}^{-1} Mpc of our position (all in the supercluster disk) are less than 100 km s^{-1}, and probably closer to 50 km s^{-1}.

5. Our Local Group is on the edge of a hole devoid of galaxies which has dimensions comparable with the dimensions of the Local Supercluster.

*This same paper was presented at the Patras General Assembly and appeared in the Highlights of Astronomy, Vol. 6, p. 747.

G. O. Abell and G. Chincarini (eds.), Early Evolution of the Universe and Its Present Structure, 239–240.

6. Almost all galaxies in the halo component lie in a small number of clouds: 56% lie in 2 clouds, 86% lie in 5 clouds, 94% lie in 7 clouds. Triaxial spheroids with axes defined by the rms separations of galaxies in these clouds contain only 4% of the available volume off the plane of the supercluster.

7. The major halo clouds are prolate, elongated 2:1, and point toward the Virgo Cluster. These shapes must be attributed to tidal distention due to the mass of the central cluster. The existence of a bound group in one of these clouds is used to set an upper limit to the epoch of cloud formation at a redshift of z = 8.

8. There is a minor feature off the plane of the supercluster but parallel to it. The plane in our vicinity and this secondary feature appear to be streaming toward each other.

Discussion

Huchra: A very quick point: I don't like differing from Sandage any more than I have to, so remember that:

$$\frac{\delta\rho}{\rho} = \frac{\rho_{interior}}{\rho_{mean}} - 1 \ .$$

For the value of $\delta\rho/\rho$, we find $\sim$ 2 from the CfA survey, and Sandage, Tammann and Yahil find $\sim$ 3, not 4.

On the Virgo cluster itself, most people analyze the dynamics assuming that all the galaxies in the 6° circle belong to a single, virialized unit. Dave Latham and I have been "drilling" this Virgo core region and now we have collected $\sim$ 300 redshifts. When we plot the distribution of galaxies in three slices in "velocity space," we find a central core which persists from minus velocities to 2000+ km/s around M87, but there also are four additional, separated clumps, including a major condensation around N4472, with much lower internal velocity dispersions. The velocity histogram for the whole sample does not resemble a gaussian. The implication is that the "core" of Virgo is not virialized, and consists of a central, much smaller, core and separate groups. The M/L must be overestimated.

Abell: Stephen Eastmond found just this same result in his thesis three years ago. In the inner 6° are several concentrations of galaxies with different mean redshifts, ranging up to more than 2000 km/s. Moreover, when Eastmond estimated relative distances to the clumps, using their luminosity functions, he found the clumps to define a linear Hubble law.

THE KINEMATICS OF THE LOCAL SUPERCLUSTER

J. R. Mould and P. L. Schechter
Kitt Peak National Observatory*
M. Aaronson
Steward Observatory, University of Arizona
R. B. Tully
Institute for Astronomy, University of Hawaii
J. P. Huchra
Center for Astrophysics

The spatial distribution of galaxies in the Local Supercluster, as described at this meeting, together with the measured anisotropy in the microwave background suggest that there exist significant deviations from a uniform Hubble flow in the velocity field of galaxies within a few thousand km/s. In principle, it is not too hard to improve on the uniform flow model: one simply needs to examine the spatial distribution of the radial velocity residuals for many nearby galaxies. In practice, the problem is non-trivial because of the coupling of velocity and distance, and the lack of a distance indicator of high precision, and the "thermal" noise in the velocity field. Our recent efforts in this direction are described in more detail in a paper in a current *Astrophysical Journal* (Aaronson *et al.* 1982*a*).

The model we fit to the local velocity field is analogous to that used to describe the motion of the Sun in the Milky Way galaxy. In both cases one decomposes the solar motion into a systematic motion of the Local Standard of Rest and a peculiar velocity with respect to the LSR. However, in the present case (in the first instance) we take the systematic motion to be a radial deceleration rather than circular motion. For an assumed spherical density enhancement centered on the Virgo cluster, it is a simple matter to predict the pattern of velocity perturbations in the Supercluster (i) assuming they are proportional to the present density distribution (the "linear model", Peebles 1976) or (ii) integrating the flow through the whole course of the expansion (the "non-linear model", Schechter 1980). We fit this model to a data set consisting of velocities and 21-cm velocity-widths for 306 spiral galaxies within 3000 km/s (Aaronson *et al.* 1982*b*). The velocity-widths are used as luminosity indicators, following Aaronson, Huchra and Mould (1979) and Tully and Fisher (1977), but it should be noted that *the current problem is totally independent of the absolute distances of galaxies*.

*Kitt Peak is operated by AURA Inc. under contract with the National Science Foundation.

G. O. Abell and G. Chincarini (eds.), Early Evolution of the Universe and Its Present Structure, 241–247.

The model is fitted by a careful χ^2 minimization technique, but as in any problem where a signal is being extracted from considerable noise, it is necessary to guard against biases in the fit. If one uses velocity-widths to predict absolute magnitude, it is possible, of course, to either underestimate or overestimate the distance to any particular galaxy, and these errors occur in a volume limited sample in a normally distributed way. However, in a magnitude limited sample, which to some extent the present sample is, this technique, in which one minimizes the *redshift* residuals, is biassed, because the galaxies with underestimated distances come from a larger volume. The problem is known as "Malmquist bias", and can be formally corrected if one knows the sample selection criteria and the true spatial distribution.

We chose to invert the problem to avoid this bias, following Schechter (1980). If one takes the distance of each galaxy from its redshift and the (iterated) flow model and uses it to predict velocity-width, there is no such volume effect. The solution is in principle unbiassed, if one minimizes *width* residuals. However, even this method is not uinbiassed, as we rapidly discovered from numerical simulations. This can be seen from the distribution of residuals in a nearby subsample in Figure 1. Suppose one is a galaxy in the "redshift plateau" in Figure 1 just "above" the Virgo cluster. Suppose also one has a redshift one or two hundred km/s different from the canonical value required by the flow model. Distance changes very rapidly with redshift in this region, and depending on the sign of the noise term, one tends to be seriously mislocated in the model either on the near or the far side of the plateau. The consequent residual, however, also enters χ^2, and the fitting routine compensates by artifically decreasing the amplitude of the flow pattern. The result is a bias in which the infall velocity is underestimated.

There is no single strategy for overcoming this bias, and we elected to (i) exclude a cone within 25° of the Virgo cluster and (ii) correct the remaining bias using Monte Carlo simulations. The one fringe benefit from this problem is that we obtain an estimate of the noise in the local velocity field of approximately 150 km/s. The local infall velocity determined by this method is 250 ± 64 km/s, and, as we shall see, there are significant peculiar motions. This value is dependent principally on the assumed power-law index of the radial density distribution. We adopted $\rho = r^{-2}$ (Yahil, Sandage and Tammann 1980*a*), and we find a change of approximately -/+65 km/s, if that index is changed by +/-0.5. We find negligible sensitivity of our result, however, to hypothesised "second parameters" in the Tully-Fisher relation, such as surface-brightness or Hubble type.

In seeking to minimize χ^2 still further, however, we did achieve a real measure of success by adding a rotational term to the flow model. We tested a rotation curve of the form $W_{rot}(r) = W_r r \exp(1-r^2)$ (de Vaucouleurs 1958). Distance (r) is in units of our distance from Virgo. We found W_r (the local rotational velocity) to be 191 ± 49 km/s, and have incorporated that in our final solution given in Table 1.

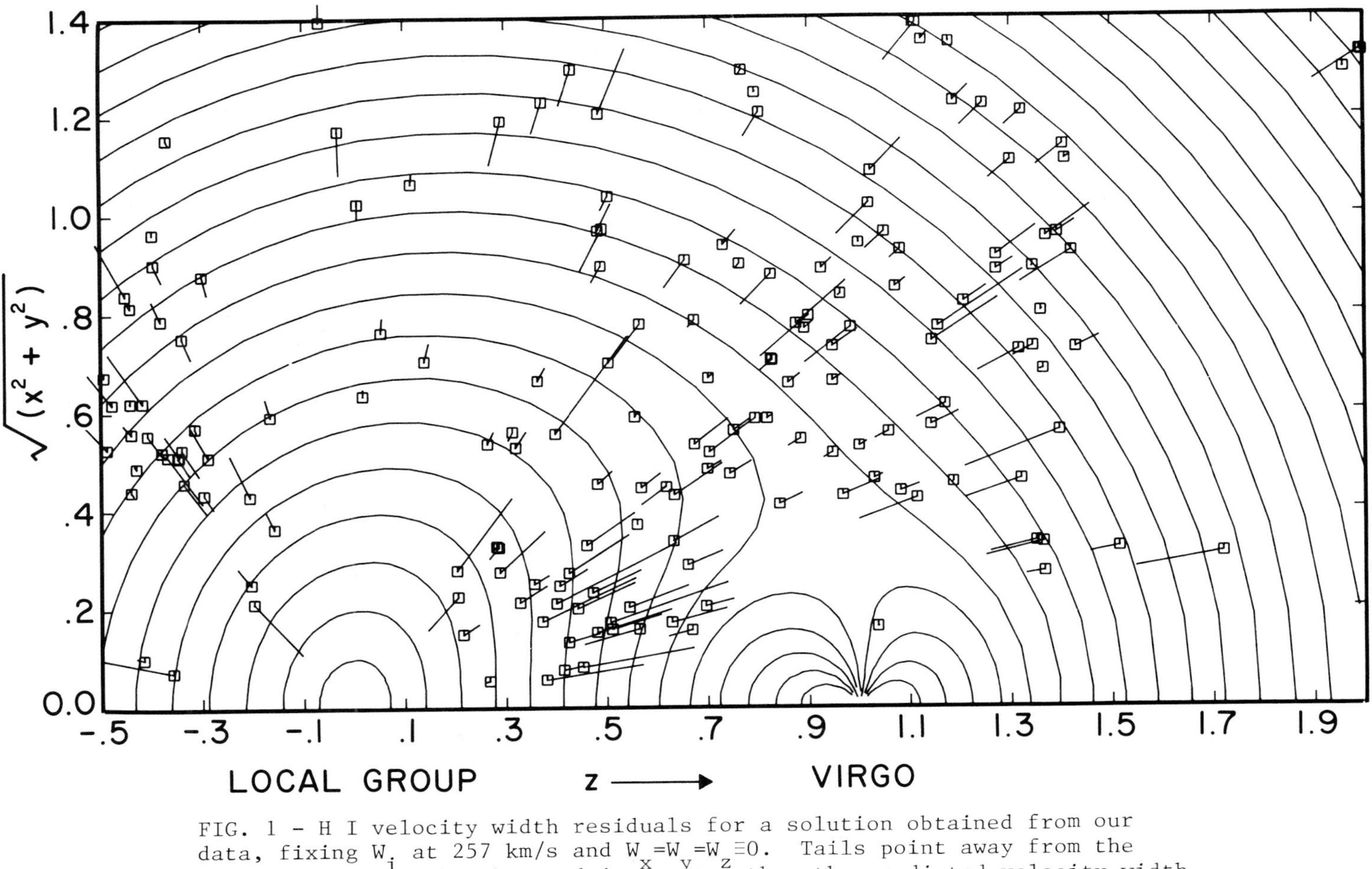

FIG. 1 - H I velocity width residuals for a solution obtained from our data, fixing W_i at 257 km/s and $W_x = W_y = W_z \equiv 0$. Tails point away from the Local Group when the observed is greater than the predicted velocity width. Contours of redshift are shown, spaced at intervals 1/10th the Virgo redshift.

Table 1 allows comparison of the present result with those of other workers. This is complicated by the assumption of different models by different groups. The second and third lines represent the most recent results from the microwave background. Here one should compare the vectors (W_x^{tot}, W_y, W_z^{tot}), where the z axis points at M87 and the x axis is in the supergalactic plane (see Aaronson *et al.* 1982*a*). The magnitude of the velocity difference is 250 ± 150 km/s. The difference appears to be significant: it is unclear to what extent it represents an additional motion relative to a distant frame or an intrinsic microwave anisotropy. The next three entries in Table 1 are pure infall solutions, and W_z^{tot} should be compared with the present result. Our value is bracketed by previous determinations. A number of other recent results are also included for comparison. It is interesting to note a significant component of our peculiar velocity (W_y) towards the south Supergalactic pole. Conceivably, this could be a result of acceleration towards the Supergalactic plane, as discussed by White & Silk (1979).

It is clear, however, that the dominant component is the infall velocity, W_i. If our kinematical model is a dynamical (gravitational) reality, and if the density contrast (ratio of interior to background densities) is 4, then the ratio of that (local) background density to the critical density in Friedmann cosmological models is approximately 0.1.

This work was partially supported with funds from the National Science Foundation.

REFERENCES

Aaronson, M., Huchra, J. and Mould, J. 1979, Ap. J., 229, 1.

Aaronson, M., Huchra, J. Mould, J., Schechter, P., Tully, R. B. 1982a, Ap. J., 258, 64.

Aaronson, M., Huchra, J., Mould, J., Tully, R. B., Fisher, J. R., van Woerden, H., Goss, W. M., Chamaraux, P., Mebold, U., Siegman, B., Berriman, G. and Persson, S. E. 1982b, Ap. J. Suppl. in press.

Aaronson, M., Mould, J., Huchra, J., Sullivan, W. T. III, Schommer, R. A., Bothun G. D. 1980, Ap. J., 239, 12. ($AHMS^2B$)

Boughn, S. P., Cheng, E. S., Wilkinson, D. T. 1981, Ap. J., 243, L113.

de Vaucouleurs, G. 1958, A. J., 63, 253.

de Vaucouleurs, G. 1972, in External Galaxies and Quasistellar Objects. IAU Symposium No. 44, ed. D. S. Evans (Dordrecht:Reidel), p353.

Hart, L. and Davies, R. D. 1982, Nature, 297, 191.

Hoffman, G. L., Olson, D. W. and Salpeter, E. E. 1980, Ap. J., 242, 861.

Peebles, P. J. E. 1976, Ap. J., 205, 328.

Rubin, V. C., Thonnard, N., Ford, W. K. and Roberts, M. S. 1976, Ap. J. 81, 719.

Schechter, P. L. 1980, A. J., 85, 801.
Smoot, G. F. and Lubin, P. M. 1979, Ap. J. (Letters), 234, 183.
Stewart, J. M. and Sciama, D. W. 1967, Nature, 216, 748.
Tonry, J. M. and Davis, M., Ap. J., 246, 680.
Tully, R. B. and Fisher, J. R. 1977, Astr. and Astrophys., 54, 661.
White, S. D. M. and Silk, J. 1979, Ap. J., 231, 1.
Yahil, A., Sandage, A. and Tammann, G. A., 1980 , Ap. J., 242, 448.
Yahil, A., Sandage, A. and Tammann, G. A. 1980 , in Cosmologie Physique, ed. R. Balian, J. Audouze and D. N. Schramm (Amsterdam:North-Holland), p127.

TABLE 1
Comparison with Selected Previous Results

Reference	W_x	W_y	W_z	W_i	W_r	W_z^{tot}	W_x^{tot}	remarks
Present	-106	-141	22	281	180	303	74	
work	±41	±47	±54	±63	±58	±39	±71	
Boughn	318	-341	411	≡0	≡0	411	318	quadrupole
et al.	±≈30	±≈30	±≈30			±≈30	±≈30	dipole; 3°K
Smoot &	178	-311	373	≡0	≡0	373	178	dipole fit
Lubin	±≈25	±≈25	±≈25			±≈25	±≈25	3°K
Yahil	≡0	≡0	≡0	230	≡0	230	≡0	V_{Virgo} = 979
et al.				±75		±75		nearby galaxies
Tonry &	≡0	≡0	≡0	440	≡0	440	≡0	V_{Virgo} = 979
Davis				±75		±75		ellipticals+S0
$AMHS^2B$	≡0	≡0	≡0	480	≡0	480	≡0	IR/H I
				±75		±75		cluster sample
de Vaucouleurs	≡0?	-250	≡0?	727	400	727	400	
(1972)		±50		±50	±50	±50	±50	
Stewart	≡0	≡0	≡0	-207	253	-207	253	
& Sciama			±230	±92				
de Vaucouleurs	-83	-106	153	≡0	≡0	153	-83	optical/H I
et al. 1981	±≈50	±≈50	±≈30			±≈30	±≈50	
Rubin	-367	-210	-138	≡0	≡0	-138	-367	ScI's
et al.								
Hoffman	≡0	≡0	≡0	250	≡0	250	≡0	2nd moment
et al.				±50		±50		
Hart &	105	-197	375	≡0	≡0	375	105	
Davies	±≈30	±≈30	±≈30			±≈30	±≈30	

Discussion

Occhionero: Does your evaluation of Ω_0 have any implication on the amount of dark matter?

Mould: I've talked about only the kinematics of the supercluster in this paper, without inquiring into the dynamics. But if one uses the linear model, for example, to determine the mass responsibl for the deceleration, one obtains a mass within the Local Group radius of 10^{14} to 10^{15} $M_\odot$, and, as the others have shown, a mass-to-light ratio of the order of 500.

Dressler: I'd like to ask Tully and Mould to clarify their conclusions about the random component, the noise, of the Hubble flow. Tully quoted a value of 50 km/sec, and Mould gave 150 km/sec, based on different analyses of similar, if not common, data. Is this difference significant?

Mould: The estimate of 150 km/s was a coarse value for the thermal noise in the Hubble flow required to produce observed systematic residuals in a very sensitive region of the supercluster. One would need a very precise distance indicator to measure the quantity Brent would like to know, namely, σ_y in the present coordinate system. The Local Group motion of 140 km/s towards the South Supergalactic Pole, seen not only in our kinematic model but also in other determinations, does give a hint about motions perpendicular to the plane, however.

Szalay: The value of Ω as inferred from the galaxy distribution is based upon two assumptions: 1) the mass distribution is spherical; 2) the galaxies represent the distribution of all matter.

It is likely that neither of these assumptions is exactly fulfilled. What deviations would you expect, were these effects taken into account?

Mould: The present work does indeed assume a spherical mass distribution in the spirit of a zeroth-order approximation to the true mass distribution in the supercluster. To test this assumption, it would be of interest to fit: 1) a flattened distribution, and 2) a clumpy or cloud model (see Tully, this meeting) to see if an improved description of the kinematics could be obtained.

Gott: We appear to be on the edge of the local supercluster.

Since a flattened disk of radius r and mass m produces a stronger acceleration at its edge than a sphere of radius r and mass m, a flattened model for the local supercluster should require somewhat less mass-to-light ratio to produce a given infall velocity. Thus, models including this refinement should lead to somewhat lower estimates of Ω than those that assume spherical symmetry.

Dekel: Dr. Gott's conclusion is valid at the post-collapse stages. If, however, we are not at the very edge of the Local Supercluster, the effect on the determined value of Ω is reversed during the one-dimensional collapse, as the density contrast interior to us grows faster due to incoming material while the dynamical effect of the flattening on the deceleration is still unimportant.

Segal: Although the motions of the Local Group estimated from CBR measurements on the one hand and by you and your colleagues by an entirely different technique on the other are in somewhat similar directions, the motion towards Virgo has the effect of impairing the fit of the theoretical (m, z) relation to, for example, the Visvanathan complete sample of E and S0 galaxies, while the former improves it. The analysis is based on an optimal nonparametric method for removal of the observational cutoff bias, and the comparative results are the same whether the Hubble or Lundmark law is used. Is your estimate of the motion of the Local Group towards Virgo at all dependent on the assumption that the galaxies in Virgo are approximately at its center and, if so, isn't this assumption fundamentally model-dependent?

Mould: In principle, we could add three further parameters to the present analysis and determine the inflow center. In practice, we have not attempted this. Davis and Huchra have, of course, considered the local "luminosity vector" from the CfA redshift survey, and we have been guided by this and the natural assumption that the mass distribution is like the light distribution.

In response to the first part of your question, I should point out that an infall field plus a local peculiar velocity is a significantly better fit to the present data than a peculiar velocity alone.

SUPERCLUSTERS AS NONDISSIPATIVE PANCAKES

Avishai Dekel
California Institute of Technology and Yale University

ABSTRACT: The formation of aspherical superclusters (SC's) is studied by 3-D N-body simulations that are confronted with the Local SC (LSC), and a simple model is developed. A nondissipative scenario, in which galaxies are formed from perturbations on smaller scales prior to the collapse of SC's, is found to be successful. It explains the disk-halo structure of SC's, their flattening and their low dispersions, which are nontransient because of the expansion along the long axes. The LSC has collapsed at $z \leq 0.5$. The large-scale velocity isotropy and the local 1-D infall indicate $\Omega < 1$. The correlation of galaxies on a few Mpc scales grows nonselfsimilarly due to recent pancaking rather than gradual clustering. This may be tested by the lack of clustering among objects at $z > 1$.

Observations of SC's have reached a stage where quantitative comparisons with formation scenarios may be attempted. The "western" picture suggests that structure that arises from small-scale peturbations (e.g. isothermal) evolves hierarchically from small to large, and that the clustering of galaxies is gradual and <u>dissipationless</u>. It is hard to see how do large aspherical SC's and holes form here. According to the "eastern" picture the structure evolves from large to small via fragmentation of SC's. Large-scale adiabatic perturbations that survive damping generate all structure, and the formation of asymmetric SC's and holes is an outcome of the large-scale velocity anisotropy (Zeldovich 1970). If the collapsing material is gaseous, <u>dissipation</u> produces thin <u>pancakes</u> (or cigars) that fragment to galaxies. This picture may allow for massive ($\sim$30eV) neutrinos which collapse nondissipatively and let the baryons collapse dissipatively in them. It is hard to explain here the presence of many galaxies ($\sim$40%) far from the LSC plane in which they are assumed to be born. Alternatively, consider a combination of the two scenarios, i.e., a <u>nondissipative pancake</u> scenario: if both small-scale and large-scale perturbations were present at recombination, galaxies or other substructures may already exist during the collapse of SC's so that they cross the SC planes (or lines) nondissipatively to form thicker pancakes, while galaxies that have not crossed the plane yet populate the SC halos. I summarize here a study of nondissipative pancakes and a confrontation with observations such as the flattening of SC's, the velocities in them and the clustering.

G. O. Abell and G. Chincarini (eds.), Early Evolution of the Universe and Its Present Structure, 249–254.

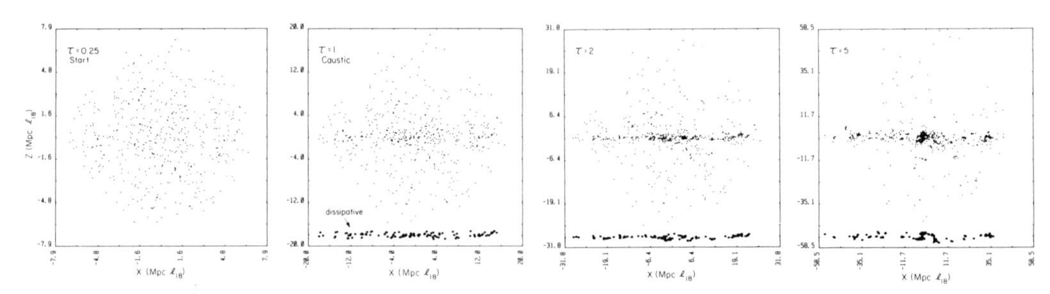

Fig. 1: Edge-on comoving snapshots.

The linear growth of nonspherical, adiabatic perturbations in a Friedman Universe is approximated by Zeldovich (1970): the position and the velocity at a time t of a particle at a comoving position $\vec{q}$ are

$$\vec{r} = a(t)[\vec{q} - b(t)\ \vec{\Psi}(\vec{q})]\ , \quad \dot{\vec{r}} = (\dot{a}/a)\ \vec{r} - a\dot{b}\ \vec{\Psi}(\vec{q}) \qquad (1)$$

where a(t) is the expansion factor, b(t) is the perturbation growth rate ($b \propto a \propto t^{2/3}$ when $\Omega=1$), and $\vec{\Psi}(\vec{q})$ is the spatial perturbation. It is likely that the perturbation is stronger in 1-D and is coherent over the critica damping length, ℓ. The velocity perturbation then leads to the formatio of a thin, dense pancake at a finite time (a=b=t=1, say) due to focusing of trajectories while the pancake is still expanding along the orthogona axes. Then the approximation breaks down. 500 particles were distribute at random in a sphere of a comoving radius ℓ, and with Hubble velocities. A 1-D adiabatic perturbation, with $\Psi_z=(\ell/\pi)\sin(\pi q_z/\ell)$ and $\Psi_x=\Psi_y<<\Psi_z$, was assumed to evolve according to eq. (1) until a=0.4. This was the start a simulation based on the Aarseth code (1972), which integrates the softened, Newtonian equations of motion. A sequence of projections of o case is shown in fig. 1 in Hubble expanding coordinates. The flattening indeed becomes apparent at focussing (when $\sim$20% of the particles have crossed the plane) but it is clearly not a transient feature! The relative flattening becomes even more pronounced later, while the absolut thickness grows. The sizes of rich clusters (that depend on the two-body interactions and on the small-scale perturbations) eventually reach the pancake thickness and affect the global flattenting (here at $t\sim5$).

The nondissipative flattening and cooling in Z is primarily due to expansion in XY. When a particle oscilates about the plane with a perio T that is smaller than the expansion time, there is an adiabatic invaria $\int_0^T \dot{Z}^2 dt \cong v^2 T \cong hv$, where h(t) and v(t) are some mean values of Z and $\dot{Z}$ at t. Out of an infinite, uniform disk, the gravitational field, $\mu(Z)$, is proportional to the surface density interior to $|Z|$. If μ is some mean value of $\mu(Z)$, $v \cong \mu T$, and therefore the thickness is expected to vary like $h \propto \mu^{-1/3}$. If r(t) is the radius of the pancake, $\mu \propto r^{-2}$, so that $h \propto v^{-1} \propto r^{2/3}$ i.e., when the pancake expands, its normal velocities are suppressed whi it thickens, but it becomes relatively flatter as $h/r \propto r^{-1/3}$ ($\propto a^{-1/3}$ if th expansion in the plane is unperturbed). One may consider two limiting cases for the matter that governs the oscilation: in one, let it all be in a thin disk at Z=0, and in the other let it be spread uniformly in Z. A simple model for the Z motions can be constructed where each particle

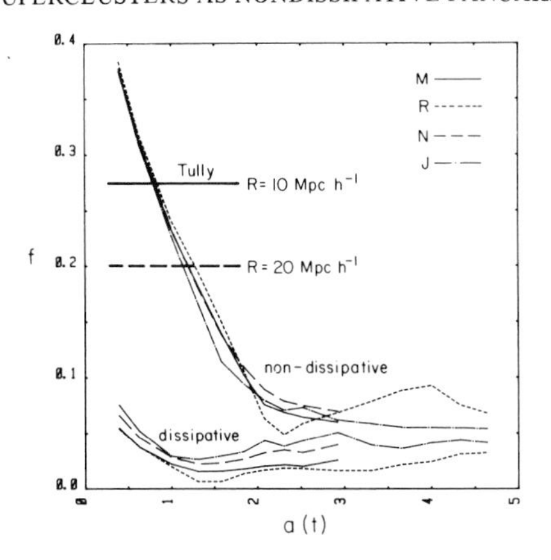

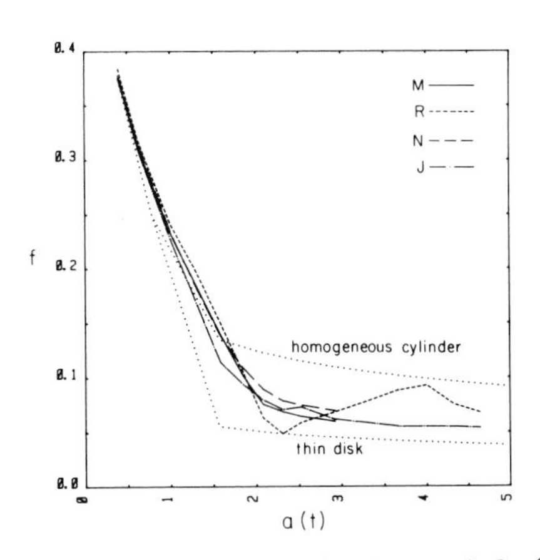

Fig. 2: Flattening in the simulations and in the analytic model (dotted).

obeys the Zeldovich approximation until it crosses the plane, and thereafter it oscilates about it subject to the adiabatic invariant in either of the above limits. The obtained limits on the flattening agree with the simulations (fig. 2b) and the model can serve as an approximation for other cases such as an open Universe (Dekel 1983a; Dekel and Aarseth 1983). The global spatial structure is found to be insensitive to Ωo.

The simulations illustrate the dissipative model when selecting pancake galaxies right after their formation (t=1), and following them evolve as N-bodies embedded in an N-body halo (neutrinos?). In fig. 1, the "dissipative" component is displaced from the center to the bottom ("*"). It remains flatter than the nondissipative pancake although it is more affected by local clustering. Note that the dissipation associated with galaxy formation is not determined by the theory and this example should be interpreted with caution.

The flattening of the LSC is estimated by Tully (1981) who counts galaxies in parallel layers of constant ΔZ, and by Yahil et al (1980) who count in bins of constant $\Delta\sin\beta$ (β is the Virgocentric latitude), both using redshifts as distance indicators. The N-body system is analyzed in both ways (Dekel 1983a). In order to compare to Tully's results we define a flattening by $f=Z(68\%)/R$, where $Z(68\%)$ is the normal width of the number count histogram. The evolution of f(a) in four of the simulations is shown in fig. 2a. The flattening of the nondissipative system resembles that of Tully near $a\sim1$ and becomes much flatter later. Neither local clustering (M vs. J) nor a slight decceleration in the plane (N), nor a rich cluster at the SC center (not shown) affect the flattening significantly. When correcting the numerical flattening according to Tully's estimate that the local thickness is roughly 2/3 of the global one, the qualitative conclusion remains the same with a preferred age $a\sim1.5$. The apparent flattening of external SC's agrees with the conclusion that a nondissipative scenario can account for the flattening of SC's in general.

The theory predicts three regions in the velocity field normal to the pancake plane, V_z: a) at large $|Z|$, the deviation from a Virgocentric

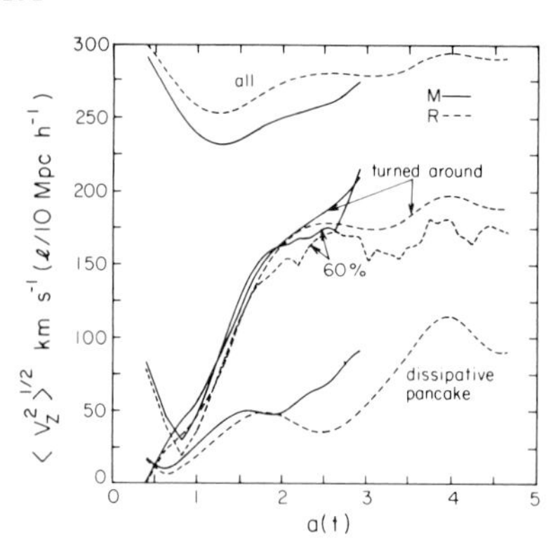

Fig. 3: Velocity dispersion

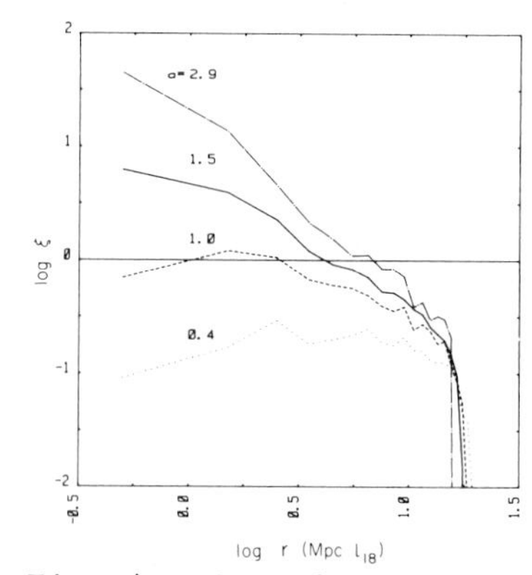

Fig. 4: Correlation function

flow is small, especially in a young SC or in an open Universe, b) interior to some $|Z|_t$, which is strongly dependent on Ωo, galaxies are falling back in, and c) in the pancake galaxies may oscilate about the plane and V_z may be virialized. Regions b and c may, in principle, provide important information about the formation of the pancake, its ag and on Ωo. The simulations (fig. 3 with $\Omega=1$) predict for the velocity dispersion in the 60% flat component $\sigma_z \sim 100$ km s^{-1} at $a\sim 1$, and $\sigma_z \sim 175$ km s^{-1} at $a \geq 2$, in agreement with the spatial flattening. Unfortunately, the observed velocities of pancake galaxies (low $|Z|$) are dominated by their fast expansion parallel to the XY plane, which makes the measure o V_z possible only for a few nearby galaxies at high $|SGB|$, and out of the Local Group. We find (Dekel 1983b), in a sample of $\sim$400 galaxies with redshift independent measured distances, no evidence for anisotropies in the Virgocentric flow at $|Z|$ larger than a few Mpc, and a weak evidence for an infall and excess V_z of both signs closer to the plane. This indicates an early stage in the 1-D collapse, in agreement with the conclusion from the flattening, and possibly an open Universe where the LSC halo is "frozen-in" the Hubble expansion.

The two-point correlation function within the simulated SC, on scal of a few Mpc, grows in a non-self-similar way due to the large-scale collapse to a pancake rather than due to local clustering (fig. 4). Fro simple geometry, a 1-D (2-D) collapse of a 3-D Poissonian distribution l to $\xi \propto r^{-1}$ (r^{-2}) without local clustering. $\xi(r) \propto r^{-1.5}$ as measured in the LSC (Rivolo and Yahil 1981, private comm.) is obtained at $a\sim 1.5$ (Dekel 1983c). Hence, the clustering of galaxies according to the nondissipati pancake scenario was still very weak at $z>1$, and is compatible with the lack of measurable clustering among the L_α absorption clouds along the lines of sight to quasars (Dekel 1982), and among the quasars themselves.

We conclude that the data is consistent with a nondissipative panca scenario which is a natural combination of the "eastern" pancakes and th "western" clustering, where both adiabatic and isothermal perturbations play a role in the formation of structure in the Universe.

REFERENCES

Aarseth, S.J. 1972, in Gravitational N-Body Problem, ed. M. Lecar (Dordrecht: Reidel), p. 373.
Dekel, A. 1982, Ap. J. L., 261, L13; 1983a, Ap. J., 264 (Jan. 15); 1983b, 1983c in preparation.
Dekel, A. and Aarseth, S.J. 1983, in preparation.
Tully, R.B. 1981, Ap. J., 255, 1.
Yahil, A., Sandage, A., and Tammann, G.A. 1980, Ap. J., 242, 448.
Zeldovich, Ya.B. 1970, Astr. Ap., 5, 84.

Discussion

Palmer: Once you allow structure on scales smaller than the damping scale of adiabatic perturbations, then a density profile focuses material into a core, which disrupts the thin pancake as it tries to form, unless the expansion velocities in the plane are large.

Dekel: An expansion in the plane is indeed required for the formation of a thin, long-lived pancake. Note further that the main reason for the rapid formation of a thin pancake is the large-scale, coherent velocity perturbation along the axis of collapse that causes the focusing of particle trajectories in the plane at a given time.

Bonometto: Your picture claims to involve an adiabatic fluctuation component, but a scenario starting from pure entropy fluctuations would present adiabatic fluctuations above M_J (Jeans mass before recombination).

If Ωh^2 is not too small, these fluctuations are large because of their collapse continuing after their entry in the horizon, while isothermal fluctuations with $M < M_J$ stay frozen.

I wonder whether the Dekel picture is not just the full picture we would expect in the primeval entropy fluctuation case. By this, I am not denying that it can also apply to a mixed adiabatic-isothermal scenario, I am only suggesting an alternative interpretation for it.

Dekel: Any truncated, adiabatic spectrum that provides coherent velocity perturbations on the scale of superclusters can do. My impression, however, is that the pre-recombination Jeans mass is too large.

Hoffman: In the simulations, how did you suppress density fluctuations on scales smaller than the damping scale in the initial conditions?

Dekel: Only the dominant wavelength was considered for the adiabatic perturbation. Perturbations on smaller scales were only Poissonian. This is still an idealistic, first approximation. Next, one should start from a generic spectrum of perturbations and investigate the structure of the individual superclusters that are formed (see the papers by Shandarin and by Efstathiou in this symposium, and an ongoing work by Aarseth and me simulating 10,000 particles and a truncated adiabatic spectrum).

Thompson: You ascribe the stability of a disk system in your N-body simulation to adiabatic cooling from the expansion parallel to the disk. Would you expect a similar cooling in a filamentary system, and do you have N-body simulations which show the stability of such filamentary structures?

Dekel: Yes. The estimates based on the adiabatic invariant are applicable to the case of a two-dimensional collapse as well. There the elongation is expected to become pronounced even faster like $R^{-2/3}$. An experiment that explores various triaxials is currently in progress (Dekel and Aarseth 1983).

IE MOTION OF THE LOCAL GROUP OF GALAXIES WITH RESPECT TO THE ACKGROUND OF GALAXIES

R.D. Davies
University of Manchester,
Nuffield Radio Astronomy Laboratories,
Jodrell Bank, Macclesfield, Cheshire, U.K.

A measurement of the motion of the Local Group of galaxies through he Universe provides an indication of their peculiar motion relative to he Hubble flow consequent upon the gravitational influence of the local arge scale mass inhomogeneities. This motion can be measured either elative to the cosmic microwave background at $z \sim 1000$ or relative to he background or nearby ($z \sim 0.01$) galaxies. The interpretation of ublished measurements is subject to some uncertainty. As an example, he Local Group motion derived from optical studies of nearby galaxies Rubin et al. 1976) differs from that derived from radio frequency easurements of the dipole anisotropy in the microwave background. Boughn et al. 1981, Gorenstein & Smoot 1981).

At Jodrell Bank, Hart & Davies (1982) have made a new determination f the Local Group motion relative to the background of galaxies which bviates many of the difficulties inherent in previous optical deter- inations. The 21 cm neutral hydrogen flux density integral of each alaxy, an indicator of the HI mass, is used as a standard candle. No orrection is required to this integral for inclination or galactic bsorption. The velocity width of the neutral hydrogen profile is used s a third parameter to indicate whether the galaxy is a giant or a warf; this is similar to the Tully-Fisher approach for optical or nfrared distance determination.

The basic data set used in our determination is the HI survey of bc (T=4 in the de Vaucouleurs morphological classification) galaxies y Davies & Johnson (1983). These galaxies were chosen so as to give good sky coverage down to Dec = -30°; they were supplemented by outhern galaxies in the HI survey by van Woerden et al. The velocity ange covered was 1000 to 5500 km s^{-1}. An independent data set was aken from the HI observations of Sc galaxies published by Rubin et al.

In the analysis we take the observed velocity of the galaxy, V_c, orrected to the centre of the Local Group, to be composed of the ollowing components

$$V_c = V_H + \Delta V_G + V_{pec}$$

. O. Abell and G. Chincarini (eds.), Early Evolution of the Universe and Its Present Structure, 255–258.

where V_H is the velocity it would have in the Universal Hubble flow, ΔV is the component of the Local Group motion in the direction of the gala and V_{pec} is the random velocity of the galaxy. We determine from a series of expressions of this form, the vector V_G giving the velocity o the Local Group relative to the backdrop of galaxies. An estimate of the true distance of each galaxy, and hence V_H, is provided by its HI flux density with a correction for its profile width.

Table I Observations of the motion of the Local Group of galaxies

Method	V_{LG} (km/s)	ℓ (°)	b (°)	L (°)	B (°)
HI fluxes: 78 Sbc galaxies	436±55	264±18	45±12	119±14	-29±12
HI fluxes: 53 Sc galaxies	580±62	245±18	35±9	111±13	-45±9
CMB: Boughn et al.	653±33	273±6	27±5	140±6	-35±5
CMB: Gorenstein & Smoot	567±60	256±9	41±7	117±8	-36±7
Average of above	546±70	261±9	39±7	122±9	-35±7

Table 1 gives the motion of the Local Group of galaxies derived from the HI and cosmic microwave background (CMB) measurements in galactic (ℓ,b) and supergalactic (L,B) coordinates. There is evident agreement between the CMB and galaxy backdrop results; the average gives the present best estimate of the Local Group motion. The disagreement with the Rubin et al. results is emphasized.

We draw the following conclusions from the data presented above.

1. The close agreement between the HI and the CMB results implies that the two types of measurement are the result of a common cause the local group motion.
2. The CMB dipole anisotropy is mainly extrinsic, $\lesssim$1 mK out of the observed 3.5 mK is likely to be intrinsic. We would expect the intrinsic quadrupole anisotropy to be <<1 mK.
3. The Local Group is moving within $\sim$30° of the direction of the centre of the Virgo cluster of galaxies; the component of infall is 450 ± 50 km s^{-1}.
4. If this infall were due to the gravitational influence of the Virgo cluster acting over the lifetime of the Universe, the local density relative to the closure density is Ω = 0.15 to 0.50.

REFERENCES

Boughn, S.P., Cheng, E.S. & Wilkinson, D.T., 1981. Astrophys.J.,243,L1
Davies, R.D. & Johnson,S.C., 1983. In preparation.
Gorenstein, M.V. & Smoot, G.F., 1981. Astrophys.J., 244, 351.
Hart, L. & Davies, R.D., 1982. Nature 297, 191.
Rubin, V.C., Thonnard, N., Ford, W.K. & Roberts, M.S., 1976. Astron.J., 81, 719.

Discussion

. *Jones:* What is the selection criterion for your samples and what causes the difference between your results and the old result f Rubin and collaborators?

avies: Our set of Sbc galaxies is composed essentially of the brightest objects in the Second Reference Catalogue. The ajority are of luminosity classes L = 1 to 4.

The reason for the difference between our HI result and the arlier Rubin et al. result is not entirely clear. It is most likely a arge-scale effect over the celestial sphere, for example, a variation f Zwicky magnitudes around the sky and/or an incorrect galactic absorp-ion correction.

anes: Could you remind us of the agreement between your study and that of de Vaucouleurs? Do these studies, which both pertain o moderately remote galaxies, agree on the existence of a (small) dis-repancy between the inferred motions and the cosmic microwave back-round? If so, what is the significance?

avies: The de Vaucouleurs et al. sample contains galaxies which are on average closer by ∿ 30% than those we have used. Thirty ercent of the de Vaucouleurs galaxies have a velocity less than 1000 m s^{-1} and ten percent are Virgo cluster members; all these galaxies ill be influenced by the Virgo-centric flow. Even so, the de Vaucouleurs ocal group motion is in a similar direction to our solution, although ignificantly smaller in magnitude.

The difference between our result and the CBR result could be ue to a small (∿ 1 mK) intrinsic component of the CBR or due to a otion of our volume of space (out to a redshift of 5000 km s^{-1}) relative o the comoving frame.

egal: We have used a nonparametric and statistically optimal tech-nique to estimate the motion of the Local Group from optical agnitudes and redshifts in the large Visvanathan E + SO sample, from he distortion such a motion would produce in the observed magnitude-edshift relation. Our cutoff-bias-removed least-squares estimate is argely independent of whether the Hubble or Lundmark law is used to upply the requisite magnitude-redshift regression, and is about 450 m s^{-1} in the direction L = 220 to 230; B = 10 to 20. An independent stimate using the same sample but based on the maximization of spatial omogeneity as measured by a V/V_m test with redshifts conservatively imited to 2250 km s^{-1} (and greater than 500 km s^{-1}), still inclusive of 200 galaxies brighter than the limiting magnitude of 12.4, also gives n estimated motion of similar magnitude almost in the galactic plane ut slightly below it. Are these estimates at least marginally consis-ent with yours? If not, do you have a suggestion regarding the dis-repancy? If they are, can you give any physical explanation for the pparent motion to be largely in the galactic plane? It seems interest-ng that if the component of this apparent motion in the plane is removed, the remaining motion is of the order of magnitude ($\lesssim$ 100 km s^{-} of the apparent peculiar velocities typical of other galaxies.

Davies: Your derivation of the Local Group velocity appears to be in the same general direction as our measurement. I believe that our results, based on samples extending to greater redshifts than your sample, show that the motion is directed significantly out of the galactic plane. This is also true of the microwave background dipole asymmetry.

Miller: The difference between your solution and the CBR solution (which presumably gives our velocity relative to comoving coordinates) could be interpreted as motion of the Virgo-centric group relative to comoving coordinates. Do you think this is reasonable?

Davies: Our solution for the Local Group motion is referred to a sphere of galaxies extending to a radius of 5000 km s^{-1} ($\sim$ 75 Mpc) around us. The difference of $\sim$ 100 km s^{-1} between our solution and the CBR solution might indicate the motion of this whole volum towards some nearby supercluster, if sufficiently massive.

NFALL OF GALAXIES INTO THE VIRGO CLUSTER

E. Shaya and R. Brent Tully
Institute for Astronomy, University of Hawaii

There are a subtantial number of galaxies in the Virgo southern xtention just beyond the Virgo Cluster which are blueshifted with re- pect to the cluster and a lesser number of galaxies slightly nearer han the Virgo Cluster which are redshifted with respect to the clus- er. These galaxies must be falling into the cluster. We are able to odel the observed infall motions by following the collapse of spher- cal shells with the Friedmann equations in a closed universe. A fam- ly of mass-age solutions provide satisfactory fits to the observa- ions.

There are two other constraints which confine our models. (a) rom a virial analysis there is an independent estimate of the mass of he Virgo Cluster. If the universe is 12-30 Gyr old then there is good greement between the mass known to exist in the central cluster and he mass required by the infall model. (b) We must also explain the bserved motion of our Galaxy with respect to the Virgo Cluster. If he age of the universe is 10-14 Gyr then the mass required to explain he infall of galaxies near to Virgo is sufficient to explain our bserved motion. There is an implication in this case that the ratio f mass to light must decrease outside the Virgo Cluster because no dditional mass is needed at large distances, at radii where 80% of the ight of the supercluster is found. If the age of the universe is less han 10 Gyr then mass is needed outside the galaxies near Virgo to xplain our motion. In this case, the mass-to-light ratio need not ecrease outside the Virgo Cluster. If the age of the universe is reater than 14 Gyr then there is already too much mass given by our nfall fits to the near Virgo data to explain the motion of our Galaxy. possible solution is to invoke the existence of a repulsive force uch as would be implied if the cosmological constant is positive.

We conclude that either (a) the universe is younger than 10 Gyr, r (b) the ratio of mass-to-light drops dramatically outside of the irgo Cluster, or (c) the cosmological constant is greater than zero.

G. O. Abell and G. Chincarini (eds.), Early Evolution of the Universe and Its Present Structure, 259–260.

There is independent evidence from the ages of globular clusters and nuclear cosmochronology against the first possibility.

The infall models permit us to calculate the present rate of galaxy accretion in the Virgo Cluster. Using distance estimators, each infalling galaxy can be located in space and its moment of arrival in the cluster can be determined. It is found that there will be a dramatic influx over the next 3 Gyr. The rate over the near future is sufficiently high that it is very plausible to suppose that all spiral and irregular galaxies now in the cluster have arrived in significantly less than a Hubble time. By contrast, the present influx of ellipticals and lenticulars is insignificant compared with the number of these types already in the cluster. Either infalling spirals are being converted into ellipticals and lenticulars or the cluster was once more extremely segregated by type than it is today. In accordance with this second possibility, we have developed a schema of galaxy formation which anticipates that first generation cluster members would be deficient in angular momentum content compared with field galaxies.

IAMETERS OF HI DISKS IN VIRGO CLUSTER - AND FIELD GALAXIES

Rein H. Warmels and Hugo van Woerden
Kapteyn Astronomical Institute,
University of Groningen
Groningen, the Netherlands

Environmental effects may be of great importance to the structure nd evolution of galaxies. Sullivan and collaborators (1981, AJ 86, 919) ind that the ratio between HI content and luminosity of galaxies tends o be smaller in regions of high galaxy density than elsewhere. It is ot quite clear what processes reduce the gas content (or enhance the uminosity!): collisions, mergers or tidal interactions of galaxies, ncounters with gas clouds, stripping by ram pressure? Detailed comparison of gas distributions and motions in cluster and field galaxies may elp answer this question.

With the Westerbork Synthesis Radio Telescope we have measured the I distributions in about 35 galaxies in the Virgo Cluster. Our observing time was limited to 2 × 2 hours, rather than 12 hours, per galaxy. ach 2-hour observation provides high (~30") angular resolution in one imension only. Thus we obtain for each galaxy the projections of its HI istribution on two different axes. From these we can derive angular iameters at a specific (face-on) surface density, σ_{HI}.

For 12 Virgo Cluster galaxies we have determined D_{HI}, the diameter at $\sigma_{HI} = 2 \times 10^{20}$ atoms/cm^2, and referred this to the face-on diameter D(0) from the Second Reference Catalogue. For the ratio $D_{HI}/D(0)$ we find an average of 1.33, with a 1σ scatter of 0.58. Taking for comparison 12 field objects with similar ranges of type and luminosity for which Bosma (1981, AJ 86, 1825, Table VI) lists diameters at $\sigma_{HI} = 1.8 \times 10^{20}$ atoms/cm^2, we find for $D_{HI}/D(0)$ an average of 2.07, with scatter 0.60.

Thus, in our 12 Virgo Cluster galaxies the HI diameters, if scaled to optical diameters, are on average ~35% smaller than in the field. We consider this result tentative, because a) the calibration of our HI surface densities is preliminary; b) the present set of 12 is only part of our final Virgo Cluster sample of ~35 galaxies, well spread in both radial velocity and distance from Cluster centre; c) our final comparison sample will be more accurately matched to the Cluster sample.

Warmels thanks ZWO for support on an ASTRON-project. The Westerbork Radio Observatory is operated by the Netherlands Foundation for Radio Astronomy, with financial support from ZWO. We are grateful to Seth Shostak for discussions.

G. O. Abell and G. Chincarini (eds.), Early Evolution of the Universe and Its Present Structure, 261.

EVIDENCE OF INTRINSIC CORRELATION BETWEEN THE LUMINOSITY AND THE VELOCITY DISPERSION IN GALAXY GROUPS

M. Mezzetti, G. Giuricin, and F. Mardirossian
Osservatorio Astronomico di Trieste, Trieste, Italy

ABSTRACT

By using the two samples of galaxy groups reselected by Rood and Dickel (1979, Astroph. J. 233, 418), we have investigated the significance of the correlations between the virial parameters of groups. Taking into account observational selection effects present in the two samples, together with statistical and observational uncertainties, we show that only an intrinsic correlation between the luminosity L and velocity dispersion V is necessarily required in order to explain all the observed correlations. There is no serious ground to state that the virial radius correlates with both velocity dispersion and luminosity. We also show that the observed slope of the Log L - Log V relation is considerably smaller than the true slope, while the contrary holds for the Log M/L - Log V relation.

An analysis similar to that described above, carried on the recent catalogue of nearby groups of galaxies prepared by Huchra and Geller (1982, Astroph. J. 257, 423),seems to reveal that also the intrinsic correlations between L and R, and R and V are significant for this new sample.

G. O. Abell and G. Chincarini (eds.), Early Evolution of the Universe and Its Present Structure, 263.

I. S. Shklovsky (left) during a scientific session. *(Courtesy, K. Brecher)*

STRUCTURE OF NEIGHBORING SUPERCLUSTERS: A QUANTITATIVE ANALYSIS

J. Einasto, A. Klypin and S. Shandarin
Tartu Astrophysical Observatory
Institute of Applied Mathematics, Moscow

So far the galaxy correlation analysis was the only quantitative method used to describe the distribution of galaxies in space. Here we consider other numerical methods to treat impersonally various aspects of the galaxy distribution.

1. CATALOGUES USED

Observational data are based on a compilation of all available redshifts by Dr. J. Huchra. Using Huchra's data, rectangular supergalactic coordinates and absolute magnitudes have been calculated for every galaxy, taking the recession velocity as a distance indicator. The relative velocity of galaxies in clusters has been reduced in order to remove the "god finger" effect. After reduction, the extent of clusters in radial and tangential direction is approximately the same.

All numerical studies have been made for galaxies within a cube of a certain size, L, and limiting absolute magnitude of galaxies, M_o. The basic observational catalogue was centered on the Virgo cluster, has cell size L = 4000 km/s in redshift space (80 Mpc for H = 50 km/s/Mpc, used in this paper), and limiting absolute magnitude $M_o = -19.5$. Other observational catalogues used have different cell sizes, limiting magnitude and center position X_o, Y_o, Z_o. Data on catalogues used are given in Table 1.

Several theoretical catalogues have been used for comparison based on the adiabatic scenario of galaxy formation, hierarchical model of galaxy distribution, and a random Poisson distribution. All simulated catalogues have approximately the same number of objects (see Table 1) and equal size, corresponding to 80 Mpc.

In the following, the catalogues are denoted as follows:
O = observed (basic variant O_2); A = adiabatic; H = hierarchical;
P = Poisson.

2. QUANTITATIVE ANALYSIS

2.1. Correlation function. Correlation analysis has been widely used by Peebles (1980). We also have applied this method to study the behavior of our catalogues. In all cases, three-dimensional initial data have been used. The results for basic catalogues are displayed in

G. O. Abell and G. Chincarini (eds.), Early Evolution of the Universe and Its Present Structure, 265–271.

Table 1

Catalogue	Name	X_o	Y_o	Z_o (Mpc)	M_o (Mpc)	L	N	ν	r_c	B (Mpc)	β	γ
Observed	O_1	0	24	0	-16.5	40	1009	1.5	2.4	0.9	0.24	0.61
	O_2	0	20	0	-19.5	80	866	1.4	4.7	0.8	0.43	0.42
	O_3	0	20	0	-21.0	80	191	1.5	12.0	2.7	0.23	0.67
	O_4	20	60	20	-21.0	160	924	1.5	12.3	1.8	0.19	0.71
	O_5	40	40	40	-21.0	200	1433	1.5	12.4	1.5	0.23	0.66
	O_6	60	60	60	-21.5	240	862	1.5	20.0	2.1	0.32	0.60
Adiabatic	A					80	753	1.5	4.8	0.7	0.00	0.51
Hierarch.	H					80	819	1.8	13:	15:	1.00	0.00
Poisson	P					80	850	0.0	7.5	3.0	0.00	0.00

Figure 1 and indicate that O, A and H catalogues are fairly similar in this respect. As expected, P catalogue has zero correlation. In Table 1 we give for catalogues studied the index ν of the correlation function:

$$\xi(r) = (r/r_o)^{-\nu}$$

2.2. Cluster analysis. New quantitative methods of the study of the galaxy distribution are based on cluster analysis. Similar methods are used in the percolation theory to study the electrical conductivity of semiconductors and in many other fields of physics (Shandarin 1982; Einasto *et al*. 1982).

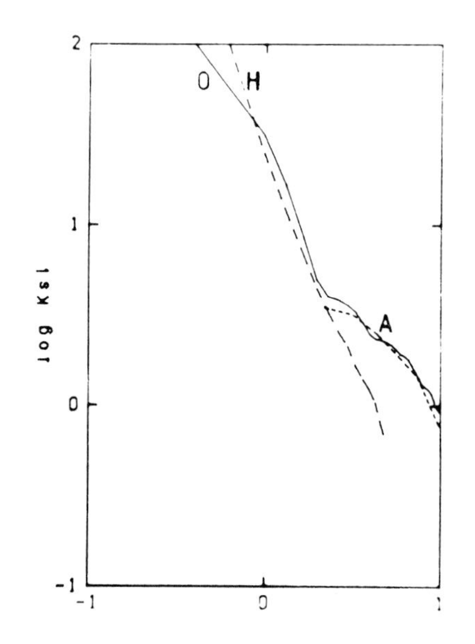

Figure 1. Correlation function for O, A and H catalogues.

In the cluster analysis the clustering tendency of test particles is studied as follows: Take a cube with size L and N test particles within the cube. The mean density of test bodies is $n = N/L^3$. Draw a sphere of radius r around each test particle. If within this sphere there are other test particles, all are considered as members of a connected system, called a "cluster" in the percolation theory. Thus, clusters consist either of single isolated test particles or of systems of test particles with each member having at least one neighbor with a distance r.

It is evident that the richness and size of clusters depend on the neighborhood radius r. For small r almost all clusters have only one member. With increasing r, the size and richness of the largest clusters increase rapidly. At some critical $r = r_c$, the size of the cluster is just sufficient to bridge two opposite sidewalls of the cube. This critical radius plays an important role in the percolation theory. Properties of clusters depend on the mean number of test particles in the sphere of radius r_c:

$$B = \frac{4\pi}{3} N(r_c/L)^3 .$$

B is a stochastic variable and changes from one kind of distribution to another. For a random distribution of test particles, the mean value is $B_P = 2.7$.

It is evident that for a random but clumpy distribution of particles the radius r_c depends not on the total number of particles but on the number of clumps. For this reason in the hierarchical case $B_H > B_P$. These expectations have been confirmed by our calculations (Zeldovich, Einasto, Shandarin 1982; Shandarin 1982; Einasto et al. 1982); see Table 1. On the other hand, if test particles are evenly spaced along strings, the parameter B can be as low as $B_S = 10^{-3}$.

In a real clumpy case, the parameter should lie somewhere between B_S and B_H, as confirmed by our calculations. For catalogues less influenced by selection effects (O_1 and O_2), B is about 15 times smaller than the respective parameter for the random hierarchical catalogue with a similar clustering parameter (H). Other observed catalogues have B that is smaller by 5 to 10 times. This difference between the observed B values in the different observed catalogues may be due to various selection effects (for example, some galaxies may be in strings connecting superclusters but are not observed). Another possibility is that in galaxy strings galaxies have lower luminosity (Einasto, Jôeveer, Saar 1980). In any case, the parameter B is much smaller than for the random hierarchical case. This indicates the presence of galaxy strings which connect all superclusters to a single network.

2.3. Cluster multiplicity. The correlation function and the percolation parameter B say little about the distribution of galaxies according to the multiplicity of systems. Thus, cluster multiplicity is an additional factor to be included in the quantitative analysis.

Let us combine galaxies into clusters using the neighborhood method outlined above. The distribution of galaxies according to cluster multiplicity depends on the neighborhood radius r. At small r most clusters are single, whereas at very large r almost all galaxies join to single huge cluster, so we do not expect large differences between various catalogues at very small and very large neighborhood radii. The differences are largest at radii where catalogues with strong tendencies toward string-like structures reach conductivity, that is, for $r = r_c$. Respective histograms for all basic catalogues under study are given in Figure 2.

Clusters can be divided into three types: poor, medium, and rich. Poor clusters are represented by the multiplicity histogram for the random P catalogue, which peaks at singles and has no clusters with multiplicity greater than $2^4 = 16$. Medium-size clusters are represented by the multiplicity histogram for the hierarchical H catalogue, which peaks at multiplicity $2^3 - 2^4$ and has practically no singles and no rich systems. Rich systems have $2^7 = 128$ or more members. Such clusters are present in all observed and in the adiabatic catalogues.

Let us denote the fractions of galaxies in poor, medium and rich systems by α, β, and γ, respectively. By definition:

$$\alpha + \beta + \gamma = 1.$$

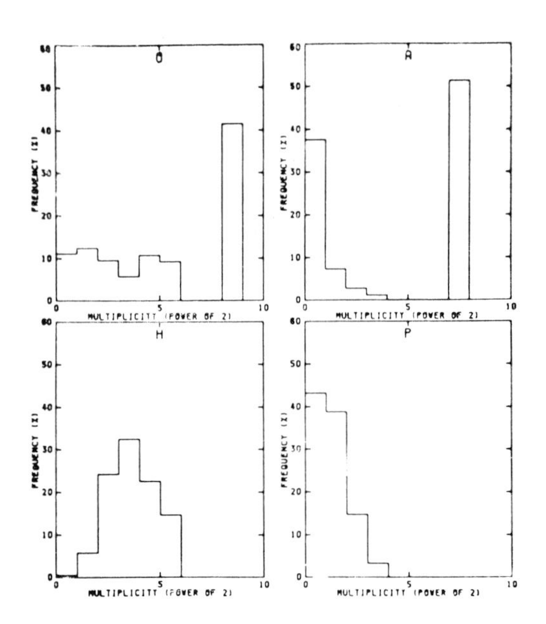

Figure 2. Distribution of galaxies according to the cluster multiplicity for neighborhood radius R = 5 Mpc.

Thus, the multiplicity distribution can be described by two independent parameters, say β and γ. Respective values are given for all catalogues in Table 1.

2.4. Shape and orientation of clusters. It is well known that clusters of galaxies are not spherical but triaxial. Statistically, the cluster shapes and orientations can be described by cluster axial ratios and by the deviation of the cluster's major axis from the direction towards nearby external clusters. It is well known (Jôeveer, Einasto, Tago 1978; Binggeli 1982) that rich clusters have a tendency to point their major axes toward neighboring clusters.

Preliminary results of the study of the shapes of clusters demonstrate the absence of large-scale sheets of galaxies. Around the Virgo cluster a small sheet has been observed (Zeldovich, Einasto, Shandarin 1982). It is too early to say whether this sheet can be identified with a Zeldovich pancake. The basic structural element seems to be a chain of galaxies or clusters of galaxies.

2.5. Morphology of galaxies in clusters. All properties studied so far represent different aspects of the geometry of galaxy distribution. It is well known (Einasto, Jôeveer, Saar 1980) that at different parts of the supercluster, galaxies have different mean absolute magnitudes, morphological types and other morphological properties. These properties play an important role for the theory of galaxy formation and evolution. However, it is difficult to describe them concisely in quantitative terms. No galaxy formation scenario is detailed enough to predict the behavior of these properties. Thus, in numerical simulations, morphological properties are neglected in most cases. Further observational and theoretical study of the morphology of galaxies in various parts of superclusters is badly needed.

3. MEAN DENSITIES

Data available allow us to derive the mean luminosity density in various parts of the universe.

As a measure of the luminosity density, we use the number of bright galaxies ($M \leq -21.0$). The total luminosity of a galaxy system can be expressed as follows:

$$L = n^* \times 3.6 \times 10^{11} L_o ,$$

where n* is the number of bright galaxies in the respective systems (Jôeveer, Einasto, Tago 1978). Here we assume that the luminosity function is the same everywhere. This is, of course, not the case, but for rough estimates it gives a good approximation.

The average luminosity density in the region bounded by X, Y = ± 5000 km/s, Z = ± 7500 km/s in redshift space is $5 \times 10^7 L_o/Mpc^3$. This may be a slight underestimate due to the exclusion of several rich clusters from the region. The density in superclusters exceeds the mean density by a factor of 3, and the density far from superclusters is about three times lower. Within galaxy strings and cluster chains, the density is 70 to 100 times higher than the mean density (Einasto et al., 1982).

In all observed catalogues the relative number of poor clusters is $\alpha \approx 0.10$. A study of the distribution of these isolated galaxies shows that at least 2/3 of them are outlying members of other systems. So we conclude that, if a field population exists at all, it cannot contain more than approximately three percent of all galaxies. Thus, the mean density of luminous matter outside galaxy systems is lower than the mean density by a factor of at least 30, and the density contrast between voids and galaxy systems is more than three orders of magnitudes (Einasto et al. 1982).

Near superclusters, galaxy systems fill about ten percent of the total volume, and in the whole region under study, only one percent. Thus, the space outside superclusters is very empty indeed (Zeldovich, Einasto, Shandarin 1982).

4. DISCUSSION

As we have seen, catalogues of different kinds behave comletely differently. All observed catalogues have similar properties. This indicates that statistical parameters found are stable and do not depend on the individual peculiarities of particular regions.

The hierarchical catalogue has only one common property with observed catalogues, the correlation function. All other parameters differ completely from observations. The present hierarchical catalogue was constructed artificially, following prescripts of Soneira and Peebles (1978). It would be interesting to compare the observed distribution with results of numerical simulations, which also involve dynamical evolution. This study is under way. If numerical simulations are close to the hierarchical catalogue studied above, then we come to the inevitable conclusion: the simple hierarchical clustering scenario contradicts observations.

The adiabatic catalogue has many properties in common with observations. However, in two points important differences are present. About one-half of all test particles of the A catalogue have multiplicity distribution like the random P catalogue. These particles can be considered a field population. The spatial density of particles of this population is lower than the average density. The observed catalogue has no appreciable field populations. This difference can be explained as follows: In regions of lower-than-average matter density, no galaxy formation takes place. The matter in low density regions remains in some pre-galactic form.

The second difference lies in the fact that the adiabatic catalogue has no systems of medium size. On the other hand, the observed catalogue has about one-half of all galaxies in systems of intermediate size. Medium-size systems are characteristic of the fine structure of superclusters. The presence of fine structure is an essential property of superclusters and should be incorporated in scenarios of galaxy formation.

Summarizing the results of the quantitative analysis of the galaxy clustering, we come to the following conclusions:

1) The basic structural element in the Universe, larger than clusters of galaxies, is a string of galaxies and clusters of galaxies;

2) No large-scale sheets of galaxies have been found so far. Small-scale sheets surround some clusters;

3) Galaxy strings connect all superclusters to a single intertwined lattice;

4) Superclusters consist of both large and medium-size strings;

5) A field population of galaxies, if it exists, contains at most three percent of all galaxies;

6) The luminosity density contrast between voids and galaxy strings exceeds three orders of magnitude;

7) Galaxy strings fill about one percent of space in the Universe; the rest is void of galaxies;

8) No theoretical scenario proposed so far explains all observed clustering properties.

REFERENCES

Binggeli, B. 1982, Astron. Astrophys. (in press).
Einasto, J., Jôeveer, M., Saar, E. 1980, M.N.R.A.S., 193, 353.
Einasto, J., Klypin, A., Saar, E., Shandarin, S. 1982 (in preparation).
Jôeveer, M., Einasto, J., and Tago, E. 1978, M.N.R.A.S., 185, 357.
Peebles, P.J.E. 1980, The Large-Scale Structure of the Universe, Princeton University Press.
Shandarin, S.F. 1982, Pis'ma v AZh (in press).
Soneira, R.M., and Peebles, P.J.E. 1978, Astron. J., 83, 845.
Zeldovich, Ya.B., Einasto, J., and Shandarin, S.F. 1982 (preprint).

Discussion

Scott: What are the observational catalogues you are referring to: 0_1, 0_2, 0_3, ... ? The results you show are very different from the distributions obtained from known catalogues by others. For example, your distribution of multiplicity is very different from that obtained from the Lick survey by Neyman, Shane and Scott (and by others), namely, a distribution continually decreasing. I know of no catalogue that will give a high probability of huge multiplicity except, of course, a catalogue of clusters. To use such would not make sense.

Einasto: All catalogues are subsamples of Huchra's (CfA) compilation. They differ in cube size, center position and absolute magnitude limit.

The reason the multiplicity function differs from those derived earlier is because a completely different method has been used to define a multiple system. In our case, the method used in the percolation theory has been applied. Rich aggregates found by us are clusters plus string systems at the percolation radius (i.e., at radius where the system joins two opposite cube sides).

Szalay: Is the Coma/A1367 complex now considered as a string?

Einasto: The Coma-A1367 supercluster consists of a number of strings. The string connecting the Coma and A1367 clusters is the strongest.

Djorgovski: Is it true that the strings intersect? If so, what is the fraction of galaxies in the knots (intersections), and what is between them?

Einasto: Yes, strings do intersect. The fraction of galaxies in knots is not yet determined. As a very crude estimate, we can take the fraction of galaxies in clusters. In any case, the exact definition of both clusters and strings is difficult due to the continuous character of the structure.

THE PERSEUS SUPERCLUSTER

RICCARDO GIOVANELLI
NAIC, Arecibo Observatory

1. INTRODUCTION

In analyzing the distribution of galaxies of a sample projected on the plane of the sky, the magnitude of a surface density enhancement produced by a clumpy structure depends on the size and magnitude of the volume density enhancement, and the depth of the sample. If the sample is too deep, or the line of sight size or volume overdensity of the clump too small, the surface enhancement may be too shallow to discern against the fore- and background objects. The Catalogue of Galaxies and Clusters of Galaxies (CGCG: Zwicky et al. 1960-68) provides a representative sample of the local universe ($cz \lesssim 15000$ km s^{-1}) and, in hindsight, possibly the one available that best enhances the inhomogeneities that appear to characterize the large scale structure of the universe. Using maps of the surface density distribution of galaxies from the CGCG, of which figure 1 is an example, Martha Haynes, Guido Chincarini and I have selected a number of filamentary structures discernable and undertaken a 21 cm redshift survey of large regions enclosing them, with the telescopes of 305 m at Arecibo and 92 m at NRAO-Green Bank. Here I shall discuss our current results from a large area extending from Pegasus to Ursa Major, which engulfs the well known Perseus supercluster (Einasto et al. 1980; Gregory et al. 1981).

2. THE SAMPLE

The Arecibo sample includes all galaxies of morphology later than S0 and angular size larger than 1′, between 22^h and 4^h in RA, 3° and 38° in Dec. In a more restricted region, we have also included spirals smaller than 1′ and brighter than m=15.7. Our partial results include approximately 1100 21 cm redshifts which, integrated with optical data from various sources, mainly the Rood and cfA catalogs, contribute to a sample of 1435 redshifts in the region mentioned. The sky distribution of galaxies in the sample is shown in figure 2. Notice that the densest part of the Perseus filament (cf figure 1),

G. O. Abell and G. Chincarini (eds.), Early Evolution of the Universe and Its Present Structure, 273–280.

located North of Dec=38°, is not yet part of our sample. The Green Bank observations discussed here refer to the region of Lynx-Ursa Major, between 6.5^h and 11.5^h in RA, 40° to 65° in Dec, and to a large section of the zone of avoidance bridging the gap between Perseus and Lynx; this sample is described in detail by Giovanelli and Haynes (1982). Our samples are incomplete in a variety of ways. Corrections for the biases introduced by the incompleteness have been applied when possible; such biases do not basically affect the conclusions that will be presented here.

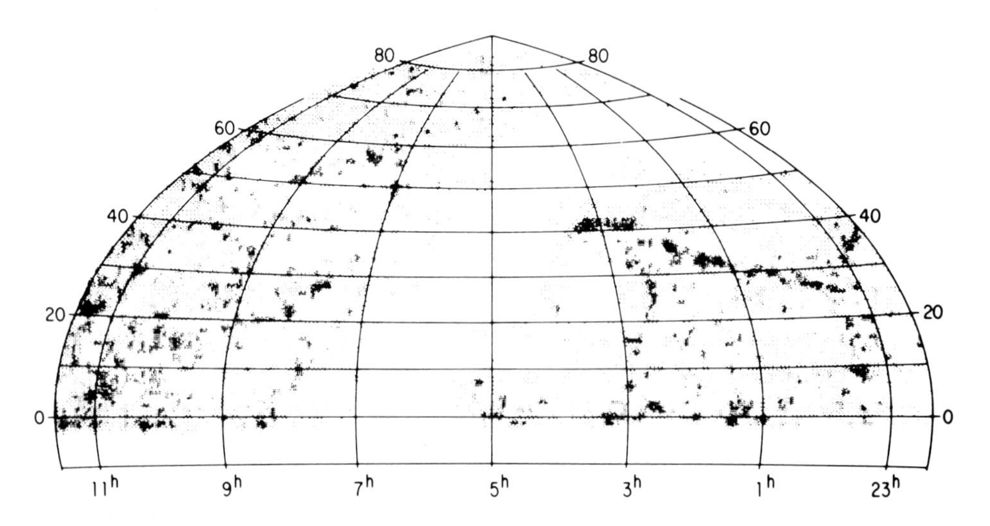

Figure 1. Surface density distribution of galaxies from the CGCG.

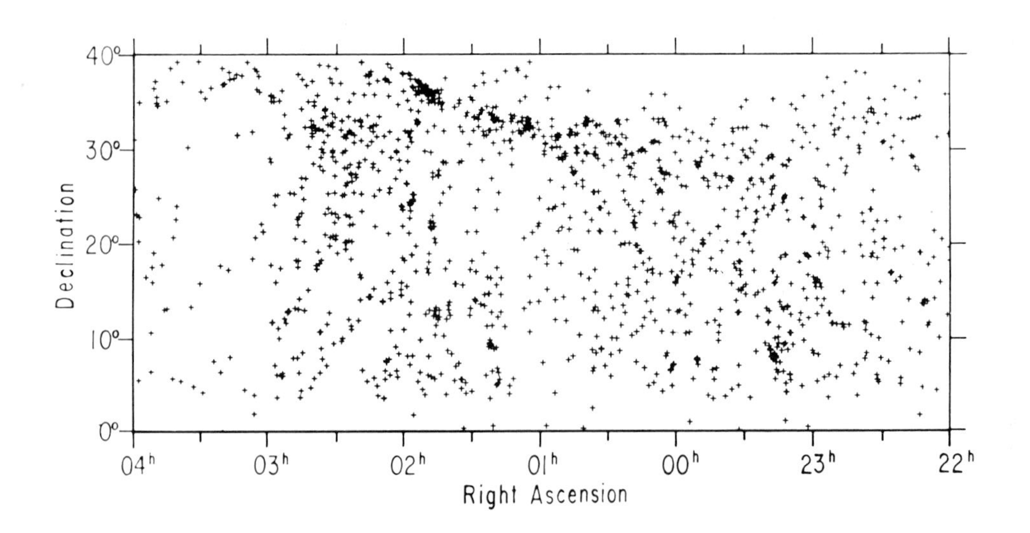

Figure 2. Distribution of redshifts of the Arecibo sample.

3. THE REGION FROM PEGASUS TO PERSEUS

The histogram in figure 3 illustrates the redshift distribution of galaxies in the Arecibo sample, while the distribution to be expected if the volume density of galaxies were uniform along the line of sight is given by the smooth curve. While the formidable enhancement near 5000 km s^{-1} represents an average over a solid angle which exceeds 2000 square degrees, the detailed structure of the supercluster appears as a maze of thin filamentary structures, which maintain a high degree of spatial coherence; sometimes they spatially merge, sometimes they merely project across each other while remaining separate along the line of sight. Unbroken, unsplit filamentary segments can extend for several tens of Mpc along one dimension (we assume H_o=50 throughout), while they are very thin in the other two, with axial ratios usually larger than 10. The table shows the characteristic parameters of filamentary segments in the region surveyed at Arecibo, obtained from numerous such structures.

Surface density contrast	......	up to 10
Volume density contrast	......	50 to 100
Length	......	45 to 90 Mpc
Width	......	3 to 8 Mpc
	......	200 to 600 km s^{-1}
Axial ratio	......	> 10
Mass	......	10^{16-17} Ω M_o

Volume densities are estimated from the observed surface density enhancement and the assumption that they are well described by a Schechter luminosity function. Masses are estimated from the derived volume overdensity and an assumed average density of matter in the universe of Ω times the critical mass. Two examples of the velocity structure are seen in figure 4. A cone diagram of galaxies within ± 3° in Dec from

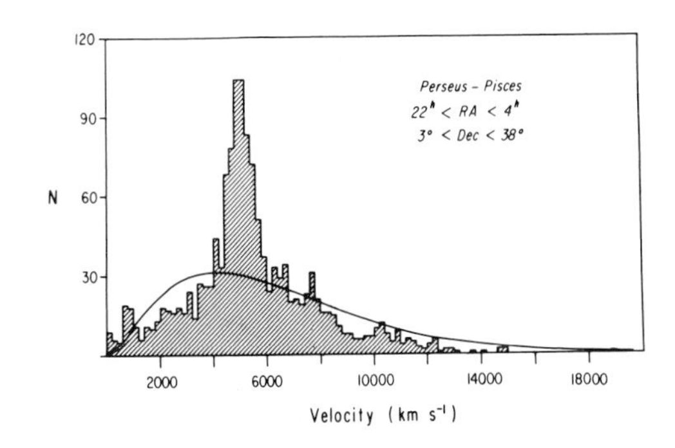

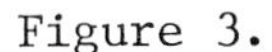
Figure 3.

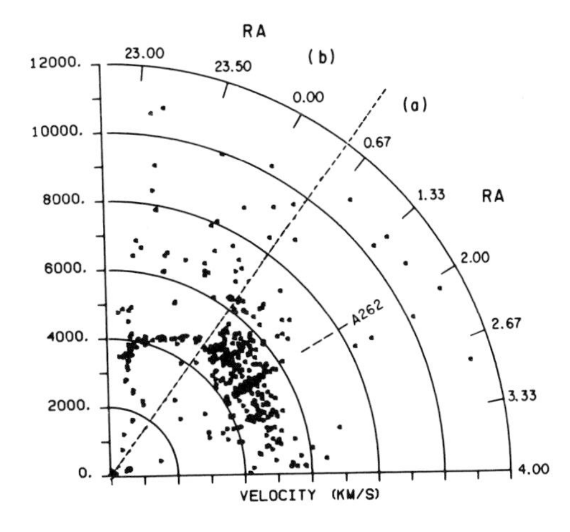

Figure 4.

$$\text{Dec} = 27.7 + 4.14\ \text{RA} \qquad 0.46 < \text{RA} < 2.00$$
$$\text{Dec} = 27.5 + 2.50\ \text{RA} \qquad 2.00 < \text{RA} < 4.00\ ,$$

the main filament, is shown in figure 4a. Notice the enhaced velocity dispersion around the dense clusters in the filament. Figure 4b presents the velocity structure of a filament running from the Pegasus cluster, near 23.3^h, 8°, to the main filament at 0.5^h, 30°. The Pegasus cluster is identified by the large velocity dispersion

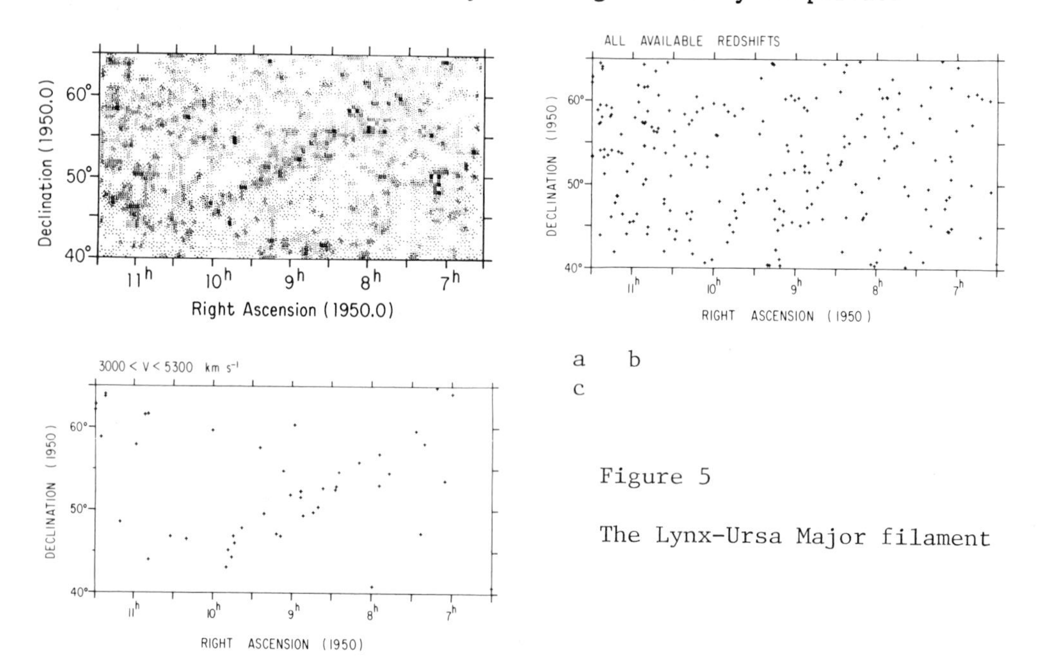

a b
c

Figure 5

The Lynx-Ursa Major filament

structure near 3500 km s^{-1}; a very narrow filament joins the cluster and the main filament shown in figure 4a. Figure 4b includes galaxies within a 4° wide band.

4. THE REGION FROM PERSEUS TO LYNX-URSA MAJOR

Inspection of figure 1 shows the main filament of the supercluster merging into the zone of avoidance near the Perseus cluster. A filamentary structure is discernible on the other side of the galactic plane, and is shown in better detail in figure 5a, as a shade plot similar to figure 1. Giovanelli and Haynes (1982) have shown the filamentary structure to be associated with a density enhancement located between 3500 and 5300 km s^{-1}, as illustrated in figures 5b and 5c. The question naturally arises of whether the Perseus and Lynx regions are connected across the zone of avoidance. Figure 6 shows the distribution of velocities in the region between the two filaments, as a histogram on which the expected redshift for a homogeneously distributed sample is superimposed as a smooth line (similarly to figure 3). A significant excess is present near 5000 km s^{-1}, confirming the suspicion of a connection, which corroborates previous suggestions put forth by Burns and Owen (1979).

5. APPLICATION TO THE DETERMINATION OF H_o

A 21 cm survey provides important additional information besides redshifts. A prominent one is the collection of line-widths which can be used, as first proposed by Tully and Fisher (1977), to determine the Hubble constant. In most application of this method, calibrators of the relationship between line width and luminosity are nearby galaxies, usually inhabitants of low density regions, while H_o has been preferentially determined using samples belonging to clusters of galaxies. The application of the method relies on the assumption that the ratio between line width (related to total mass) and luminosity is an environment independent quantity. We have investigated this question using subsamples of supercluster galaxies ($4000 < cz < 6000$ km s^{-1}) of various morphological types (in order to single out the effects of morphological segregation in the supercluster), analyzing the dependence of the total mass luminosity ratio as a function of local galaxian density. The total mass was determined from the velocity width and the customary assumption for the shape of the rotation curve (assumed flat) using a Brandt formula (cf Roberts 1975); the luminosities are from the CGCG, corrected for reddening and for the irregularities discussed by Bothun and Schommer (1982). The Local galaxian density is defined as

$$\rho = \Sigma_i L_i \exp(-0.5(r_i/\sigma)^2)$$

where L_i, r_i are the luminosity and projected distance (on the plane of the sky) of the i-th neighbor found in the CGCG, assumed at the same redshift as the sample galaxy, and $\sigma = 0.5$ Mpc. The summation is carried on over all neighbors within 4 Mpc. Figure 7 shows the behavior of the total mass to luminosity ratio as a function of

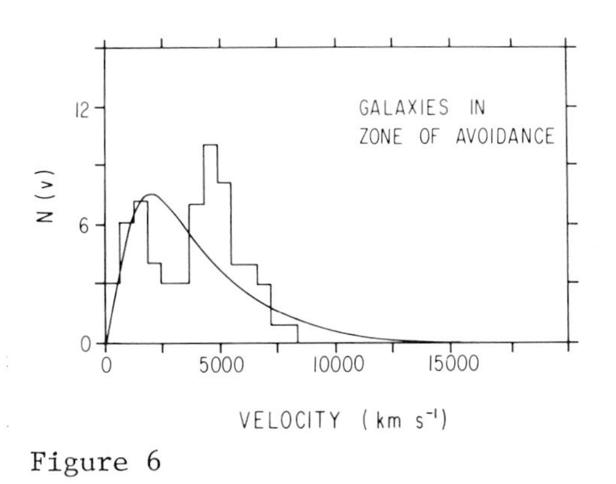

Figure 6

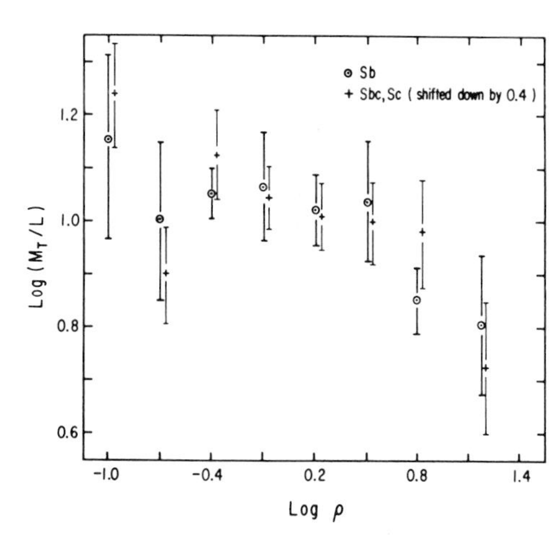

Figure 7

ρ separately for two samples of Sb and Sbc, Sc galaxies. In both cases a trend is discernible for galaxies in higher density environments to have lower mass to luminosity ratios than those in low density environments. The difference, on the order of a factor of two, could produce an overestimate of the value of H_o derived from cluster galaxies by about $\sqrt{2}$, unless it results from a bias in Zwicky magnitudes or UGC morphological types which differentially affect regions of different density.

6. CONCLUSIONS

The distribution of galaxies between Pegasus and Ursa Major, analyzed mainly on the basis of 21 cm redshifts, resembles a network pattern of thin filamentary segments. The segments, typically extending several tens of Mpc, represent density enhancements of 30 to 100 times the average volume density, and have masses of $10^{16-17}\,\Omega\,M_o$. The network pattern can be followed uninterruptedly for at least several hundreds of Mpc, as indicated by the suggestive connection between the Perseus and Lynx regions across the zone of avoidance. The large 21 cm sample of supercluster galaxies enables us to analyze the environmental dependence of integral properties of galaxies, suggesting that the Hubble constant as derived previously via the Tully-Fisher method may have been overestimated, if the difference in the mass to luminosity ratio between high and low density regions is not the effect of so far unknown observational biases.

The Arecibo Observatory is part of the National Astronomy and Ionosphere Center, operated by Cornell University under contract with the National Science Foundation. The National Radio Astronomy Observatory is operated by Associated Universities, Inc., under contract with the National Science Foundation.

REFERENCES

Bothun, G.D. and Schommer, R.A. 1982, Ap.J.(Letters) 255, L23.
Burns, J.O., and Owen, F.N. 1979, A.J. 84, 1478
Einasto, J., Joeveer, M. and Saar, E. 1980, M.N.R.A.S. 193, 353.
Giovanelli, R. and Haynes, M.P. 1982, A.J., in press.
Gregory, S.A., Thompson, L.A. and Tifft, W.G. 1981, Ap.J. 243, 411.
Roberts, M.S. 1975 in Galaxies and the Universe, U. of Chicago Press.
Tully, R.B. and Fisher, J.R. 1977, A.A. 54, 661.

Discussion

Szalay: What was the method by which the masses were determined of the objects plotted on Figure 7?

Giovanelli: They were determined in the traditional way from the 21-cm line width, corrected for inclination, and assuming that the rotation curve is flat and that the maximum rotational velocity is reached at a fixed fraction of the optical diameter.

Szalay: As far as I know, there are no data on whether spiral galaxies in rich environments also have flat rotational curves. Is that true?

Giovanelli: I would agree. I am not aware of any systematic mapping of rotation curves in the denser parts of clusters.

Thompson: If Dr. Dekel is correct and filamentary structures are stabilized by streaming motions along the filaments, then any filament oriented with its major axis out of the plane of the sky should show evidence for such streaming motions. For the filaments which appear in your study, can you test Dr. Dekel's hypothesis?

Giovanelli: The clearly discernible filament with the largest angle to the plane of the sky is located between 10 and 30 degrees declination, 2.4 and 3 hours right ascension. It exhibits a gradual velocity change of about 3000 km s^{-1} from one extreme to the other, or an inclination of about 45 degrees to the plane of the sky. It could provide an ideal case to test Dr. Dekel's hypothesis. At present, however, the sampling of that filament is rather coarse, and it is premature to say whether the relatively large redshift width is an effect of streaming or just of poor sampling associated with the large velocity gradient due to differential Hubble flow. Fifty more redshifts in the filament could help to answer less ambiguously.

Aaronson: I just wanted to repeat a comment I made in Patras regarding your last transparency (Figure 7). We do not see any evidence for environmental effects on the IR Tully-Fisher method. In particular, we now have a sample of ten distant clusters ranging from high-density, spiral-poor objects like Coma to low-density, spiral-rich objects like Pisces. All these yield a similar Hubble ratio. Furthermore, we also have a sample of distant field Sc galaxies drawn from the studies of Sandage-Tammann and Rubin et al. These yield a Hubble ratio which agrees with the cluster data. We generally use a circle of only 3° radius for the clusters to that a large range in density contrast does exist between the cluster and field samples.

Giovanelli: I wish to underscore again the caution with which I am mentioning the result. It is possible that the effect is milder or disappears if one uses infrared magnitudes; we don't have them for our sample, yet. As for the difference in local density between galaxies in clusters like Coma and Pisces, I think that the detected galaxies at 21 cm may be less representative of the difference in density among clusters than you imply. As you know, spirals in cores of clusters are usually very gas-poor and not chosen for Tully-Fisher samples because of the difficulty in measuring a 21-cm line width. It is likely that the Coma spirals used for FT studies are in lower density regions than cluster cores, even if projected onto them; similarly in other dense clusters. Hence, detected cluster spirals used in your samples are likely to inhabit regions that bridge a very narrow dynamic range in local densities; thus, the similarity of the inferred results is not inconsistent with the point made in this paper. The results for your field Sc sample, on the other hand, appear to indicate disagreement

Scott: Can you give some details as to how your filaments were determined? As seen by me sitting here, the dog had a lot more legs.

Giovanelli: The redshift information helps to disentangle surface density features.

Tarenghi: How was your sample chosen, what is your detection rate and what is the velocity range of search for unknown redshifts?

Giovanelli: We observed all spirals of type Sa or later, with optical sizes greater than one arcminute (UGC objects), except in a few fields where we have observed smaller Zwicky galaxies. Our overall detection rate is of about 80 percent. Our range of search is between zero and about 14,000 km s^{-1}.

THE URSA MAJOR SUPERCLUSTER

N.J. Schuch
Observatório Nacional, CNPa/ON, Rio de Janeiro, Brasil

The Ursa Major supercluster is an association of ten Abell clusters which lie within 7° of each other. The center of the system has equatorial coordinates RA $11^h\ 39^m$, Dec 54°5 (1950.0) and galactic coordinates $\ell = 143°$, $b = 60°$.

The observational work consists of an optical and a radio survey. The optical observations consist of a spectroscopic survey in which redshift data for cluster galaxies and optical identifications of radio sources were obtained with the 98-inch Isaac Newton telescope at the Royal Greenwich Observatory, and the 200-inch Hale telescope; the photographic survey in B, V and R colors was made with the 48-inch Schmidt telescope at Palomar. The reduction of these plates is under way with the use of a computer-controlled PDS which has just become operational at CNPq-ON. For the analysis of the two-dimensional galaxy distribution counts of the new statistically-corrected Shane and Wirtanen Catalogue, plate numbers 1066 and 1067 were used.

The main conclusions of these surveys may be summarized as follows: the two-dimensional distribution of galaxies over the supercluster region is very complex; the rich clusters of galaxies are highly asymmetrical, the cluster A1377 possibly being the most massive cluster in the system. There are many groups of galaxies scattered between the rich clusters of galaxies and in many of the regions between the rich clusters and groups of galaxies. The surface density of galaxies with $m_{pg} < 19$ is higher (~ 2) than the average background. Peebles' model to describe the overall distribution of galaxies in the universe is compatible with the analysis of the counts of the galaxies. The observed redshifts segregate into at least four distinct ranges: i) the Local Supercluster and foreground groups ($\sim$ 2500 km s^{-1}); ii) the major association of matter of the Ursa Major Supercluster ($\sim$19,700 km s^{-1}); iii) background groups and clusters ($\sim$ 30,000 km s^{-1}); iv) very distant groups and clusters ($\sim$ 41,000 km s^{-1}). This result shows that the remarkable association of galaxies observed is partly produced by projection on the sky of physically independent systems, confirming the existence of large "voids" or gaps in the three-dimensional galaxy distribution, also well-demonstrated for regions around other superclusters such as Coma/A1367, Hercules and Perseus/Pisces. It is likely that the Ursa Major

G. O. Abell and G. Chincarini (eds.), Early Evolution of the Universe and Its Present Structure, 281–283.

Supercluster is a physical system of four clusters, shown in the following table.

Cluster		Mean Velocity	Velocity Dispersion	Virial Mass $M_\odot$	No. of Observed Galaxies
	$\langle Z \rangle$	$\langle V \rangle$ km.s^{-1}	$\langle V^2 \rangle^{1/2}$ km.s^{-1}		
A1291	0.0530	15885	975	1.4×10^{15}	7
A1318	0.0564	16923	284	1.2×10^{14}	4
A1377	0.0514	15431	488	3.5×10^{14}	8
A1383	0.0603	18102	395	2.3×10^{14}	5

There are also groups and field galaxies. The entire system has a mean velocity of 17,250 km.s^{-1}, velocity dispersion of 1193 km.s^{-1}, and a virial mass of 5.1×10^{15} $M_\odot$. It is 345 Mpc away, and has a projected radius of $\sim$ 10 Mpc. The crossing times are greater than the Hubble time though there is observational evidence for strong gravitational interaction between the clusters and groups. The supercluster probably is a bound system that has not yet reached a stationary state. The system is embedded in a large association of matter with projected radius -24 Mpc. This association is evidently expanding with the Hubble flow.

The 5C10 survey is the tenth survey with the Cambridge One-Mile telescope at 408 and 1407 MHz simultaneously, and known collectivel as the 5C surveys. The main conclusions of this survey may be summarize as follows: At 408 MHz a flux limit (before envelope correction) of 9.8 mJy (= 6σ, where σ = 1.63 mJy) was adopted, and the catalogue contains 265 sources above this level. At 1407 MHz, a limit of 1.7 mJy (= 5σ, where σ = 0.34 mJy) was adopted, and 48 sources are listed. Twenty-six sources appear extended and 12 of these are identified with galaxies. Four of the sources have radio-tail structure. At 408 MHz, 27 (11 percent) of the unresolved sources have been identified. When resolved sources are included, the total identification rate is 18 percent. At 1407 MHz, 41 percent (17 sources) of all sources have possibl identifications. Six out of ten Abell clusters surveyed in 5C10 probabl have radio sources in them. Sixteen 5C10 sources are associated with galaxies which are apparently in groups of galaxies. Of these, eight ar in the magnitude range of 15 to 17, and presumably two may be related to the supercluster. About three percent of the galaxies in the Ursa Major supercluster have been detected as radio sources. The overall statistic of the 5C10 survey are not substantially biased by the presence of the supercluster. The distribution of radio spectral index for a combined sample of 5C surveys differs significantly from that for bright (2 Jy) sources. In the central cluster (A1318), 9 percent of the galaxies are detected as radio sources; its luminosity distribution is compatible wit that of the Coma cluster. Intrinsically bright galaxies are more likely to be radio sources, and those in rich clusters are more likely than those in the general field of the supercluster or in groups.

I am very grateful to Professors Sir M. Ryle, A. Hewish, M.S. Longair, J.E. Gunn, and G.O. Abell and to Drs. J.M. Riley, J. Wall,

.nd M. Seldner. I am grateful to the Mullard Radio Astronomy Observa-
:ory, Royal Greenwich Observatory, and the Hale Observatories, Palomar
Mountain, for the use of the telescopes and computer facilities.

DYNAMICS OF SOME REMOTE SUPERCLUSTERS

R. J. Harms[1], H. C. Ford[2], and R. Ciardullo[3]
1. University of California, San Diego, USA
2. Space Telescope Science Institute, USA
3. University of California, Los Angeles, USA

ABSTRACT

Redshifts of clusters of galaxies have been obtained in four regions of the sky selected as suspected rich superclusters at intermediate redshift ($0.1 < z < 0.3$). Measurements to date have detected the existence of several superclusters with dimensions up to 50 h^{-1} Mpc ($h = H_o/100$ k/s/Mpc), irregular shapes, and containing as many as seven rich clusters in the richest supercluster. The velocity dispersions suggest some slowing of the Hubble flow internal to the superclusters. However, the inferred mass densities are low enough to prefer an open universe if the mass-to-luminosity ratios within the best-studied superclusters are comparable to the universal ratio.

1. OBSERVATIONS

Our most nearly complete redshift measurements have been obtained in the regions 1451+22 = Abell #11 (Abell 1961) = MFJG #18 (Murray et al. 1978) and 1615+43 = MFJG #19. Details of these observations are (or will be soon) published elsewhere (Ford et al. 1981, Paper I; Harms et al. 1981, Paper II; Ciardullo et al. 1983, Paper IV). Figures 1-4 summarize the locations on the sky and in redshift of the clusters of galaxies which form these two rich superclusters. Work in progress has provided less complete redshift data for several other candidate superclusters. Figures 5-8 present positions and redshifts of clusters in the directions of 0138-10 = MFJG #2 and 2306-22 = Abell #16 = MFJG #20. Note that the observations are not yet complete; candidate superclusters do contain more clusters and some may also contain additional Abell cluster members.

G. O. Abell and G. Chincarini (eds.), Early Evolution of the Universe and Its Present Structure, 285–290.

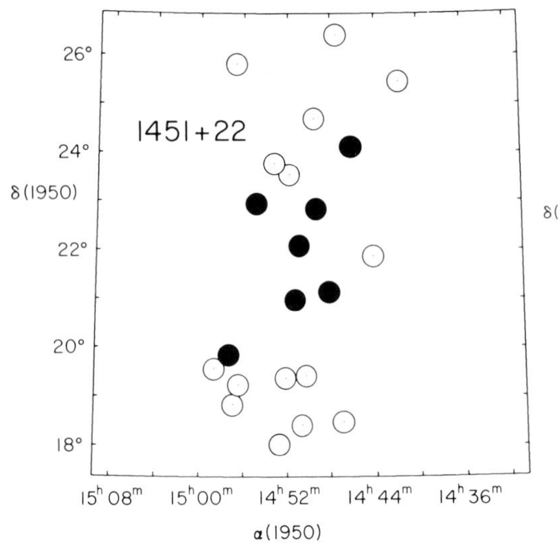

Fig. 1. Abell cluster members (filled circles) & other cluster members (open circles) of 1451+22A.

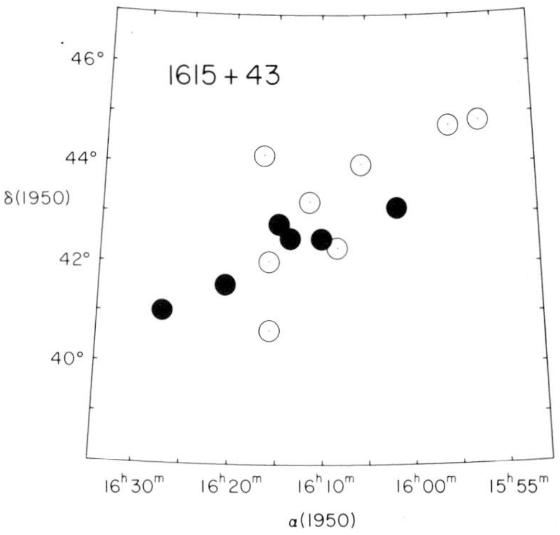

1615+43
46°
44°
42°
40°
δ(1950)
16h 30m
16h 20m
16h 10m
16h 00m
15h 55m
α(1950)

Fig. 2. Abell and other cl ter members of 1615+43A.

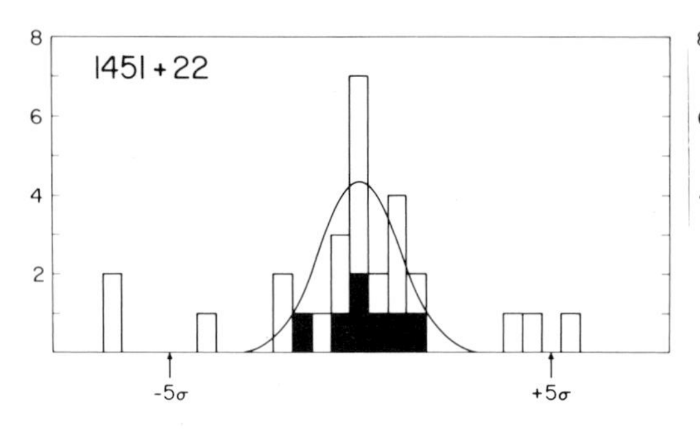

Fig. 3. No. vs redshift bin of Abell (solid) & other cluster (open) members and near-members of 1451+22A.

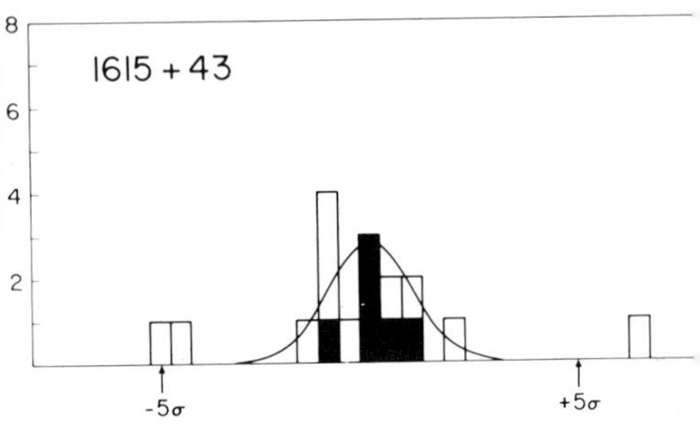

Fig. 4. No. vs redshift bi of member & near-member clusters of 1615+43A.

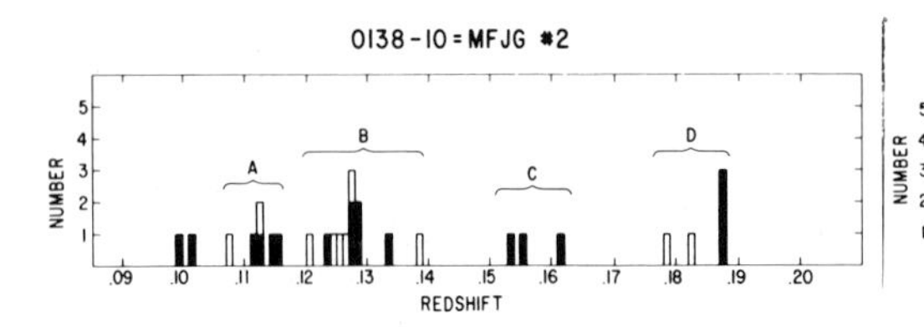

Fig. 5. Distribution of cluster redshifts toward 0138-10.

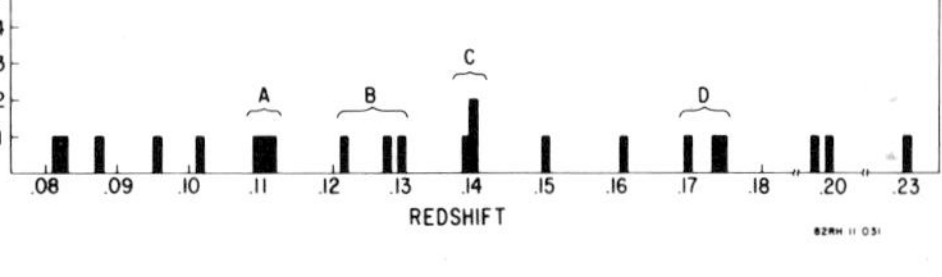

Fig. 6. Cluster (all Abell redshifts toward 2306-22.

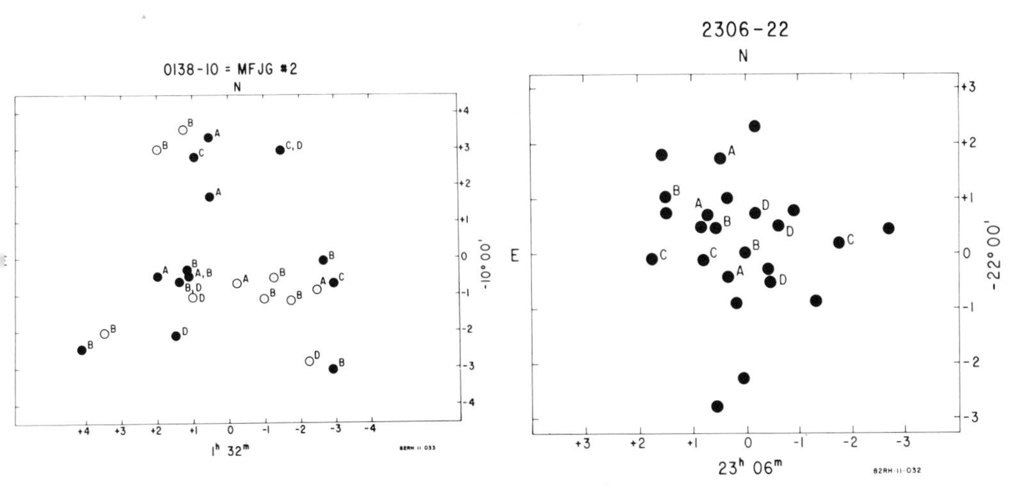

Fig. 7. Locations of clusters in 0138-10 A-D.

Fig. 8. Locations of Abell clusters toward 2306-22.

2. SUPERCLUSTER PROPERTIES

Table 1 lists some of the properties of the superclusters. The well-sampled superclusters 1451+22A and 1615+43A are elongated on the sky with dimensions on order of $50h^{-1}$ Mpc, while 0138-10B appears irregular but not elongated. The maps also suggest a core-halo structure in the superclusters, with Abell clusters concentrated toward the center (but often elongated) surrounded by less rich clusters. Two-color photography with crude redshift resolution capable of determining supercluster membership is underway to define the geometry and structure of the superclusters.

For both of the two well-sampled superclusers, 1451+22A and 1615+43A, the apparent thickness ΔZ (refer to Table 1) is less than the dimensions ΔX and ΔY on the sky in the α and δ directions, suggesting possible slowing of the Hubble flow within the superclusters. Table 2 summarizes the results of applying a t-test of significance to the means in ΔX, ΔY, ΔZ pairs for the four cases analyzing all or best-sampled superclusters using Abell or all cluster members. In every case, the mean ΔZ is less than the mean values for ΔX and ΔY, but always only at marginally significant levels. Two effects bias these results. We preferentially sample the full range of ΔZ but not ΔX and ΔY. Also, the measurement uncertainty in ΔZ is larger than in ΔX and ΔY due to galaxy motions in their clusters and any random cluster motions. Because both these effects lead to

TABLE 1

Super-cluster		# Clusters Abell	Total	Redshifts ⟨z⟩	σ	Dimensions(h^{-1}Mpc) ΔX	ΔY	ΔZ
1451+22	A	7	22	.1161	.0019	24.0	47.4	21.3
	B	1	6	.1019	.0045	25.8	13.2	37.5
	C	2	4	.1526	.0010	7.2	42.3	6.3
	D	3	4	.1820	.0056	34.2	65.4	39.0
	E	2	2	.2464	.0049	6.0	103.2	20.7
1615+43	A	6	14	.1354	.0027	46.2	29.7	28.2
	B	2	2	.0797	.0001	15.0	3.3	0.3
	C	3	3	.1845	.0021	32.7	17.1	11.7
0138-10	A	4	6	.1124	.0028	26.1	24.6	24.6
	B	6	12	.1275	.0046	46.2	43.8	52.8
	C	3	3	.1571	.0041	31.8	30.0	23.7
	D	3	5	.1845	.0041	35.7	56.1	27.9
2306-22	A	3	3	.1105	.0013	2.4	12.9	7.5
	B	3	3	.1262	.0041	9.9	6.9	24.0
	C	3	3	.1391	.0007	24.9	1.8	3.9
	D	3	3	.1725	.0023	3.6	10.8	12.9

Supercluster properties of measured member clusters; dimensions are from extreme values of α, δ, z:

$\Delta X = c\langle z\rangle \Delta\alpha \cos \langle|\delta|\rangle/100h$

$\Delta Y = c\langle z\rangle \Delta\delta /100h$

$\Delta Z = c\Delta z/100h$ with angular differences α, δ in radians

TABLE 2

Sample Description	Sample Size	ΔX m±σ	ΔY m±σ	ΔZ m±σ	Null Hypothesis Probability(%) ΔX-ΔY	ΔX-ΔZ	ΔY-ΔZ
Best Superclusters Abells only	2	24.9±12.6	20.4±7.8	16.5±0.3	71	45	55
Best Superclusters all clusters	2	35.1±15.6	38.4±12.6	24.6±4.8	84	46	29
All Superclusters Abells only*	15	19.8±14.1	26.4±24.2	15.0±11.1	39	31	13
All Superclusters all clusters	16	23.1±14.4	31.8±27.3	21.3±14.1	27	72	19

*Excluding 1451+22B, which contains only one Abell cluster

overestimating ΔZ, we conclude that the observed smallness of ΔZ probably does indicate slowing of the Hubble flow within at least some of the superclusters, but the issue is by no means resolved beyond doubt.

Neither 1451+22A nor 1615+43A show significant variation in redshift or in redshift variance with position on the sky. The less-studied 0138-10B may contain a redshift gradient in the north-south direction, but more measurements will be necessary for a reliable result.

3. COSMOLOGICAL IMPLICATIONS

The deceleration of the clusters within each supercluster is small at best. This is in qualitative agreement with the analysis in Papers I and II implying $\Omega \lesssim 0.3$ (and preferring $\Omega \simeq 0.1$), assuming that the mass-to-light (or mass-to-luminous galaxy) ratios in the superclusters are typical of the universe. In any case, the seemingly inexorable increase in M/L with increasing scale appears to have halted for 1451+22A and 1615+43A, as depicted in Figure 9.

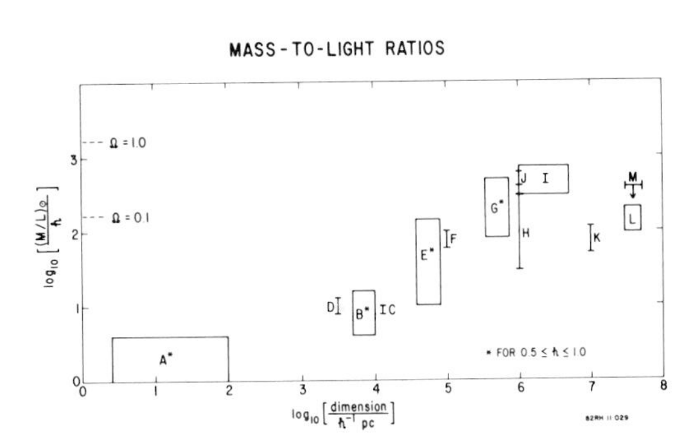

Fig. 9. Variations of M/L with scale. Samples A-K from Table 5 of Peebles (1980). L&M from Papers I and II.

REFERENCES

Abell, G. O. 1961, A.J., 66, 607.
Ciardullo, R., Bartko, F., Ford, H. C., and Harms, R. J. 1983, submitted to Ap. J. Supp. (Paper IV).
Ford, H. C., Harms, R. J., Ciardullo, R., and Bartko, F. 1981, Ap. J. (Letters), 245, L53 (Paper I).
Harms, R. J., Ford, H. C., Ciardullo, R., and Bartko, F. 1981, in The Tenth Texas Symposium on Relativistic Astrophysics, ed. R. Ramaty and F. C. Jones, (New York: New York Academy of Sciences), p. 178 (Paper II).
Murray, S. S., Forman, W., Jones, D., and Giacconi, R. 1978, Ap. J. (Letters), 219, L89.
Peebles, P. J. E. 1980, in Physical Cosmology, ed. R. Balian, J. Audouze, and D. N. Schramm, (Amsterdam, New York, Oxford: North Holland Publishing Co.), p. 213.

Discussion

Szalay: Your values of the Hubble flow seem to indicate a dynamic overdensity of 2 to 3. You mentioned a volume density contrast of 20 to 50; this would imply a highly nonlinear infall. Which is the case?

Harms: The models used were nonlinear, but easily solvable due to the simplifying approximation of spherically symmetric noncrossing mass shells for the superclusters. The initial conditions involved matching the supercluster expansion rate to the Hubble flow at "initial" redshifts ranging from 2 to 1000. The nonlinear equations for the supercluster were solved iteratively to obtain that mass within each supercluster which slows its expansion rate to the measured values at the observed epoch. Models were obtained for both the cases $\Omega \to 0$ and $\Omega = 1$ in each case the derived values for mass densities within the superclusters range from about 1 to 4 times the closure density. Note, as yet, no reference to overdensity has been made.

The luminosity contrast within the superclusters has been estimated from counts of Abell clusters, and in a preliminary fashion, from galaxy counts assuming a Schechter luminosity function. Such counts yield overdensities from about 15 to 70. If we assume these number or luminosity overdensities are similar to the mass overdensities then we can conclude that $\Omega < 1$.

A2197 + A2199 SUPERCLUSTER REGION

Laird A. Thompson
Institute for Astronomy, University of Hawaii
Stephen A. Gregory
Physics & Astronomy Dept., Bowling Green State University

A2197 and A2199 are among the nearest rich clusters of galaxies. They are worthy of special attention because (i) they form a close binary system, the analysis of which provides an independent estimate of the clusters' mass to light ratio, and (ii) these two clusters fall along a bridge of interconnected clusters that stretches at least 50 Mpc. We report 78 new redshifts in the A2197+A2199 region, thus tripling the number of known redshifts.

The A2197+A2199 group is located 25° north of the Hercules cluster group. Because both groups have similar mean redshifts ($z=0.0309$ for the former and $z=0.0360$ for the latter), Chincarini, Rood, and Thompson (1981) searched the intervening area of sky for galaxies with $z \sim 0.030$. These observations showed that a bridge of galaxies links the two regions. While making new observations in the A2197+A2199 area, we probed a 72 square degree region further to the north. Five additional galaxies with redshifts $z \sim 0.030$ were identified, as many as would be expected if the bridge continues northward at least another 10°.

The binary nature of A2197+A2199 provides the means to determine the system's total mass to light ratio (M/L) independent of virial techniques. We use a formalism which was introduced by Peebles (1974) and Gunn (1974) and originally applied to the Local Group. Since the equations contain one too many unknowns to fully specify a solution, we calculate the binary cluster's M/L as a function of the angle α between the plane of the sky and the line connecting the cluster centers. While high values of M/L are permissible if $\alpha < 5°$ or $\alpha > 75°$, all intermediate values of α require $M/L < 200$. If $13° < \alpha < 62°$ then $M/L < 50$. This stands in contrast to results commonly obtained for rich cluster cores via the virial theorem (for the A2199 core we find $M/L \sim 185-225$). In all likelihood, high M/L ratios are characteristic only of cluster cores, and the outskirts of rich clusters may show lower M/L ratios.

Chincarini, G., Rood, H., and Thompson, L. 1981, Ap.J.Lett., 249, L47.
Gunn, J. 1974, Comments Ap. & Space Phys., 6, 7.
Peebles, P.J.E. 1971, Physical Cosmology (Princeton Univ. Press).

G. O. Abell and G. Chincarini (eds.), Early Evolution of the Universe and Its Present Structure, 291–292.

Discussion

Shandarin: Why do you think these clusters are gravitationally bound?

Thompson: If the double cluster is modeled with the equations appropriate for an unbound system, the solution requires a very low mass for the double cluster. In that case, $M/L < 5.1$.

Salpeter: If massive neutrinos are dominant, one could have both a large M/L and the small deviation ΔV from the Hubble expansion which you observe: ΔV would be small if the neutrinos in the super cluster extend beyond the region where the cluster cores are, with smaller density contrast than for galaxies.

Thompson: I agree that this is one possible explanation. However, before making any wide-ranging conclusions, other close binary clusters must be observed to insure that the A2197+A2199 system is telling us the right value for M/L.

Peacock: It may well be that by selecting close pairs on the sky you are biased towards angles close to the line of sight. High M/L is then not so improbable.

Thompson: If the clusters selected for study are widely separated from one another in the line of sight ($\alpha \gtrsim 75°$), then their galaxy luminosity functions should show the difference in distance. Preliminary data for A2197 and A2199 indicate that these two clusters do not suffer from this effect. However, anyone who selects other pairs for similar analysis must take your advice into account.

Einasto: This double cluster is a part of a very long cluster filament (Hercules supercluster) which is seen almost in the plane of the sky. The method used is very sensitive to small changes in projection angle, α, in this particular case, and therefore the result has low weight.

Thompson: In fact, the filament you speak of is tilted off the plane of the sky by 26°. If the angle α for the A2197 + A2199 system is the same as the filament (i.e., $\alpha = 26°$), then the solution has high weight and $M/L < 50$. However, I think it is not advisable to presume that such a close binary would necessarily fall along the filament.

THE INDUS SUPERCLUSTER

Harold G. Corwin, Jr.*
Department of Astronomy
University of Edinburgh

A survey of rich galaxy clusters, redshifts for many of those clusters, and galaxy counts by eye to B = 19.0 show the Indus Supercluster to be an annular (in projection) configuration of nine rich clusters at $0.073 < z < 0.080$ apparently connected by bridges of galaxies.

Photoelectrically calibrated photographic photometry of galaxy images on six U.K. Schmidt plates using the COSMOS machine at the Royal Observatory, Edinburgh, gave photometric information for about 150,000 galaxies. From this, the luminosity function of the Indus Supercluster was extracted. To B = 21.5, the Supercluster includes about 25,000 galaxies, its estimated total luminosity is 7×10^{13} $L_\odot$, and -- if its mass-to-light ratio is typical -- its total mass is $\sim 1 \times 10^{16} \mathfrak{M}_\odot$. Its diameter is about 40 Mpc. These parameters make it similar to other known superclusters.

In addition, the integrated apparent field luminosity function for galaxies, derived from the 140 square degrees of sky scanned on Schmidt plates by COSMOS, agrees with most previous determinations.

The general picture of a sponge-like cellular distribution of galaxies as developed by Einasto et al (1980 and references therein) is confirmed. Though there are some indications that this structure is primordial, neither data nor theories are yet sufficient to allow an adequate explanation of the development of such structure in a big bang universe. Details have been presented by Corwin (1981).

I thank Guido Chincarini and George Abell for their help with this presentation.

REFERENCES

Corwin, H.G. (1981). Ph.D. thesis, University of Edinburgh.
Einasto, J., Joeveer, M. and Saar, E. (1980). M.N.R.A.S. 193, 353.

*Present address: Dept. of Astronomy, RLM 15.220
University of Texas
Austin, TX 78712, USA

G. O. Abell and G. Chincarini (eds.), Early Evolution of the Universe and Its Present Structure, 293.

THE SOUTHWEST EXTENSION OF THE PERSEUS SUPERCLUSTER

P. Focardi[1], B. Marano[1], G. Vettolani[2]
[1]Istituto di Astronomia, Università di Bologna
[2]Istituto di Radioastronomia, CNR, Bologna

The main structure of the Perseus supercluster is defined by a long chain of clusters and rich groups of galaxies with very similar radial velocities ($\bar{v}$ = 5200 ± 200 km/s). The redshift survey performed by Gregory *et al*. (1981) has shown that most of the "field" galaxies are also members of the supercluster.

We have considered an area of 270 square degrees, between R.A. 23^h30^m and 1^h0^m and between Dec. 21°30' and 33°30', which lies along the axis of the Perseus supercluster beyond its previously supposed limits. The projected distribution of galaxies suggests a further extension of the supercluster in this direction. Ninety-nine galaxies brighter than M = 14.5 are present in this area. Almost all of them are "field" galaxies. Only four are members of the clusters A2634 and A2666. New redshifts have been obtained for 44 galaxies. With the data available in the literature (Palumbo *et al*. 1982), the radial velocities of 93 galaxies out of 99 are known.

We find that: a) A well-defined population of galaxies with $\bar{v}_o$ = 5240 km/s and σ_r = 312 km/s is present; this proves the further extension of the supercluster. b) As found in different fields by other authors, foreground galaxies are mainly grouped; only a few, if any, isolated galaxies are found. c) Background galaxies with $6000 < v < 10{,}000$ km/s are mostly organized in an elongated structure encompassing the clusters A2634 and A2666, having similar velocities. A trace of filamentary structures, with bends and bifurcations, suggestive of the "cell" model (Einasto *et al*. 1980), seems to emerge.

REFERENCES

Einasto, J., Jôeveer, M., Saar, E. 1980, *M.N.R.A.S.*, 193, 353.
Gregory, S.A., Thompson, L.A., Tifft, W.G. 1981, *Ap. J.*, 243, 411.
Palumbo, G.C., Tanzella-Nitti, G., Vettolani, G. 1982, "Catalogue of Radial Velocities of Galaxies," Gordon and Breach, in press.

G. O. Abell and G. Chincarini (eds.), Early Evolution of the Universe and Its Present Structure, 295.

THE HOROLOGIUM SUPERCLUSTER

G. Chincarini[1,2], M. Tarenghi[2], H. Sol[2], P. Crane[2],
J. Manousoyannaki[1] and J. Materne[3]
[1]University of Oklahoma; [2]European Southern Observatory
[3]Technische Universität, Berlin

We measured redshifts for a random subsample of 286 objects, m (Shapley) $\leq$ 16.5, drawn from the catalogue of the Horologium region published by Shapley. The distribution of the sample objects on the celestial sphere is shown in Fig. 1, the characteristics of the redshift distribution are represented in Figs. 2 and 3.

The Shapley magnitudes correlate fairly well with blue magnitudes derived from photoelectric observations, Fig. 4. The two point angular correlation function of the subsample is similar to that of the whole catalogue. The clumpy distribution in depth, Fig. 2, bias, however, the determination of Ω by Peeble's $\xi(\sigma, \pi)$ method. The clump at $<v>$ $\simeq$ 13450 km/sec has a velocity dispersion of $\simeq$480 km/sec and is consistent with the idea that some of these structures are rather thin and filamentary rather than pancake-structures (Chincarini 1982, Chincarini, Giovanelli and Haynes 1982, and Giovanelli, these Proceedings).

REFERENCES

Chincarini, G., 1982. Summer School of Cosmology and Relativity. Rio de Janeiro, January 1982, preprint.
Chincarini, G., Giovanelli, R. and Haynes, M., 1982, submitted to A.A.
Shapley, H., 1935, Annals of Harvard College Observations, Vol. 88, No. 5.

G. O. Abell and G. Chincarini (eds.), Early Evolution of the Universe and Its Present Structure, 297–298.

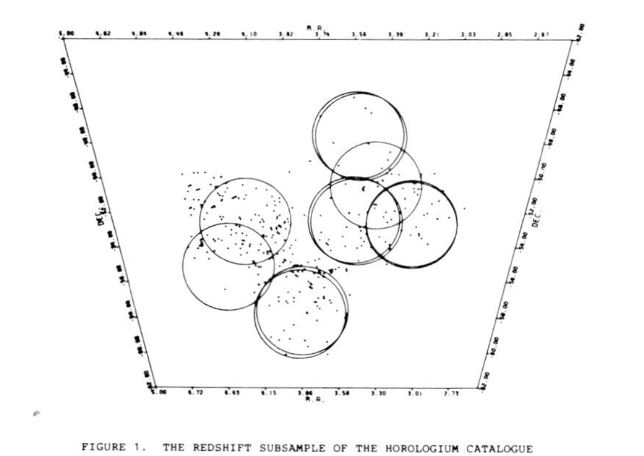

FIGURE 1. THE REDSHIFT SUBSAMPLE OF THE HOROLOGIUM CATALOGUE BY SHAPLEY. THE CIRCLES INDICATE THE AREAS OVER WHICH THE SURVEY BY SHAPLEY IS PROBABLY COMPLETE, SMAG < 16.5.

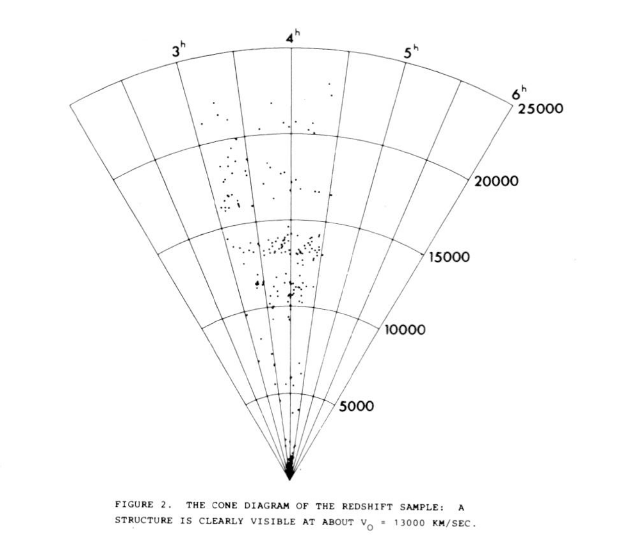

FIGURE 2. THE CONE DIAGRAM OF THE REDSHIFT SAMPLE: A STRUCTURE IS CLEARLY VISIBLE AT ABOUT V_O = 13000 KM/SEC.

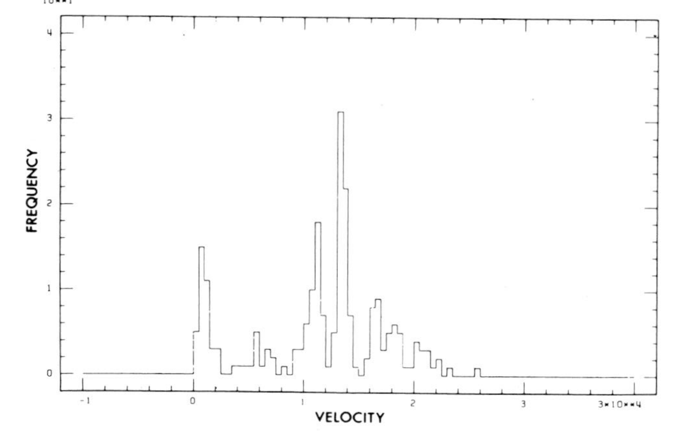

FIGURE 3. HISTOGRAM OF THE DISTRIBUTION OF REDSHIFTS FOR THE HOROLOGIUM SAMPLE.

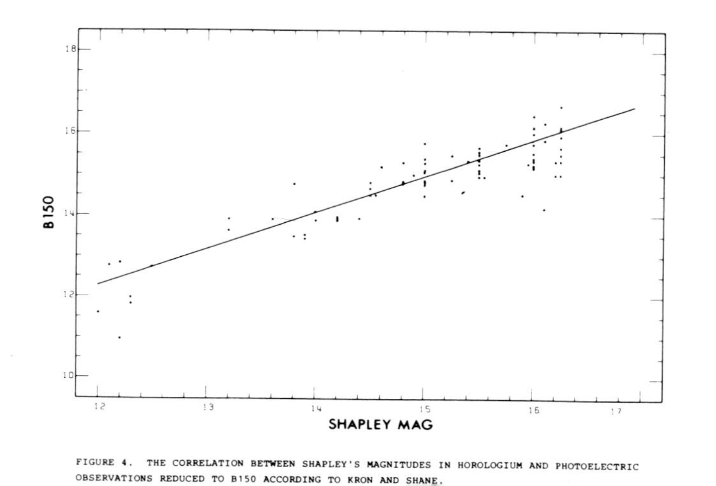

FIGURE 4. THE CORRELATION BETWEEN SHAPLEY'S MAGNITUDES IN HOROLOGIUM AND PHOTOELECTRIC OBSERVATIONS REDUCED TO B150 ACCORDING TO KRON AND SHANE.

UNSEEN MASS

Martin J. Rees
Institute of Astronomy, Madingley Road,
Cambridge CB3 OHA, England

INTRODUCTION

The arguments for "unseen mass", and the evidence on its distribution, are still somewhat controversial as far as the details are concerned. The following four statements would, however, be widely accepted:

(i) Baryons cannot contribute much less than 1% of the critical density - i.e. $\Omega_{baryon} \gtrsim 0.01$. This limit comes from the "luminous" content of galaxies, and from the inferred amount of X-ray emitting gas in clusters.

(ii) $(M/L)_B$ values in the range around 200 h_{100} are derived from studies of virialised clusters and from the cosmic virial theorem. These refer to length scales of $(1 - 2)h_{100}^{-1}$ Mpc.

(iii) $(M/L)_B$ appears to be a non-decreasing function of length scale.

(iv) If $\Omega \simeq 1$ — the value favoured by some theorists, especially the advocates of "inflationary" cosmology — then $(M/L)_B$ must continue to increase out to at least $\sim 10\ h_{100}^{-1}$ Mpc. Otherwise, a high density Universe would be incompatible with the low random velocities which lead to conclusion (ii) above.

The straightforward relationship

$$\langle M/L \rangle_B \simeq 2300\ (\Omega h_{100}) \tag{1}$$

shows that 90% (or even, if $\Omega \simeq 1$, as much as 99%) of the mass-energy in the Universe must be in some form with much greater M/L than the objects that astronomers normally investigate.

G. O. Abell and G. Chincarini (eds.), Early Evolution of the Universe and Its Present Structure, 299–305.

2. CANDIDATES FOR "UNSEEN MASS"

Neutrinos or Other "...inos"

If neutrinos have non-zero masses, then

$$\Omega_\nu = 0.03\, h_{100}^{-2} \left(\frac{n_\nu}{n_\gamma}\right) \sum_{\text{species}} (m_\nu)_{ev} ,$$

where n_γ is the photon number density in the microwave background (and two-component neutrinos are assumed). In the "standard" hot big bang theory, $n_\nu/n_\gamma = 3/11$ (provided that the rest masses are not so large that mc^2 is comparable with kT at the temperature (~ 1 Mev) when neutrinos decouple); so one species with mass 100 h_{100}^2 suffices to give $\Omega_\nu = 1$.

Theoretical physicists have other particles in reserve - photinos or gravitinos, for instance - which could (if they existed) contribute to Ω in an analogous way. The only difference would be that (n/n_γ) could be less than for neutrinos because the other "...inos" may have decoupled before muons (or even hadron pairs) annihilated - the latter would then boost the neutrinos but not the still more weakly coupled "...inos". For gravitinos, we may have $(n/n_\gamma) \lesssim 10^{-2}$. Szalay will discuss this topic in more detail, so I will make just two remarks about neutrinos, etc.:

(a) If neutrinos dominate the dynamics of galaxy formation, then there is a characteristic mass-scale of $\sim 2 \times 10^{18} (m_\nu)_{ev}^{-2}\, M_\odot$, which is of "supercluster" size for $m_\nu \simeq 10$ ev. (This is the horizon mass at the stage when $kT = m\,c^2$, and can also be expressed as $m \sim m_{Planck} (m_\nu/m_{Planck})^{-2}$.) For other "...inos" this mass may be of galactic order.

(b) Considerations of primordial nucleosynthesis yield a well-known general argument favouring non-baryonic "unseen mass". The observed deuterium abundance cannot be produced in the big bang unless $\Omega_b < 0.015\, h_{100}^{-2}$. A baryon density within a factor of two of this limit fits best with the data on other light elements (^{3}He, ^{4}He, ^{7}Li). Advocates of a high Ω_b must appeal to an "astrophysical" origin for D, or else to the "escape clauses" permitted in non-standard big bang models (e.g. David and Reeves 1979).

Baryonic "Unseen Mass"

If galaxies and clusters evolved hierarchically from sub-units that condensed earlier, most of the initial baryons might have been incorporated in a pregalactic Population III. Ideally, one would like to be able to calculate what happens when a cloud of $10^6 - 10^8\, M_\odot$ condenses out soon after recombination - does it form one (or a few) supermassive objects, or does fragmentation proceed efficiently down to low-mass stars? Our poor understanding of the initial mass function (IMF) for stars forming now in our own Galaxy gives us little confidence that we can predict the nature of pregalactic stars, forming in an environment differing from our (present-day) Galaxy in at least four significant ways:

(i) The initial cloud masses may be larger than any dense clouds in our Galaxy - the maximum scale on which $\delta\rho/\rho \simeq 1$ at recombination depends on the initial fluctuation spectrum, but may be as high as 10^8 $M_\odot$.

(ii) There may be no coolants apart from H and He.

(iii) The microwave background prevents cooling below $\sim 3(1+z)^{\circ}$K.

(iv) The energy density in background radiation ($\propto(1+z)^4$) is so high that "Compton drag" may inhibit free-fall collapse if the material is partially ionized.

Kashlinsky and I have (1982) tried to investigate these processes in some detail, but we reach the depressing conclusion that one cannot yet confidently pin down the masses within even ten orders of magnitude (10^{-2} - 10^8 $M_\odot$)! Starting with a post-recombination bound cloud of 10^6 - 10^8 $M_\odot$, there are two extreme possibilities. Fragmentation may be so ineffectual that a single supermassive object results. On the other hand, if fragmentation were maximally efficient, we could end up with 10^{-2} $M_\odot$ stars, this being the Jeans mass if the material compressed to the highest density ($\sim 10^{10}$ cm^{-3}) permitted by the clouds' likely initial rotation, and turned into H_2 at $\sim 10^3$ $^{\circ}$K.

Astrophysical and Observational Constraints on the Masses of Population III Remnants.

Despite our inability to "predict" what Population III should be like, there are several constraints which together allow us to conclude that the masses must either be $\lesssim 0.1$ $M_\odot$ or else in the range 10^4 - 10^6 $M_\odot$.

The (M/L) ratio. Relation (1) obviously rules out masses above 0.1 $M_\odot$ unless the stars have all evolved and died, leaving dark remnants.

Nucleosynthesis, background light etc. A severe constraint comes from the requirement that Population III should not overproduce heavy elements. If this population predates all Population II, the fraction of heavy elements produced must be $\lesssim 10^{-4}$; if Population II and Population III are coeval, maybe up to 10^{-3} is permissible. This sets strict limits on the mass fraction going into the upper mass range for "ordinary" stars (15 - 100 $M_\odot$). Limits on the range 100 - 400 $M_\odot$ are uncertain because only ^{4}He may be ejected, the "heavies" in the core collapsing into a black hole remnant. An uncertainty in the evolution of massive or supermassive stars is the amount of mass loss during H-burning; however an IMF such that most mass goes into very massive objects (VMOs) of $\gtrsim 10^4$ $M_\odot$ is compatible with the nucleosynthesis constraints. A further consideration favouring these high masses is that they are likely to terminate their evolution by a collapse which swallows most of the mass - if most of the material were ejected we would need to invoke "recycling" through several generations in order to end up with most of the material in black holes rather than gas.

The background light constraint depends on the redshift at which the VMOs form, and on whether the energy can be thermalised (see Rowan-Robinson's contribution to these proceedings).

Detailed discussions of Population III are given by Carr, Bond and Arnett (1982) and by Tarbet and Rowan-Robinson (1982).

Dynamical friction, etc. Carr (1978) has reviewed the effects of massive black holes on the dynamics of our galactic disc and the effects of accretion, showing that our Galactic halo cannot be composed of objects whose individual masses exceed $\sim 10^6$ $M_\odot$. This limit cannot however be applied to the more diffusely-distributed objects that might contribute $\Omega \simeq 1$.

Gravitational Lenses

If a remote source (e.g. a quasar) is sufficiently compact, the probability that our line of sight passes close enough to one of the Population III objects for significant lensing to occur is $\sim \Omega$. Note that the individual lensing masses do not affect this probability, only their total contribution to Ω. The characteristic scale of the images is

$$\theta \simeq 10^{-6} \left(\frac{M_{lens}}{M_\odot} \right)^{\frac{1}{2}} \text{ arc sec.} \quad (2)$$

The precise coefficient in (2) is, of course, a function of the source and lens redshift and of the cosmological model, but is ~ 1 for sources with $z \gtrsim 1$ and lenses $\sim \frac{1}{2}$ way along the line of sight.

The structure on milli-arc second scales predicted by (2) for 10^6 $M_\odot$ holes could be detected by VLBI. For "Jupiters" of $\lesssim 0.1$ $M_\odot$ there is no short-term hope of achieving the angular resolution required ($< 10^{-6}$ arc seconds); however there is then a chance of detecting variability on timescales of years due to transverse motions of the sources themselves (Gott 1981, Young 1981). Such variability would be found only in source components whose intrinsic angular size was less than the θ given by (2). For $z \simeq 1$ this requires linear dimensions of around a light day or less. The region emitting the quasar optical and X-ray continuum probably fulfils this requirement, but its rapid intrinsic variability would be hard to disentangle from effects due to lensing. Canizares (1982) argues that one can already exclude $\Omega \gtrsim 1$ in "Jupiters" because the apparent magnitude of the typical quasar continuum region would then be altered by a factor ~ 2 relative to that of the (larger) line-emitting component, thereby introducing an unacceptably large scatter in the equivalent widths of quasar emission lines. (For a related argument, see Setti and Zamorani's contribution to these proceedings.)

3. LARGE-SCALE SEGREGATION OF LUMINOUS AND "UNSEEN" MATERIAL

We only observe the 1 - 10% mass fraction of the Universe that is in "luminous" galaxies. To what extent is this a valid tracer of the total mass distribution on various scales? The answer to this question, for scales (2 - 100)h_{100}^{-1} Mpc, is relevant to three key issues: (i) Is $\Omega \simeq 1$ possible? (ii) What is the nature of the "voids"? (iii) Can the observed clustering of galaxies be the outcome of purely gravitational forces?

One cannot yet give confident answers to any of these questions. Szalay's paper discusses galaxy formation in neutrino-dominated cosmology, where dissipation can separate baryons and "unseen" neutrinos on scales at least up to the characteristic damping mass; on the other hand, for an initial fluctuation spectrum with a high wave number cut-off, purely gravitational effects can perhaps adequately reproduce the "linear" features of the observed galaxy clustering (Klypin and Shandarin 1982, Davis, Frenk and White 1982). The figure shows schematically how the "mass fraction" of the Universe may be apportioned between "unseen mass" gas and galaxies. Gas $\leftrightarrow$ galaxy processes continue up until recent epochs; moreover, the differing effects of dissipation, dynamical friction, etc. on the three components will cause their ratio to vary spatially.

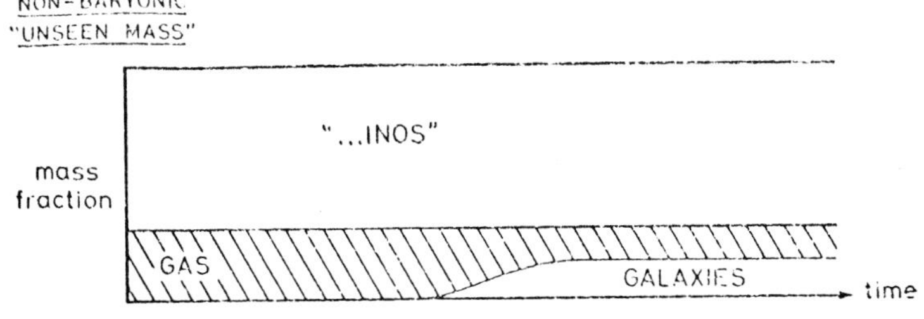

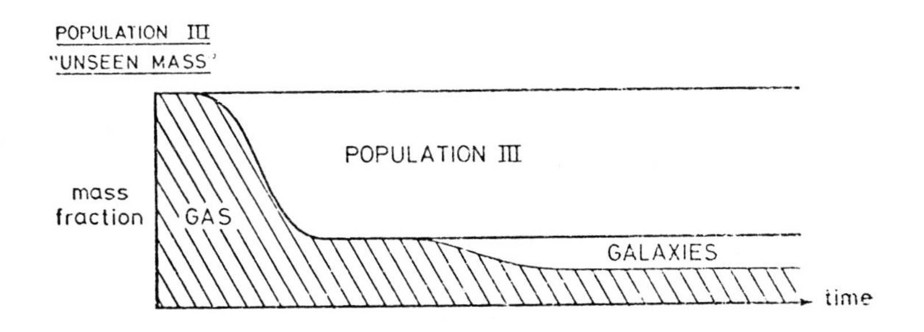

Figure 1. Cosmic 'mass fractions' as a function of time.

I should like to draw attention to two ways in which galaxy formation may be inhibited in large volumes, without the baryon content of those volumes being depleted.

(a) Heating of pregalactic gas to $\sim 10^6$ ^oK would make M_{Jeans} higher than a galactic mass, and inhibit condensation of protogalaxies.

This would require pregalactic energy input as envisaged in the Ostriker-Cowie (1981) scheme. However, it is much less extravagant in terms of energy to heat up a large volume than to evacuate it: 10^6 ^oK corresponds to only $\sim$ 100 ev per proton, whereas to evacuate a "void" as large as that discovered by Kirshner *et al.* (1981, and these proceedings) would require $\sim 10^5$ ev per proton.

(b) Several models for the evolution of primordial density fluctuations lead to a post-recombination spectrum of slope $n \simeq -3$ on mass-scales up to superclusters. This spectrum has the property that $\langle(\delta\rho/\rho)^2\rangle$ is independent of scale - it is not the case that smaller scales have larger amplitude and condense out at earlier epochs. The important consequence ensues that few galactic-mass perturbations in incipient voids would be gravitationally bound, and even purely gravitational effects would inhibit galaxy formation in regions destined to lie outside clusters.

Either of the above schemes (or some combination of the two, whereby galaxies form first (for reason (b)) in incipient superclusters, and their power output heats the intervening volume) would give rise to voids which could contain *more* gas than apparent clusters, and where even the baryonic component could be much more homogeneous than the distribution of galaxies might indicate.

4. CONCLUSIONS

The acceptable forms for "unseen" mass can be listed as follows:

Non-baryonic	$\sim$ 10 ev neutrinos	(10^{-32} gm)
	Other elementary particles (photinos, gravitinos, etc.)	
Baryonic	"Jupiters" of $< 0.1\ M_\odot$	
	$10^4 - 10^6\ M_\odot$ black holes	($\sim 10^{37} - 10^{39}$ gm)
	Diffuse gas in "voids"	

Observations may soon reduce the range of options. Alternatively, they may show that more than one type of unseen mass exists: there may, for instance, be a widely-diffused non-baryonic component, as well as a large number of Population III remnants concentrated in the halos of individual galaxies. But at the moment our ignorance is encapsulated by the statement that there are still > 70 powers of ten uncertainty in the masses of the entities that constitute > 90% of the content of the Universe!

REFERENCES

Canizares, C. 1982. Astrophys.J. (in press).
Carr, B.J. 1978. Comments on Astrophys., 7, 161.
Carr, B.J., Bond, J.R. and Arnett, W.D. 1982. Astrophys.J. (submitted).

David, Y. and Reeves, H. 1979. Phil. Trans. R. Soc., 296, 381.
Davis, M., Frenk, C. and White, S.D.M. 1982. Astrophys.J. (submitted).
Gott, J.R. 1981. Astrophys.J., 243, 140.
Kashlinsky, A. and Rees, M.J. 1982. Mon.Not.R.astr.Soc. (submitted).
Kirshner, R.F., Oemler, A., Schechter, P.L. and Sheckman, S.A. 1981. Astrophys.J., 248, L.57.
Klypin, A. and Shandarin, S. 1982. Sov. Astron. (submitted).
Ostriker, J.P. and Cowie, L.L. 1981. Astrophys.J. (Lett.), 243, L.127.
Tarbet, P.W. and Rowan-Robinson, M. 1982. Nature, 298, L. 127.
Young, P.J. 1981. Astrophys.J., 244, 756.

Discussion

Salpeter: The formation of low-mass stars out of pure H and He may select against high-velocity turbulence (which can dissociate H_2, which in turn prevents radiative cooling) and for high gas density. These "Jupiters" might then be concentrated to the innermost cores of rich clusters.

Rees: In the scheme Kashlinsky and I have discussed, the best chance of obtaining dense H_2 would occur if a rotationally-supported disk can form (maybe around a central massive object). There are, however, some effects which would make it easier for H_2 to form (or to re-form after shock heating) at early epochs. At, say, $z \simeq 300$, the radiation temperature is $\sim 10^3$ K. Moreover, its energy density is high enough that an electron can Compton-cool to the radiation temperature before recombining. A high density of free electrons or negative ions at a low temperature provides very propitious conditions for molecule formation.

NEUTRINO MASS AND GALAXY FORMATION

Alexander S. Szalay
Astronomy and Physics Dept., Univ. of California, Berkeley
Dept. of Atomic Physics, Eotvos Univ. Budapest, Hungary

J. Richard Bond
Physics Department, Stanford University
Institute of Astronomy, Cambridge University

The adiabatic theory of galaxy formation in neutrino-dominated universes is reviewed. Collisionless damping leads to a density fluctuation spectrum with a cutoff, the nonlinear evolution of which naturally results in the formation of pancakes, strings and voids.

INTRODUCTION

The concensus of the observational papers presented in this volume (see for example Oort, Chincarini and Einasto) is that the structure of the universe is dominated by distinctive large scale features: superclusters, with lengths of order 25-100 Mpc, that are often more string-like than sheet-like. Rather than being isolated, there are hints of a network structure to the superclusters with large voids almost free of galaxies in between. The announcement of Lubimov et al. (1980), that their experiment on the beta decay of tritium indicates the mass of the electron neutrino lies in the range 16 eV $< m_\nu <$46 eV was in large part responsible for the resurgence of interest in the neutrino-dominated universe. In this paper we review how the presently observed distribution of galaxies may arise if the neutrino is endowed with a rest mass of order 10-100 eV.

Gershtein and Zeldovich (1966) first suggested that even a small neutrino mass may have important cosmological consequences, by contributing a larger fraction of the overall density of the universe than baryons. Marx and Szalay (1972), Cowsik and McClelland (1972) and Schramm and Steigman (1981) have refined their arguments: the present limits on the Hubble constant and the deceleration parameter of the universe result in an upper bound to the sum of the neutrino masses: Σm_ν < 100 eV. Calculations of the primordial He and D abundance indicate baryons can only contribute a small fraction of the critical density. The likely value of the baryon density parameter at the time of primordial nucleosynthesis lies in the range of $0.01<\Omega_B< 0.1$ (Olive et al. 1981). This suggests that if baryons dominate the mass density then the universe

G. O. Abell and G. Chincarini (eds.), Early Evolution of the Universe and Its Present Structure, 307–312.

is open by a wide margin. In such a universe the growth of the initially adiabatic fluctuations is highly suppressed,as we shall see later: the presence of nonlinear structure now is incompatible with the tight upper bounds on small scale density perturbations arising from observations of microwave background temperature fluctuations, and effectively rules out such a case. Szalay and Marx (1976) first noted that the way fluctuations behave is quite different in a neutrino-dominated universe than in a baryon-dominated one. Below T~1 MeV neutrinos are collisionless particles and are thus not affected by radiation drag as ordinary matter is. This enables much larger fluctuation growth to occur, without strongly influencing the microwave background. On the other hand, neutrinos are subject to phase mixing of their orbits,which results in a characteristic Jeans mass scale corresponding to superclusters, as was realized by many authors recently (Doroshkevich et al. 1980abc, Klinkhamer and Norman 1981, Bond et al. 1980, Sato and Takahara 1980). The collapse of such systems was shown by Zeldovich (1970) to lead to highly anisotropic structures, the pancakes. These are not isolated: a cellular structure would form with huge voids in between, filaments would appear at the intersection lines of pancakes. The neutrino mass explains this structure in a simple and elegant way, as we now detail.

LINEAR PERTURBATIONS IN A NEUTRINO DOMINATED UNIVERSE

Small fluctuations may emerge naturally near the big bang itself. In the presently popular grand unified theories of particle physics the most important fluctuations are those which preserve the baryon/photon ratio, namely the adiabatic ones. These fluctuations are presumed to have a smooth scale free spectrum existing over a wide range of mass scales. As the expansion proceeds, larger and larger masses come within the horizon. There are two competing effects: gravity attracts the particles towards the highest densities, while the pressure due to thermal motion tries to prevent this. On large scales gravity always wins, matter condenses in some regions, and is rarefied in others. On small scales pressure is more important. The perturbations behave like acoustic waves; excess density is accompanied by excess pressure, and the local density oscillates.

Until the temperature has dropped to a few thousand degrees, the radiation is still sufficiently energetic to keep the matter ionized. The growth of baryon fluctuations by gravitational instability is inhibited in the ionized phase because the radiation provides a strong source of viscosity. The radiation is scattered mostly by free electrons, so once the electrons recombine into H atoms, radiation streams independently of the matter. There is no longer any resistance to fluctuation growth, and gravitational instability proceeds. This happens at a relatively late stage, at $z \sim 1000$. During recombination the smallest scale fluctuations are subject to viscous damping (Silk 1968). The compressed radiation tends to diffuse and thereby smooth out all baryon fluctuations below $\sim 10^{13} - 10^{15}\ M_{\odot}$.

Even when pressure becomes negligible, the rate of fluctuation growth is small in low density universes for redshifts $< \Omega^{-1}$, when the curvature dominates the expansion. This can lead to a total growth factor from decoupling to today as small as 15, to be compared with the total growth of 1000 if $\Omega=1$. Since the photons were not scattered after decoupling, the present value of the temperature fluctuations of the CBR reflects the amplitude of the baryon density perturbations at that epoch. Upper limits on small scale fluctuations, such as given by Davies in this volume (10′ corresponds to $\sim 10^{15}$ $M_\odot$) implies growth by even that factor of 1000 will not be enough to give nonlinear structure by now in baryon-dominated universes.

As Doroshkevich et al. (1980a) and Bond et al. (1980) emphasized, one of the principle triumphs of a universe dominated by neutrinos or other weakly interacting particles such as gravitinos or photinos is that they can beautifully sidestep this difficulty. This occurs for two reasons. Firstly, Ω can be larger, hence the curvature-dominated era, if it exists at all, will have occured for a much shorter time. More importantly, once neutrinos become non-relativistic, which occurs before recombination, their fluctuations become gravitationally unstable on sub-horizon scales, as long as these scales are above the instantaneous value of the neutrino Jeans mass. Consequently, the amplitude of the neutrino density fluctuations greatly exceeds those of the baryon perturbations by recombination. Once decoupling occurs, the baryon fluctuations experience an accelerated growth until they equal the neutrino fluctuations. This new feature implies one can get nonlinearity occuring at z ~5, and still have the induced temperature fluctuations just smaller than the upper limits on small scales. Though safely lower than the upper bounds set by observers so far, the $\Delta T/T$ structure on all scales larger than ~10′ is predicted to be rich on the 10^{-5} to 10^{-4} level in a neutrino-dominated universe.

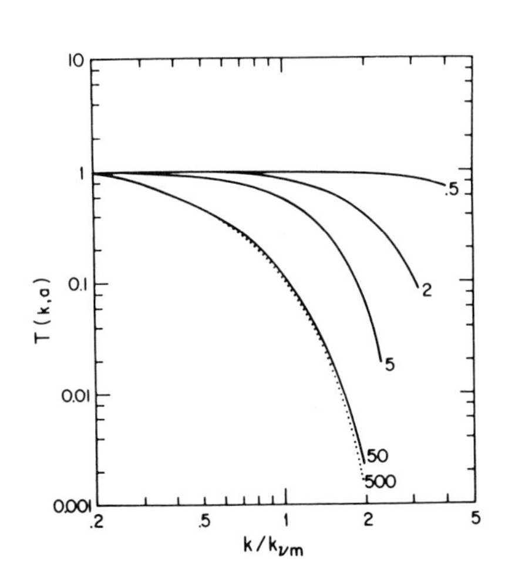

Figure 1. The transfer function of neutrino fluctuations (see text).

The occurrence of a damping cutoff in the neutrino fluctuation spectrum determines the nature of large scale structure formation in neutrino-dominated universes. Since momentum suffers the cosmological redshift, once neutrinos go nonrelativistic their velocity slows from the speed of light down to a present rms value of 6 km/s (30 eV/m_ν). The total comoving displacement the typical freely streaming neutrino traverses converges to a finite time-independent value given by the mass scale ~ 0.5 $M_{\nu m}$ (Bond and Szalay 1982), where the damping scale is the maximum value the neutrino Jeans mass attains,

$$M_{\nu m} = 2\, m_p^{\,3}\, m_\nu^{-2} = 3 \times 10^{15} \quad (30\ \mathrm{eV}/m_\nu)^2\ M_\odot$$

(Bond et al. 1980, Bisnovaty-Kogan and Novikov 1980, Doroshkevich et al.

1980c). Below this scale orbits of neutrinos streaming in different directions and/or with different speeds phase mix, effectively erasing any inhomogeneities initially present. Doroshkevich et al.(1980a) and Wasserman (1982) give a quantitative discussion of this limiting regime. Above this scale, neutrinos cannot move from one lump to another, and the self-gravity of compressions cause them to grow in magnitude via the usual Jeans instability. Bond and Szalay (1981, 1982) and Peebles (1982) have numerically integrated the coupled Einstein-Boltzmann equations to determine the detailed shape of the fluctuation spectra, which ties together these two regimes. The temporal evolution of the transfer function (amplitude of neutrino perturbations normalized to a fluctuation with mass much above the critical scale) is illustrated in Fig.1., taken from Bond and Szalay (1982). The scale factor has a=1 when the neutrinos just become nonrelativistic. It is apparent that most damping occurs between a=0.5 and a=50, which is near recombination, after which the spectral shape down to $\ll M_{\nu m}$ is frozen in.

NONLINEAR STRUCTURE IN THE UNIVERSE

Once the first mass scale in a spectrum with a damping cutoff such as that obtained from Fig.1. reaches nonlinearity, neutrino trajectories cease expanding away from each other and begin converging, resulting in the temporary formation of caustics. Just as in elasticity, we can envisage this behaviour by considering a cubical volume smaller than the damping scale. The cube suffers deformation due to the particle motion with contractions occuring along some axes, expansions along others. A spherically symmetric contraction is a special, highly degenerate case. Generally, the distribution of deformations in each of the principal directions favours asymmetry (Doroshkevich 1970). Gravitational attraction further amplifies this strain, and a highly flattened quadrangle which is still expanding or mildly contracting in the other directions results. The mass inside the cube is preserved, so when both the thickness and the volume of the cube approaches zero, its density becomes very high and a flat 'pancake' is formed, as was originally suggested by Zeldovich (1970). At first they form at isolated spots where the initial velocity perturbations had the largest gradient. Soon these regions grow, turning into huge thin surfaces which intersect, tilt and form the walls of a cell-structure which is itself unstable gravitationally. The universe may be at this cellular stage today as detailed numerical calculations indicate.

In the nonlinear phase, mode-mode coupling among Fourier components sends power to short wavelengths, and correlates phases even though the initial fluctuation spectrum may have had random phases. The evolution of structure is then best calculated in real space. As Shandarin discusses in this volume, Arnold, Zeldovich and Shandarin (1982) have applied the methods of catastrophe theory to analyze structure that develops in potential motion, i.e. with no vorticity, which, if present, should be small by the end of the linear phase anyway. They found that the two-dimensional pancakes are only the lowest order singularities, and one-dimensional strings (superclusters?) and zero-dimensional points (rich clusters?) should also appear. These features can be seen in the

N-body simulations of Doroshkevich et al. (1980d), Melott (1982), Clypin and Shandarin (1982), Frenk et al. (1982).

When the intersection of trajectories takes place, gas pressure builds up, the velocity of the collapsing gas exceeds the speed of sound, and a shock wave is formed (Sunyaev and Zeldovich 1972, Doroshkevich et al. 1978). The gas heats up to more than a million degrees, and emits radiation over a broad spectrum, cooling the gas, especially in the central layers, where the density is higher. Recently, Bond, Centrella, Szalay and Wilson (1982) calculated the cooling of collapsing neutrino-baryon pancakes, the details of which are considerably different from those in a pure baryon pancake: the baryon density is lower, but the infall velocities are higher, and thus the cooling rate is much slower. We found that the fraction of cooled baryons is a very sensitive function of the total mass of the collapsing object. Above $\sim 10^{15}$ $M_{\odot}$, no more than about ten percent of the baryons can cool before significant transverse flows take place. This cooling is required, since only cool gas is able to form smaller lumps, the seeds of galaxies. The details of this cooling may be important: a transverse flow in the pancake towards the line singularities will increase the local density, thus enhance cooling. Strings may well be the locations of the most efficient galaxy formation. The UV and soft X-ray emission can photoionize the intergalactic medium, making galaxy formation in regions that have not yet formed pancakes more difficult, which would accentuate the contrast in galaxy density between the strings and pancakes vs. voids, even though the density contrast may be only ~3-10 (Zeldovich and Shandarin 1982). One of the principal difficulties in this picture is how to drive fragmentation at all, since isolated pancakes have no power on small transverse scales. This question was adressed by Doroshkevich (1980), but has not yet been satisfactorily answered.

Cosmic neutrino pancakes may lead to an attractive explanation of the dark halos of galaxies. In the pancake collapse most of the neutrinos acquire large velocities. Some, however, move only slowly at first, since these neutrinos were initially closer to the midplane of the pancake. Around this midplane, we assume a thin gas layer condenses and fragments. Bond, Szalay and White (1982) have constructed a simple one-dimensional model which demonstrates that the slowest moving neutrinos will add on to the baryonic seeds first, followed by progressively faster ones, resulting in a halo which has the total-to-baryonic mass ratio and velocity dispersions needed to describe the halos of spirals. Further, the one-at-a-time addition makes it more likely an r^{-2} halo will form than the r^{-3} found in violent relaxation studies, which would be expected if the dark matter and the baryons collapsed together. Finally, the fraction of neutrinos captured and their velocity dispersion are found to rise with the baryonic mass. The more difficult problem of neutrino capture on galaxies formed along strings has not yet been addressed.

In conclusion, we hope it has become clear that the neutrino-dominated universe seems capable of explaining most features of the large scale structure. At several points in the development of the theory (CBR fluctuations, gas cooling, the phase space constraints of Tremaine and Gunn (1979), the formation of galaxy halos) <u>it just works</u>, which may be the best argument of all for taking it seriously.

REFERENCES:

Arnold,V.I., S.F.Shandarin, Ya.B.Zeldovich. 1982. Geophys. Astrophys. Fluid Dynamics 20,111.
Bond,J.R., J.Centrella, A.S.Szalay, J.R.Wilson. 1982. Preprint.
Bond,J.R., G.Efstathiou, J.Silk. 1980. Phys.Rev.Lett. 45,1980.
Bond,J.R., A.S.Szalay. 1981. Proc Neutrino'81. 1,59
Bond,J.R., A.S.Szalay. 1982. submitted to Ap.J.
Bond,J.R., A.S.Szalay, S.D.M.White. 1982. submitted to Nature
Clypin,A.A., S.F.Shandarin. 1982. submitted to M.N.R.A.S.
Cowsik,R., J.McClelland. 1972. Phys.Rev.Lett. 29,669
Doroshkevich,A.G. 1970. Astrofisika, 6,581.
Doroshkevich,A.G., S.F.Shandarin, E.Saar. 1978. M.N.R.A.S. 184,643.
Doroshkevich,A.G.1980. Sov.Astron. 57,259.
Doroshkevich,A.G., M.Yu.Khlopov, R.A.Sunyaev, A.S.Szalay, Ya.B.Zeldovich. 1980a. Proc. Xth Texas Symposium on Relativistic Astrophysics, New York Acad. Sci. p.32.
Doroshkevich,A.G., M.Yu.Khlopov, R.A.Sunyaev, Ya.B.Zeldovich. 1980b. Pisma Astr. Zh. 6,457.
Doroshkevich,A.G., M.Yu.Kholpov, R.A.Sunyaev, Ya.B.Zeldovich. 1980c. Pisma Astr. Zh. 6,465.
Doroshkevich,A.G., E.V.Kotok, I.D.Novikov, A.N.Polyudov, S.F.Shandarin, U.S.Sigov. 1980d. M.N.R.A.S. 192,321.
Gershtein,S.S., Ya.B.Zeldovich. 1966. JETP Lett. 4,174.
Klinkhamer,F.R., C.A.Norman. 1981. Ap.J.Lett. 243,1.
Lubimov,V.A., E.G.Novikov, V.Z.Nozik, E.F.Tretyakov, V.S.Kozik. 1980. Phys.Lett. 943,266.
Marx,G., A.S.Szalay. 1972. Proc. Neutrino'72. 1,123.
Melott,A. 1982. M.N.R.A.S. in press.
Olive, K.A., D.N.Schramm, G.Steigman, M.S.Turner, J.Yang.1981. Ap.J. 246,557.
Peebles,P.J.E. 1982. Ap.J. 258,415
Sato,H., F.Takahara. 1980. Preprint RIFP-400. Kyoto University
Schramm,D.N., G.Steigman. 1981. Ap.J. 243,1.
Silk,J. 1968. Ap.J. 151,459.
Sunyaev,R.A., Zeldovich,Ya.B. 1972. Astron.Astrophys. 20,189.
Szalay,A.S., G.Marx. 1976. Astron.Astrophys. 49,437.
Tremaine,S., J.E.Gunn. 1979. Phys.Rev.Lett. 42,407
Wassermann,I. 1981. Ap.J. 248,1.
Zeldovich,Ya.B. 1970. Astron.Astrophys. 5,84.
Zeldovich,Ya.B., S.F.Shandarin. 1982. Pisma Astr.Zh. 8,259.
Frenk,C., S.D.M. White, M. Davis. 1982. Preprint.

MASSIVE NEUTRINOS AND ULTRAVIOLET ASTRONOMY

D.W. Sciama
International School for Advanced Studies, Trieste
and Department of Astrophysics, Oxford University

ABSTRACT

Massive neutrinos (or photinos) dominating galactic halos may decay into less massive particles by emitting ultraviolet photons. The lifetime for this process can be calculated from particle physics in a model-dependent way. The observed ultraviolet background constrains this lifetime to exceed about 10^{24} seconds. If the photon energy exceeds 13.6 ev, the existence of HI structures from the galactic plane imposes a similar constraint. The existence of Si IV and C IV in the halos of our own and other galaxies could be due to $\sim$ 50 ev photons emitted by neutrinos or photinos of rest-mass $\sim$ 100 ev if their lifetime $\sim 10^{27}$ seconds. This lifetime could be in agreement with the theoretical value for 100 ev particles.

INTRODUCTION

The idea that massive neutrinos might be unstable was suggested by Tennakone and Pakvase (1971) and by Bahcall, Cabibbo and Yahill (1972). The production of photons in such a decay process was considered by Pakvase and Tennakone (1972). The crucial idea that if massive neutrinos dominate galactic halos such photons would lie in the ultraviolet and might be detectable was introduced by de Rujula and Glashow (1980). Their calculations of the decay rate have since been improved by Pal and Wolfenstein (1982), who give references to earlier work.

An alternative possibility is that massive photinos play some or all of the astronomical roles which have been attributed to massive neutrinos. These photinos are spin 1/2 partners of photons in super-symmetric theories, in which bosons and fermions can be transformed into one another, and can belong to the same (super) multiplet (for a review, see Fayet and Ferrara 1977). Their role in cosmology, and in particular their decay into photons, has been considered by Cabibbo, Farrar and Maiani (1982).

Observations of the ultraviolet background can be used to place a lower limit on the radiative lifetime τ of a decaying particle of rest-mass lying in the range of 20 to 100 electron volts, which is assumed to dominate a) the universe, b) clusters of galaxies, and c) the halo of our Galaxy.

G. O. Abell and G. Chincarini (eds.), Early Evolution of the Universe and Its Present Structure, 313–320.

If the photon energy exceeds 13.6 ev, the ionization of hydrogen becomes possible, and the existence of neutral hydrogen structures away from the galactic plane (in the plane absorption of such photons would dominate) can also be used to derive a lower limit on τ.

Finally, various observed or inferred ionization stages of hydrogen and helium in the intergalactic medium, and of silicon and carbon in galactic halos, including our own, if attributed to photons from decaying neutrinos or photinos, would lead to a mass ($\sim$ 100 ev) and a radiative lifetime ($\sim 10^{27}$ secs) for these particles which would be acceptable both cosmologically and in terms of elementary particle physics. These ideas could be tested by searching for a narrow line at $\sim$ 50 ev in the radiation background of the galaxy.

ELEMENTARY PARTICLE ASPECTS

The decays envisaged are $\nu_1 \rightarrow \gamma + \nu_2$, where ν_2 is a neutrino of lower mass m_2 than ν_1, and $\bar{\gamma} \rightarrow \gamma + \bar{g}$, where $\bar{\gamma}$ is a photino and $\bar{g}$ is goldstino (the spin 1/2 partner of the Goldstone boson whose existence results from the spontaneous breaking of supersymmetry).

If the parent particle is at rest relative to the observer, then conservation of energy and momentum in the decay process results in a photon energy E_γ given by:

$$E_\gamma = \frac{m_1^2 - m_2^2}{2m_1} .$$

If $m_2 << m_1$ we can simplify this to:

$$E_\gamma \sim \frac{1}{2} m_1 .$$

Thus, if m_1 lies in the range 20 to 100 ev, which is relevant both for cosmology (Cowsik and McClelland 1972) and for the domination of galaxy clusters and individual galaxiès (Tremaine and Gunn 1979), we would have

$$E_\gamma \sim 10 - 50 \text{ ev} ,$$

(unless $m_2 \approx m_1$), so that the decay photons would lie in the ultraviolet.

The lifetime for neutrino decay depends critically on whether a cancellation process called GIM suppression is operating (GIM = Glashow, Iliopoulos, Maiani). If one assumes that there are only the three standard neutrino flavors (e-type, μ-type and τ-type), then GIM suppression would occur and theoretical values of τ would range from:

$$\frac{10^{36}}{\sin^2 2\beta_1} \left(\frac{30 \text{ ev}}{m_1}\right)^5 \text{ sec}$$

to ten times this quantity (Pal and Wolfenstein 1981). Here β_1 is the mixing angle between ν_1 and ν_2 (which also influences their oscillation rate).

This theoretical lifetime turns out to be too long to be of astronomical interest for the foreseeable future, but various mechanisms exist for eliminating the GIM mechanism. These have been discussed by de Rujula and Gashow and by Pal and Wolfenstein. Examples are introducing a fourth generation with a much heavier charged lepton (analogous to e, μ, and τ) or a fourth neutrino without an associated charged lepton. In the latter case, Pal and Wolfenstein obtain for $m_1 >> m_2$:

$$\tau \sim \frac{6 \times 10^{29}}{\sin^2 2\beta_2} \left(\frac{30 \text{ ev}}{m_1}\right)^5 \text{ sec},$$

where β_2 is the mixing angle between ν_1 and the fourth neutrino. The resulting decrease in τ would bring it into the realm of astronomical interest if $\sin^2 2\beta_2$ is not much less than unity, as we shall see.

The photino lifetime has been calculated by Cabibbo, Farrar and Maiana (1982). They obtain:

$$\tau \sim 6.8 \times 10^{22} \left[\frac{d}{(100 \text{ Gev})^2}\right]^2 \left(\frac{30 \text{ ev}}{m_\gamma}\right)^5 \text{ sec},$$

where d is a parameter associated with the breaking of supersymmetry (roughly speaking, this breaking occurs at an energy $\sim d^{1/2}$). There is no question of GIM suppression here, but the value of d is unknown. Some particle physicists consider that it is likely to exceed the weak interaction value of 10^5 Gev^2. For the lower part of this range ($d \leq 3 \times 10^7$ Gev^2), the resulting decay rate would be faster than even the enhanced neutrino rate, and again could be astronomically relevant.

Finally, we note that values of τ which would arise if leptons and quarks are composite have been considered by Stecker and Brown (1982).

ASTRONOMICAL LOWER LIMITS ON τ

We note first the coincidence that the photon flux at the Earth coming from the proposed galactic neutrinos has the same order of magnitude as that coming from cosmological neutrinos (ignoring absorption effects for the moment). This follows from the fact that the proposed galactic enhancement in the neutrino concentration over its cosmological value is of the same order ($\sim 10^5$) as the ratio of the radius of the universe ($\sim$ 3000 Mpc) to the scale-height of the galactic halo ($\sim$ 30 kpc). The main difference is that the galactic flux is nearly monochromatic (since the velocity-dispersion of the galactic neutrinos $\sim$ 200 km/sec << c), whereas the cosmological flux would be drawn out into a continuous spectrum by the differential redshift associated with the expansion of the universe.

In fact, this spectrum would have the form:

$$\underset{(\lambda \geq \lambda_0)}{I_\lambda} = \frac{c n_\nu(z=0)\, \lambda_0^{3/2}}{H_0\, 4\pi\tau\, \lambda^{5/2}} \left[1 + (2q_0 - 1)\left(1 - \frac{\lambda_0}{\lambda}\right)\right]^{-1/2},$$

where λ_0 is the rest wavelength of the decay photon, λ the observed wavelength, τ the neutrino lifetime and q_0 the deceleration parameter. Any absorption effects would have to be added to this relation. (There is a numerical error $\sim 10^4$ in equation [7c] of de Rujula and Glashow [1980] which has been corrected by Kimble, Bowyer and Jakobsen [1981]).

It has been pointed out by Stecker (1980) and Kimble, Bowyer and Jakobsen (1981) that one can use this spectrum to obtain lower limit on τ from the observed ultraviolet background even if λ_0 corresponds to a wavelength at which the galaxy is opaque. Since the observed background is probably due to other sources, it has been used to limit τ rather than to determine it. From an observed background $\sim$ 200 to 300 photons cm^{-2} sec^{-1} $ster^{-1}$ A^{-1}, they deduce that:

$$\tau > 10^{22} - 10^{23} \text{ sec}; \quad 10 \text{ ev} < \frac{hc}{\lambda_0} < 50 \text{ ev}$$

if the intergalactic medium is transparent out to the largest redshifts involved ($z \sim 6$). We return to the question of this transparency later.

Stecker also discussed the possibility that a reported increas in the background at 1700 Å might represent a photon flux from galactic neutrinos and he suggested that $\tau \sim 3 \times 10^{24}$ sec. According to Kimble, Bowyer and Jakobsen, this increase has not been confirmed and is best treated as an upper limit to the actual intensity.

These authors also derived a lower limit on τ for galactic neutrinos, from the general observed background in the 30 to 50 ev range (which corresponds to the mass range in which neutrinos of cosmological origin could dominate the Galaxy). Their limit depends on the uncertain opacity of the Galaxy at these photon energies, and on the photon energy itself, but lies in the range:

$$\tau > 10^{20} - 10^{22} \text{ sec}; \quad 30 \text{ ev} < E_\gamma < 50 \text{ ev} .$$

More stringent limits have been derived by Shipman and Cowsik (1981) and Henry and Feldman (1981) from optical and ultraviolet observations of the Virgo and Coma clusters of galaxies. If these clusters are dominated by neutrinos of appropriate mass, the derived limits are:

$$\tau > 10^{23} - 10^{25} \text{ sec}; \quad 1 \text{ ev} < E_\gamma < 10 \text{ ev} .$$

Shipman and Cowsik consider that with existing or proposed instruments one could improve these limits up to the range 10^{26} to 10^{27} sec.

All these limits are derived from direct observations of photon fluxes. One can also use arguments derived from the ionizing effects of photons, as pointed out by Melott and Sciama (1981). These differ from the previous arguments in that they can lead to limits on (or perhaps evaluations of) photon fluxes at positions distant from the galactic plane, where the effects of absorption by neutral hydrogen or dust will be different (and in general less). For example, Melott and Sciama (1981) demanded that these photon fluxes should not completely ionize the High Velocity Clouds. These clouds are neutral hydrogen features observed at 21 cm to be predominantly approaching us with velocities of a few hundred kilometers per second. Various arguments suggest that some clouds lie at least a kiloparsec above the galactic plane, while a recent estimate puts them at tens of kiloparsecs away. In fact, the limits obtained for τ are valid so long as the clouds lie within the Local Group of galaxies, but not within the galactic plane. One finds in this way that:

$$\tau > 10^{24} \text{ sec}; \qquad E_\gamma > 13.6 \text{ ev} .$$

These astronomical lower limits on τ are several orders of magnitude less than the expected theoretical values, except for photinos with a small value of d. The next step should therefore be to search for more sensitive ionization processes which would lead either to an actual effect being identified, or at least to a much more stringent limit on τ in the relevant photon energy range. Some possible ionization processes will be considered in the next section.

POSSIBLE IONIZATION PROCESSES

We consider first the intergalactic medium, which is generally believed to be highly ionized in H I and He I (this is inferred from the absence of absorption troughs in quasar spectra (Gunn and Peterson 1965; Green *et al.* 1980; Wampler *et al.* 1973; Ulrich *et al.* 1980). The ionization source is unknown. Perhaps the most plausible possibility is ultraviolet radiation from quasars (Sherman 1981).

The ionization of intergalactic H I by photons from decaying cosmological neutrinos was suggested by Raphaeli and Szalay (1981) and by Sciama (1981a). The first authors took $\tau < 10^{25}$ sec, which would lead to ionization at relatively early cosmic epochs, but is in disagreement with most of the particle physics expectations. Sciama took $\tau \sim 10^{27}$ sec, which may avoid this disagreement, but would lead to ionization of H I only after a cosmic epoch corresponding to a redshift ~ 4 and only for a low density intergalactic medium. He also pointed out that if He I is ionized in this way, one would need $m_\nu >> 50$ ev in order that the decay photons be energetic enough. On the other hand, if appreciable quantities of He II are present in the IGM (which might be testable by observations with IUE [Gondhalekar 1981] or Space Telescope) one might be able to infer that $m_\nu \lesssim 110$ ev.

Another possible ionization process involves Si IV and C IV in galactic halos. The presence of these high ionization stages is known for the Milky Way halo from the IUE observations (*e.g.*, Savage and de Boer 1979, 1981; Bromage, Gabriel and Sciama 1980; Ulrich *et al.* 1980; Pettini and West 1982). Their presence in other galactic halos would follow from the now widely-held view that many of the absorption line systems in quasar spectra arise in intervening galactic halos (Weymann, Carswell and Smith 1981; Young, Sargent and Boksenberg 1982).

The ionization involved could be collisional in origin ($T \sim 10^5$°K) or due to photons from hot stars, hot gas, or the integrated ultraviolet radiation from quasars, etc. The recent observations of Pettini and West (1982) which suggest that in our Galaxy N(C IV)/N(Si IV) = 4.5 ± 1.5 throughout the region between 1 and 3 kpc from the galactic plane, put in doubt both the collisional explanation (which would require an abnormally constant temperature throughout this region [but see

Hartquist 1982]), and the hot stars (whose radiation would be too soft to give the observed C IV/Si IV ratio).

The other conventional explanations are numerically reasonable but may be ineffective because of absorption of H I, He I and He II near the plane of the Galaxy, in the outer regions of the halo, in intergalactic space, and in quasars. Accordingly, it is of interest to examine the possibility that the ionization is due to photons from decaying neutrinos (Sciama and Melott 1982) or photinos (Sciama 1982b,c) which are assumed to dominate galactic halos. Such particles would then exist close to any observed Si IV and C IV, and this might overcome the absorption problem. Evidence against this ionization hypothesis has been adduced by Feldman, Brune and Henry (1981), who observed an emission line component of the background radiation field at high galactic latitudes. They attributed this to hot gas in the halo, which they also related to the observed Si IV and C IV. However, their observed emission lines could have been produced in warm gas close to the galactic plane (Deharveng, Joubert and Berge 1982). This seems the likely explanation in view of the results of Pettini and West on the constancy of the C IV/Si IV ratio noted above.

The photon energy required to produce C IV (47.9 ev) would imply that the parent particle has a mass $\sim$ 100 ev, and the abundance of C IV then implies that $\tau \sim 10^{27}$ sec. These results are cosmologically reasonable, are compatible with some of the particle physics estimates of τ and are comparable to the values required by the IGM ionization hypothesis. However, if we demand that the universe has the critical density (Guth 1982), its age would be rather low ($\sim 7 \times 10^9$ years) for neutrinos, and photinos might be a better choice. The reason is that for photinos we could tentatively take $d \sim 5 \times 10^7$ Gev^2. They would then decouple earlier in the big bang than do neutrinos ($T_d \sim 200$ Mev instead of 1 Mev) (Sciama 1982c) and so would have a lower cosmological number density by a factor ~ 4 (since muons and pions annihilating after photinos decouple would feed the 3°K background but not the photinos). A critical density in photinos would then correspond to a higher age for the universe ($\sim 13 \times 10^9$ years) which would be more compatible with other estimates of this quantity. Moreover, with this value of d and $m_{\tilde{\gamma}} \approx 100$ ev, $\tau_{\tilde{\gamma}}$ would be close the value $\sim 10^{27}$ sec required by our ionization hypothesis. Finally, we note that this suppression of the photino abundance would just be compatible with primordial nucleosynthesis, which appears to permit three "neutrino" types but not four so long as they were relativistic at the time of nucleosynthesis (Schramm 1982).

These ideas could be tested by searching for a narrow line ($\Delta\lambda/\lambda < 10^{-3}$) in the galactic background at ~ 50 ev with a flux $\sim 10^3$ photons cm^{-2} sec^{-1}. It is hoped that such a measurement can be carried out in the near future.

REFERENCES

Bahcall, J.N., Cabibbo, N., and Yahill, A. 1972, Phys. Rev. Lett., 28, 316.

Bromage, G.E., Gabriel, A.H., and Sciama, D.W. 1980, Proc. 2nd European IUE Conf. Report No. ESA SP-157.

Cabibbo, N., Farrar, G.R., and Maiani, L. 1981, Phys. Lett., 105B, 155.
Cowsik, R., and McClelland, J. 1972, Phys. Rev. Lett., 29, 669.
Deharveng, J.M., Joubert, M., and Berge, P. 1982, Astron. Astrophys., 109, 179.
De Rujula, A., and Glashow, S. 1980, Phys. Rev. Lett., 45, 942.
Fayet, P., and Ferrara, S. 1977, Phys. Reports 32C, 249.
Feldman, P.D., Bruce, W.H., and Henry, R.C. 1981, Ap. J. Lett., 249, L51.
Gondhalekar, P. 1981, private communication.
Green, R.F., Pier, J.R., Schmidt, M., Estabrook, F.B., Lane, A.L., and Wahlquist, H.D. 1980, Ap. J., 239, 483.
Gunn, J.E., and Peterson, B.A. 1965, Ap. J., 142, 1633.
Guth, A.H. 1982, Proc. Roy. Soc. A., to be published.
Hartquist, T.W. 1982, to be published.
Henry, R.C., and Feldman, P.D. 1981, Phys. Rev. Lett., 47, 618.
Kimble, R., Bowyer, S., and Jakobsen, P. 1981, Phys. Rev. Lett., 46, 80.
Melott, A.L., and Sciama, D.W. 1981, Phys. Rev. Lett., 46, 1369.
Pakvase, S., and Tennakone, K. 1972, Phys. Rev. Lett., 28, 1415.
Pal, P.B., and Wolfenstein, L. 1982, Phys. Rev. D, 25, 766.
Pettini, M., and West, K.A. 1982, Ap. J., 260, 561.
Rephaeli, Y., and Szalay, A.S. 1981, Phys. Lett., 106B, 73.
Savage, B.D., and de Boer, K. 1979, Ap. J. Lett., 230, L77.
Savage, B.D., and de Boer, K. 1981, Ap. J., 243, 460.
Schramm, D.N. 1982, Proc. Roy. Soc. A., to be published.
Sciama, D.W. 1982a, Mon. Not. Roy. Astr. Soc., 198, 1P.
Sciama, D.W. 1982b, Phys. Lett., 112B, 211.
Sciama, D.W. 1982c, Phys. Lett., 114B, 19.
Sciama, D.W., and Melott, A.L. 1982, Phys. Rev. D, 25, 2214.
Sherman, R.D. 1981, Ap. J., 246, 365.
Shipman, H.L., and Cowsik, R. 1981, Ap. J. Lett., 247, L111.
Stecker, F.W. 1980, Phys. Rev. Lett., 45, 1460.
Stecker, F.W., and Brown, R.W. 1982, Ap. J., 257, 1.
Tennakone, K., and Pakvase, S. 1971, Phys. Rev. Lett., 27, 757.
Tremaine, S., and Gunn, J.E. 1979, Phys. Rev. Lett., 42, 407.
Ulrich, M.H., et al. 1980, Mon. Not. Roy. Astr. Soc., 1982, 561.
Wampler, E.J., Robinson, L.B., Baldwin, J.A., and Burbidge, E.M. 1973, Nature, 243, 336.
Weymann, R.J., Carswell, R.F., and Smith, M.G. 1981, Ann. Rev. Astron. Astrophys., 19, 41.
Young, P., Sargent, W.L.W., and Bokensberg, A. 1982, Ap. J. Suppl., 48, 455.

Discussion

Stecker: It should be pointed out that some recent analyses based on observations of He in less evolved galaxies give a very low primordial He abundance ($\sim$ 21.6%) as compared with the prediction of the standard model with $N_\nu \gtrsim 3$ ($\gtrsim$ 25%). This has important complications for all arguments trying to limit N_ν, since they may be basically self-contradictory (Stecker, 1980, Phys. Rev. Lett., 44, 1237; 46, 517; Rana, 1982, Phys. Rev. Lett., 48, 209; Rayo et al., 1982, Ap. J., 255, 1).

Sciama: A recent critical analysis of the data on helium abundances by Pagel at the Royal Society meeting on cosmology yields a preferred value of 0.25. This is partly based on the recent work of Kunth and Sargent, shown on a poster at this meeting. This value, together with an estimate of n_b/n_γ of 3×10^{-10}, would lead to an allowed number of effective neutrino types lying between 3 and 4. This result would be important if correct, as it would require photinos to be suppressed (unless the tau neutrino is more massive than 1 Mev, so that it was nonrelativistic when neutrons froze out).

WHAT DOES THE DYNAMICAL ANALSIS OF CLUSTERS OF GALAXIES TELL US ABOUT MASSIVE NEUTRINOS?

G. des Forêts, D. Gerbal, G. Mathez, A. Mazure,
E. Salvador-Solé*
Observatoire de Meudon
*Universidad de Barcelona

It is often claimed that massive neutrinos (ν's) can solve the "missing mass" problem, but it is not so clear in the particular case of clusters of galaxies (C.O.G.). Let us assume that the unseen matter is composed by massive ν's only. If they are cosmological, the ν's should obey Fermi-Dirac statistics with a density of $\sim$ 100 ν/cm^3/species. But if "relic," the ν's would be so slow (1) that they cannot exist in this form (because of the previous Jeans instability or because they are trapped in wells generated by baryonic matter). Since the time when the ν's decoupled from the primeval mixture (T $\sim$ 3-1 MeV), the ν's can be considered as a "gravitational plasma," so that violent relaxation occurs in inhomogeneous systems, leading to a Lynden-Bell distribution defined by three parameters: n_ν (numerical density), V_ν (r.m.s. velocity) and the ν-mass, m_ν, all unknown. All three of these parameters are, in fact, necessary to define a state of ν-matter.

Dynamical analyses of C.O.G. with a single baryonic component model lead, for the mass defect ratio (Virial mass/luminous mass), to values running from 3 (2) to 20.

Two interpretations are possible: 1. Galaxies have dynamical masses 3 to 20 times their luminous ones and ν's are located in massive haloes; then $m_\nu \gtrsim$ 22 eV (4) and V_ν of the same order as the star velocity ($\sim$ 100 km/s). The total ν-mass is then 3 to 20 times the visible one. This solution is very simple, but it has been argued that if galaxies are too massive, the two-body relaxation time (too small) is in contradiction with the nonrelaxation state observed in the C.O.G. (3). Moreover, the "questionable" dependence of M/L on scale cannot be explained.

2. The ν's have the same density profile as the galaxies, and the virial masses of galaxies are equal to their luminous ones. This occurs if $V_\nu \equiv V_{gal}$, and suggesta that $n_\nu \propto$ n(baryonic) everywhere. From the dynamical value M/L we then obtain:

$$m_\nu \simeq \frac{(M/L)}{30} h_{50} \qquad \begin{array}{l} m_\nu = 1.5 - 9 \text{ eV } (H = 50); \\ m_\nu = 3 - 18 \text{ eV } (H = 100). \end{array}$$

These values have to be compared with the result $m_\nu \gtrsim 4$ eV of Tremaine and Gunn (4). Once again this explanation cannot elucidate the dependence of M/L on the scale of the system.

G. O. Abell and G. Chincarini (eds.), Early Evolution of the Universe and Its Present Structure, 321–323.

In fact, the previous considerations are obtained from one-component analysis. If massive ν's are present, they are dynamically influential and a two-component analysis is necessary. It then turns out that the predominance is governed by the ratio V_ν/V_{gal} (6 and references therein). a) In the very hot case ($V_\nu >> V_{gal}$), a baryonic cluster is located in the center of the ν-cluster. The gravitational potential is smooth (except at the center) and the modeling of the baryonic cluster is then not affected by the ν-cluster; the usual analysis and the missing mass problem are unaffected! b) There is, in principle, a possibility of detecting tepid ν's: i) If they exist, the asymptotic slope of the density profile of galaxies depends only on the properties of neutrinos (6). ii) If they do not exist, the galaxy density profile depends on the velocity dispersion profile of galaxies.

In the "cold" case, two possibilities exist: a) $V_\nu \lesssim V_*$ (100 km/s): ν's are confined in the galaxies of the cluster so that it is equivalent to the first case discussed. b) $100 \lesssim V_\nu \lesssim 500$ km/s (500 is the velocity of the most massive galaxies): Cold ν's are then located at the center of the cluster, deepening the potential so that a central enhancement must appear in the density profile. What is the real situation? Quintana (5) has noted such an enhancement in the center of the Coma cluster. We confirm that result by comparison of small- and large-scale maps of the Coma cluster (2). Are neutrinos then really detected in the Coma cluster? Unfortunately, other physical explanations are possible. For instance, the previously noticed existence of massive binary systems of galaxies (7) may also deepen the potential, as well as taking into account the anisotropy of the dispersion velocity tensor.

In conclusion: Although neutrinos are exciting from a cosmological point of view, they do not solve the missing mass problem so easily, at least for clusters of galaxies. From dynamical consideration it seems to us that if massive ν's (or other inos) exist, they are probably located in galaxy haloes and then are distributed by mising in the centers of rich clusters, leading to possible observational effects.

REFERENCES

1. Doroshkevich, A.G., Khlopov, Y., Sunyaev, R.A., Szalay, A.S., and Zeldovich, Ya.B. 1981, Ann. N.Y. Acad. Sci., 375, 32.
2. Gerbal, D., Mathez, G., Mazure, A., and Salvadore-Solé, E. 1982, Moriond Second Astrophysics Meeting, p. 299, Edition Frontières, Paris. des Fôrets, G., Dominguez-Tenreiro, R., Gerbal, D., Mathez, G., Mazure, A., and Salvadore-Solé, E. 1982, submitted to Ap. J.
3. White, S.D.M. 1977, M.N.R.A.S., 179, 33.
4. Tremaine, S., and Gunn, J.E. 1979, Phys. Rev. Lett., 42, 407.
5. Quintana, H. 1979, A. J., 84, 15.
6. Capelato, H.V., Gerbal, D., Salvador-Solé, E., Mathez, D., Mazure, A. and Roland, J. 1979, Astron. Astrophys. Suppl., 38, 295.
7. Valtonen, M.J., and Byrd, G.G. 1979, Ap. J., 230, 655.

Discussion

Schallwich: I should mention that we have detached diffuse optical emission in a cluster of galaxies (A1146) which could provide as much mass as there is in the galaxies of the cluster.

Dressler: As you pointed out, observations of clusters of galaxies rule out a large two-body relaxation, which in turn implies that the missing mass is not associated with individual galaxies. This has not been considered a problem, however, since calculations such as those done by Richstone predict that the halos would be tidally stripped in rich clusters.

MULTIMASS MODELS FOR CLUSTERS OF GALAXIES

A. Mazure*, G. des Forêts*, D. Gerbal*, G. Mathez*,
E. Salvador-Solé†

*Observatoire de Meudon
†Universidad de Barcelona

It is now a widely spread opinion that a ratio of 10:1 between dark and luminous matter exists. Supported by the existence of flat rotation curves at large radii for spirals, this fact reinforces cosmological scenarios with, for instance, massive neutrinos. This content of dark matter is often estimated from the dynamical analysis of clusters of galaxies based essentially on the application of the Virial theorem or the monomassive Emden sphere or deduced from numerical simulations. However, a careful examination shows crucial failures in such approaches[1], at least the lack of a mass spectrum and/or of a dynamically influent Intra Cluster Medium. This has been included in simple models[1] together with other realistic features such as temperature gradient, isovelocity and/or isothermicity of the gravitational plasma. Our aim is thus to account simultaneously for all the available data concerning both galaxies and ICM; namely, the Nonisothermal Multimass Models[1] allow us to fit jointly the numerical density profiles of galaxies, the luminosity function, the velocity dispersion profiles versus magnitude or radius, the luminosity segregation[2], the X-ray temperature, luminosity and surface brightness profiles.

Applied to the Coma cluster, these models give two main results. First, it appears that the content of dark matter is only 3 ± 1 times the luminous one. Second, a central enhancement already underlined by Quintana[3] is exhibited in the numerical density profile of galaxies. Is this peak due to the presence of massive neutrinos or a massive bound system at the center of the cluster is discussed in another paper by D. Gerbal.

REFERENCES

[1] des Forêts, G., Dominguez-Tenreiro, R., Gerbal, D., Mathez, G., Mazure, A., Salvador-Solé, E. 1982, submitted to Ap. J.

[2] Capelato, H.V., Gerbal, D., Mathez, G., Mazure, A., Salvador-Solé, E., Sol, H. 1980, Ap. J., 241, 521.

[3] Quintana, H. 1979, A. J., 84, 15.

G. O. Abell and G. Chincarini (eds.), Early Evolution of the Universe and Its Present Structure, 325.

PERTURBATIONS IN THE UNIVERSE WITH MASSIVE "...INOS"

G.S. Bisnovatyi-Kogan, V.N. Lukash, I.D. Novikov
Space Research Institute, Moscow

Weak interacting particles (WIPs): neutrinos (ν_e, ν_μ, ν_τ), the hypothetical photino ($\bar{\gamma}$), the gravitino ($\bar{g}$), etc., may have nonzero rest mass. This fact is extremely important for cosmology. WIPs do not annihilate in the very early Universe and their number is preserved. If they have a rest mass, their mass density may dominate in the Universe (1).

The present number density of some kinds of WIPs depends mainly on the number of kinds of ordinary matter fermion pairs (e^+e^-, p^+p^-, $n\bar{n}$, etc.), N_d that existed in the thermal bath at the moment of decoupling of the WIPs from the matter and radiation. N_d is the greater the earlier the WIPs decoupled and, during the cosmological expansion, these pairs' annihilation results in the temperature difference of WIPs and relict radiation. Their present density numbers are related as (2): $n_\gamma/n_W \approx N_d^{3/4}$ for $N_d >> 1$. More rigorous calculations are given in the following table:

	ν_e, ν_μ, ν_τ	?	?	?
N_d	1	10	10^2	10^3
n_γ/n_W	2.85	12	64	360

At the present time $n_\gamma \simeq 450$, so for $m_\nu \simeq 30$ eV (3), we have $\rho_\nu \simeq 160\ m_\nu \simeq 10^{-29}$ g/cm^3. This value may close the Universe. The clustering of the neutrinos together with matter may solve the problems of the small scale angular isotropy of the relict radiation and of the "hidden" mass in the rich galaxy clusters.

The problems connected with the heavy ...inos in the Universe have been considered in many recent publications (4,9). Here we want to stress that all the results obtained for neutrinos are valid also for other WIPs, if we take into account the different present number densities, according to the table. For example, the characteristic mass of a WIP cluster for large N_d is of the order of $M_{W(min)} \simeq 10^{15}\ M_\odot\ N_d^{-3/2}$ $(m_W/30\ \text{eV})^{-2}$ and the ratio of the WIP mass, M_W, to the ordinary mass M

G. O. Abell and G. Chincarini (eds.), Early Evolution of the Universe and Its Present Structure, 327–329.

in the cluster is $M_w/M \simeq N_d^{-3/4} m_{w,eV}$ for $N_d^{3/4}$ eV < m_w < 1 MeV, where $m_{w,ev}$ is m_w in eV-units (4,5).

There are some peculiarities in the law of the growth of the perturbations of the relativistic ordinary matter in the gravitational field of the WIP cluster. The equation describing this growth at the stage when WIPs are nonrelativistic and dominate the expansion has the form (5):

$$t^2\ddot{\delta}_m + \frac{2}{3} t\dot{\delta}_m + \kappa^2(\delta_m - A) = 0 .$$

Here δ_m and δ_w (= $27A\kappa^2/8$) are Fourier-components of the relativistic matter and WIP density perturbations, $\kappa = ct/\lambda\!\!\!^{-}\sqrt{3} \approx t^{1/3}$, A = constant (gravitational potential of the WIP condensation), and $\lambda\!\!\!^{-}$ is a perturbation scale. The solution of this equation is a sine wave oscillating around the constant value: $\delta_m = A + B \cdot \sin(3\kappa + \zeta)$; B, ζ = const. Let us remember that in the case of nonrelativistic plasma the equation for δ_m is analogous, but the second term is 4/3 t $\dot{\delta}_m$ (10), which leads to the growth of perturbations instead of pure oscillations.

In a Universe with heavy neutrinos $\delta_{m1} \sim \delta_{\nu 1}$ at $t \sim t_1$, when neutrinos first become nonrelativistic. Then δ_m grows to $\delta_{\nu 1}\kappa_{\nu 1}^{-2} \geq \delta_{\nu 1}$, while δ_ν increases continuously as $t^{2/3}$. (Note that for $\kappa_{1} >> 1$, perturbations damp [11].) After the hydrogen recombination at $t \sim t_2$, ρ_ν/ρ_m is constant and the radiation pressure no longer prevents the growth of δ_m, which thus grows rapidly to $\sim \delta_\nu$ (12). For $m_\nu \simeq 30$ eV: $t_1 \simeq 300$ years, $\delta_{m2} \simeq \delta_{\nu 1}$ $(M/3 \cdot 10^{13} M_\odot)^{2/3} << \delta_{\nu 2}$, the characteristic mass of the usual galaxy cluster $\sim 3 \cdot 10^{13} M_\odot$ and the hidden mass of the neutrino halo $\sim 10^{15} M_\odot$, in agreement with observations.

The inequality $\delta_m << \delta_\nu$ at the epoch of the last scattering of relict quanta lowers the amplitude of small scale fluctuations of the relict radiation temperature: $\Delta T/T \simeq 10^{-5}$ $(1 + Z_0)$ for $m_\nu \simeq 30$ eV; the coefficient is 3 (10) times more for $m_\nu \simeq 3$ eV($m_\nu = 0$), and less in the presence of heavy neutral leptons ($m_L \gtrsim 1$ GeV) or primordial black holes. Z_0 is the moment of the beginning of the nonlinear evolution ($\delta_{m0} = 1$) (13).

REFERENCES

[1] Gershtein, S.S., Zeldovich, Ya.B. 1966, Pisma Zh. E.T.F., 4, 179.
[2] Landau, L.D., Lifshitz, E.M. 1976, Statistical Physics, "Nauka."
[3] Lyubimov, V.A., Novikov, E.G., Nozik, V.Z., Tretyakov, E.F., Kosik, V.S. 1980, preprint ITEF-62, Moscow.
[4] Bisnovatyi-Kogan, G.S., Novikov, I.D. 1980, A. Zh., 57, 899.
[5] Bisnovatyi-Kogan, G.S., Lukash, V.N., Novikov, I.D. 1980, Proc. of 5-Regional Meeting (IAU/EPS), Liege, Belgium, 29-31 July.
[6] Doroshkevich, A.G., Zeldovich, Ya.B., Syunyaev, R.A., Khlopov, M.Yu. 1980, Pisma A. Zh., 6, 457.
[7] Klinkhamer, S., Norman, C.A. 1981, Ap. J. Lett., 243, L1.
[8] Melott, A. 1982, Nature, 296, 721.
[9] Bond, J., Szalay, A., Turner, M. 1982, Phys. Rev. Lett., 48, 1636.
[10] Zeldovich, Ya.B., Novikov, I.D. 1975, The Structure and Evolution of the Universe, "Nauka."

[11]Bisnovatyi-Kogan, G.S., Zeldovich, Ya.B. 1970, A. Zh., 47, 942.
[12]Chernin, A.D. 1981, A. Zh., 58, 1.
[13]Zabotin, N.A., Naselskii, P.D. 1982, submitted to A. Zh.

Discussion

Bonometto: What is meant by $\Delta T/T$ in the results projected? Do you consider the significance to be due to an actual amplitude of the antenna beam? Which angles are being considered?

Novikov: The estimate $\Delta T/T$ in the paper means the average $\sqrt{<\Delta T^2/T>}$ (over all sky) for an angle of the order of the 10 arcmin.

Contopoulos: Does anyone know what are the present experimental limits for the mass of the neutrino?

Stecker: Recently R.W. Brown and I have compiled what we feel are the best present limits on τ_ν, which is really a function of photon energy $E \sim m_\nu/2$. We also point out a possible neutrino decay signal in the UV background which would give $m_\nu \sim 14$ eV and $\tau_\nu \sim 6 \times 10^{24}$ s. Such a short lifetime may imply that the neutrino is a composite particle (F.W. Stecker and R.W. Brown, 1982, Ap. J., 257, 1).

Novikov: The question was replied to by Szalay and Stecker.

[Szalay also responded to Prof. Contopoulos' question; we regret that his answer was not received. Eds.]

BLACK HOLES AND THE FATE OF A CLOSED UNIVERSE

Demosthenes Kazanas
NASA/GSFC, Code 665, Greenbelt, MD. 20771 and
University of Maryland, College Park, MD. 20742

To date observations have not yet unequivocally determined whether the universe is open or closed. Although for the luminous matter $\Omega \lesssim 0.1$, the possibility of existence of non-luminous matter (especially if the neutrino has a non-zero mass) leaves ground for considering that the universe may indeed be closed. In this case the universe is expected to recollapse and become again, at a time t_o, radiation dominated. Hence $R(t) \sim (t_f - t)^{1/2} \sim 1/T\ (t)$, where t_f is the time of collapse to the singularity and T is the temperature. During this new radiation dominated era, a black hole of mass M_o will accrete at a rate

$$\frac{dM}{dt} \simeq 4\pi\ r^2(t)\ \rho(t)\ c \tag{1}$$

where $\rho(t) \simeq \alpha T^4(t)$ and $r(t) = 2GM(t)/c^2$. Eq (1) upon integration yields

$$\frac{M}{M_o} = \left[1 - \frac{4\pi r_o^2 \rho_o c\ \tau_o}{M_o} \left(\frac{\tau_o}{\tau} - 1\right)\right]^{-1} \simeq \left[1 - \frac{4\pi r_o^2 \rho_o c\ \tau_o}{M_o} \frac{\tau_o}{\tau}\right]^{-1} \tag{2}$$

(All subscript zero quantities are taken at $t = t_o$). Eq (2) shows that the mass of the black hole diverges when $\tau_o/\tau \sim M_o/4\pi r_o^2\ \rho_o\ c\tau_o$ where $\tau_o = t_f - t_o$.

For $M_o \simeq 1\ M_\odot$, and since $\rho_o \simeq 10^{-19}$ g cm^{-3} ($= aT_o^4$ for $T_o \simeq 1$ ev) and $\tau_o \simeq 10^{13}$ sec one obtains $\tau_o/\tau \sim 10^{16}$. Hence the divergence temperature will be $T \simeq 10^8\ T_o$. Since $T_o \simeq 1$ eV is then $T \simeq 100$ MeV - 1 GeV, i.e the mass diverges at a finite, modest (by early universe standards) temperature at which the micro-physics is well understood. The source of this devergence is the prescribed time dependence of the temperature during recollapse. This certainly breaks down when the radius of the black hole becomes comparable to R. At this point however the universe has irrepairably departed from homogeneity and isotropy, and its reemergence into a new Robertson-Walker cycle seems quite unlikely.

G. O. Abell and G. Chincarini (eds.), Early Evolution of the Universe and Its Present Structure, 331.

THE [AN]ISOTROPY OF THE X-RAY SKY

R.A. Shafer, NASA/Goddard Space Flight Center and Dept. of Physics and Astronomy, University of Maryland, U.S.A. (currently at Institute of Astronomy, Cambridge, U.K.)
A.C. Fabian, Institute of Astronomy, Cambridge, U.K.

1. Introduction

In this presentation we show how the study of the isotropy of the X-ray sky contributes to our understanding of the structure of the universe at moderate redshifts ($1 \lesssim z \ll z_{recombination}$). Actually, the anisotropy of the sky flux provides the information, much as the microwave sky anisotropy does for earlier epochs. [See reports in this volume.] Though we are currently unable to make measurements with the precision and small solid angles typically achieved in the microwave, comparatively crude limits from the X-ray fluctuations place limits on the largest scale structure of the universe. We first outline the measurements of the X-ray sky and its anisotropies made with the HEAO 1 A-2 experiment. Detailed presentations are found elsewhere [Shafer 1982; Marshall *et al*. 1980; Piccinotti *et al*. 1982; Iwan *et al*. 1982; Shafer *et al*. in prep.]. We then show how the anisotropies place limits on the origin of the X-ray sky and on any large scale structure of the universe, following the example of previous analyses which used earlier anisotropy estimates [see *e.g.* Fabian and Rees 1978; Rees 1980; Fabian 1981].

2. The X-ray Sky

In Figure 1 we present the extragalactic sky spectrum. Several properties of the X-ray portion of the spectrum are noteworthy:

(1) It is bright. A spectrum with slope -1 in Figure 1 has equal energy per decade; thus the energy density in 3-100 keV X-rays is second only to the density of the microwave region.

(2) It is easily detected and nearly isotropic. The region from about 3 keV to ~1 MeV is the only well studied portion of the sky spectrum other than the microwave background that is not dominated by a strong galactic component.

(3) It has a well determined, if not well understood, spectrum. In terms of accuracy and bandwidth the measurement of the X-ray spectrum surpasses even determinations of the microwave spectrum [de Zotti

G. O. Abell and G. Chincarini (eds.), Early Evolution of the Universe and Its Present Structure, 333–343.

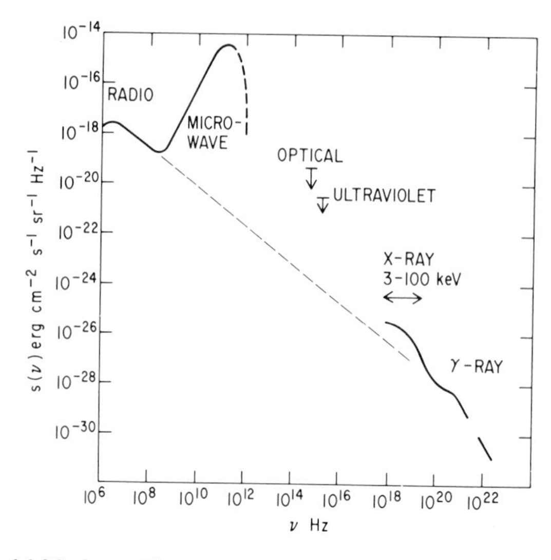

Figure 1. The Cosmic Extragalactic Spectrum. The light dashed line extends the radio discrete sources to higher energies assuming a 0.7 energy spectral index. [Radio and microwave: Longair 1978; Optical: Dube, Wickes and Wilkinson 1979; UV: Paresce, McKee and Bowyer 1980; X-ray: Marshall et al. 1980, Rothschild et al. 1983; γ-ray: Fichtel & Trombka 1981].

1982a]. The spectrum from 3 to 100 keV is well represented by a thin thermal bremsstrahlung model with a temperature of 40±5 keV [Marshall et al. 1980; Rothschild et al. 1983]. No population of sources with a single index power law spectra provides the right shape. A suitably constrained model of an evolving population with a sharp break in their spectral index is consistent with the 3-50 keV data [de Zotti et al. 1982b].

A variety of models have been presented to account for parts of the sky flux, but no coherent picture exists that satisfies all the observations. Particular models may be consistent with the data but all leave open questions or await particular observational confirmation. Known and possible fractions of the X-ray sky flux are:

(1) A galactic component. Because of the observed isotropy of the sky, any flux associated with the galactic disk can contribute no more than ~2-10% of the high galactic latitude flux, depending on the latitude of the observation [see e.g. Iwan et al. 1982]. In addition, known populations of galactic sources do not generally have the same spectrum as the sky.

(2) Well observed extragalactic sources. Based on the HEAO 1 A-2 all-sky survey the extragalactic sources resolved at high flux are predominantly either clusters of galaxies or active galactic nuclei (Seyferts, N galaxies, etc.) [Piccinotti et al. 1982]. The derived local luminosity function of these two populations can be used to estimate contributions to the total sky flux of about 4% and 18% respectively, assuming no significant evolution of the luminosity function. In addition, the low temperature thermal spectra of cluster sources [Mushotzky et al. 1978] and the power law spectra with a photon index near 1.7 typical of active galaxies do not correspond to the sky spectrum [Mushotzky et al. 1980; Rothschild et al. 1983]. In fact, no source or population has been observed to have the proper spectrum to provide the bulk of the sky emission.

(3) An evolving population of sources. This could be the evolution of

the above known populations or the introduction of a new population of sources. QSOs, which are undergoing apparent evolution at other wavelengths, have been shown to be strong emitters in the softer 0.5 to 3.5 keV band covered by the *Einstein* observatory [see *e.g.* Zamorani *et al.* 1981]. Unfortunately, there are too few high quality broad band X-ray spectra of QSOs to make an unambiguous estimate of their contribution at the higher energies typical of the bulk of the X-ray sky flux. Avni [1978] pointed out that active galaxies may make up the total sky flux if they undergo only moderate evolution, in comparison to the amount of evolution suggested for QSOs in the optical. However, the evolution must involve the spectral form of the objects as well as their luminosity function [*e.g.* Leiter and Boldt 1982]. Suggestions for new populations of X-ray sources have included hot gas associated with the initial generation of stars of young galaxies [Bookbinder *et al.* 1980] and primordial black holes [Carr 1980]. Though the proposed spectra are in accordance with the X-ray sky spectrum, there have been no identifications of these new objects with observed X-ray sources.

(3) A totally diffuse component, such as a hot intergalactic medium. This model has the correct spectral form, but there are possible difficulties providing the energy to heat the medium [Field and Perrenod 1977; Fabian 1981].

Though the exact origin of the sky flux is still an open question, we have a better understanding of the principal sources of the observed anisotropy. We classify the variations in the X-ray sky intensity as large angular scale anisotropies or as fluctuations (small scale variations).

The galaxy does not dominate the 2-10 keV sky flux, but the dominant large scale *variation* is associated with the galactic disk. An early model of this variation was the cosecant |b| law of an infinite plane of emission [Warwick, Pye and Fabian 1980]. Other studies have noted a longitudinal component associated with the galaxy [Protheroe, Wolfendale and Wdowczyk 1980; Iwan *et al.* 1982]. An expected large scale anisotropy, of smaller magnitude, is a cosine or dipole anisotropy. Such a signal is expected for the same reason as the dipole variation seen in the microwave sky, *i.e.* motion of the observer with respect to the rest frame of the emission, the Compton-Getting effect.

The dominant contribution to the small scale fluctuations is a continuation of known source populations to lower flux levels where sources are no longer individually detectable. The size and shape of the frequency distribution of the fluctuations is a function of the number of sources versus flux relation, N(S), at those fluxes. We can explain all the fluctuations in terms of the known populations, without evolution, and place an upper bound on the size of any other small scale variations, hereafter referred to as the excess variance. This bound constrains all other sources of anisotropy. Possible origins of additional variation would be an evolving or new population of sources, or a clumping of the sources that make up the background. At the largest angular scales source clumping indicates large scale structure in the universe, such as a global perturbation in the density $\delta\rho$. An upper bound on the excess variance limits the allowed strength of such structure, $\delta\rho/\rho$, with no assumption about the origins of the X-ray sky flux other than presuming that variations in the X-ray volume emissivity are proportional to $\delta\rho$. (For general reviews of the X-ray sky see e.g. Boldt [1981], Fabian [1981].)

3. The Data

Our results are based on measurements taken with a xenon proportional counter, one module of the A-2 experiment on the HEAO 1 satellite [Rothschild et al. 1979], taken during an all-sky survey. For the X-ray sky spectrum, 90% of the counts in this detector originate in the 2.5-13.3 keV band. We measure flux, S, in units of counts s^{-1} cm^{-2}. For typical extragalactic spectra 1 count s^{-1} cm^{-2} is equivalent to 1.35×10^{-8} ergs s^{-1} cm^{-2} (2-10 keV). The all-sky flux, S_{as}, is 58 counts s^{-1} cm^{-2}. The measured count rate depends on detector area, collimator solid angle, and integration time. For our measurements the mean sky intensity, I_{sky}, is 17.06 counts exp^{-1} (one exposure is 1.28 s). The mean count rate of the internal, non-X-ray, background is 3.5 counts exp^{-1}. The average uncertainty due to counting statistics was 0.23 counts exp^{-1}. The angular size of the measurements is fairly large, over 100 square degrees, but much of this area contributes little to the total count rate. 90% of the total comes from a rectangle of $11.2^o\times4.4^o$, covering 49 square degrees. The central area of ~26 square degrees contributes 71% of the sky intensity but 90% of any excess variance in the intensity.

The fluctuation data were restricted to high galactic latitudes, $|b|>20^o$, and free from contamination by X-ray sources in the Magellanic Clouds and bright high latitude galactic sources. When looking for large scale structure, we included data down to latitudes of 10°, excluding all contamination from any resolved source cataloged in the complete all-sky sample of Piccinotti et al. [1982].

The A-2 detectors had several unique features for the continuous monitoring of internal background and determining the X-ray sky flux. The performance of the detectors, as monitored by repeat scans of the same area of the sky six months apart, was very stable. The internal background was also very stable. After selection of data to avoid noisy periods, the variation in the background was roughly 0.05 counts exp^{-1}, corresponding to a sigma 1.3% of the non-X-ray count rate and only 0.25% of the total intensity.

4. Large scale variations: Galactic and Dipole anisotropies

Using a similar set of HEAO 1 A-2 data, Iwan et al. [1982] showed that there was a variation associated with galactic longitude in addition to the latitude variation. Following that paper we fit the galactic component with a disk of finite radius and an exponential scale height. These parameters are strongly correlated and their upper bounds poorly determined. The best fit radius was 1.8 R_{gc}, where R_{gc} is the distance from the sun to the galactic center, roughly 10 kpc. The best fit value for the scale height was 0.4 R_{gc}, with the 90% lower limit of 0.1R_{gc}, corresponding to 1 kpc, a scale typically larger than most galactic X-ray source populations. A discussion of this model and its implications is given in Iwan _et al_. [1982].

After removal of the best fit galactic model, we fit a dipole model, $\delta I = I_{CG} \cos \theta$, where θ is the angle between the observation and the signal maximum. The addition of this new model to the fit produced a drop in χ^2 significant at the 95% level. The strength of the signal, I_{CG}, is 0.09±0.03 counts $\exp^{-1}$, about 0.5% of the sky intensity. The best fit direction in galactic coordinates, (ℓ,b), is (282°,+30°). However the 90% confidence region for the direction is very large, covering about one eighth of the total sky. A result of similar direction, magnitude, precision, and confidence was found by Protheroe, Wolfendale and Wdoczyk [1980] using _UHURU_ data.

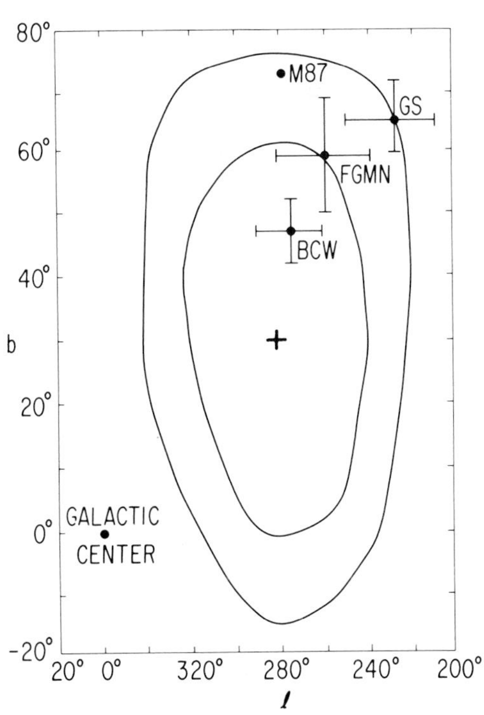

Figure 2. Position of Dipole Maximum. Contours show 70% and 90% confidence regions. The center + marks the best fit position. Also shown are the one-sigma error bars for measurements of the dipole maximum in the microwave. [BCW: Boughn, Cheng & Wilkinson 1981; FGMN: Fabbri, Guidi, Melchiorri and Natale 1980; GS: Gorenstein and Smoot 1981.]

One possible interpretation of this statistically marginal result is in terms of the Compton-Getting effect, where the size of the dipole signal is related to the observer's velocity by

$$I_{CG} = \overline{I}\ (2+\Gamma)\ v/c. \qquad [1]$$

Γ is the photon index of the sky flux, ~1.4 for the band we are interested in. The derived value for the velocity, 475±165 km s^{-1}, and the direction are consistent with observations of the dipole signal in the microwave sky. The microwave directions in the literature are shown in Figure 2 along with the X-ray confidence regions. A synthesis performed by Wilkinson at this symposium of the different experimental results gave a best fit direction of $(265^o, +50^o)$ and a velocity of 372 ± 25 km s^{-1}. The origins of the Compton-Getting velocity may be responsible for an additional component of the observed X-ray dipole signal. If the velocity observed in the microwave is the integral of the acceleration caused by a large scale overdensity ("lump") at the same redshifts range at which the X-ray sky emission originates, then the lump should produce an excess in the X-ray emissivity, producing an enhancement in the direction of the lump in addition to the Compton-Getting velocity signal. Comparisons of the X-ray and microwave large-scale anisotropies help to decouple the two effects of such a lump, providing constraints on the lump's properties. [see e.g. Warwick, Pye and Fabian 1980; Fabian 1981].

The total dipole signal may be due to variations other than the Compton-Getting effect. The form of the underlying structure may not be a pure dipole. For instance, there may be second order anisotropies associated with the galaxy but not included in the finite disk model. Also, it is intriguing that the contour for the direction of the dipole maximum includes a large fraction of the local supercluster, which could provide a large scale enhancement to the sky flux. The best fit value of I_{CG} corresponds to a maximum surface brightness of 0.02 counts s^{-1} cm^{-2} sr^{-1}. The volume emissivity and total luminosity of the local supercluster required to dominate the "dipole" signal are dependent on geometry. Assuming a disk of emission centered on the dipole maximum with a radius of 4 Mpc, an estimate of the total luminosity is 2×10^{42} erg s^{-1} $(H_o = 50)$. Previous attempts to correlate X-ray surface brightness with the local supercluster have yielded only upper limits larger than the above estimates. [see e.g. Schwartz 1980]. To test if the local supercluster is in part responsible for the fit dipole signal requires direct testing of models for the supercluster, a project now in progress. The tidal "12 hour" signal reported by Warwick, Pye and Fabian [1980] in the Ariel V data was not observed in the A-2 data.

5. Small Scale Anisotropies: The Fluctuations

Point sources have an impact on measurements of the sky flux even if the sources are too numerous to be individually resolved. The actual number of sources, and hence their total intensity, varies from one part of the sky to another. The process of extracting information about the sources from the size and shape of the intensity distributions was pioneered by radio astronomers [Scheuer 1957; Condon 1974; see e.g. Condon and Dressel 1978]. Given a model for the differential number of sources as a function of flux, N(S) dS, the distribution of intensities, $P_{I-\bar{I}}(I)$ dI, can be predicted, assuming the distribution of sources is completely random and unclumped. Standard statistical tools are used to evaluate the N(S) models by comparing the predicted distributions to observations, extending our knowledge of the X-ray source counts beyond what was directly accessible from resolved sources [Fabian 1975; Schwartz 1976; Pye and Warwick 1976].

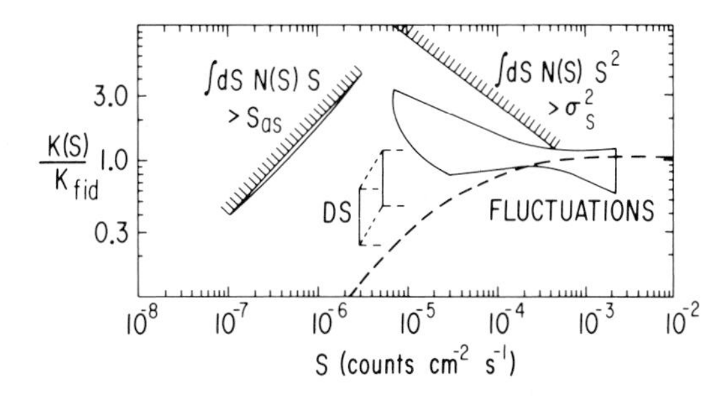

Figure 3. Number of Sources Versus Flux Derived from the Fluctuations. The plotted quantity is the ratio of the number to that expected for a fiducial Euclidean model, where $N \propto S^{-5/2}$. See text for details.

In Figure 3 we present the results of such model fitting using the HEAO 1 A-2 data. We restricted our model to a single power law form,

$$N(S)\, dS = 4\pi\ K\ S^{-\gamma}\ dS\ , \qquad [2]$$

with a sharp cutoff imposed at the low flux where the intensity contributed by the sources equals the total sky flux:

$$\int dS\ N(S)\ S = S_{as}\ . \qquad [3]$$

A population of sources distributed randomly through Euclidean space will follow a power law model with $\gamma = 5/2$. We compare an N(S) relation to the Euclidean form by defining a function K(S)

$$N(S) \equiv 4\pi\ K(S)\ S^{-5/2}. \qquad [4]$$

Figure 3 plots K(S) with respect to K_{fid}, the K value for the best fit Euclidean model, $1.48\text{x}10^{-3}$ (counts s^{-1} cm^{-2})$^{1.5}$. Other Euclidean models would appear on Figure 3 as horizontal lines, K(S) constant. The trumpet-shaped region on the right shows the behavior of power law models acceptable at the 90% level. The hatched line at the far left is where the power law models must be terminated to avoid exceeding the total sky flux (equation [3]). With our data, an acceptable power law model that is truncated between the hatched line and the left hand edge of the trumpet shaped region is statistically indistinguishable from the model as continued to the hatched line. The right hand edge shows the limits of the HEAO 1 resolved source counts.

The greatest constraint placed by the fluctuations is on sources roughly an order of magnitude in flux below resolved sources. However care is required in interpreting the limits on N(S). The formal validity of the confidence region rests on the assumption that the actual N(S) is well modelled by a single power law continuing without change of index past the lower limits of the region. More complicated models that do not lie wholly within the indicated region may be acceptable, *e.g.* the dashed line of Figure 3. This line is a schematic representation of the N(S) behavior of sources observed directly at higher fluxes and extrapolated without evolution to the lower values of

S using the appropriate luminosity functions.

To assess the degree of source evolution at low fluxes, direct measuments of N(S) would be the easiest to interpret. The deep surveys performed with the *Einstein* observatory [*e.g.* Giacconi *et al.* 1979] provide such information. However information from the deep survey, indicated by the two bars labeled DS in Figure 3, also have problems of interpretation. One difficulty is in transforming source fluxes from the *Einstein* 1-3 keV band to the higher energy band measured in the A-2 data without good spectral information. The upper-right bar of the pair uses the published presumption that the sources have a 1.4 index power law photon spectrum, typical of the unresolved sky flux in the 2-10 keV band. The lower-left bar instead assumes that the deep survey sources have the 1.7 index spectra characteristic of active galactic nuclei. Both bars assume that N(S) is Euclidean at the deep survey flux limit. If γ were nearer that of the unevolved population's models at that flux, $\gamma \sim 1.8$, or if the sources were strongly evolving so that γ were near 3, the bars would be adjusted to 50%-130% of their indicated value. Conclusions drawn from the deep survey results must explictly consider the impact of these assumptions. Results from the *Einstein* medium survey [see *e.g.* Maccacaro *et al.* 1982] show source evolution less indirectly.

6. Excess Variance

The fluctuations can be totally described by models of non-evolving sources, such as the dashed line in Figure 3. The 90% upper bound to any additional variance, added as a pure Gaussian, is $\sigma^2_I \lesssim 0.057$ (counts exp^{-1})^2. If we assume that the unevolved populations account for 20% of the sky intensity then σ_I is $\lesssim$1.7% of the remaining intensity. Any other source of variation is constrained by this limit.

Any evolving populations's distribution, $N_{ev}(S)$, is a source of fluctuations. It must satisfy the integral constraint

$$\sigma^2_S = \int dS\ S^2\ N_{ev}(S)\ . \qquad [5]$$

σ^2_S is a measure of the excess variance that is independent of the measurement solid angle, $\sigma^2_S \lesssim 7\times10^{-4}$ (counts s^{-1} cm^{-2})^2. The impact of this limit depends on the particular form of $N_{ev}(S)$ but if the evolved sources are to make up the remainder of the sky flux we place a lower limit on the number of sources at 75 per square degree and an upper limit on their mean flux at 1.5×10^{-5} counts s^{-1} cm^{-2} (2.0×10^{-13} erg s^{-1} cm^{-2}). If we assume that $N_{ev}(S)$ is of the form $4\pi\ K_{ev}\ S^{-3}$ we can set a limit $K\lesssim1.7\times10^{-5}$ (counts s^{-1} cm^{-2})^2, indicated by the right-hand hatched line in Figure 3. A wide latitude for the behavior of $N_{ev}(S)$ is allowed.

These limits all assume that the sources that make up the background are not clustered, that is, their distribution among the measurements is Poisson. We can estimate that the total allowed variation on a scale of 26 square degrees, roughly the size of a Schmidt survey plate, is 2.3%. If QSOs contribute all of the remaining sky flux, then this limit means that the total $(\delta N/\bar{N})_{QSO} \lesssim 0.023$, where N is the mean number of QSOs on a scale of 26 square degrees, and δN is the total variation including Poisson statistics and clumping. If we assume the number of QSOs is ~200 per square degree in order to estimate the Poisson noise portion, the additional variation due to clumping can be at most 1.9%. If QSOs are observed to have clustering with a larger value of δN on these scales, then the excess variance places an upper bound on their contribution to the total sky flux.

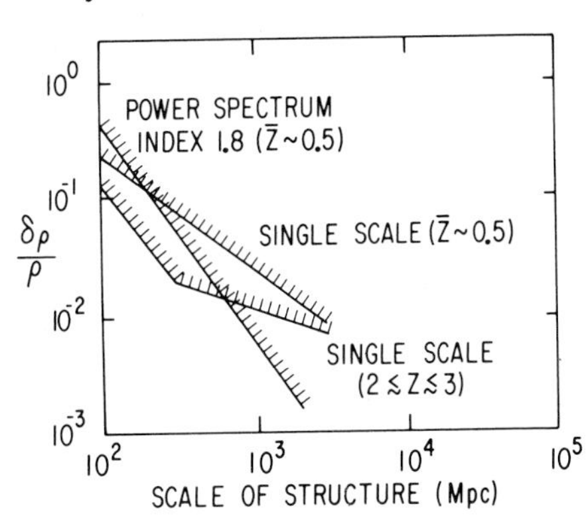

Figure 4. Preliminary upper bounds on magnitude of large scale structure from limits of X-ray excess variance.

Figure 4 shows estimates of the limits on the large scale structure of the universe placed by a previous upper bound on the excess variance. The three curves illustrate the dependence of the limits on models of the structure as well as on the origins of the X-ray sky. Two of the curves compare the case for an unevolved origin of the sky flux, $\bar{z}$~0.5, and the density variation restricted to a single scale or defined by a power spectrum with index 1.8, typical of the inferred distribution at smaller scales [see *e.g.* Peebles 1980]. The third curve is for a single scale of variation where the sky flux is dominated by an evolved component with $2 \lesssim z \lesssim 3$.

7. Conclusions

All small scale fluctuations are consistent with what we expect from known, unevolved populations of sources resolved by HEAO 1. The large scale variation, dominated by a galactic anisotropy, also admits a dipole signal which may be interpreted as a Compton-Getting dipole signal consistent with the microwave results, although other interpretations are possible. The bound on any excess variance places limits on structures at moderate redshift and of large scale otherwise not easily accessible.

We would like to thank our colleagues at GSFC, particularly C.M. Urry, for their contributions and critical advice on this presentation. ACF aknowledges the financial support of the Royal Society of London.

References

Avni, Y., 1978, Astron. Astrophys., 63, L13.
Boldt, Elihu, 1981, Comments Astrophys., 9, 97.
Bookbinder, J., L.L.Cowie, J.H.Krolik, J.P.Ostriker & M.J.Rees, 1980, Ap.J., 237, 647.
Boughn, S.P., E.S.Cheng& D.T.Wilkinson, 1981, Ap.J.(Lett.), 243, L113.
Carr, B.J., 1980, Nature, 284, 326.
Condon, J.J., 1974, Ap.J., 188, 279.
Condon, J.J., & L.L.Dressel, 1978, Ap.J., 222, 745.
de Zotti, G., 1982a, Acta Cosmol., 11, 65.
de Zotti, G., E.A.Boldt, A.Cavaliere, L.Danese, A.Franceschini, F.E.Marshall, J.H.Swank & A.E.Szymkowiak, 1982, Ap.J., 253, 47.
Dube, R.R., W.C.Wickes & D.T.Wilkinson, 1979, Ap.J., 232, 333.
Fabbri, R., I.Guidi, F.Melchiorri & V.Natale, 1980, Phys.Rev.Lett., 44, 1563. (Erratum in 45, 401).
Fabian, A.C., 1975, M.N.R.A.S., 172, 149.
Fabian, A.C., & M.J.Rees, 1978, M.N.R.A.S., 185, 109.
Fabian, A.C., 1981, in Ramaty and Jones (eds.), 10th Texas Symposium on Relativistic Astrophysics, Ann.N.Y.Acad.,Sci., 375, 235.
Fichtel, C.E. & J.I.Trombka, 1981, Gamma Ray Astrophysics, NASA SP-453, (Government Printing Office:Washington).
Field, G.B., & S.C.Perrenod, 1977, Ap.J., 215, 717.
Giacconi, R., J.Bechtold, B.Branduardi, W.Forman, J.P.Henry, C.Jones, E.Kellogg, H.van der Laan, W.Liller, H.Marshall, S.S.Murray, J.Pye, E.Schreier, W.L.W.Sargent, F.Seward & H.Tananbaum, 1979, Ap.J.(Lett.), 234, L1.
Gorenstein, M.V., & G.F.Smoot, 1981, Ap.J., 244, 361.
Iwan, D., F.E.Marshall, E.A.Boldt, R.F.Mushotzky, R.A.Shafer & A.Stottlemyer, 1982, Ap.J., 260, 111.
Leiter, D. & E.Boldt, 1982, Ap.J., 260, 1.
Longair, M.S., 1978, in Gunn, Longair & Rees Observational Cosmology', (Geneva Observatory:Sauverny, Switzerland).
Maccacaro, T., Y.Avni, I.M.Gioia, P.Giommi, J.Liebert, J.Stocke, J.Danziger, 1982, Ap.J., subm.
Marshall, F.E., E.A.Boldt, S.S.Holt, R.Miller, R.F.Mushotzky, L.A.Rose, R.Rothschild & P.Serlemitsos, 1980, Ap.J., 235, 4.
Mushotzky, R.F., P.J.Serlemitsos, B.W.Smith, E.A.Boldt & S.S.Holt, 1978, Ap.J., 194, 1.
Mushotzky, R.F., F.E.Marshall, E.A.Boldt, S.S.Holt & P.J.Serlemitsos, 1980, Ap.J., 235, 377.
Paresce, F., C.F.McKee & S.Bowyer, 1980, Ap.J., 240, 387.
Piccinotti, G., R.F.Mushotzky, E.A.Boldt, S.S.Holt, F.E.Marshall, P.J.Serlemitsos & R.A.Shafer, 1982, Ap.J., 253, 485.
Peebles, P.J.E., 1980, The Large-Scale Structure of the Universe, (Princeton University Press:Princeton).
Protheroe, R.J., A.W.Wolfendale & J.Wdowczyk, 1980, M.N.R.A.S., 192, 445.
Pye, J.P. & R.S.Warwick, 1979, M.N.R.A.S., 187, 905.
Rees, M.J., 1980, in Abell and Peebles (eds.), Objects at High Redshifts, I.A.U. Symp. 92, 209, (D. Reidel:Dordrecht).
Rothschild, R., E.Boldt, S.Holt, P.Serlemitsos, G.Garmire, P.Agrawal, G.Riegler, S.Bowyer & M.Lampton, 1979, Space Sci.Inst., 4, 269.
Rothschild, R.E., R.F.Mushotzky, W.A.Baity, D.E.Gruber & J.L.Matteson, 1983, Ap.J., submit.
Scheuer, P.A.G. 1957, Proc.Camb.Phil.Soc., 53, 764.
Schwartz, D.A., S.S.Murray, H.Gursky, 1976, Ap.J., 204, 315.
Schwartz, D.A., 1980, Physica Scripta, 21, 644.
Shafer, R.A., 1982, Ph.D. dissertation, University of Maryland.
Shafer, R.A., et al., 1983, in prep.
Warwick, R.S., J.P.Pye & A.C.Fabian, 1980. M.N.R.A.S., 190, 243.
Zamorani, G., J.P.Henry, T.Maccacaro, H.Tananbaum, A.Soltan, Y.Avni, J.Lieber, J.Stocke, P.A.Strittmatter, R.J.Weymann, M.G.Smith, J.J.Condon, 1981, Ap.J., 245, 357.

Discussion

Tyson: How, how often, and how well do you recalibrate the detector, and what limits does this set to the contribution of systematic errors to your claimed dipole signal?

Shafer: The data used to fit the dipole component were taken during the first nine months of satellite operations during which the experiment scanned the entire sky 1 1/2 times. The region of the dipole maximum was included in the region that was scanned twice. The detector was continually calibratable as far as pulse height gain. Gain variations were small and in any case not expected to be a major problem for the wide bandwidth measurements used to fit the dipole. The absolute sensitivity of the detector was a larger problem. The internal background was continuously monitored, while the sensitivity of the detector to X-rays could best be checked by using the sky as a reference point. By comparing measurements of the north ecliptic pole (which was measured every scan) as well as measurements of the same patch of sky (six months apart), we were able to detect a slight linear secular drift in X-ray sensitivity. In conjunction with a measured linear decrease in internal background, the total linear drift in our intensity was $\sim$ -0.06 counts $\exp^{-1}$ in six months. Higher order variations did not have significantly better fits. We have physical models to explain the sensitivity drift. Even if we did not remove the secular drift, it would have an all-sky amplitude of about 0.1%, in comparison to the dipole measured strength of 0.5%. This is not to say that "systematics" are not important, but our current understanding of our detector, along with the independent indication from UHURU of a dipole in the same direction, indicates that the systematics associated with the actual X-ray sky (i.e., higher order galaxy contamination and/or the local supercluster) are of a greater concern than the physical performance of the detector.

HOT ACCRETION DISKS AND THE HIGH ENERGY BACKGROUND

M. Kafatos
Department of Physics, George Mason University, Fairfax, VA and Laboratory for Astronomy and Solar Physics, NASA, Goddard Space Flight Center, Greenbelt, MD, U.S.A.
Jean A. Eilek
Physics Department, New Mexico Tech, Socorro, NM, U.S.A.

1. INTRODUCTION

The origin of the high energy (X-ray and gamma-ray) background may be attributed to discrete sources, which are usually thought to be active galactic nuclei (AGN) (cf.Rothschild et al. 1982, Bignami et al. 1979). At X-rays a lot of information has been obtained with HEAO-1 in the spectral range 2-165 keV. At gamma-rays the background has been estimated from the Apollo 15 and 16 (Trombka et al. 1977) and SAS-2 (Bignami et al. 1979) observations. A summary of some of the observations (Rothschild et al. 1982) is shown in Figure 1. The contribution of AGN to the diffuse high energy background is uncertain at X-rays although it is generally estimated to be in the 20-30% range (Rothschild et al. 1982). At gamma-rays, in the range 1-150 MeV, AGN (specifically Seyfert galaxies) could account for all the emission.

2. HOT ACCRETION DISKS

Accretion disks around massive black holes can be a copious source of X-rays, gamma-rays and relativistic electron-positron pairs (Eilek and Kafatos 1982). The X-ray emission arises from the Comptonization of seed photons by a hot, thermal gas ($T_e \sim 10^9$ K). The hot electron gas is accreting onto a non-rotating or rotating black hole. The relevant parameter is the so-called Comptonization parameter y defined as equal to the product of fractional energy change of the photons per scattering and the mean number of scatterings or

$$y = 4kT_e/m_e c^2 \cdot \max(\tau_{es}, \tau_{es}^2) \qquad (1)$$

(cf. Shapiro, Lightman and Eardley 1976). A large amplification of the incoming soft flux occurs when $y \simeq 1$ (unsaturated process) and Shapiro, Lightman and Eardley (1976) show that the condition $y \simeq 1$ is a relatively stable condition. The spectral index of the energy flux (keV/cm^2 sec keV) is related to y through the approximate relation $y \sim 4/(3\Gamma+\Gamma^2)$ and, therefore, $\Gamma \sim 1$ when $y \sim 1$.

Rotating (Kerr) black holes are important for the production of gamma-

G. O. Abell and G. Chincarini (eds.), Early Evolution of the Universe and Its Present Structure, 345–346.

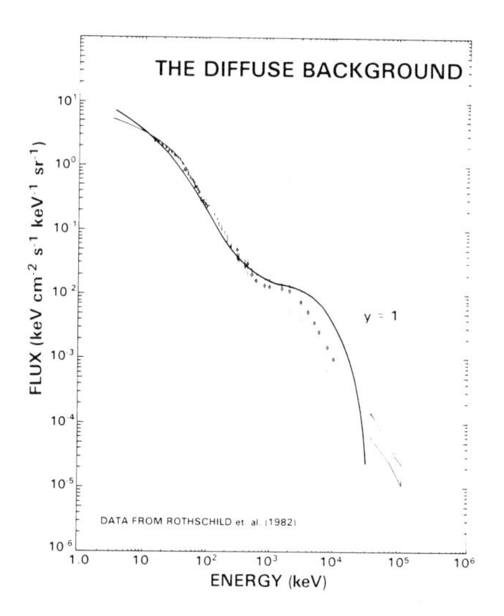

Figure 1. The high energy background and a hot accretion disk fit with y = 1

rays and relativistic electron-positron pairs. The ion temperature T_i is much larger than T_e. We find that in order for the hot, inner region to exist one needs a luminosity of the disk at least a few percent of the Eddington luminosity or $\dot{M}_*/M_8 \gtrsim 0.03\ M_8^{-1/32}\ (\alpha/0.1)^{-17/32}$, where α is the usual viscosity parameter estimated to be in the range 0.01-1 (Shapiro, Lightman and Eardley 1976), $\dot{M}_*$ is the accretion rate in $M_\odot$/yr and M_8 is the mass of the black hole in $10^8\ M_\odot$. We find that for α~0.1 the disk is thick The ion temperature is assured then to be greater than 10^{12} K without any other requirement. Models that are near the Eddington limit produce 10-20% of the bolometric luminosity in gamma-rays of energy greater than 1 MeV, and a similar energy in the form of 50-100 MeV e^+e^- pairs. Figure 1 sh a particular fit for $\dot{M}_*/M_8 = 1$, α =.1 y = 1 and a Kerr metric. The general agreement is striking. Other good fit are obtained for y in the approximate range 1-3 as long as the accretion rate remains within a factor of a couple near the Eddington limit. The need to evoke different mechanisms or sources to explain both the X-ray and gamma-ray backgrounds is not present. The gamma-rays and pairs arise from pions produced due to the high ion temperatures. Due to the high $\gamma\gamma \rightarrow e^+e^-$ opacity gamma-rays of energy higher than ~ 3-10 MeV do not escap from the disk, in agreement with both the background and gamma-ray spect of AGN (Bignami et al. 1979). We obtain about 3×10^{-4} sources/Mpc^3 with a average mass of the central object of about $5\times10^4\ M_\odot$, producing on the average 10^{42} erg/sec in the 2-10 keV range. In these models y≳1.

REFERENCES

Bignami,G.F.,Fichtel,C.E.,Hartman,R.C.,and Thompson,D.J.:1979,Ap.J. 232, pp. 649-658.
Eilek,J.A.,and Kafatos,M.:1982,Ap.J.,submitted.
Rothschild,R.E.,Mushotzky,R.F.,Baity,W.A.,Gruber,D.E.,and Matteson,J.L.: 1982,Ap.J.,submitted.
Shapiro,S.L.,Lightman,A.P.,and Eardley,D.M.:1976,Ap.J. 204,p.187.
Trombka,J.I.,et al.:1977,Ap.J. 212,p.925.

X-RAY OBSERVATIONS OF ACTIVE GALACTIC NUCLEI

C. Megan Urry*, Richard F. Mushotzky, Allyn F. Tennant**,
Elihu A. Boldt, and Stephen S. Holt
NASA/Goddard Space Flight Center, Greenbelt, MD, USA
*also Johns Hopkins University, Baltimore, MD, USA
**also University of Maryland, College Park, MD, USA

ABSTRACT. HEAO 1 A2 and Einstein SSS spectral observations of Seyfert galaxies and BL Lac objects suggest that in both cases, the X-ray emission is due to relativistic particles. The five BL Lac objects have very soft spectra and at higher energies (above 10 keV) may have hard tails. Combining our X-ray data with radio, infrared, optical, and ultraviolet observations, we can fit the BL Lac spectra with the familiar synchrotron self-Compton model if we allow for relativistic beaming (Urry and Mushotzky 1982, Urry et al. 1982). We show that Doppler beaming of an underlying (Seyfert-like) source population flattens the observed luminosity function, and we emphasize that the relative numbers of BL Lacs and quasars in given spectral intervals are strong functions of selection effects, the degree of Doppler beaming, and the form of the intrinsic luminosity function.

The twenty-eight X-ray spectra of Seyfert galaxies are remarkably homogeneous: all are well-fit by power laws with mean energy spectral index $\alpha = 0.65 \pm 0.13$, where the latter number indicates the dispersion (Mushotzky et al. 1980, Mushotzky 1982). This power law, taken with the IR-optical-UV emission, suggests that Compton scattering is the dominant X-ray production mechanism. No changes in spectral form are seen on short (10-10^4 s) or long (0.5-1.5 yr) timescales in either the A2 (2-40 keV) or SSS (0.5-3.5 keV) data even when the intensity changes. With one major exception (NGC 6814), no variability with $\Delta I/I>0.1$ is seen for timescales of 5 seconds to 6 hours (Tennant and Mushotzky 1982). For 6 month timescales, at most one third of the galaxies are variable in intensity, and these tend to be the lower luminosity objects.

Mushotzky, R.F.: 1982, Ap.J. 256, pp. 92-102.
Mushotzky, R.F., Marshall, F.E., Boldt, E.A., Holt, S.S., and Serlemitsos, P.J.: 1980, Ap.J. 235, pp. 377-385.
Tennant, A.F. and Mushotzky, R.F.: 1982, to be pub. in Ap.J. 264.
Urry, C.M. and Mushotzky, R.F.: 1982, Ap.J. 253, pp. 38-46.
Urry, C.M., Mushotzky, R.F., Kondo, Y., Hackney, K.R.H., and Hackney, R.L.: 1982, to be pub. in Ap.J. 261.

G. O. Abell and G. Chincarini (eds.), Early Evolution of the Universe and Its Present Structure, 347.

QSO ABSORPTION LINES

BRUCE A. PETERSON
Mount Stromlo and Siding Spring Observatories
The Australian National University

ABSTRACT

The evidence for the evolution of the intergalactic clouds that produce the Lyman-alpha absorption lines in the spectra of QSOs is reviewed. The recent detection of OVI in clouds that produce both Lyman-alpha and Lyman-beta absorption lines is discussed. The identification of molecular hydrogen in the spectrum of 1442+101 (OQ172) is not confirmed by high resolution spectra.

1. INTRODUCTION

The absorption lines seen in the spectra of QSOs are produced by gas clouds along the line of sight to the QSO. There are three major sites for these absorbing clouds: 1) near the QSO, 2) the interstellar medium of intervening galaxies, and 3) intergalactic space. The absorbing material associated with the QSO forms large absorption troughs on the short wavelength side of the emission lines. These absorption troughs are deepest nearest the emission line, and decrease in optical depth at shorter wavelengths (which correspond to the larger ejected velocities). In many cases, the smooth absorption troughs break up into discrete absorption line systems (Clowes et al. 1979, Wright et al. 1979, Turnshek et al. 1980). The absorption line systems produced by the interstellar medium in intervening galaxies are similar to the absorbtion line systems seen in satellite ultraviolet spectra of stars in our own galaxy. The column densities measured relative to HI and compared to solar show that OI, NI, SiII, SII, and FeII are down by a factor of 10, typical of HI clouds in the halo of our own galaxy (Morton et al. 1980, Savage et al. 1981). The intergalactic clouds produce narrow absorption lines that are due to Lyman-alpha. Recent work on the absorption lines produced in the intergalactic clouds is described in the next two sections.

For a comprehensive review of QSO absorption lines, see Weymann, Carswell and Smith (1981) and references therein.

G. O. Abell and G. Chincarini (eds.), Early Evolution of the Universe and Its Present Structure, 349–357.

2. THE EVOLUTION OF THE LYMAN-ALPHA CLOUDS

Consider the distribution of the Lyman-alpha absorption lines that would be produced by intergalactic clouds with invariant cross-sections, and with a uniform space distribution. The number of clouds in a unit redshift interval is given as a function of redshift by Equation 1.

$$dN/dz = \frac{c}{H_o}\sigma\rho_o\frac{(1 + z)}{(1 + 2q_o z)^{\frac{1}{2}}} \tag{1}$$

Here sigma is the cloud cross-section, rho is the number of clouds per unit volume at the present epoch, H and q are the Hubble constant and the deceleration parameter at the present epoch, c is the speed of light, and z is the redshift (lambda, the cosmological constant, is taken to be zero.)

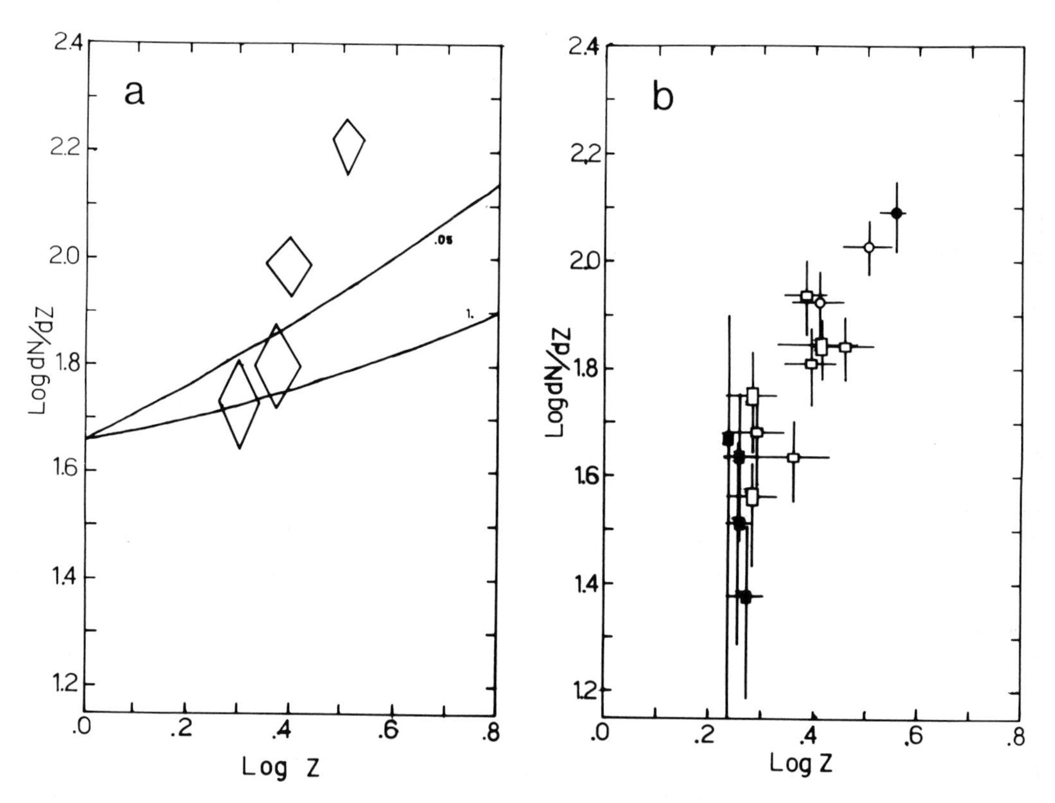

Figure 1. Line density (the number of absorption lines in a unit redshift interval) vs. redshift on log-log scales. The horizontal extent of the symbols indicates the redshift interval. The vertical extent of the symbols indicates the r.m.s. error ($N \pm N^{\frac{1}{2}}$) where N is the number of observed lines in the interval. Panel a shows the first evidence for the evolution of the intergalactic clouds (Peterson 1978). The diamonds are from counts of absorption lines in the four QSO's 1442+101, 0805+046, 0329-255, 1448-232. The curved lines represent the expected relation between line density (given by Equation 1) and redshift for non-elvolving clouds and $q_o = 0.05$ and $q_o = 1.0$ Panel b shows the current evidence. The line counts include unidentified lines with rest frame equivalent widths greater than 0.32. The data are from (●) 2000-330 Peterson (1982); (○) 1442+101, 0805+046 Peterson, Chen, Morton, Wright, Jauncey (1982), Chen, Morton, Peterson, Wright, Jauncey (1982); (□) 0453-423, 2126-158, 0002-422, 1225+317, 0100+130 Sargent, Young, Boksenberg, Tytler (1980); (▯) 0420-388, 0122-380, 1104-264 Smitn (1978), Carswell, Whelan, Smith, Boksenberg, Tytler (1982); (■) 1115+080, 0119-046, 0002+051 Young, Sargent, Boksenberg (1982). The data on which the counts for 2000-330 were made were supplied by Hunstead, Murdoch, Blades, and Pettini.

The first evidence for evolution of the intergalactic clouds came from counts of absorption lines in the spectra of four QSOs at various redshifts (Peterson 1978). These counts are shown in Fig. 1a along with two lines which represent the relation between the number of clouds in a redshift interval and the redshift of the interval for q = 1.0 and for q = 0.05 as given by Equation 1 for non-evolving clouds. On the basis of the data in Fig. 1a., Peterson suggested that the rapid increase in the number of absorption lines with redshift could be understood in terms of a progressive ionization of the intergalactic medium that started with the turn on of QSOs at z of about 4, or in terms of a change in cloud cross-section produced by collapse, perhaps to form galaxies. These conclusions and the significance of this newly discovered effect were disputed by Sargent et al. (1980) who studied five QSOs at various redshifts. They state

> "We have found that the overall Lyman-alpha absorption line density is statistically the same in all five QSOs. This is in sharp contrast to Petersons (1978) conclusion that in four QSOs the line density increases systematically by a factor of 3 with redshift over the range Z(em)=2.21 to Z(em)=3.53."

However, inspection of the data given by Sargent et al. in their Table 9 reveals that their data is in better agreement with the rapidly increasing line density discovered by Peterson. Using their Table 9 data only, a least squares fit gives

$$dN/dz = K(1 + z)^{\gamma}, \quad \gamma = 1.6 \pm 1.3.$$

The current status regarding the significance of the evidence for evolution of the intergalactic clouds is shown in Fig. 1b. Using all the data shown in Fig. 1b, gamma = 2.2 with an RMS error of 0.4. The data are given for the number of lines with rest frame equivalent widths greater than 0.32, and are taken from the work of several groups, as indicated. With this data, the relation between the number of lines per unit redshift interval and redshift as given by Equation 1 for non-evolving clouds is excluded at the 3.5 sigma level for the most favorable case of q = 0.05. Larger values of q are excluded with greater significance.

Thus the current evidence supports Petersons original conclusion that the increase in the numbers of lines seen in the spectra of high redshift QSOs is more rapid than allowed by the cloud model with no evolution of the cloud properties. It may be that the clouds evaporate or increase their ionization as the Universe expands, or that the clouds are self-gravitating and collapse after a time.

3. ABUNDANCES IN THE LYMAN-ALPHA/BETA CLOUDS

For most of the intergalactic Lyman-alpha clouds, the column density is too low for any other lines to be seen. In some cases, the

Lyman-alpha column density approaches logN(HI) = 15, and the Lyman-beta line can be identified, but at this neutral hydrogen column density, no lines of heavier ions are strong enough to be observed in a single redshift system. In order to increase the detecability of other lines, Norris et al. (1982) added the rest frame spectra of 65 clouds that produced both Lyman-alpha and Lyman-beta lines in the spectra of the two QSOs 0805+046 (Chen et al. 1982) and 1442+101 (Peterson et al. 1982). Norris et al. were able to detect OVI and measure a column density of logN(OVI) = 13.8 in the Lyman-alpha/beta clouds with logN(HI) = 14.9. Upper limits of 13.2 for logN(CIV) and 13.5 for logN(NV) were obtained. Fig.2 shows the composite rest frame spectra of each QSO, and for both QSOs added together. In Fig. 2, the two components in the NV doublet and in the OVI doublet have been added together. The curves in the panel on the right in Fig. 2 are calculated line profiles for logN(HI) = 14.9 and for logN(OVI) = 13.5 and 14.0 with b = 30 km/s and a Gaussian instrumental profile of 0.5 A FWHM.

No detection of NV was claimed. The significance of the OVI detection was tested by generating sets of random redshifts and requiring that the equivalent widths of both components of the OVI doublet in the composite spectrum obtained from the random redshift sets be greater than or equal to that observed. This occurred 2.7% of the time for 0805+046, 0.6% of the time for 1442+101, and 0.4% of the time for both QSOs added together.

The oxygen to hydrogen abundance ratio was calculated by assuming that the clouds were ionized by the integrated QSO background flux. With a cloud temperature of 40000K and a volume density of log n(H) = -3 to -4, the following (log) abundances (relative to solar) were obtained: O/H = -1.8 to -1.9, C/H less than -1.1, N/H less than -0.8. For a wide range of cloud model parameters, no values for an abundance less than values found in halo objects in our galaxy are deduced.

The abundance of the clouds is similar to the heavy element enrichment found in isolated extragalactic HII regions and in Population II material in our galaxy. Thus the clouds are not primeval, and it may be necessary to consider a pre-galactic enrichment process (e.g. see Peebles and Dicke 1968).

4. MOLECULAR HYDROGEN ABSORPTION IN QSO SPECTRA?

Molecular hydrogen absorption lines are produced in the ultraviolet spectra of galactic stars by the interstellar medium in our galaxy (Morton 1975, Morton and Dinerstein 1976). Molecular hydrogen absorption lines have been identified in the spectra of various QSOs (Carlson 1974, Aaronson et al. 1974, Varshalovich and Levshakov 1982 and references therein). In the particular case of 1442+101, Levshakov and Varshalovich (1979) have identified molecular hydrogen with two redshift systems at z = 2.651 and 3.092 using a spectrum obtained by Baldwin et al. (1974). Peterson et al. (1982) have obtained a spectrum of 1442+101 at higher resolution, and it has been examined for

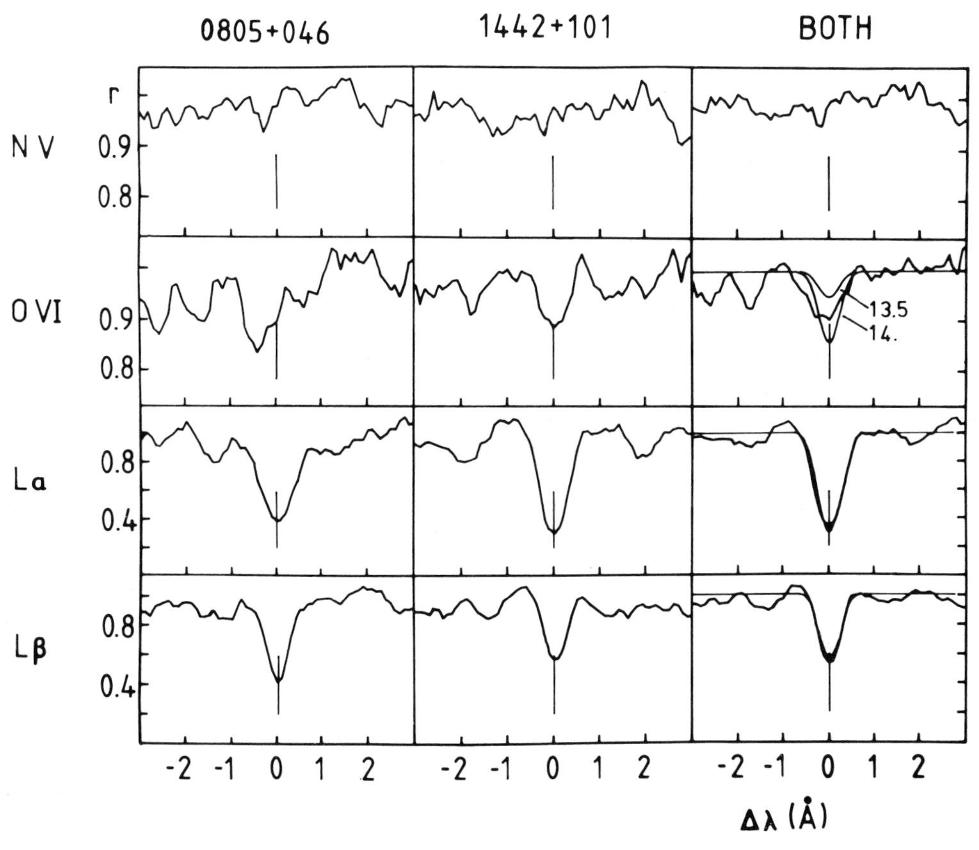

Figure 2. Composite spectra of the QSO's 0805+046 and 1442+101 in the regions of NV OVI Lα and Lβ (Norris et al. 1982). The abcissa represents the distance from the line center. The composite spectra were produced by adding (in the rest frame) the spectra of 27 absorption redshift systems identified in the spectrum of 0805+046 by Chen et al. (1982) and 38 systems identified by Peterson et al. (1982) in the spectrum of 1442+101 with Lα/Lβ pairs that have rest frame equivalent widths greater than 1.0 A.

redshift systems containing the strongest interstellar lines seen in the spectra of zeta-Oph (Morton 1975) and zeta-Pup (Morton 1978) (plus CIV, NV, OVI, and SiIV) and for redshift systems containing the lines of molecular hydrogen at various temperatures. Fig. 3 shows the result of cross-correlating the ion lines (upper panel) and the molecular hydrogen lines (lower panel) with the high resolution spectrum of 1442+101. The two redshift systems identified by Peterson et al. at z = 2.0701 and 2.5631 are seen in the ion line cross-correlation. There is no significant ion line cross-correlation amplitude for the two redshift systems of Levshakov and Varshalovich, and no significant molecular hydrogen cross-correlation amplitude at their redshifts or at any other redshift for molecular hydrogen in the temperature range 3K to 2000K. A detailed comparison of the QSO spectrum with the spectrum of molecular hydrogen at the redshifts of Levshakov and Varshalovich shows that at least one of the three strongest lines is missing.

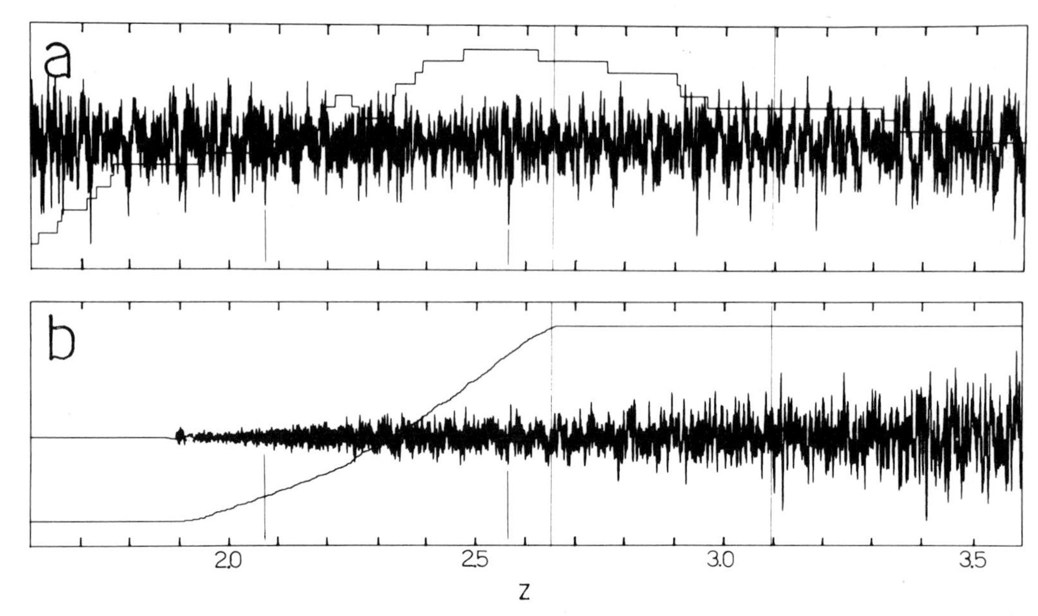

Figure 3. Cross-correlation amplitude per line (together with the number of lines used in the cross-correlation) vs. redshift. The locally averaged cross-correlation amplitude has been subtracted. Panel a shows the cross-correlation of atomic ions [C IV N V O VI Si IV plus the ions with the largest column densities through the interstellar medium of our galaxy in the direction of ζOph (Morton 1975) and ζPup (Morton 1978) all given equal weights] with the spectrum of 1442+101 obtained by Peterson et al. (1982). Panel b shows the cross-correlation of the lines of H_2 [weighted according to their absorption strengths at 1000° as calculated from the molecular data given by Morton & Dinerstein (1976)] with the spectrum of 1442+101 obtained by Peterson et al. (1982). The atomic ion absorption line systems at $Z = 2.0701$ and $Z = 2.5631$ found by Peterson et al. (1982) and the redshifts of the H_2 and atomic ion absorption line systems at $Z = 2.651$ and $Z = 3.092$ identified by Levshakov & Varshalovich (1979) in the spectrum of 1442+101 obtained by Baldwin et al. (1974) are marked in both panels.

Thus with increased resolution, there is no significant evidence for molecular hydrogen in the spectrum of 1442+101.

5. SUMMARY

The intergalactic clouds which produce Lyman-alpha absorption lines in QSO spectra evolve as the Universe expands, in the sense that there were more absorption lines produced at earlier epochs. The clouds may become more ionized and evaporate with time, or they may be gravitationally bound and collapsing.

These clouds are not primeval, but have undergone enrichment similar to Population II objects in the galaxy.

The identification of molecular hydrogen absorption lines in the spectrum of 1442+101 is not confirmed by high resolution spectra.

REFERENCES

Aaronson,M., Black,J.H., and McKee,C.F. 1974, Ap. J. (Letters), 191, L53.
Baldwin,J.A., Burbidge,E.M., Burbidge,G.R., Hazard,C., Robinson,L.B., and Wampler,E.J. 1974, Ap. J., 193, 513.
Carlson,R.W. 1974, Ap. J. (Letters), 190, L99.

Carswell,R.F., Whelan,J.A.J., Smith,M.G., Boksenberg,A., and Tytler,D. 1982, M.N., 198, 91.
Clowes,R.G., Smith,M.G., Savage,Ann, Cannon,R.D., Boksenberg,A., and Wall,J.V. 1979, M.N., 189, 175.
Chen,J.-S., Morton,D.C., Peterson,B.A., Wright,A.E., and Jauncey,D.L. 1982, M.N., 196, 715.
Levshakov,S.A., and Varshalovich,D.A. 1979, Ap. Letters, 20, 67.
Morton,D.C. 1978, Ap. J., 222, 863.
Morton,D.C. 1975, Ap. J., 197, 85.
Morton,D.C., Chen,J.-S., Wright,A.E., Peterson,B.A., and Jauncey,D.L 1980, M.N., 193, 399.
Morton,D.C., and Dinerstein,H.L. 1976, Ap. J., 204, 1.
Norris,J., Peterson,B.A., and Hartwick,F.D.A. 1982, Ap. J., (submitted).
Peebles,P.J.E., and Dicke,R.H. 1968, Ap. J., 154, 891.
Peterson,B.A. 1982 (in prep.).
Peterson,B.A. 1978, in "The Large Scale Structure of the Universe", (IAU Symposium No. 79) eds. M.S.Longair and J.Einasto (Dordrecht:Reidel), pp. 389-392.
Peterson,B.A., Chen,J.-S., Morton,D.C., Wright,A.E., and Jauncey,D.L. 1982, M.N., (in prep.).
Sargent,W.L.W., Young,P.J., Boksenberg,A., and Tytler,D. 1980, Ap. J. Suppl., 42, 41.
Savage,B.D., and Jeske,N.A. 1981, Ap. J., 244, 768.
Smith,M.G. 1978, Vistas in Astronomy, 22, 321.
Turnshek,D.A., Weyman,R.J., Leibert,J.W., Williams,R.E., and Strittmatter,P.A. 1980, Ap. J., 238, 488.
Varshalovich,D.A., and Levshakov,S.A. 1982, Comments Astrophys., 9, 199.
Weymann,R.J., Carswell,R.F., and Smith,M.G. 1981, Ann. Rev. Astron. Astrophys., 19, 41.
Wright,A.E., Morton,D.C., Peterson,B.A., and Jauncey,D.L. 1979, M.N., 189, 611.
Young,P., Sargent,W.L.W., and Boksenberg,A. 1982, Ap. J., 252, 10.

Discussion

Boksenberg: I have two comments: 1) I do not accept that the statement concerning evolution of the hydrogen clouds made by Sargent, Young, Boksenberg and Tyler is controversial as you have stated. Our full study showed that a non-evolving population was consistent within the carefully derived errors of our result. In this, we accounted for the equivalent width spectrum including blending effects and imposed a cutoff in the rest frame equivalent in deriving the number density as a function of redshift for all six objects. You did not do this for your data in your earlier note. Subsequently, with more data on a broader redshift baseline, Young, Sargent and Boksenberg did show positive evolution in the number density of the Lyman α clouds, with a hint of a break in number density at redshift near 2. More data we now have seem to confirm this, but they have not yet been fully processed.

2) The systems of large column density in H I you have picked to look for the "Lyman α" systems and heavy element systems merge: the heavy element systems more commonly have high column density, while the "Lyman α" systems more commonly have low.

Peterson: I disagree with your suggestion that I should not have seen in 1978 what you have only just seen in 1982.

M. Burbidge: There is an alternative hypothesis for the origin of the narrow Lα absorptions; that is, they really belong to the QSOs and do not arise in unconnected intergalactic clouds. Pointers in this direction are that the numbers and strengths of Lα absorptions in the <u>same</u> wavelength band (say 3300 - 3600 Å) appear to depend on the emission-line redshifts and luminosities of the objects. In looking at that wavelength band in QSOs with z_{em} between 1.9 and about 3, one shoul be sampling the <u>same</u> redshift range of intergalactic gas and the Lα should be randomly distributed, independent of z_{em} or luminosity of the QSO. However, one finds QSOs at the low end of the 1.9 - 3 range of z_{em} which have really few and relatively weak Lα narrow absorptions.

Peterson: If one considers the same wavelength region in the observer's frame, then as the redshift of the QSO increase the absorption lines seen in this observing window first consist of Ly-alpha, then Ly-beta and Ly-alpha, then Ly-gamma, Ly-beta, and Ly-alph so that as the redshift of the QSO increases, absorption lines higher in the Lyman series, which are produced by the more distant absorbing clouds, are seen along with the Ly-alpha absorption lines produced by the nearby clouds. This produces the correlation between the number of absorption lines in a fixed wavelength interval in the observer's frame and the QSO redshift.

If one considers the same wavelength region in the emitter's frame, say the region between the Ly-alpha and the Ly-beta emission lines as I have done here, then there should be no change in the number of absorption lines as a function of the QSO redshift if the absorption lines are associated with the QSO. The observed increase, in the number of absorption lines in this interval as a function of the QSO redshift, is evidence that the absorption lines are not associated with the QSO but with intergalactic clouds.

Boksenberg: 1) I have observed two QSOs of similar redshifts, but one having more than ten times the luminosity of the other and I have found no difference in H I line density.

2) If the H I absorption redshifts are interpreted in terms of velocities of ejection from the QSOs: a) the number of densities of clouds is uniformly distributed in "ejection velocity"; b) the number density is statistically the same in all objects measured in a consistent way; c) there is no correlation between line strength and "ejection velocity."

3) I believe it inherently unlikely that an ejection theory can be constructed to account for these observations, but no difficulty arises when we interpret the systems as being cosmologically distributed intervening material not associated with the QSOs.

Wolfe: I question your estimate of the oxygen abundance for two reasons: First, in order to know the total column density of hydrogen, you have to know the fractional ionization which is likely to be quite high. The latter depends on the cloud density and the mean intensity of ionizing radiation, both of which are uncertain by many orders of magnitude. Second the O VI column density you quote should be treated as a lower limit. The reason is that the absorption profile is likely to break up into narrow components which would be undetectable at the resolution you used. These narrow components may contain most of the oxygen, yet they would contribute little to the equivalent widths. So I believe that the O/H ratio which you quoted has little meaning.

Peterson: Of course, the oxygen-to-hydrogen abundance ratio is model dependent, but it varies within a range restricted by limits that can be placed upon the model parameters. The table below illustrates the sensitivity of the abundance ratio to the cloud density, the spectral index of the ionizing continuum, and the intensity of the ionizing continuum.

Sensitivity of [O/H] to the Ionizing QSO Background Flux

$$I_\nu = 10^{-21} I_{\nu-21}(\nu/\nu_o)^{-\alpha}$$

		α	0	1	2
		T	60000	40000	30000
$I_{\nu-21}$	log n			[O/H]	
0.1	-3.6 to -4.6		-1.2 to -0.9	-1.4 to -1.9	-0.2 to -2.2
1	-3.0 to -4.0		-1.0 to -0.9	-1.8 to -1.9	-1.4 to -2.7
10	-2.4 to -3.4		-0.9	-1.9	-2.2 to -2.9

The cold, high density clouds associated with 21-cm and Mg II absorption line systems have low velocity dispersions which result in narrow, saturated absorption lines that cannot be resolved optically. I agree that abundance determinations for these clouds suffer from the problems that you mention. However, the O VI clouds discussed here are hot clouds that have thermal velocity dispersions which correspond to our observed line profiles. Therefore, our O VI column density measurement does not suffer from the saturation and blending effects that you mention and is more reliable than you believe.

ABSORPTION STRUCTURE IN THE BL LAC OBJECT 0215+015 AT 20 km s^{-1} RESOLUTION

R. W. Hunstead and H. S. Murdoch
School of Physics, University of Sydney

M. Pettini
Royal Greenwich Observatory

and J. C. Blades
Rutherford Appleton Laboratory

The origin of the narrow metal absorption lines in the spectra of QSOs remains uncertain despite the large amount of high-quality data obtained at high resolution over the past decade. Recently, statistical tests of a uniform sample of C IV absorption systems have shown that their redshift distribution is consistent with the view that these lines arise in randomly distributed intervening galaxies in a Friedmann Universe (Young, Sargent and Boksenberg 1982, hereafter YSB). In a complementary approach we consider in detail the physical properties of the absorbing material in cases of particular interest. We have chosen the high-redshift BL Lac object 0215+015 (Blades et al. 1982, hereafter BHMP) for detailed spectroscopic study because it is currently in a bright phase (V ~ 14.5 - 16.5) and because it has several strong absorption systems with differing ionization structure. From our medium-resolution spectra of 0215+015 (BHMP) it appears that the density of C IV systems per unit redshift in this object is somewhat higher than, but not inconsistent with, the average density in the YSB sample.

Following the high-resolution (0.65 Å FWHM) observations obtained by BHMP in 1979 and 1980, we have now reobserved the C IV systems at z_a = 1.549 and 1.649 at a resolution of 0.27 Å FWHM (20 km s^{-1}). These observations were made in 1981 using the IPCS and RGO spectrograph on the 3.9m Anglo-Australian Telescope. A red-blazed 1200 line mm^{-1} grating was used in second order to provide a dispersion of 5 Å mm^{-1} with the 82cm camera. The z_a = 1.549 C IV system is shown in Figure 1 with the corresponding "low-resolution" (1.5 Å FWHM) profile superimposed. It is interesting to note that what appears as a simple C IV doublet at 1.5 Å resolution breaks up into many narrow components at 0.27 Å. There is also a clear detection of Galactic Ca II K absorption with W_λ = 0.12 Å. The z_a = 1.649 C IV system shows even greater complexity. We have

G. O. Abell and G. Chincarini (eds.), Early Evolution of the Universe and Its Present Structure, 359–364.

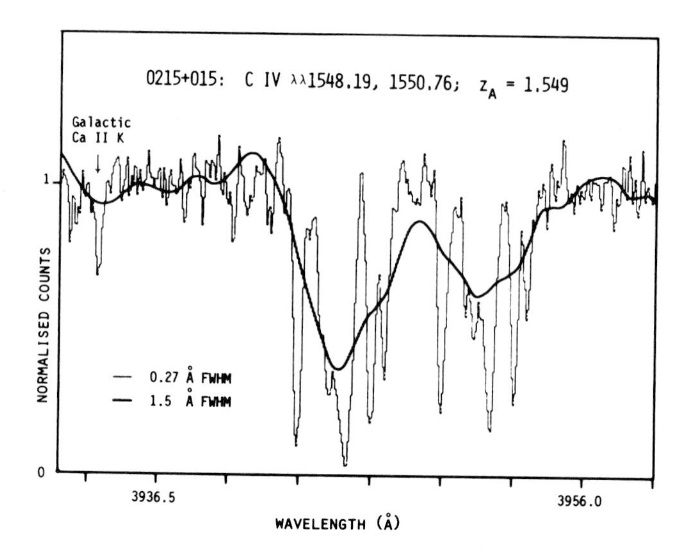

Figure 1. C IV absorption in 0215+015 at high and "low" resolution.

applied a model-fitting procedure to the profiles to determine the absorption parameters of individual clouds. Since there is only minimal blending in the profiles at this very high resolution, the velocity dispersions and column densities are much better defined than is usually the case. For the $z_a = 1.649$ system we fit 9 components over a velocity range of 910 km s^{-1} with b values from 8 to 25 km s^{-1} and column densities N ranging from 1 to 13 x 10^{13} cm^{-2}. For the $z_a = 1.549$ system there are 7 components spread over 300 km s^{-1} with b ranging from 6 to 15 km s^{-1} and N from 3 to 20 x 10^{13} cm^{-2}. It is worth noting that for the majority of components the b values are now directly comparable with that due to thermal broadening alone for gas at T = 10^5K, namely b = 12 km s^{-1}.

Such complex absorption can hardly be due to a single intervening galaxy, but can it be explained by a cluster of galaxies? To test whether this is plausible we have considered as an example the Coma cluster for which the luminosity function and surface density of galaxies versus radial distance are known (Godwin 1976, Abell 1977). By assuming galactic C IV cross-sections implied by the QSO absorption-line statistics of YSB and an average of two C IV velocity components per galaxy, we can reproduce the observed multiplicity of C IV components in 0215+015 with a line of sight passing within 0.3 Mpc(H_0=100 km s^{-1} Mpc^{-1}, q_0 = 0) of the cluster centre. This estimate only includes galaxies brighter than M_V = -16.9 and the impact parameter could be larger if fainter galaxies are considered. The likelihood of two such encounters with a rich cluster (corresponding to z_a = 1.549 and 1.649) is difficult to assess because the distribution of matter at these redshifts is not known. On the other hand, there are serious difficulties with interpreting these complex C IV systems as due to material ejected by the BL Lac object. The redshift of 0215+015 is likely to be of order 1.7, based on the highest absorption redshift identified (z_a = 1.719) and the lack of a Ly α forest (Blades, Hunstead, Murdoch and Pettini, in preparation). Thus, ejection velocities ≳ 5000 and 16000 km s^{-1} are required for the 1.649 and 1.549 systems, respectively. Current models for the intrinsic formation of narrow absorption-line systems in QSOs (Falle et al. 1981) cannot at present accommodate such large ejection velocities.

As part of the programme to obtain high-resolution spectra of 0215+015 we have also observed ions from the rich mixed-ionization system at z_a = 1.345 with resolution of 20 - 30 km s^{-1} FWHM. In BHMP it was pointed out that the derived ion column densities matched very closely those observed in Galactic halo sight-lines, making this system a highly plausible candidate for an intervening galaxy. At high resolution the line profiles are complex with at least four components spanning 250 km s^{-1}. The pair of lines Fe II λλ 2586, 2600 was observed on 1981 Nov 6.7 UT and again on 1981 Dec 1.5 in order to improve the signal to noise ratio. Clear differences in the absorption profiles are apparent when the spectra are superimposed, indicating substantial changes in the column density of some components. The two spectra are shown in Figure 2(a). Since changes in absorption structure on such a short time-scale (11 days at z_a = 1.345) are completely unexpected, a third spectrum of the same region was obtained on 1982 Jul 16.8. This spectrum is shown in Figure 2(b) superimposed on the 1981 Dec 1.5 spectrum and again there has been a clear change. As an analogous change is also seen in Fe II λ2586, there can be little doubt that the variations are real. Since this system is unlikely to be intrinsic (ejection velocity ~ 40,000 km s^{-1}) we have explored some of the more obvious mechanisms for producing variations. Using typical sizes for interstellar clouds in the Galaxy (0.1 - 10 pc), it is impossible to explain the variations by transverse motion of either the source or the absorber, unless by chance the line of sight has intersected a transient event in the absorber, such as a supernova.

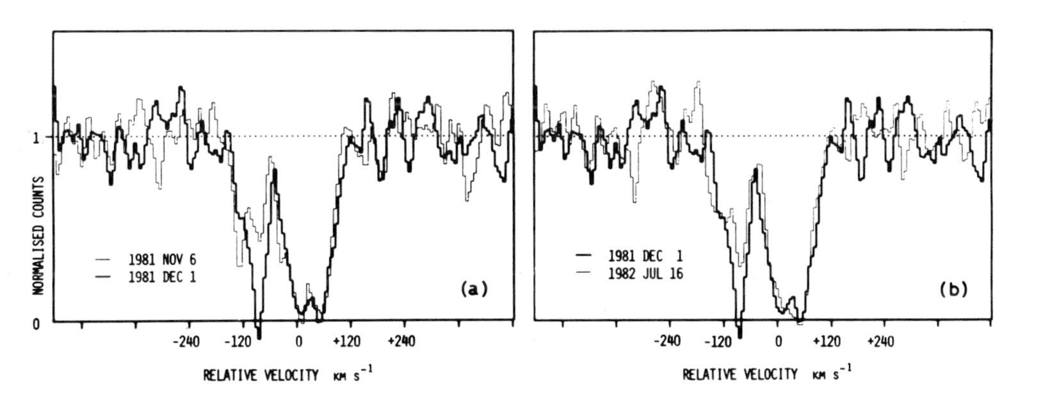

Figure 2. Time variation in the Fe II λ2600 absorption profile in the z_a = 1.345 system; the resolution at each epoch was 27 km s^{-1} FWHM (after smoothing with a gaussian of 10 km s^{-1} FWHM).

We are led therefore to a scenario in which the source is composed of several components of varying brightness and the absorption variability arises from the integrated effect of multiple sight lines through the absorbing galaxy or galaxies. The multiple source components may be intrinsic to the BL Lac object or alternatively may be gravitational lens images, perhaps formed by one of the absorption regions detected. In the latter case, the known variations in the BL Lac object combined with path differences through the lens galaxy will lead naturally to non-coherent variations in the images. It should be possible to test this hypothesis using VLBI mapping and by looking for correlations between the apparent magnitude of 0215+015 and the characteristics of the absorption profile.

REFERENCES

Abell, G.O.: 1977, Astrophys. J. 213, pp. 327-344.
Blades, J.C., Hunstead, R.W., Murdoch, H.S. and Pettini, M.: 1982. Mon. Not. R. Astr. Soc. 200, pp. 1091-1111.
Falle, S.A.E.G., Perry, J.J. and Dyson, J.E.: 1981, Mon. Not. R. Astr. Soc. 195, pp. 397-427.
Godwin, J.G.: 1976, Ph.D. Thesis, University of Oxford.
Young, P., Sargent, W.L.W. and Boksenberg, A.: 1982. Astrophys. J. Suppl. Ser. 48, pp. 455-505.

Discussion

Boksenberg: I would like to comment on the observed frequency of multiple heavy element absorption systems seen in QSO and BL Lac spectra. I believe these are much rarer than the literature would lead us to believe. Most published absorption systems have been discovered in low-resolution spectral surveys, when the complex systems, of large equivalent width, are more evident than simple systems. The discovered systems then are studied at high resolution and the complex multiple structure is revealed. An unbiased survey I have begun with Sargent, in which we do not pre-select "interesting" QSOs for study at high resolution, so far suggests that most heavy-element systems in fact are single, not multiple. This is entirely consistent with these being due to intervening galaxies.

Wolfe: Your detection of variations in Fe II line strength in 0215+015 is very exciting. The only other case in which such variability is observed is in the $z = 0.5$, 21-cm absorption spectrum in another BL Lac object, AO 0235+164 (Wolfe, Davis, and Briggs, 1982, Ap. J., 259, 495). The latter system also exhibits strong Fe II absorption, and I suggest that you look for Fe II variations there. This is important since it is highly unlikely that the 21-cm variations are due to passages of gravitational lenses across the beam because the radio beam diameter is at least $\sim$ 20 pc compared to optical beam sizes of less than $\sim$ one light week. So I believe that the detection of Fe II variations in 0235+164 would eliminate the lens model.

Hunstead: The H I variability in AO 0235+164 certainly calls for high-resolution optical observations at several epochs.

Baldwin: Are the differences seen between the three observations outside the absorption line to be taken as typical? Some look as large as those visible in the absorption line itself. Are they confined to one of the days?

Hunstead: In general, the continuum fluctuations are consistent with photon statistics. We believe the claimed variation is highly significant, especially in view of the correlated variation seen in the Fe II λ2586 line.

Dressler: The interpretation that the various absorption components occur in intervening galaxies in rich clusters like Coma seems inconsistent with the small velocity spread you find of several hundred km/sec. A typical spread of velocities in a dense cluster is of order 2000 - 3000 km/sec.

Hunstead: We would not necessarily expect a single sight of line intersecting a small number of galaxies to sample the full velocity spread of the cluster.

Wampler: I would like to emphasize the unusual nature of 0215+015. First, in August of 1981 it was approximately 14th magnitude. With a redshift of $z = 1.7$ it would be the brightest known object in the universe if it is at the Hubble distance.

Second, the polarization is very high 20% - 30%. I don't know the effects of gravitational lensing on polarization, but perhaps time delays in the light path or some other effect would give additional constraints on the system.

Peterson: How many photon counts do you have in the continuum bins in each of the three spectra that you used to detect variability?

Hunstead: About 50 counts, with a sky count of about 20.

Schallwich: On your question "Can such absorption arise in intervening clusters of galaxies?" mentioning the Coma cluster, Sholomitskii, Sunyaev, Wielebinski and Schallwich observed a QSO with absorption line systems with the 100-meter radiotelescope to search for hot gas (clusters of galaxies) via the Sunyaev-Zeldovich effect, which adds up on the line of sight.

We could not detect such an effect at λ 6 cm or at λ 2.8 cm on a level of $\Delta T/T \sim 10^{-4}$ (I have a positive detection on this level for the cluster Abell 2218).

So no "Coma clusters" (I mean rich clusters) should be expected to be responsible for the absorption.

Hunstead: This question bears on the general question of the origin of QSO absorption lines and not simply on 0215+015. We put forward the cluster absorption hypothesis simply to see whether it is at all plausible.

HIGH-REDSHIFT MOLECULAR CLOUDS AND ABSORPTION-LINE SPECTRA OF QUASARS

D. A. VARSHALOVICH and S. A. LEVSHAKOV
A.F. Ioffe Physical-Technical Institute,
USSR Academy of Sciences, Leningrad, USSR

The optical spectra of distant quasars (OQ 172, PHL 957, PKS 0237-233 and 11 others) were reanalysed with the purpose of searching molecular lines /2, 4/.

Several systems of absorption lines redshifted by Z_a= 2-3 which include lines of the H_2 and/or CO molecules together with atoms and ions were found among the preliminary unidentified absorption features. The probability of accidental line coincidences was estimated for each of the systems; in some cases the reliability of the identification of molecules is high enough.

For example, a system at Z_a = 2.651 in the OQ 172 spectrum contains about a dozen of the H_2 lines of the Lyman and Werner bands: L4-0, L5-0, L6-0, L7-0, L8-0, W1-0, W2-0, W3-0 and others (Fig.1). In addition to molecular lines a series of atomic and ionic lines can be identified at the same Z_a: H I (L_α, L_β, L_γ); C I (1328.8, 1277.2, 945.5); C II (1334.5, 1036.3); N I (1199.6); O I (1302.2); S I (1295.7); S II (1259.5); Ca II (1649.9, 1341.9, 1342.5); Fe II (1096.9). It gives a strong argument for the H_2 identification. The column density of H_2 is approximately $10^{20}cm^{-2}$ and the absorbing material is essentially molecular. Hence it is a high-redshift molecular cloud.

Consequently, molecular clouds existed at the early cosmological epoch Z_a = 2-3, when the age of the Universe was only 10-30 percent of the present one. The basic properties of these clouds - the column density of the material, its composition, its ionization and excitation - are similar to those of the Galactic interstellar clouds with $N_H \geq 3\cdot10^{19}cm^{-2}$ ($A_V \geq 0^m.5$). It is most likely they belong to some distant, unseen intervening galaxies.

Investigation of high-redshift molecular clouds may give an important information on the physical conditions and the chemical and isotopic composition of the matter at the early stages in the evolution of galaxies.

G. O. Abell and G. Chincarini (eds.), Early Evolution of the Universe and Its Present Structure, 365–366.

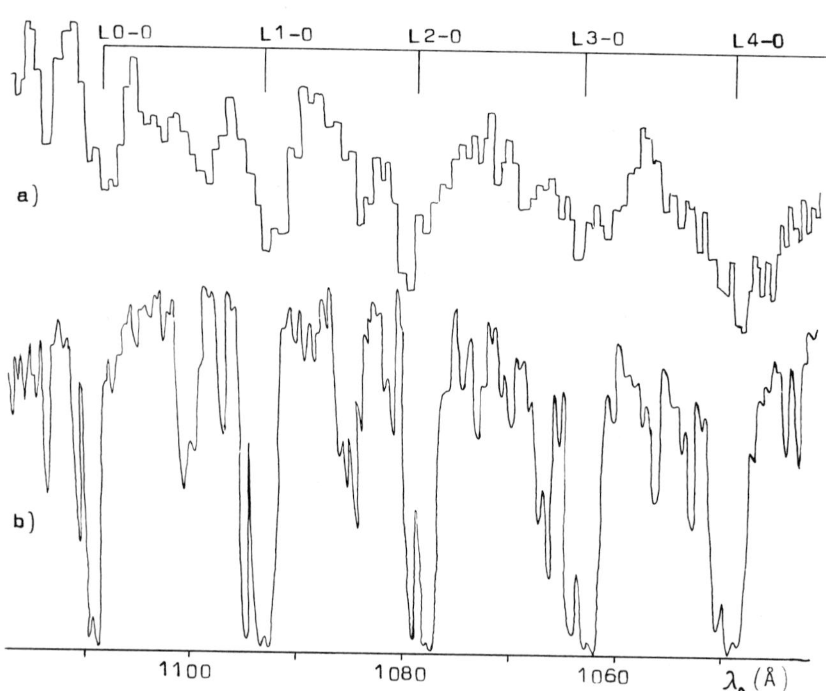

Fig.1 a) A fragment of the 00172 spectrum (Baldwin et al. 1974); the absorption features marked by Lv-0 can be identified with the H_2 Lyman band red-shifted $\lambda = \lambda_0(1+z_a)$ at $z_a = 2.651$.

b) The UV absorption spectrum of the interstellar H_2 in the direction to ξ Per (Savage et al. 1977).

REFERENCES

1. Baldwin J.A., Burbidge E.M., Burbidge G.R., Hazard C., Robinson L.B., and Wampler E.J.: 1974 Astrophys.J. **193**, p. 513.
2. Levshakov S.A., and Varshalovich D.A.: 1979 Astrophys.Letters 20, p. 67; 1982 Astrofizika **18**, p. 49.
3. Savage B.D., Bohlin R.C., Drake J.F., and Budich W.: 1977 Astrophys.J. **216**, p. 291.
4. Varshalovich D.A., and Levshakov S.A.: 1978 Sov.Astron.Letters **4**, p. 61; 1979 Sov.Astron.Letters **5**, p. 199; 1981 Sov.Astron.Letters **7**, p. 204; 1982 Comments on Astrophys. **9**, No. 5.

Dr. Varshalovich had intended to present this paper at the Symposium. At the last minute, he was unable to attend. The paper was summarized at the Patras General Assembly. Eds.

THE STATISTIC STUDY OF Lα ABSORPTION LINES

Chen Jian-Sheng*
(European Southern Observatory) and
(Peking Astronomical Observatory, Academia Sinica)

* This work was done in collaboration with Zou Zhen-Long, Bian Yu-Lin, Tang Xiao-Ying and Cui Zhen-Xing of Peking Astronomical Observatory.

The statistic studies of the absorbing forest in the spectra of high redshift QSOs will provide an important approach to the understanding of the physical properties of the absorbing materials. For this purpose a collection of uniform samples would be very important and urgent. It is very time consuming of the large telescopes to take the high resolution, high ratio-to-noise and wide coverage of wavelength of the QSO's spectra, therefore, an international collaboration in a uniform way would be very desirable.

We added here 4 spectra of high redshift QSOs, namely PKS 0528-250, PKS 0805+046 (4C 05.34), PKS 1448-232 and PKS 1442+101 (OQ 172), to SYBT's sample (Sargent, Young, Boksenberg and Tytler, 1980 (Ap. J. Suppl. 42, 81), all of which were taken at RGO spectrograph and IPCS attached to the Cassegrain focus of 3.9^m AAT with 1200/m grating and 25cm camera resulting in 1-2 Å resolution, so can be matched to the SYBT's sample. This larger sample nearly doubles the number of SYBT's and extends the upper limit of the redshift to 3.543 (OQ 172). Since the strong correlation between Lα and Lβ, we have asked the Lα sample satisfying the condition: $1025.72(1+Zem) \leqslant \lambda \leqslant 1215.67(1+Zem)$. This makes a slight modification for SYBT in taking off the lines outside this window and leaves altogether 350 lines for 9 QSOs. A series statistical test, which is similar to that of SYBT, leads to the conclusion that: 1) There is no significant difference in the number density of Lα lines among these 9 QSOs; 2) There is no significant variation in line density with redshift. The number variation is consistent with the Friedman cosmological models (with $0 \leqslant q_0 \leqslant 1$); 3) The rest equivalent width spectrum shows no significant variation with redshift; 4) There is no difference in the properties of Lα absorption lines in the wing of the Lα emission and those in the continuum. This means the size of the clouds D^c 10^{20} cm; 5) The two-point correlation function of the Lα clouds is flat, unlike that of galaxies. In summary, the results strengthen the point of view: the Lα clouds are probably intergalactic clouds.

G. O. Abell and G. Chincarini (eds.), Early Evolution of the Universe and Its Present Structure, 367.

ABSORPTION-LINE SPECTROSCOPY OF CLOSE PAIRS OF QSOs

P.A. Shaver[1] and J.G. Robertson[2]

[1]European Southern Observatory, Munich
[2]Anglo-Australian Observatory, Sydney

Close pairs of QSOs (separations $\lesssim$1-5 arcmin) provide a powerful approach to the study of narrow absorption lines in QSO spectra. By looking for absorption in the spectrum of the higher-redshift QSO at the redshift of the other ("associated absorption"), or absorption in both spectra at the same redshift ("common absorption"), one may address several issues: the cosmological nature of QSO redshifts, the origin of the narrow absorption lines of high redshift and excitation (intrinsic or intervening), the clustering of absorbing systems (with each other, and with QSOs), the sizes of the absorbing regions (for metal absorption lines and for Lyα lines; absorption cross-sections of individual galaxies and of clusters of galaxies), and the presence and nature of gaseous halos around QSOs.

The table summarizes work that has been done to date on close pairs of QSOs. In 3 pairs out of the 4 for which appropriate data exist, absorption has been found in the spectrum of the higher-redshift QSO within 2000 km/s of the emission-line redshift of the other. This result provides strong support for the cosmological interpretation of QSO redshifts (if further support is needed), and shows directly that at least some narrow absorption lines of high z and excitation arise in intervening matter. It also demonstrates that QSOs are located in regions of high matter density - plausibly clusters of galaxies. In the Q0028+003/Q0029+003 pair the absorption may arise in an extended ($\gtrsim$400 kpc) gaseous halo associated with the lower-redshift QSO itself.

Conclusive evidence of absorption in both spectra of a pair at the same redshift has only been obtained from the gravitational lens QSO Q0957+561, for which the projected separations are very small. No coincidences (within 2000 km/s) are present in 5 CIV systems in the spectra of the Q0254-334 and Q1623+268 pairs. This suggests that typical CIV absorbing regions are not strongly clustered together and are smaller than $\sim$400 kpc. A cross-correlation study of the narrow Lyα lines in the Q1623+268 pair showed no excess at small velocity splittings, indicating that the Lyα absorbers are also not strongly clustered and probably smaller than 1 Mpc.

G. O. Abell and G. Chincarini (eds.), Early Evolution of the Universe and Its Present Structure, 369–370.

QSO Pair	z_{abs}	Separation			Relevant Lines	Ref.
		ang. (')	proj. (kpc)	ΔV (km/s)		
a) Absorption at z_{em}(lower) ("associated absorption")						
0028+003/0029+003	1.7	1.0	390	190	CIV	SBR
0254-334/0254-334	(1.9)	1.0	380	$\gg$2000	CIV	WMPJ
1228+076/1228+077	1.9	3.4	1300	1400	CIV	RS
1548+114/1548+114	0.4	0.1	19	1400	MgII	BSWW
b) Absorption at Common z ("common absorption")						
0254-334/0254-334	1.7	1.0	380	$\gg$2000	CIV	WMPJ
"	0.2	1.0	140	100?	MgII?	WMPJ
0957+561/0957+561	1.4	0.1	$\leqslant$0.3	<10	CIV	WCW
"	1.1	0.1	$\leqslant$5	<10	CIV	YSBO
1623+268/1623+269	1.9-2.2	2.9	1120	>2000	CIV	SYS
"	2.2-2.5	2.9	1150	-	Lyα	SYS

(H_o=100 km $s^{-1}Mpc^{-1}$, q_o=0)

REFERENCES

Burbidge,E.M., Smith,H.E., Weymann,R.J., Williams,R.E. 1977, Ap.J. 218,1 (BSWW)

Robertson,J.G., Shaver,P.A. 1982, Nature (submitted) (RS)

Sargent,W.L.W., Young,P., Schneider,D.P. 1982, Ap.J. 256,374 (SYS)

Shaver,P.A., Boksenberg,A., Robertson,J.G. 1982, Ap.J.Letts.(in press) (SBR)

Walsh,D., Carswell,R.F., Weymann,R.J. 1979, Nature 279,381 (WCW)

Wright,A.E., Morton,D.C., Peterson,B.A., Jauncey,D.L. 1982, Mon. Not. Roy. astr. Soc. 199,81 (WMPJ)

Young,P., Sargent,W.L.W., Boksenberg,A., Oke,J.B. 1982, Ap.J. (in press) (YSBO)

Discussion

Braccesi: Should you not have to see reabsorption in the emission line itself of quasar 1, if this quasar is surrounded by a gas cloud able to produce an absorption line in quasar 2?

Shaver: The absence of such absorption could be explained if the "halo" is in the form of a thin disk. Alternatively, it could be an indication that the observed absorption does not arise in gas physically associated with the foreground quasar, but rather in a galaxy located in the same cluster as the quasar.

EVOLUTIONARY EFFECT IN QUASARS AS A CONSEQUENCE OF GALAXY FORMATION PROCESS

I.S. Shklovsky

Many years ago M. Schmidt established the presence of a strong evolutionary effect in quasistellar objects (QSOs). Though this important problem is not yet fully resolved, there can be no doubt that in earlier epochs of the Universe evolution of the spatial density of QSOs (in the comoving coordinates) was much higher than it is now. Thus, for instance, the spatial density of the optically selected QSO is:

$$\rho = \rho_0 \ e^{10\tau}$$

according to Schmidt (1977). Here $\tau = z/(1+z)$, whence, for instance, if $z = 3$, $\rho \sim 1000 \ \rho_0$, where ρ_0 is the local density.

On the other hand, the density of QSOs ceases to grow at a redshift somewhere between $3 < z < 4$. According to Osmer (1981), the spatial density of QSOs decreases considerably for $3.7 < z < 4.7$. The available, though unfortunately still insufficient, data make it possible to state that the spatial density of QSOs has its maximum near $z \sim 3$, and the "effective width" of their distribution is $|\Delta z| \sim 0.5$.

That there is an "epoch of quasar formation" at about $z \sim 3 \pm 0.5$ in the history of the Universe is the most natural conclusion from the above distribution. The interval

$$T = (1.3 \text{ to } 1.9) \cdot 10^9 \text{ years}$$

of the Universe age corresponds to that epoch. Here the age of the Universe (for the closed model) is given by:

$$T = T_0(1+z)^{-3/2}; \ \Delta T = 3/2 \ T_0(1+z)^{-5.2} \ \Delta z \ ,$$

where $T_0 = 2/3 \ H$. For a Hubble constant of 50 km/s, $T_0 \doteq 13 \cdot 10^9$ years. Thus, the evolutionary effect in the case of QSOs is most naturally explained by the fact that the overwhelming majority of QSOs formed at a certain stage of the Universe's evolution, when it was about two billion ($2 \cdot 10^9$) years old. What stage is that?

With a good degree of certainty we may now state that QSOs are galactic nuclei with an extremely high level of activity. Therefore,

G. O. Abell and G. Chincarini (eds.), Early Evolution of the Universe and Its Present Structure, 371–373.

the epoch of QSO formation is the epoch when galactic nuclei were forming. The almost-normal (that is, solar-like) chemical composition of the plasma-radiating QSO emission lines is a convincing argument in favor of the concept that QSOs are closely associated with the nuclei of galaxies. As far back as 1964, we emphasized that the phenomenon was not trivial (Shklovsky 1964).

It is most evident that a protogalaxy forming from diffuse matter needs time -- and a long time, indeed -- to develop a nucleus (and a disk, too) with the "normal" chemical composition. The suggested interpretation of the evolutionary effect implies that it should be $T_1 \sim (1.6 \pm 0.5) \cdot 10^9$ years. Therefore, we may estimate the epoch when protogalaxies started forming from the initial inhomogeneity of the matter in the Universe. The beginning of contraction of such inhomogeneities under the effect of gravitational instability is at

$$T_2 = T_1/2 ;$$

whence: $z_2 \sim 5$.

The estimates of T_1, T_2 and z_1 derived above are obviously most tentative. We have used simplified cosmological relationships, valid for the case $\rho = \rho_\alpha = 3H^2/8\pi G$. Since at present it is not yet clear which model, open or closed, is valid for the Universe, the above estimates may differ much from the real values. Our aim is to emphasize the explicit connection between the evolution of the Universe and the statistics of observational data for QSOs.

In the framework of the interpretation suggested here for the QSO evolutionary effect, a specific problem of comparatively near ("local") objects arises. The point is that the QSO phenomenon should be fairly short-lived. Hence, it is wrong to assume that local QSOs have been emitting over cosmological periods of time. There are two possibilities to explain the phenomenon of local QSOs:

a) Local QSOs are associated with quite recently formed nuclei of galaxies. It implies an assumption of a continuing formation of galaxies from the intergalactic gas in the expanding Universe;

b) Local QSOs are associated with the "rejuvenation" (activation) of the nuclei formed long ago. "Cannibalism" processes in the world of galaxies, gaining enhanced attention in recent years, may be mentioned as a cause of such "rejuvenation." Not abandoning, in principle, "cannibalism" as a possible reason for a temporary increase in the activity of galactic nuclei, we believe the phenomenon of local QSOs should be explained by a continuing process of the formation of the galactic nuclei in the Universe. The process should be regarded as slowing down, for at z = 2 to 3, its rate was hundreds of times higher than in our epoch.

It might be argued that the "cannibalism" hypothesis is supported by the recent discovery of three companions near QSOs (Stockton 1980). In our view, however, this is but another fact favoring the concept of QSO localization in poor clusters. As is known, similar companions are observed at about the same distance in the vicinity of M31 and other near galaxies, M87 among them. The fact of galaxy interaction alone does not imply a likelihood of their "rejuvenation."

On the contrary, new supporting arguments have recently been offered in favor of the hypothesis of a continuing process of galactic nuclei formation. The galaxy NGC 5128 has been considered by many authors to be a classical example of "cannibalism." However, recent thorough studies of NGC 5128 have shown no traces of a collision of two galaxies there. Everything indicates that a newly (about 10^9 years ago) formed disk -- and therefore a nucleus -- is observed in this galaxy. Note that Cygnus A and some other objects belong to the same class of galaxies. Another argument is provided by recent studies of the age of the disk and nucleus populations in our Galaxy; the age of the oldest stars in the disk appears to be at most $\sim 7 \cdot 10^9$ years (O'Connel 1980). Apparently, the nucleus of our stellar system formed only when the age of the Universe was already about ten billion ($10 \cdot 10^9$) years, that is, $z \sim 0.5$, whereas the overwhelming majority of nuclei of other galaxies in the Universe formed at $z \sim 2$ to 3.

Observations of recent years lead us to the idea that even within one cluster the time required for nuclei of constituent galaxies to form may vary over a large scale. A natural assumption is that galactic nuclei form more rapidly in the clusters with a higher initial density, since they have a shorter free-fall phase of gravitational contraction. The conditions for the condensation of more rarefied clouds might possibly be more favorable at cluster peripheries.

Attention should be drawn to the fact that the gas of which our protogalaxy formed was not yet enriched by heavy elements -- in particular, iron. It may mean that the condensation started prior to the formation of an intergalactic hot plasma in the cluster, the plasma being rich in heavy elements. We may also imagine the formation of protoclusters of galaxies at $z < z_1$, which, having a comparatively low mean density, should evolve much slower. It may explain the observed situation whereby local QSOs mainly occur in poor clusters with lower densities.

REFERENCES

Schmidt, M. 1977, Physica Scripta, 7, 135.
Osmer, P. 1981, Ap. J., 253, 28.
Shklovsky, I.S. 1964, Soviet A.J., 41, 801.
Stockton, A. 1982, Ap. J., 257, 33.
Marcelin, M., Boulesteix, J., Courtes, G., Milliard, B. 1982, Nature, 297, 38.
O'Connel, R. 1980, Ap. J., 236, 430.
Stocke, J., and Perrenod, S. 1981, Ap. J., 245, 375.

ARE QSOs GRAVITATIONALLY LENSED?

J. A. Tyson
Bell Laboratories

There are now four known cases of multiply imaged QSOs, one with a detected foreground object at roughly half the affine distance to the QSO: 0957+561 (17 mag, z=1.4, separation =6"), 1115+080 (17 mag, z=1.7, s1=1.8", s2=2.3"), 2345+007 (19 mag, z=2.1, s=7") and 1635+267 (19 mag z=2, s=5"). In addition, 1548+115 (19 mag, z=1.9) is a probable lens event with a foreground QSO, but no secondary image has been found. Perhaps 500 candidate QSOs have been surveyed optically for multiple images by all observers. 0957+561 is the only catalogued QSO shown to be multiple. Of the remaining 1548 QSOs currently catalogued, any secondary image is masked by atmospheric scattering of the QSO light. Typically, this sets detection limits of $\gtrsim$ 3 mag fainter and $<$ 2 arcsec separation from the bright component, for any secondary image. Objective prism and grism surveys look directly for multiple QSOs with identical emission lines and have surveyed 1500 QSOs. The remaining three lensed QSOs come from these more efficient surveys. Although the exciting search for multiply imaged QSOs has only begun, sufficient data already exist to test two hypotheses: (A) QSOs are intrinsically luminous and occasionally are multiply imaged through a chance alignment with a foreground galaxy of sufficient mass gradient and (B) all QSOs are the result of gravitational lens magnification of a distant Seyfert nucleus by foreground galaxy(s). I will first address hypothesis B, then A. I assume that mass (seen and unseen) clusters with galaxies and/or clusters of galaxies.

Case B: This appeals to some because QSOs would be no more intrinsically luminous than Seyferts. It also follows that the log N/m relation for case B QSOs has twice the galaxy log N/m slope, independent of other parameters. The data support this 2:1 slope relation, out to m_V = 19 mag for QSOs. Nevertheless, there are several problems with this hypothesis: (1) If foreground galaxies do the lensing, their average velocity dispersion would be 500 km/sec to explain the QSO number density - much larger than observed, since 80% of galaxies are spirals with $\sigma \sim$ 200 km/sec. Furthermore, a large fraction of QSOs would thus appear as multiple images with identical spectra separated by many arcsec. This is not found in optical QSO surveys, and it is not found in a VLA 0.1 arcsec resolution

G. O. Abell and G. Chincarini (eds.), Early Evolution of the Universe and Its Present Structure, 375–378.

survey of 400 QSOs (Perley). Finally, deep CCD multicolor surface photometry and spectroscopy of two bright QSOs shows no evidence for a superposed galaxy other than that one for which the QSO is the nucleus. (2) If galaxy clusters lens every QSO, we would still expect the observed 2:1 slope ratio in the log N/m plots, the required velocity dispersion in the lensing clusters would be acceptable ($\sim 10^3$km/s), but multiple images would not be expected in most cases. However, clusters of this 'Abell' size are not seen superposed on low redshift QSOs, where they can be seen by visual inspection of the Palomar Sky Survey. One might argue that since most QSOs have redshifts > 1, perhaps only higher redshift QSOs are all lensed. I have a deep CCD survey to 27 R mag arcsec^{-2} of 30 higher redshift QSOs ($1<z<1.5$) with few showing foreground clusters. Thus, the case for all QSOs arising from gravitational lensing appears to be unlikely in the extreme, unless my assumption that mass clusters with galaxies is violated. One would need a cosmological density of dark, compact objects of galactic mass and size, spatially uncorrelated with the observable galaxies. It should be reemphasized (Press and Gunn) that counts of gravitational lenses are one of the few cosmological tests for compact matter.

Case A: Under this hypothesis we treat QSOs as a separate family of objects with their own surface density $N_Q(m_Q)$ on the sky, unrelated to $N_G(m_G)$ for foreground galaxies. What do we then expect for the lensed QSO surface density $N_L(m_Q,m_G)$? Foreground galaxies within the critical angle for lensing, $\Theta_c = 2\sigma_{250}^2$ arcsec, will initiate a lens event. σ_{250} is the galaxy's velocity dispersion/250 kms^{-1}. By definition, a QSO appears stellar: $m_G - m_Q \geq 3$. Other arguments imply $m_G - m_Q \approx 3$ mag. Thus, N_L is simply the product of known surface densities: $N_L = \pi\Theta_c^2 f N_G(m_G) N_Q(m')$, $m' = m_Q + 2.5 \operatorname{Log} A$, with $m_G - m_Q = 3$, where f is the fraction of mass in galaxies and A is the average amplification in lens events giving detected secondary images ($A \approx 2$). Using the data for galaxy counts $N_G = 1.5 \times 10^4$ dex.44(m_G-24) deg^{-2} and QSO counts $N_Q = 3.2 \times 10^{-17}$ dex.9m_Q deg^{-2} for $m_Q < 19$ mag, I get

$$N_L = 2.5 \times 10^{-28} \ f \ \sigma_{250}^4 \ A^{2.2} \ 10^{1.3m_Q} \ \deg^{-2}$$

for the surface density of lensed QSOs as a function of their apparent magnitude. Note the strong dependence on m_Q. Taking f = 1 and A = 2 this leads to $N_L(19) = 230\sigma_{250}^4$/sky to 19th mag. Taking σ = 210 kms^{-1} gives $N_L(19) = 115$/sky. Since 6.3 x 10-3 of the sky was surveyed to 19th in the current QSO catalogue (1040 QSOs <19 mag vs 164,000 predicted over the entire sky by complete surveys over small areas), I expect 0.7 lensed QSOs to be found in the current catalogue to 19th mag out of 1000. In fact, one was found (0957+561). The grism (700) and objective prism (800) surveys are more efficient, and account for the remaining three out of 1500. Thus, surprisingly good agreement is obtained using canonical values for σ, f, A. Since the prediction is that we should eventually find $\sim$ 100 lensed QSOs over the sky to 19th V mag, there will then be sufficient data to test the steep m_Q dependence. Counting lensed QSOs survives as one of the best ways to test cosmology by sampling the mass distribution at $z \sim .3 - 1$.

Discussion

G. Burbidge: Searches for close pairs of QSOs using the grism technique will tend to find objects with the same redshift. However, what we really need to find out is how many close pairs there are with arbitrary redshifts. Close pairs with very different redshifts need to be explained.

Tyson: What I have done is to compare two hypotheses for the origin of close multiple QSO images with identical spectra. In obtaining lens count data for <u>both</u> of these hypotheses, we want to use an observing technique with the highest efficiency for detection of multiple QSO images with identical spectra. Grism or objective prism surveys complete in a given area of sky and to a given apparent flux have higher efficiency for this purpose than any inhomogeneous compilation of QSOs based on BSO or radio properties. I agree that to address a third hypothesis that somehow discrepant redshift QSOs are correlated on the sky, we should choose another observation technique, but I am not addressing that hypothesis here. If by invoking nothing more radical than general relativity and the observed galaxy and QSO surface densities we arrived at the observed lensed QSO surface density, there is no compelling reason to consider more complicated models to explain this surface density.

J. Barnothy: The B) solution was, as you well know, proposed by me in 1965. I think in your calculations you have forgotten three factors: 1) A gravitational lens with distributed mass can have 2 - 10 times greater intensification than a lens of the same compact mass; 2) Your inference that very few double images have been observed is not a valid argument against lensing in general. The few magnitude intensification needed to render a Seyfert nucleus visible through lensing can be achieved with medium mass intervening galaxies. But then the spacing between the two crescent images will be merely a fraction of an arcsec, not resolvable with optical telescopes. To see a double QSO, a separation of at least 5 arcsec is needed. This would mean that the mass of the lens has to be 100 - 1000 times larger, which, of course, would make it a rather rare event; 3) If our universe is not an expanding universe, and thus a Doppler effect is not present in the luminosity distance, a much lower intensification is needed, so that the nucleus of a Seyfert galaxy should become observable by lensing. For example, in the FIB cosmology, 90% of the QSOs seem to be lensed, while the remaining 10% of not-lensed Seyfert galaxies are to be found at very low z values and around the antipode at $z = 3.81$.

Tyson: I did not consider alternative cosmologies, nor did I intend to imply that QSO counts themselves are inconsistent with your suggestion. The problem is with the multiple QSO counts. I disagree with your statement that > 3 mag lens-intensified QSOs would have multiple images separated by a fraction of an arcsec. The inferred velocity dispersion, mass, and the luminosity of the lensing galaxy in 0957+561 are consistent with the 6-arcsec image separation.

I must remark that it is a testimonial to Zwicky, the Barnothys, and, ultimately, to Einstein that gravitationally-lensed objects were predicted before they were discovered!

Gorenstein: On a related topic, I would like to announce that N.L. Cohen, I.I. Shapiro, E.E. Falco, A.E.E. Rogers and I have obtained high-resolution and high-sensitivity radio maps of 0957+561 A,B using VLBI techniques. With these data we have also detected a new compact radio component which may be either a third image of the quasar, or the radio core of the elliptical galaxy situated near the B image. The preliminary maps of the A and B images appear consistent with the gravitational light-bending hypothesis.

CAN ALL QUASARS BE GRAVITATIONALLY LENSED SY's NUCLEI ?

G. Setti[1] and G. Zamorani[2]
1 European Southern Observatory, Garching bei München, FRG.
2 Istituto di Radioastronomia CNR, Bologna, Italy.

There has been a good deal of discussion in recent literature about the hypothesis, first put forward by Barnothy and Barnothy (1968), that quasars could be gravitationally lensed nuclei of Seyfert's galaxies (Turner, 1980; Tyson, 1981). Large amplifications (> 3-4 magnitudes) are needed to account for the widespread distribution in the intrinsic optical luminosities of quasars. A direct verification of this hypothesis is difficult to achieve due to the limited angular resolution of ground based telescopes. However, this hypothesis may be tested in a global sense by referring to radiation properties of Sy 1 nuclei which must be preserved, or must change in any predictable way, through the lensing process.

One such test is already provided by the X-ray emission properties of Sy 1 nuclei and quasars. It is known that the X-ray to optical emission ratio for radio quiet quasars depends on their intrinsic optical luminosity and that Sy 1 type nuclei ($M_B > -23.8$) are relatively stronger X-ray emitters than "true" quasars. This in itself already tells us that not all quasars can be magnified Sy 1 nuclei. To make the argument more quantitative we have considered a sample of 70 optically selected quasi stellar objects and Seyfert 1 nuclei ($M_B \lesssim -20$) for which Zamorani (1982) has derived a best fit line

$$\alpha_{ox} = 0.129 \text{ Log } L_{2500} - 2.427 \qquad (1)$$

where α_{ox} is the nominal spectral index between the optical (2500Å) and the X-ray (2keV) emissions, and L_{2500} is the intrinsic luminosity at 2500 A° in units of erg s^{-1} Hz^{-1}. In this sample there are no objects such that $\Delta\alpha = \alpha_{ox}$ (best fit line) $- \alpha_{ox}$ (observed) > 0.395, which according to Poisson statistics implies that at most 3 objects with $\Delta\alpha > 0.395$ could have been found at a 95% confidence level. This finding can be used to set an upper limit to the fraction of gravitationally lensed objects since, according to the relationship (1), one would expect that large amplifications would result in larger $\Delta\alpha$'s. By adopting a power law dist-

G. O. Abell and G. Chincarini (eds.), Early Evolution of the Universe and Its Present Structure, 379–380.

ribution of the amplification factors with an exponent in the range 2-3 in agreement with the quasar source counts (Peacock, 1982; Canizares 1982), we find that no more than 20% of the objects in the sample could result from amplifications larger than $\sim$2.5 magnitudes.

In this derivation we have tacitly assumed in agreement with present observational and theoretical knowledge that the X-ray photons are produced within the same region from which the optical continuum radiation is emitted.

Our upper limit is consistent with the results of Canizares (1982) for a flux limited sample and adds further evidence to the exclusion of the existence of a closure density of compact objects with masses $\gtrsim 1\ M_{\odot}$ if the radiation is emitted within 10^{-3} pc.

REFERENCES

Barnothy, J. and Barnothy, M.F. 1968, Science, 162, 348.
Canizares, C.R. 1982, Ap. J., in press.
Peacock, J.A. 1982, M.N.R.A.S., 199, 987.
Turner, E.L. 1980, Ap. J. (Letters), 242, L139.
Tyson, J.A. 1981, Ap. J. (Letters), 248, L89.
Zamorani, G. 1982, Ap. J. (Letters), in press.

Discussion

Rees: These limits are interesting because they constrain the amount of lensing due to individual compact masses (which may contribute $\Omega \simeq 0.01$). Canizares has used the equivalent width of QSO emission lines in a similar way: the continuum source of quasars may be lensed, but the line-emitting region may be too large to be affected, so the effect of lensing would be to introduce too large a spread in the equivalent widths.

STATISTICAL ANALYSIS OF OPTICALLY VARIABLE QSOs AND BRIGHT GALAXIES: A HINT FOR GRAVITATIONAL LENSES?

R. Bacon
J.-L. Nieto
Observatoire du Pic-du-Midi
F. 65200 Bagneres-de-Bigorre

Nieto (1979) found an excess of optically variable QSOs (OV) near bright galaxies ($m < 15.7$): 6 observed versus 1.6 expected for $r < 5'$. The probability involved was $p = 5 \times 10^{-3}$. Because of the small number of OV QSOs in this sample ($N = 41$, sample 1), this result needed a confirmation. So the same analysis was repeated with a sample of 112 QSOs (sample 3) from Hewitt and Burbidge (1980). Eleven objects were observed at $r < 5'$ versus 4.4 expected, so $p = 4 \times 10^{-3}$, the sample made up with 71 objects (sample 2) supporting slightly the result found with the first 41 objects. A notable difference between these two samples 1 and 2 is that the objects included in sample 2 are fainter than the objects included in sample 1. Repeating then the same analysis on samples of QSOs at different brightness levels suggests that the excess is related to the apparent brightness of the QSOs.

Concerning the output of statistical studies of QSO-galaxy associations, our results can be summarized in the following fashion:
-- the strong excess presented at first by the brightest radio QSOs is not confirmed by much larger QSO samples (see Nieto, 1978);
-- the same analysis repeated for various classes does not yield particular feature of the QSO distribution with respect to galaxies (Nieto 1978, 1979), except for: 1) bright and 2) optically variable QSOs, as shown by our sample 3 confirming the preliminary study made with sample 1.

We are tempted to believe that these statistical results can have a physical explanation which need not call into question a cosmological nature of QSOs. They could perfectly fit in the following scenario independently suggested by Canizares (1981), namely gravitational lens effects by stars located in the halos of intervening galaxies.

Then, a random distribution of QSOs with respect to galaxies would appear unrandom to us if some QSOs being (by chance) on the line of sight of a galactic halo (and their cosmological distances) are submitted to a gravitational lens effect coming from a halo star. Such a phenomenon would affect the luminosity function of QSOs in the neighborhood of galaxies and would produce some variabilities in the lensed images of the QSOs (already variable or not).

G. O. Abell and G. Chincarini (eds.), Early Evolution of the Universe and Its Present Structure, 381–382.

REFERENCES

Canizares, C.R. 1981, *Nature*, 291, 620.
Hewitt, A.H., Burbidge, G.R. 1980, *Ap. J. Suppl.*, 43, 57.
Nieto, J.-L. 1978, *Astron. Astrophys.*, 70, 219.
Nieto, J.-L. 1979, *Astron. Astrophys.*, 74, 152.

NATURE OF 'UNSEEN' GALACTIC ENVELOPES

W. H. McCrea
Astronomy Centre, University of Sussex, Brighton BN1 9QH
England

It is suggested that unseen matter in a galactic envelope or in a group of galaxies may consist of substellar bodies originating as the first permanent 'stars' in the formation of a very massive galaxy according to a model previously proposed by the author.

The writer has described a model for galaxy-formation (McCrea 1979, 1982). It is on the basis of simple big-bang cosmology: since this includes no means for initiating condensations, some additional assumption is unavoidable: we assume that after the epoch of decoupling the material of the cosmos splits into clouds of all sizes. The universe continues expanding, but neighbouring clouds may fall together under mutual gravitation. In an encounter of interest the relative motion is supersonic; it produces a layer of shocked material which breaks up into primary condensations. Each first collapses on itself forming short-lived supermassive stars which end in outbursts that (a) produce the *first heavy elements*, (b) induce in the rest of the condensation the formation of the *first normal stars*. These disperse or remain as a *globular cluster*, the aggregate of all such stars and clusters forming the *halo of a galaxy* composed of the material involved in the encounter. Any left after forming the halo goes to produce a nucleus or disk stars which mainly determine the optical appearance; here we are concerned with the halo population which on the model contains most of the mass. For a galaxy of mass about that of our Galaxy $M_G \simeq 10^{11}\mathcal{M}_\odot$ the model correctly predicts a mass $m_G \simeq 10^6\mathcal{M}_\odot$ for a globular cluster. At present it does not predict the mean mass $\mathcal{M}_{*G}$ of a star formed as in (b); so we shall adopt (c) $\mathcal{M}_{*G} \simeq 0.5\mathcal{M}_\odot$ as an estimated mean mass of halo stars in the Galaxy.

For a galaxy of mass M formed according to the model, if m is the mass of a primary condensation and ρ its mean density when formed, we have $m \propto M^{-1/3}$ and $\rho \propto M^{2/3}$. When the first stars of average mass $\mathcal{M}_*$ are formed, the mean density ρ_* is different, but it is natural to assume $\rho_* \propto \rho \propto M^{2/3}$. When stars are formed in material of density ρ_*, according to almost any theory of a critical mass, we have $\mathcal{M}_* \propto \rho_*^{-1/2}$. Combining these results $\mathcal{M}_* \propto M^{-1/3}$, or writing $M = \mathcal{M} M_G$ and using (c) we have $\mathcal{M}_* \simeq 0.5\mathcal{M}^{-1/3}\mathcal{M}_\odot$. Also we see that $m/\mathcal{M}_*$ is independent of M i.e on

G. O. Abell and G. Chincarini (eds.), Early Evolution of the Universe and Its Present Structure, 383–385.

the model the number of stars in a globular cluster is about the same fo all galaxies. Finally, the halo has diameter $\propto M^{1/3}$. Taken literally w have inferred that the halo of a 'galaxy' say 100 times as massive as ou Galaxy would be composed of stars of average mass about $0.1\mathcal{M}_\odot$, which ar about the faintest stars observed; the diameter would be about 5 times that of our Galaxy. Presumably this halo would be undetectable by direc observation since it would be out-shone by the rest of the galaxy. Agai a galaxy about 1000 times as massive as our Galaxy would have a halo of 10 times the diameter that would be composed of bodies of about $0.05\mathcal{M}_\odot$. According to usual estimates, nuclear burning would not be ignited in these, so we should have an enormous completely dark envelope.

Well-known reasons show that surrounding certain actual galaxies there is an amount of dark matter much exceeding the visible parts in mass; also in some groups of galaxies there is evidence of much dark matter. Such matter must be the dominant constituent of the system concerned. Here I consider the hypothesis that it is *baryonic*, i.e. neither neutrinos nor black holes. Because of cosmic abundances it must then be mainly hydrogen and helium, but not gas which would be too impermanent. The only known permanent state is in bodies between about $0.1\mathcal{M}_\odot$ and $0.001\mathcal{M}_\odot$; more massive bodies would not be dark and less massive would probably evaporate. Thus we want extensive 'galaxies' tha are composed of such *substellar bodies* and relatively little else. The only way to provide such a structure is to have the bodies made througho the required region; once they have been formed and are moving under th gravitation of the system itself the state is long-lasting. These requirements regarding actual dark matter appear to be met in a rather convincing manner by the halos of very massive galaxies predicted by our model. Thus our purpose is to propose the hypothesis:

> Unseen galactic envelopes and unseen matter in groups of galaxies consist of substellar bodies formed as the first permanent 'stars' in very massive galaxies.

The problem of unseen matter is so formidable that a drastic soluti is demanded. Therefore a suggestion that some galaxies possess masses large enough to furnish them, according to the model, with halos composed of bodies of mass not exceeding about $0.1\mathcal{M}_\odot$ may be not too fantastic. On the other hand, our numerical examples show that the hypothesis become of interest for masses exceeding about 100 times the mass of our Galaxy, which are normally considered more applicable to groups of galaxies. In the case of such a mass it may therefore be better to regard the model as applying to the first stage in forming a whole group rather than as necessarily leading to the production of a single enormous galaxy. This possibility of a unified treatment of dark matter in both cases would be an additional recommendation. The suggestion that dark matter consists of faint stars is not novel, but its derivation from a model for galaxy formation that offers even tentative predictions about numbers and masses appears to be new.

REFERENCES

McCrea, W.H.: 1979, Irish Astr. Jl. 14pp. 41-49.
McCrea. W.H.: 1982, *Progress in Cosmology* (A.W. Wolfendale, ed.) D. Reidel Publ. Co., Dordrecht, Holland, pp. 239-257.

Discussion

Rees: I would just like to emphasize that your proposed correlation between the masses of stars and the scale of the bound system they belong to is expected only if the stars form at a relatively "recent" epoch. If Population III objects condensed at Z $\gtrsim$ 100 (as is envisaged in models where their luminosity gives rise to distortions in the microwave background), they would not at that stage "know" what scale of bound system they are destined to be in.

McCrea: I should have done better not to mention Population III objects! The halo "stars" -- or "jupiters" if the mass is small enough -- about which I spoke are not the Population III stars of other models. On the present model the halo objects are formed when a galaxy as a whole (or possibly a group of galaxies as a whole) is formed. So they "know" from the outset the scale of the bound system to which they belong; there is no paradox in their mean mass being dependent upon that system's total mass.

In giving a brief oral description of the model, I referred to Population III stars because other speakers had mentioned them; in the written version I refer to "short-lived supermassive stars" and in my spoken account I said that these may resemble Population III stars of certain other models. But in my model such stars are a (temporary) part of the galaxy being formed; so they are not pre-galactic in the sense of those other models. In that sense, as Professor Rees says, my processes may be somewhat more "recent" than postulated pre-galactic processes. However, as stated in the cited references, I contemplated an epoch z $\simeq$ 100, which may not be a lot more recent, but I appeal to no similar pre-galactic happenings.

THE THEORY OF LARGE-SCALE STRUCTURE OF THE UNIVERSE: LOCAL PROPERTIES AND GLOBAL TOPOLOGY

A.G. Doroshkevich, S.F. Shandarin, Ya.B. Zeldovich
Keldysh Institute of Applied Mathematics
USSR Academy of Sciences, Moscow 125047, USSR

ABSTRACT

Properties of the large-scale distribution of galaxies are considered. Particular attention is paid to properties of the large-scale structures such as anisotropy of superclusters and the existence of large regions practically devoid of galaxies. Another question discussed in detail is the link between superclusters and formation of a network or cellular structure. An explanation of the latter is proposed in the frame of the fragmentation scenario. The role of the neutrino rest mass is discussed.

1. INTRODUCTION

The theory that will be considered here is based on the big bang model and the idea that physical processes before decoupling leave us with small amplitude longwave perturbations in cold gravitating matter. These conditions are fulfilled in the flat universe because of photon viscosity. Later, arguments concerning the small MBR temperature fluctuations and the matter density appearing to be much less than the critical value are shown to reveal a quantitative discrepancy, leading to the idea of a neutrino-dominated universe.

In this case the wavelength of surviving perturbations is determined by the neutrino travel distance, which is limited because the neutrinos are no longer relativistic at a temperature before decoupling.

Luckily, the two scenarios (baryon- or neutrino-dominated Universe), very different physical ideas, are leading to very similar structural properties. Many ideas worked out in 1970-1978 in the frame of baryon dominance remain valid for the neutrino model. Of course, the exact value of the neutrino mass and even its nature remain to be confirmed by particle physicists.

We will discuss only the rather late epoch, ($z < 10$), when density perturbations become nonlinear, $\delta\rho/\rho \sim 1$ and the structure acquires features, many of which exist at present. The long previous evolution of small perturbations guarantees that only growing modes existed. Therefore, perturbations must be random and smooth and of potential type (rot $v = 0$).

G. O. Abell and G. Chincarini (eds.), Early Evolution of the Universe and Its Present Structure, 387–391.

2. OBSERVATIONS

Observational data about the large-scale structures are discussed in detail by many others and are well summarized by Oort (see this volume). Perhaps the most striking features of the large-scale galaxy distribution are the following: Most galaxies belong to superclusters; superclusters occupy only about 10% of the volume; the other space is practically devoid of galaxies; superclusters are highly asymmetric unrelaxed structures; superclusters are not isolated from each other but probably form a network of cellular structure. Typical sizes of the structure reach 100 Mpc or even more. The two-point correlation function for galaxies is much less than 1 on these scales (Peebles 1980).

3. SCENARIO FOR STRUCTURE FORMATION

In the scenario under consideration, perturbations were small for a long time after decoupling. The smallest perturbed scale was about $M_C \approx (10^{14} - 10^{15})\ M_\odot$, which is much greater than Jeans mass M_J ($M_C \gg M_J$) of neutral gas and/or cold neutrinos. This means that gas pressure is unimportant for evolution of the perturbations prior to formation of the first objects. Zeldovich (1970) has argued that the beginning of collapse in a medium with low pressure is one-dimensional. This results in the formation of so-called pancakes with masses of about $M_C \approx (10^{14} - 10^{15})\ M_\odot$.

The pancakes are bounded by the shock waves (gas component), the multistream regions with high neutrino density. Once formed, the pancakes remain gravitationally bound.

Later, it became clear that pancakes are the large-scale objects formed first but they are not the only ones. Catastrophe theory provides a full list of all structures of the generic types (Arnold, Shandarin, Zeldovich 1982) forming at the nonlinear stage. There are two-dimensional (pancakes), one-dimensional (filaments or strings) and zero-dimensional (clusters) structures in the list. Structures of the generic types are those elements from which the large-scale structure is built. They are extremely anisotropic, so in this scenario anisotropy of superclusters is explained quite naturally.

Numerical simulations of the structure formation in this scenario were done in two-dimensional (Doroshkevich et al. 1980) and three-dimensional cases (Klypin and Shandarin 1981). They have known that at some stage soon after pancake formation regions of large density form a single network structure.

Galaxies are formed inside pancakes and higher order singularities in a rather complicated process. Gas heated by shock waves first cools to ~ 10^4 K (Sunyaev, Zeldovich 1972); thereafter, it fragments (Doroshkevich, Shandarin, Saar 1978). The rarefied gas outside pancakes is heated and ionized by pancake radiation. Probably no galaxies are formed in this gas, which would explain the voids (Zeldovich, Shandarin 1982).

4. GLOBAL TOPOLOGY OF THE STRUCTURE

Let us consider this question in detail. Recently, many observers (see this volume) pointed out that many superclusters link each other in a single network structure; however, the lack of quantitative technique to study this problem makes it difficult to establish this objectively. The widespread, two-point correlation analysis gives much interesting information about the galaxy distribution (Peebles 1980); however, it gives little help in answering the question concerning pattern recognition. The question about the global topology of the structure was raised by Zeldovich (1982). If superclusters really occupy only about 10% of the total volume and the majority of galaxies belong to them, it is surprising that superclusters form a connected system.

To explain it, let us consider an epoch prior to pancake formation. At this stage perturbations are small and one easily can find that there are four types of fluid element behavior: i) contracting along all three directions; ii) contracting along two directions but expanding along the third one; iii) contracting along only one direction; and iv) expanding in all three directions. Here contraction or expansion mean peculiar motion in comoving coordinates with the mean Hubble flow; i.e., it is superimposed on the general expansion. To form a pancake, it is enough to contract along only one direction and as was shown by Doroshkevich (1970) about 92% of the matter contracts along at least one direction. At the stage of small perturbations, this matter forms one connected region because it occupies also about 92% of the volume.

The eight percent of the "to be rarefied gas" is occupying disconnected islands in the sea of "to be compressed gas." These islands, devoid of galaxies, remain disconnected when their volume has increased to more than 50%, just due to their expansion. Topology is preserved. It is a cell structure if galaxy formation is strong enough after one-dimensional compression; it is a net if higher order singularities are needed. In all cases the cell and/or net structure remains gravitationally unstable, it is an intermediate asymptote, which will be disrupted after a time of the order of (several) ages of the Universe.

Recently, Shandarin (1982) proposed a new quantitative method to recognize the patterns in observations. The method is based on ideas of percolation theory (see, for example, B.I. Shklovski, Efros, 1979) and is particularly sensitive in distinguishing between network or cellular structure on the one hand and isolated clumps of galaxies on the other. It can be used for both two- and three-dimensional distributions.

Cut out in space a cube with size L, containing N galaxies within it. Draw a sphere of radius r around each galaxy. If there is another galaxy within the sphere, we shall call the two galaxies connected. All connected galaxies form a cluster. The number and sizes of clusters depend on the radius r. At small r, clusters are numerous but small; at large r, the number of clusters decreases and sizes grow. There is a critical radius r_c when a cluster is linking the opposite sides of the cube. In terms of percolation theory, percolation arises. The condition of percolation is usually expressed in terms of a dimensionless parameter $B = 4/3\ \pi\ N(r_c/L)^3$, which is a mean number of galaxies within a sphere of percolation radius r. If galaxies were distributed independently with a Poisson distribution, percolation would arise at $B_{Poisson} \simeq 2.7$ (Kurkijarvi 1974; Pike and Seager 1974; Skal and B.I. Shklovski 1973). It is clear that if the galaxy distribution tends towards a network, percolation would arise along the network more easily.

than in the Poisson case: $B_{network} < B_{Poisson}$. In the opposite case, if galaxies are concentrated towards isolated clumps, percolation would be impeded between clumps: $B_{clumps} > B_{Poisson}$.

Below we give the first estimates of percolation parameters for four different distributions. For details see Shandarin (1982) and Einasto _et al_. (this volume). The first distribution (A) is a real distribution of galaxies within a cube with size 80 Mpc (H=50), taken from the Huchra sample. Three others are model distributions. One of them (B) is taken from the Klypin and Shandarin (1981) three-dimensional numerical simulation in the adiabatic scenario; the second (C), is a realization of Poisson distribution; and the third (D) is hierarchical distribution constructed in accord with the prescription by Soneira and Peebles (1978).

	Type of Distribution	Number of Objects	Percolation Parameter
A	Galaxies (observational sample)	866	0.8
B	Adiabatic model	753	0.7
C	Poisson	850	3
D	Hierarchical model	819	7

These values definitely show that in both observational and adiabatic samples there are filamentary structure. The hierarchical sample, where isolated clumps dominate, gives a value of B far from that observed. This analysis, in its present form, disproves clumps but was not discriminate between cellular or network structure. However, in principle, this method can distinguish between these two possibilities if one will also study percolation problems in "empty" regions. Summarizing, the observations favor the adiabatic scenario. The new ideas of Ostriker and Cowie about the role of explosions remain to be analyzed.

REFERENCES

Arnold, V.I., Shandarin, S.F., Zeldovich, Ya.B. 1982, _Geophys. Astrophys Fluid Dynamics_, 20, 111.
Doroshkevich, A.G. 1970, _Astrofisica_, 6, 581.
Doroshkevich, A.G., Shandarin, S.F., Saar, E. 1978, _M.N.R.A.S._, 184, 643.
Doroshkevich, A.G., Kotok, E.V., Novikov, I.D., Polyudov, A.N., Shandarin, S.F., Sigov, Yu.S. 1980, _M.N.R.A.S._, 192, 321.
Klypin, A.A., Shandarin, S.F. 1981, Pr-t No. 136, _In-t Appl. Math. Acad. of Sciences USSR_.
Kurkijarvi, J. 1974, _Phys. Rev_., B9, 770.
Peebles, P.J.E. 1980, _The Large-Scale Structure of the Universe_, Princeton University Press.
Pike, G.E., Seager, C.H. 1974, _Phys. Rev_., B10, 1421.
Shandarin, S.F. 1982, _Pis'ma v Astron. Zh_.
Shklovski, B.I., Efros, A.L. 1979, _Electronic Properties of the Alloyed Semiconductors_, Moscow.

Skal, A.S., Shklovski, B.I. 1973, FTP, 7, 1589.
Soneira, R.M., Peebles, P.J.E. 1978, Astron. J., 83, 845.
Sunyaev, R.A., Zeldovich, Ya.B. 1972, Astron. Astrophys., 20, 189.
Zeldovich, Ya.B. 1970, Astron. Astrophys., 5, 84.
Zeldovich, Ya.B. 1982, Pis'ma v Astron. Zh., 8, 195.
Zeldovich, Ya.B., Shandarin, S.F. 1982, Pis'ma v Astron. Zh., 8, 131.

Discussion

Thompson: For your three-dimensional simulation which produced a filamentary structure, what did you assume for the model's initials conditions?

Shandarin: In our simulations we started from random but smooth (on short scales) perturbations of a potential type. These initial conditions are typical for the adiabatic or fragmentation scenario of structure formation.

Bonometto: How many points were used in your simulation and which numerical technique did you use?

Shandarin: We used a fast Fourier technique with a total number of particles of $32^3 \approx 3.277 \times 10^4$. The number of cells was the same.

B. Jones: The Lick catalogue is two-dimensional; whereas, the numerical simulations of both the pancake theory and isothermal theory are three-dimensional. Is it possible to use the percolation coefficients to compare these or is it essential to have a three-dimensional picture of our universe such as is provided by the redshift surveys?

Shandarin: Yes, it is possible. However, in the two-dimensional case, a critical value of the percolation parameter, B_c, is different. $B_c^{(2)} \approx 4.1$, but $B_c^{(3)} \approx 2.7$.

SELF-SIMILAR GRAVITATIONAL CLUSTERING

G. Efstathiou
Institute of Astronomy,
Madingley Road,
Cambridge, England.

The nature of the distribution of galaxies poses a challenging problem for theorists. It seems reasonable, as a start, to suppose that galaxies and clusters arose from small perturbations by gravitational instability. However, one still has the problem of the choice of initial conditions, for example, the shape of the fluctuation spectrum and the cosmological density parameter Ω. A considerable simplification is to assume that the clustering pattern obeys some simple similarity scaling, so that the clustering at some early time, apart from a change in length scale, is statistically indistinguishable from the pattern observed today. The power-law shape of the two-point correlation function and the simple forms of higher order correlation functions (Peebles, 1980) have provided some evidence that such a simplifying assumption may be relevant - just how relevant is the subject of this article.

1. THE TWO-POINT CORRELATION FUNCTION

Self-similar gravitational clustering requires that the initial conditions do not possess any characteristic length scales and that the expansion of the universe does not present any characteristic timescales. Thus, A $\Omega = 1$ ($\Lambda = 0$, pressure = 0). B The initial power-spectrum of the matter distribution must be a power law $|\delta_k|^2 \propto k^n$. C The clustering must be due to gravity alone - non gravitational forces are ignored. With these assumptions the two-point correlation function obeys a scaling relation $\xi(x,t) \rightarrow \xi(s)$ with $s = x/t^{\alpha}$. The 'similarity' parameter α can be fixed from the initial conditions using the growing mode from linear theory, $\alpha = 4/3(3+n)$. The slope of the correlation function in the non-linear regime $\xi >> 1$ may be fixed by the requirement of small-scale stability (i.e. that clusters once formed are bound and stable with no subsequent evolution),

$$\xi(s) \propto s^{-\gamma}, \quad \gamma = 3(n+3)/(n+5), \tag{1}$$

(Davis & Peebles, 1977). The observed non-linear slope $\gamma = 1.8$ implies white noise initial conditions, $n = 0$. Together with the stability

G. O. Abell and G. Chincarini (eds.), Early Evolution of the Universe and Its Present Structure, 393–399.

assumption, the similarity solution for the three- and four-point correlation functions is $\zeta \propto s^{-2\gamma}$, $\eta \propto s^{-3\gamma}$ (Peebles, 1980, ¶ 73) and these relations are in excellent agreement with the observations.

Several approaches have been used in an attempt to derive the shape of the two-point function in the regime where $\xi \sim 1$. Davis & Peebles (1977) used the BBGKY equations truncated in a manner consistent with the observations of the three-point function ζ. Fry & Peebles (1980) have used a Monte-Carlo N-body technique, but the most widely used method has been the direct integration of Newton's equations in an expanding universe (e.g. Miyoshi & Kihara, 1975; Gott et al., 1979; Efstathiou & Eastwood, 1981; Frenk et al., 1982).

Figure 1 shows results for ξ from the N-body calculations of Efstathiou & Eastwood. The models give a correlation function which is much steeper than the observations and are in disagreement with eq. (1) because the stability assumption does not apply on scales corresponding to $\xi \lesssim 50$. Instead, one observes a radial streaming as clusters collapse in order to generate enough kinetic energy to satisfy the virial theorem.

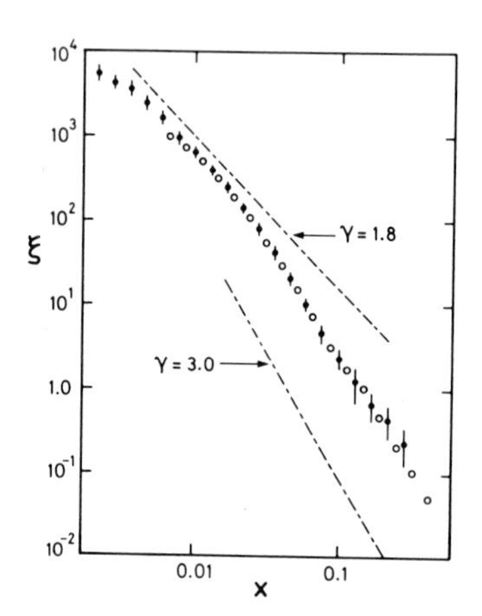

Figure 1. Results for the two-point correlation function using Poisson initial conditions and $\Omega = 1$ after expansion by a factor of 9.9. Filled circles show results for 6 models with N = 1000 and the open circles show results for one 20000 body model.

Of course there are difficulties in the interpretation of the N-body results; one certainly does not expect the similarity solution to apply on scales smaller than the mean inter-particle separation and for small N systems this is uncomfortably close to the size of the whole system. The N-body results do deserve to be taken seriously because A different N-body schemes yield results which are in good agreement (Figure 1). B ξ does evolve in a manner consistent with a similarity solution (Efstathiou et al., 1979; Frenk et al., 1982).

There is an additional important discrepancy between the N-body models discussed above and observations. If the models are scaled so that $\xi(r_o) = 1$ with $r_o = 4h^{-1}$ Mpc (h is Hubble's constant H_o in units of 100 km sec^{-1} Mpc^{-1}), the one-dimensional r.m.s. peculiar velocity between particle pairs of separation r_o is $<w^2>^{1/2} \sim 900$ km sec^{-1}. As discussed below this is much larger than the peculiar velocities between galaxy pairs.

2. PECULIAR VELOCITIES

The mean square peculiar velocity between galaxy pairs of separation r can be calculated from an integral over the three-point correlation function. For $\xi(r) = (r_o/r)^{1.8}$ the relation is (Peebles, 1976),

$$<w^2(r)>^{\frac{1}{2}} \sim 870\ Q^{\frac{1}{2}}(r_o/4h^{-1}\mathrm{Mpc})^{0.9}(r/1h^{-1}\mathrm{Mpc})^{0.1}\Omega^{\frac{1}{2}}\mathrm{km\ sec}^{-1}, \quad (2)$$

where Q is the ratio of the amplitude of the three-point correlation function to the square of the two-point function. There are two main assumptions necessary in deriving eq. (2) - the stability assumption and that the galaxy correlation functions measure the mass distribution.

In order to derive $<w^2>$ from redshift data I use a method similar to that of Peebles (1979). Consider two galaxies with observed radial velocities v_1 and v_2 separated on the sky by an angle θ_{12}. The approximate separation of the pair parallel to the line of sight is $\pi = (v_1-v_2)/H_o$ and the separation perpendicular to the line of sight is $\sigma = \sqrt{v_1 v_2}\,\theta_{12}/H_o$. The clustering pattern is distorted in the π direction because of peculiar velocities (and velocity errors) hence the two-point function ξ_v measured using the coordinates σ and π will differ from the true correlation function. This is modelled as,

$$\xi_v(\sigma,\pi) = \int \xi\{[(\sigma^2+(\pi-w/H_o)^2]^{\frac{1}{2}}\}\ f(w)dw, \quad (3)$$

where w is drawn from the distribution function f(w).

Two recent magnitude limited redshift surveys have been analysed; the Kirshner et al. (1978) survey containing ~ 160 galaxies and the Anglo-Australian survey (Peterson et al., 1982) containing ~ 340 galaxies. Figure 2 shows estimates of $\xi_v(\sigma,\pi)$ for the combined sample. The method used to estimate ξ_v is described in detail by Bean et al. (1982). The distribution function f in eq. (3) is assumed to be

$$f(w) \propto \exp(-0.7966\,|w|^{3/2}<w^2>^{-3/4}), \quad (4)$$

which has been found to be a good fit to the N-body models. The power-law model for ξ is assumed with $\gamma = 1.8$ and the best values of $<w^2>$ and r_o are found by a least squares fit of the model (eq. 3) to the observed histograms of $\xi_v(\sigma,\pi)$ over the interval $0 < \pi < 10\ h^{-1}$Mpc. These methods have been thoroughly tested using simulated catalogues similar to the AAT survey generated according to the prescription of Soneira & Peebles

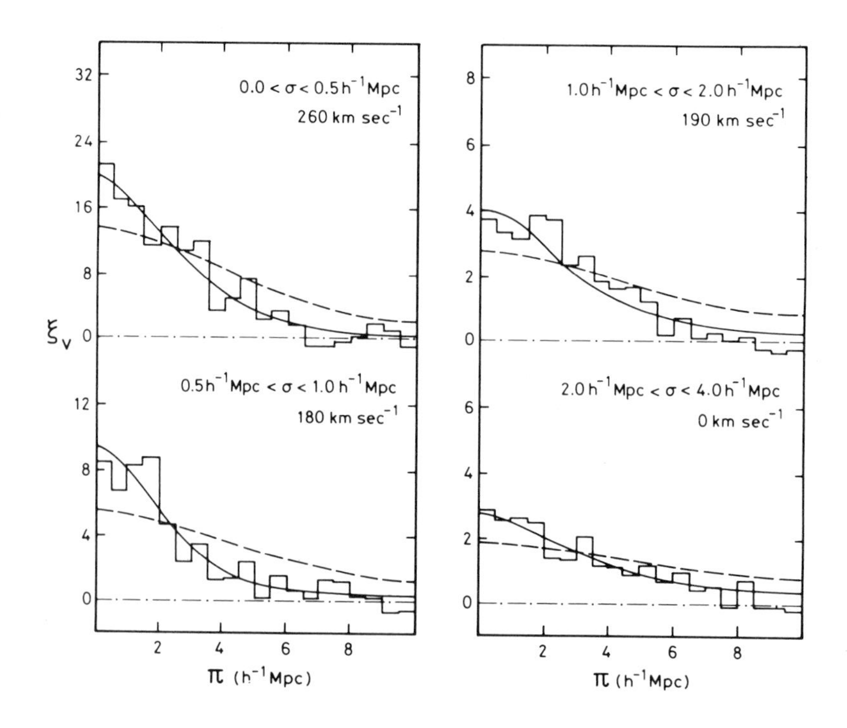

Figure 2. Estimates of ξ_v for the Anglo-Australian survey combined with the Kirshner et al. sample. The solid lines give the best least squares fit using eq. (3) and (4) corresponding to the quoted value of $\langle w^2\rangle^{\frac{1}{2}}$. The dashed line shows the best fit obtained using $\langle w^2\rangle^{\frac{1}{2}} = 500$ km sec^{-1}.

Table 1. Monte-Carlo results and error estimates for $\langle w^2\rangle^{\frac{1}{2}}$

	200 km sec^{-1}		500 km sec^{-1}	
σ (h^{-1}Mpc)	$\langle w^2\rangle^{\frac{1}{2}}$	s.d.	$\langle w^2\rangle^{\frac{1}{2}}$	s.d.
0.0 - 0.5	210 ± 6	(40)	490 ± 13	(83)
0.5 - 1.0	203 ± 8	(50)	463 ± 13	(79)
1.0 - 2.0	205 ± 13	(85)	463 ± 15	(92)
2.0 - 4.0	189 ± 37	(233)	430 ± 34	(215)

The columns denoted by $\langle w^2\rangle^{\frac{1}{2}}$ give the mean values measured from two sets of 40 models, one set having $\langle w^2\rangle^{\frac{1}{2}}$ set to be 200 km sec^{-1} and the other with $\langle w^2\rangle^{\frac{1}{2}} = 500$ km sec^{-1}. The columns denoted s.d. give the standard deviation of $\langle w^2\rangle^{\frac{1}{2}}$ expected from one model.

(1978) modified to include peculiar velocities. The results for two sets of 40 Monte-Carlo simulations are summarised in Table 1 which shows that the methods are essentially unbiased. From Figure 2 and Table 1, I conclude $\langle w^2 \rangle^{\frac{1}{2}} = 240 \pm 60$ km $\sec^{-1}$ at $r \sim 1\ h^{-1}$Mpc. The surveys also give $r_o = 4.1 \pm 0.3\ h^{-1}$Mpc and $Q = 0.6 \pm 0.1$, thus eq. (2) yields $\Omega = 0.1 \times (2^{\pm 1})$ (see also Davis & Peebles, 1982).

3. CONCLUSIONS

The results of Sections 1 and 2 suggest more complex initial conditions than those required for a similarity solution. The observed peculiar velocities imply an open universe unless the mass density is dominated by dark material which is much more uniformly distributed than galaxies on scales $\lesssim 5\ h^{-1}$Mpc. There are many other possibilities worthy of further work, e.g. schemes in which galaxies form before clusters but which give rise to non-power law fluctuation spectra (Bond et al., 1982; Peebles, 1982; Hogan & Kaiser, 1982) or schemes in which all fluctuations on scales smaller than clusters or superclusters are erased by Silk damping or the decay of neutrino perturbations (Bond et al., 1980). In the latter cases the main goal is to account for $r_o \sim 4\ h^{-1}$Mpc without obtaining large peculiar velocities; results from N-body simulations are discussed by Klypin & Shandarin (1982) and Frenk et al. (1982).

REFERENCES

Bean, J., Efstathiou , G., Ellis, R.S., Peterson, B.A. and Shanks, T.: 1982, in preparation.
Bond, J.R., Efstathiou, G. and Silk, J.: 1980, Phys.Rev.Lett., 45, 1180.
Bond, J.R., Szalay, A.S. and Turner, M.: 1982, Phys.Rev.Lett., 48, 1636.
Davis, M. and Peebles, P.J.E.: 1977, Ap.J.Suppl., 34, 425.
Davis, M. and Peebles, P.J.E.: 1982, preprint.
Efstathiou, G. and Eastwood, J.W.: 1981, M.N.R.A.S., 194, 503.
Efstathiou, G., Fall, S.M. and Hogan, C.: 1979, M.N.R.A.S., 189, 203.
Frenk, C.S., White, S.D.M. and Davis, M.: 1982, preprint.
Fry, J.N. and Peebles, P.J.E.: 1980, Ap.J., 236, 13.
Gott, J.R., Turner, E.L. and Aarseth, S.J.: 1979, Ap.J., 234, 13.
Hogan, C. and Kaiser, N.: 1982, in preparation.
Kirshner, R.P., Oemler, A. and Schechter, P.L.: 1978, A.J., 83, 1549.
Klypin, A.A. and Shandarin, S.F.: 1982, preprint and this volume.
Miyoshi, K. and Kihara, T.: 1975, Pub.Astr.Soc.Japan, 27, 333
Peebles, P.J.E.: 1976, Ap.J.Lett., 205, L109.
Peebles, P.J.E.: 1979, A.J., 84, 730.
Peebles, P.J.E.: 1980, "The Large-Scale Structure of the Universe", Princeton University Press, Princeton, N.J.
Peebles, P.J.E.: 1982, preprint.
Peterson, B.A., Bean, J., Efstathiou, S., Ellis, R.S., Shanks, T. and Zou, 1982, in preparation.
Soneira, R.M. and Peebles, P.J.E.: 1978, A.J., 83, 845.

Discussion

Palmer: One of the most damning arguments against the self-similar clustering model is the velocity dispersion argument. The estimate of the observed dispersion depends upon your choice of the weight function, f(w). Could you say how sensitive your result is to the choice of this function, and why you chose the one you did?

Efstathiou: I chose a function which is a good fit to the distributions from the N-body models. If instead one uses a gaussian, then $\langle w^2 \rangle^{1/2}$ decreases by $\sim$ 10 km sec^{-1} and if one takes an exponential, the velocities increase by $\sim$ 10 km sec^{-1}. I would not seriously consider the possibility that the velocities are wrong by several hundred km sec^{-1} because of an inappropriate choice of the distribution function. Also, the quality of the model fit to ξ_v gives some indication of whether the distribution function is reasonable. Fitting a model to the observations is preferable to taking moments of ξ_v.

Shandarin: Did you estimate the two-point correlation function of clusters of particles in your simulations?

Efstathiou: Joshua Barnes has devised a fast algorithm for locating groups of particles. We found that the two-point correlation function for groups is the same as the two-point function for individual particles. Thus, rich groups are good tracers of the mass distribution. Apparently, this result is in conflict with the results of Hauser and Peebles which show that the two-point correlation function for Abell clusters has an amplitude $\sim$ 10 times larger than that for the galaxies. Recent work by Bahcall and Soneira agrees with the results of Hauser and Peebles.

B. Jones: Doesn't the fact that the cluster-cluster correlation is so much higher than the galaxy-galaxy correlation on the same scale simply suggest that either galaxy or cluster light is not a good tracer of the mass distribution?

Efstathiou: Perhaps neither galaxies nor clusters trace the mass distribution! I would have expected that the cluster-cluster correlation function should agree with the galaxy-galaxy correlation function, and this issue deserves more attention. One check comes from the way in which the peculiar velocities between galaxy pairs scale with pair separation. The results, particularly those from the CfA survey, are consistent with the scaling expected from the galaxy correlation functions.

Szalay: If one would invoke dissipation, wouldn't that change some of these results?

Efstathiou: Yes. The shape of the two-point function on small scales could be very different from that found in the purely gravitational N-body simulation of Klypin and Shandarin and Frenk, White and Davis. If the dissipation was recent, one might also obtain lower peculiar velocities than those expected from the virial theorem.

Now in the self-similar model, dissipation is ignored but if dissipative effects are important only on scales corresponding to individual galaxies, then the conclusions should be unaltered.

G. Burbidge (right) replying to a question. M. Schmist presiding.
(Courtesy, K. Brecher)

MONTE-CARLO SIMULATIONS OF THE DISTRIBUTION OF FAINT GALAXIES

H.T. MacGillivray[1] and R.J. Dodd[2]
1. Royal Observatory, Blackford Hill, Edinburgh, Scotland, UK.
2. Carter Observatory, PO Box 2909, Wellington 1, New Zealand.

Simulations of static (non-evolving) fields of galaxies are carried out by means of a Monte-Carlo computer technique which models all physical effects acting on the light from distant galaxies, from emission at the source up to the final detection and measurement on the photographic plate by means of a high-speed measuring machine. Parameters for the model galaxies are computed in order to match the main types of measurement made on astronomical photographs (e.g. image centroids, isophotal magnitudes and colours, orientations and shapes).

In the simulations, galaxies are generated according to any desired clustering scenario in three dimensional (3-D) space, and then projected onto the plane of the sky. Wherever possible, observationally determined parameters are used as input to the simulations. For example, we use the cluster luminosity function of Schechter (1976), power-law radial distribution of galaxies within clusters from Peebles and Groth (1975), the shapes of clusters from Binggeli (1982), etc. The light from a galaxy is increased for evolution and cosmology and decreased for internal absorption, effects of distance and K-dimming, Galactic obscuration, telescope vignetting, scattering by the point-spread function and finally for detection above an isophotal threshold. The simulation proceeds until a specified number of galaxies in the model field is detected above the limiting magnitude.

The model fields may be compared directly with observational data from scans, using the COSMOS automatic plate-measuring machine at the Royal Observatory Edinburgh (MacGillivray, 1981), of deep photographs taken with the UK Schmidt (UKST) and Anglo-Australian (AAT) Telescopes. Images are detected on these plates down to thresholds in the range 2-10% of night sky intensity level (corresponding approximately to the B = 26-27 mag/sq arcsec isophotes). Star/galaxy separation is performed by computer down to $m_J \sim 21$-22 (MacGillivray and Dodd, 1980b) and the distribution of the galaxy images investigated by means of suitable clustering algorithms.

G. O. Abell and G. Chincarini (eds.), Early Evolution of the Universe and Its Present Structure, 401–404.

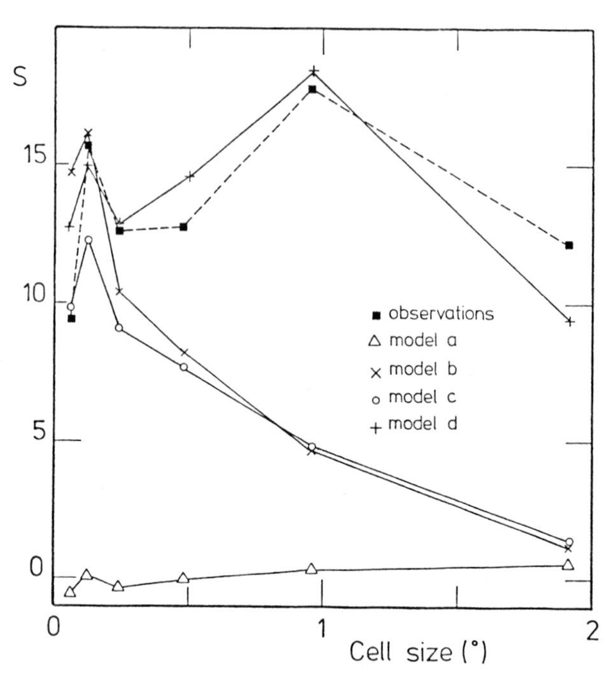

Figure 1. The result of applying Meads analysis to observational data described in the text and various simulated fields involving galaxies distributed at random (triangles), within clusters of fixed and continuous scales (crosses and open circles respectively) and galaxies in clusters which are in turn clustered into high-order clusterings (plus-signs).

Figure 1 shows the result of applying the method of Meads analysis (see Shanks, 1979) to the galaxy field of MacGillivray and Dodd (1980a), involving some 30000 galaxies in a field of 4 × 4 degrees down to a limiting magnitude of B ∿ 22 near the South Galactic Pole. If the estimated plate selection function for these data is correct, then at the typical sample redshift of z' ∿ 0.15, the clustering detected at angular scales of 0.12° and 1.0° corresponds to linear scales of ∿ 1.0 and 8.0 Mpc respectively (with H_0 = 75 km s^{-1} Mpc^{-1} and q_0 = +1). The larger clustering scale compares with the size of typical groups of clusters, e.g. the Hercules supercluster (Tarenghi et al., 1979) and the Horologium-Reticulum supercluster (Dawe et al., 1979).

The results for various simulated fields (with similar parameters for the plate/machine combination - i.e. seeing of 2 arcsec and isophotal threshold at the 9.6% night sky level, and no luminosity evolution included) are shown also in Figure 1. The models shown include a completely random model (model a), clustering on a fixed scale (model b) with cluster radii of 2.5 Mpc, clustering on continuous scales (model c) with all cluster radii in the range 1-5 Mpc being allowed, and a model involving second-order clustering of the galaxies, with supercluster diameters of 10 Mpc (model d). As can be seen, only models incorporating second order clustering can reproduce the peak at angular size of 1°, thus supporting the presence of second order clustering in the observational data. However, the exact nature of this second order clustering cannot be unambiguously ascertained from these data.

REFERENCES

Binggeli, B.: 1982, Astron. Astrophys. 107, pp. 338-349.

Dawe, J.A., Dickens, R.J., Lucey, J., Mitchell, R.J., and Peterson, B.A.: 1979, New Zealand Journal of Science 22, pp. 369-370.

MacGillivray, H.T.: 1981, in Astronomical Photography 1981, eds. J.L. Heudier and M.E. Sim, pp. 277-289.

MacGillivray, H.T., and Dodd, R.J.: 1980a, Mon. Not. R. astr. Soc. 193, pp. 1-6.

MacGillivray, H.T., and Dodd, R.J.: 1980b, Astrophys. Sp. Sci. 72, pp. 315-318.

Peebles, P.J.E., and Groth, E.J.: 1975, Astrophys. J. 196, pp. 1-11.

Schechter, P.: 1976, Astrophys. J. 203, pp. 297-306.

Shanks, T.: 1979, Mon. Not. R. astr. Soc. 186, pp. 583-602.

Tarenghi, M., Tifft, W.G., Chincarini, G., Rood, H.J., and Thompson, L.A.: 1979, Astrophys. J. 234, pp. 793-801.

Discussion

Giovanelli: I would like to make a comment on your remark on Martian canals. If superclusters have redshift depths of several hundred km/s, even volume density enhancements as high as 100 would not appear conspicuous in galaxian distributions as deep as the ones you showed. For example, for a sample as shallow as the Zwicky catalog, an overdensity of 100, about 600 km/s deep will appear with a surface density contrast of at best (depending where it is centered in redshift) of only 7 or 8. If the depth of the sample is many times higher, as yours, superclusters will wash away.

MacGillivray: I agree that any large-scale structures would be completely swamped by foreground and background contamination in these very deep galaxy surveys. This makes it even more interesting when examining isoplethal plots of the distribution of the galaxies, in which "ridges" are clearly seen connecting the rich clusters. The consequences for the interpretation of ridges seen in other galaxy samples cannot be ignored.

J. Jones: In working on the calibration of Schmidt plates, I have found a great deal of patchiness and nonuniformity over the plates at these low light levels. Have you done any tests to ensure that this is not affecting your conclusions about large-scale structure?

MacGillivray: The problem of nonuniformities over plates is one which we are examining with great care at Edinburgh. Ultimately, one can be absolutely certain only when two plates of the same field are examined and in such instances our results remain valid.

On a similar note, we have also attempted to model the effect of patchy galactic obscuration on simulated galaxy fields using the most recent observations of H I clouds. We find that although patchy obscuration can indeed influence the galaxy counts, it cannot produce features of the magnitude observed in our data at large ($\sim 1°$) angular scales.

NEIGHBORING SUPERCLUSTERS AND THEIR ENVIRONS

J. Einasto[1] and R.H. Miller[2]
[1]Tartu Astrophysical Observatory
[2]European Southern Observatory

Recently finished redshift surveys make it possible to study the large-scale environment of superclusters and their mutual relationship.

Figure 1 shows the distribution of nearby clusters in the sky in supergalactic coordinates at two redshift intervals. Nearby clusters in the distance interval 75 to 150 Mpc form a belt around us which is close to the supergalactic equator; its inclination is only 20°. The following superclusters belong to this belt: Ursa Major-Lynx (Giovanelli and Haynes 1982), Coma, Hydra-Centaurus, Pavo-Corona Australes, and Perseus-Pisces. Coordinates and redshifts for a number of previously unknown southern clusters have been derived by Dr. H. Corwin and Dr. M. Tarenghi (Einasto et al. 1982).

Clusters in the distance interval 150 to 250 Mpc are found at much higher supergalactic latitudes. Clusters in this distance interval form a number of superclusters: Hercules, Ursa Major-Leo, Pegasus and several southern superclusters.

All these superclusters belong to cells which can be called the Northern Local Cell and the Southern Local Cell (Einasto et al. 1982). Nearby superclusters form together with our Local Supercluster a disk about 250 Mpc in diameter and 50 Mpc thick, which is located between both local cells. The Hercules supercluster is located between the Northern Local Cell and The Bootes cell, studied by Kirsher et al. (1981). The Perseus-Pisces and Pegasus superclusters are located between the Northern Local Cell and the Perseus cell.

In the sky, the Northern Local Cell covers the whole northern supergalactic hemisphere and the Southern Local Cell covers the whole southern hemisphere. These are probably the largest objects in the Universe which can be considered "local" ones.

Figure 2 shows the distribution of the same clusters in rectangular supergalactic coordinates, X, Y, and Z, where X = V cos SGB cos SGL, Y = V cos SGB sin SGL, and Z = V sin SGB.

The Sandage-Tammann survey covers the whole sky and is complete to $13^m.2$, which corresponds for galaxies brighter than -21.0 to a distance 70 Mpc (H = 50/km/s/Mpc). The CfA survey is complete to $14^m.5$ in the northern sky; the corresponding distance limit is 130 Mpc. Within the

G. O. Abell and G. Chincarini (eds.), Early Evolution of the Universe and Its Present Structure, 405–409.

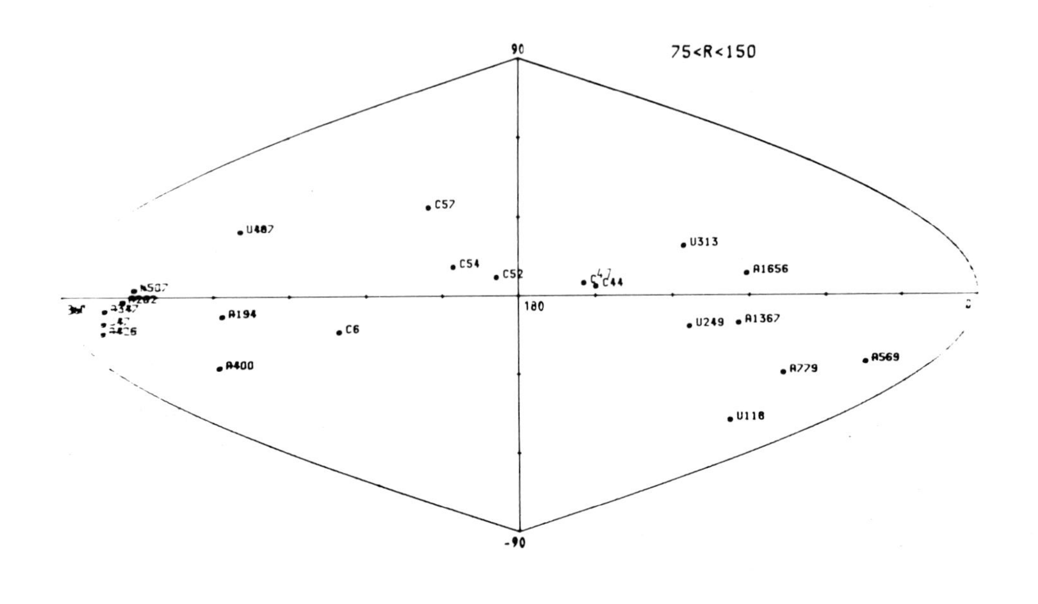

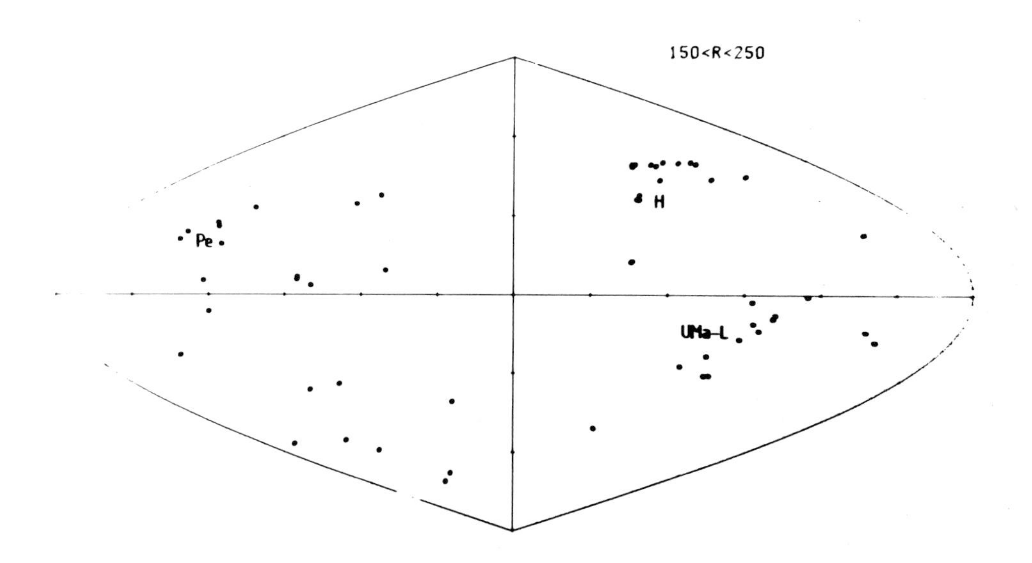

Figure 1. Distribution of clusters at two distance intervals in equal area supergalactic coordinates.

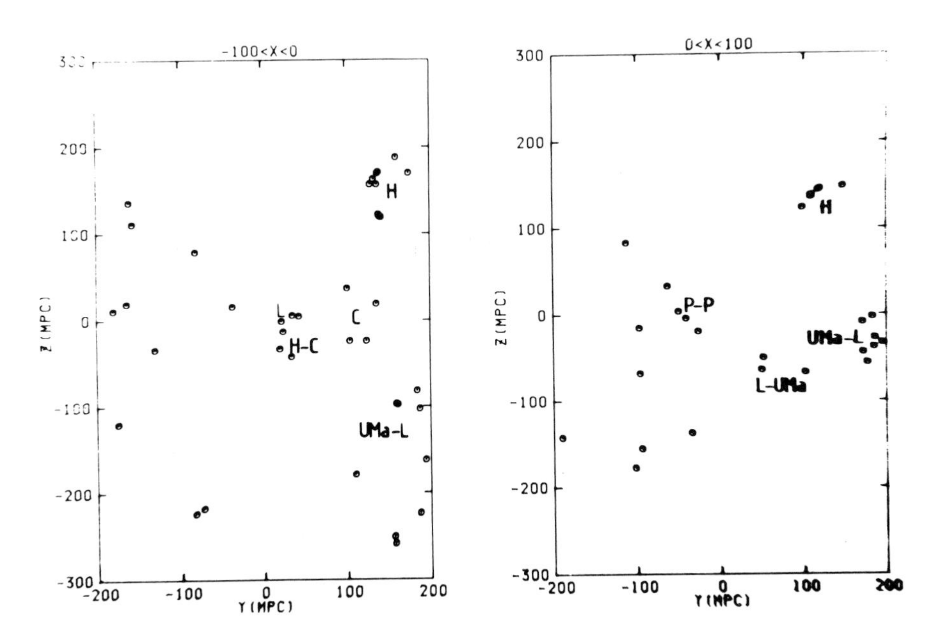

Figure 2. Distribution of clusters in rectangular supergalactic coordinates.

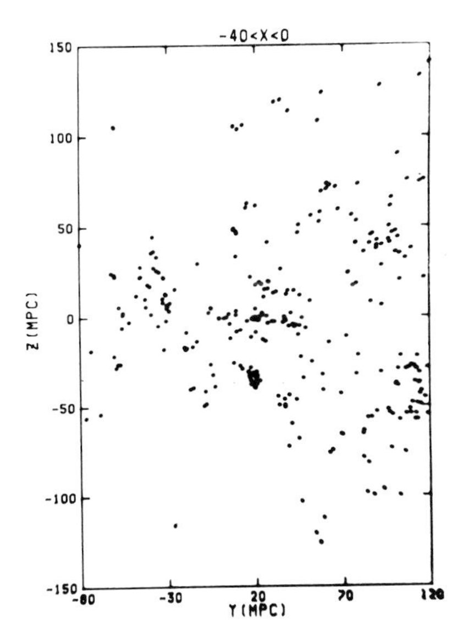

Figure 3. Distribution of bright galaxies ($M \leq -21.0$) in rectangular supergalactic coordinates ($H = 50$ km/s/Mpc).

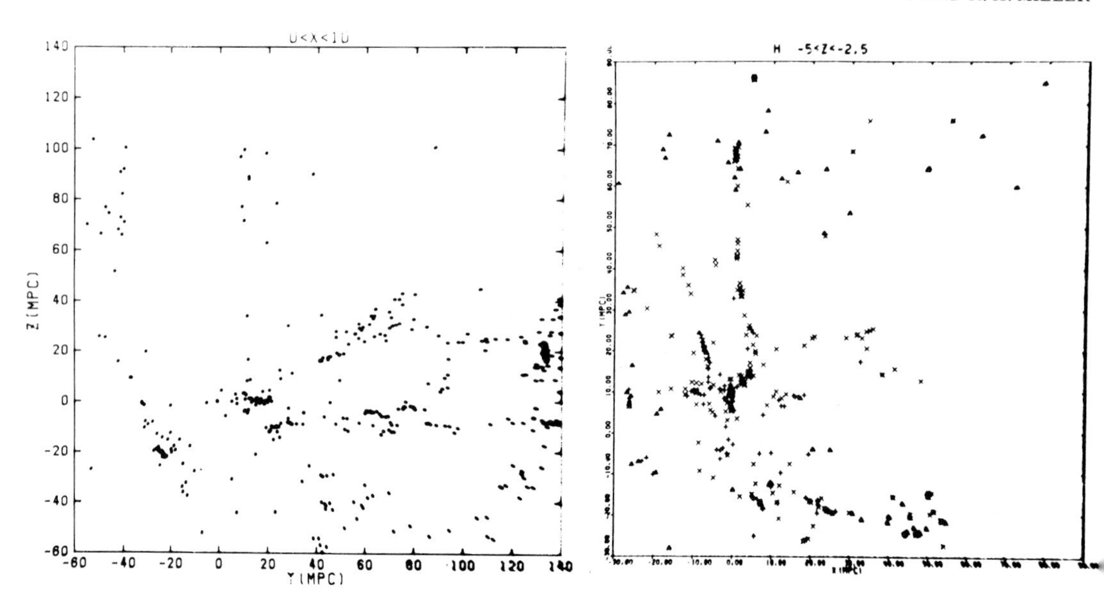

Figure 4. Thin slices through the Local Supercluster, in SG coordinates, corresponding to Hubble constant 100.

Local Supercluster it is possible to use the Fisher-Tully and other surveys of dwarf galaxies. Here much fainter galaxies can be studied.

These surveys provide a unique opportunity to study in detail the galaxy distribution in cells and cell walls.

A rectangular plot of galaxies brighter than -21.0 absolute magnitude is given in Figure 3. The density has a peak at the plane Z = 0 and decreases on each side. Figure 3 shows that the galaxy density does not drop to zero near the center of the cell interior. In other words, voids defined by rich clusters of galaxies are not completel empty. The mean density of galaxies in the void center is about ten time lower than in cell walls and three times lower than the mean density of the Universe.

Figure 3 also shows the presence of some smaller voids in the galaxy distribution. In thin slices through the Local Supercluster, plotted in Figure 4, these small voids are seen much better. We see also that the whole disk consists of a lattice of intertwined galaxy strings, located both in horizontal (X,Y) and vertical (Z) directions.

The galaxy chains seen in Figure 4 do not differ in principle from cluster chains. The difference lies only in the scale and population. Within superclusters, chains are rich; in cell interiors, they are poor. The geometry is the same.

To visualize the three-dimensional distribution of galaxies, a series of motion pictures has been prepared. To see the third dimension, rotation is artificially introduced. The string-like distribution of galaxies, the concentration of galaxies into a disk of superclusters, and the very low galaxy density in cell interior, are apparent in the films showing the distribution centered near the center of the Northern Local void.

REFERENCES

Einasto, J., Corwin, H.G., Jr., Huchra, J., Miller, R.H., and Tarenghi, M. 1982, preprint.

Giovanelli, R., and Haynes, M.P. 1982, preprint.

Kirsher, R.P., Oemler, A., Jr., Schechter, P.L., and Shectman, S.A. 1981, *Ap. J.*, 248, L57.

NUMERICAL EXPERIMENTS ON GALAXY CLUSTERING

R. H. Miller
European Southern Observatory and Max Planck Institut fur Astrophysik, Garching bei Muenchen, West Germany

ABSTRACT

A study of the way observable clustering depends on expansion history is reported. Observable shapes that result from evolving otherwise identical systems are intercompared to show differences due to different expansion histories. Four cases are compared: non-expanding, Ω =1, and two open universes with 0.10 and 0.03 as final values of Ω There is remarkably little difference in observable forms for the expanding cases. The 0.03 universe expanded by a factor 500 during the experiment. This study is an example of the way numerical experiments can be used is studies of galaxy clustering.

1. INTRODUCTION

Numerical experiments yield configurations that look remarkably like the observed structure of galaxy clustering when started from fairly smooth initial conditions. Several groups have run calculations of this kind. Doroskevich and Shandarin (1983) and Efstathiou (1983) have reported this kind of work at this meeting. Their papers provide references to earlier work. A remarkable feature of all this work is that everyone gets structures that he says look a lot like the observed Universe even though the details of the calculations differ substantially. This suggests that structures like those observed are easy to get through simple dynamical processes based on gravitational forces. The observational properties that catch our attention do not depend much on details of the dynamics, the cosmological history, or boundary and initial conditions. This is reassuring from the standpoint of trying to simulate the Universe but it is distressing from the alternative point of view of trying to distinguish among various

G. O. Abell and G. Chincarini (eds.), Early Evolution of the Universe and Its Present Structure, 411–416.

theoretical models. This situation might have been foreseen: the fact that many different theoretical models each gave some measure of agreement with observation already indicated that observations do not constrain the models very much. We need a uniqueness proof--but there seems to be no such thing in astronomy.

Numerical experiments seem to be a popular indoor sport, but what part can they play in helping toward an understanding of the physical processes that caused the Universe to look the way it does? Numerical experiments can be a keen analytical tool in showing precisely how different models or different initial or boundary conditions affect observable properties. This in turn can help to identify observational features that will provide some leverage on the key questions of what our Universe looked like at earlier times and of what physical processes dominated galaxy clustering. We need some qualitative guides before detailed quantitative studies can be of much use. Some steps in that direction are reported here. It turns out that some of the features we had hoped would produce significant observable differences actually produce very subtle--almost indistinguishable--effects.

Two matters of viewpoint are essential. (1) The notion of numerical experiments, and (2) Heavy reliance on direct comparison with observation.

(1) A numerical experiment is the closest analogue we have to a laboratory experiment in the dynamics of galaxies and larger systems. One tries to include as much of the essential physics as possible in a kind of laboratory setup (the computer program), in ways such that possible instrumental effects (numerical errors, grid effects, different boundary conditons) can be calibrated and controlled. A series of experiments is run in this environment, varying one parameter at a time to "pick the picture apart," to identify the important physical effects. The approach is the same as in laboratory physics--it is not theory in the conventional sense. A strength of the method lies in the intercomparison of results obtained in a systematic search of the parameter space. Instrumental effects affect the various experimental runs in much the same way so differences can safely be attributed to changes in the parameters.

(2) Three dimensional forms as complex as the Universe are difficult to visualize both in experimental results and in observations. It is safest to compare experimental results and observations directly at a level as close to the basic observational material as possible. The motion picture showing a three-dimensional representation of the observed Universe shown by Einasto (1983) at this conference was prepared as part of this effort. A film showing the experimental results presented in the same way will be shown as part of the presentation of this paper [1]. The film is a vital part of the paper. We stress that apparent similarity of observed and experimental structures is necessary for the experiments to be convincing. It is not sufficient. On the other hand, while more abstract summaries of the

observational data (e.g., correlation functions) can be difficult to interpret because of different boundary conditions in the experiment and in the real Universe, the visual comparison of the two kinds of results has an unambiguous immediacy.

2. THE EXPERIMENTAL METHOD

The basic experimental setup is described in a paper which we hope will soon be published (Miller 1981) to which we refer for details. We summarize it briefly.

An n-body system in which the motions of 100,000 particles are followed self-consistently as they move under forces generated by their own self-gravitation is the basic tool. Particles and forces have periodic boundary conditons in which the periodic cell partakes of the general (externally specified) expansion. The forces that act on individual particles are derived from potentials that are obtained by solving Poisson's equation on a grid, here 64 active grid points in each direction of a cartesian lattice. This technique is often referred to as a "Fourier method," but Fourier is only a computational trick to speed up the numerical solution of the Poisson equation, and it has nothing whatever to do with the match between the computation and the physics of the problem.

The paper mentioned describes some of the first results, which include several checks on the consistency and robustness of the results. The principal features found were (1) disturbances grow at the rate given by linear perturbation theory to an accuracy of 1-2% to clumping strengths greater than those in the present-day Universe with no evident saturation. (2) The dominant visual impression on watching the dynamical development in a motion picture is one of growing voids--of opening holes that sweep material before them and pile it up in a structure that looks like superclusters with huge voids.

3. DEPENDENCE OF OBSERVABLE CLUSTERING ON EXPANSION HISTORY

These experiments were all started from identical runs of pseudorandom numbers in the initial load. All particles were nearly at rest. They were given small initial velocities to suppress the dying mode of linear perturbation theory. Disturbances grow at the expected rates in all experiments. Motion pictures will be shown for the expanding cases. The forms are virtually identical at equal clumping strengths. "Snapshots" of these systems are rotated to show the three-dimensional forms at stages of equal clumping strength. The similarities are remarkable.

But the similarities are even more remarkable at microscopic levels (see accompanying figures). Little particle aggregates can be recognized in each of the configurations.

Forms seen at equal clumping strengths are similar for each of these three expansion histories. This implies that values of Ω are not easily determinable from particle locations. However, since the time-dependence of clumping strength depends on Ω, there may well be Ω-dependent information in the particle velocities.

A careful look shows slightly sharper clumps at $\Omega = 0.03$ than at $\Omega = 1$. This results from a nonlinear effect whose nature is clarified by the non-expanding (NE) experiment. Features are not as tightly clumped in the NE experiment as in any of the others. A study of the time-dependence of clumping strength shows the cause: disturbances at high wavenumber (small wavelength) start to grow and then stop in the NE case, while they continue to grow in the remaining experiments. Particle velocities build up in the NE experiment to where the Jeans instability is suppressed at short wavelengths. This is a characteristic nonlinear effect that appears at larger clumping strengths. In the expanding cases, redshifting of the velocities diminishes the effect until, at $\Omega = 0.03$, clumping can continue at short wavelengths, producing tighter clumps.

The use of identical starting conditions, with only the expansion history varying, is the key to the sharp result obtained.

Films shown with this presentation were produced at the NASA-Ames Research Center largely through the efforts of Dr. Bruce F. Smith. Dr. Smith has been an equal partner in the galactic dynamics program since its inception. Computations leading to these results were carried out at the Max Planck Institut fur Astrophysik, whose support is gratefully acknowledged. The writer has enjoyed the generous hospitality of the European Southern Observatory as joint visitor at ESO and MPA, where this work was done while on leave from The University of Chicago. This work was partially supported by Interchange No. NCA2-OR108-902 between NASA-Ames and the University of Chicago.

REFERENCES

Doroskevich, A., and Shandarin, S. 1983, this volume, p.387.

Efstathiou, G. 1983, this volume, p. 393.

Einasto, J. 1983, this volume.

Miller, R. H. 1981, Astrophys. Journ.

FOOTNOTE

(1) Prints of these and other films are available at cost. Write Astronomy Department, University of Chicago, requesting ordering information for Miller's films.

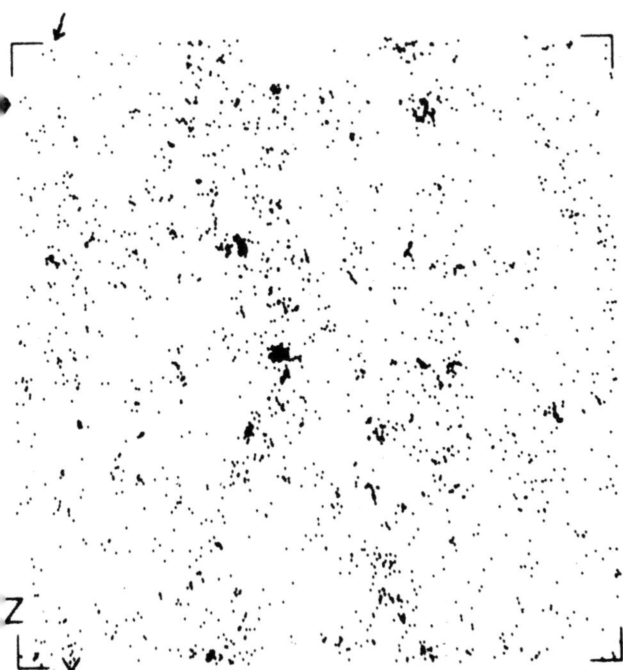

One view of the configurations from each of two different experiments at the same clumping strength. The upper figure shows the Ω =1 experiment at an expansion of 35 past the initial condition. The lower figure shows the same view for the Ω =0.03 experiment at an expansion of 500 past the initial condition.

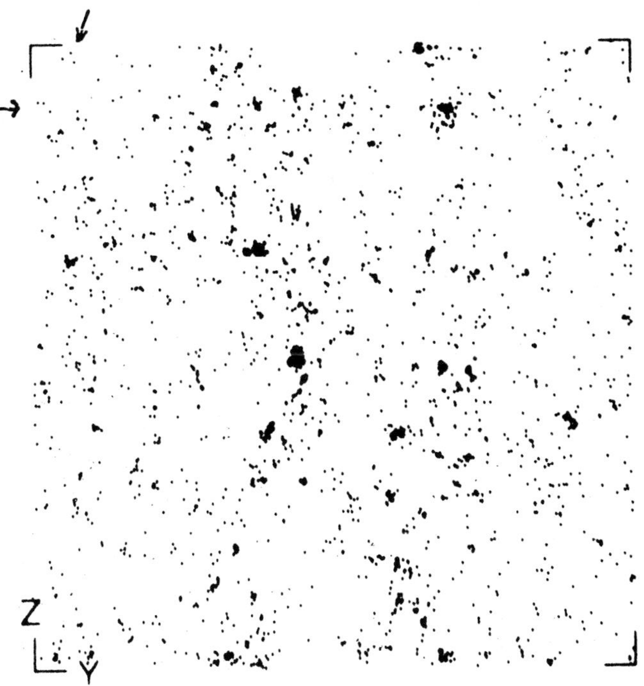

Notice the nearly identical particle aggregates like the marked triples in the upper left-hand corner or the pair near the upper right-hand corner brackets. Many more such associations can easily be found. Clumps are somewhat tighter in the Ω =0.03 experiment.

Discussion

Efstathiou: I object to the statement that the two-point correlation function does not tell you much about the clustering pattern. The numerical models with Ω = 1 and n = 0 give a correlation function which is too steep compared to the observations. The disagreement is more apparent if $\Omega << 1$ and n = 0.

Miller: I said correlation functions are not sensitive to the features that catch the eye in the observed clustering. They do not describe the filamentary pattern that is so striking in the observations. I agree with you that they provide one statistic that is useful in numerical comparisons with observed clustering.

Dekel: What is the cell size in your simulations, in comparison with the size of the clusters?

Miller: Clusters typically have diameters of at least 8 - 12 of the cells used for the potential calculation. We don't trust features only 1 - 2 cells across.

Bhavsar: One must remember when comparing the N-body simulations with observations that we are comparing the mass distributions of the simulations with the light distribution of the real universe. The extent to which light is a good tracer of mass will depend on the uniformity of the mass-to-light ratio.

Miller: Good point. Our motion pictures and analyses trace the active mass distribution. This need not be the same as the observable luminosity distribution. They do, however, trace the stuff that controls the dynamics.

RELATIVISTIC STELLAR DYNAMICS

G. Contopoulos
University of Athens, Greece

Abstract: Three main areas of relativistic stellar dynamics are reviewed: (a) Relativistic clusters, (b) Systems containing a massive black hole, and (c) Perturbed expanding Universes. The emphasis is on the use of orbit perturbations.

There are three main areas of research that belong to the field of Relativistic Stellar Dynamics, namely:

(a) The dynamics of clusters, or nuclei of galaxies, of very high density,

(b) The dynamics of systems containing a massive black hole, and

(c) The dynamics of particles (and photons) in an expanding Universe.

(a) Up to now the emphasis has been put on the first subject. It is known that clusters evolve by condensing their central parts while expanding their outer parts. If the density contrast between the center and the envelope is sufficiently high the evolution produces a collapse of the center to theoretically infinite density. This is what Lynden-Bell calls a "gravothermal catastrophe" (Lynden-Bell and Wood, 1968).

Although infinite densities cannot be realized in practice, one expects that the core of the cluster will become relativistic. Relativistic effects become significant for densities of the order of $10^{12} M_{\odot}/pc$, in which case the stars are at distances similar to those of the planets in the solar system.

A review of the dynamics of relativistic clusters was provided by Ipser (1975). The main topic in this review was the stability of clusters. It seems that highly relativistic clusters are unstable in general.

A recent development in this direction was provided by Vandervoort

G. O. Abell and G. Chincarini (eds.), Early Evolution of the Universe and Its Present Structure, 417–423.

and Ipser (1982). These authors proved that a large class of clusters are unstable because of gravitational radiation. This result led them to the conjecture that gravitational radiation makes all clusters unstable.

A similar theorem in the case of fluid systems (stars) was formulated a few years ago by Friedman (1978; see also Friedman and Shutz, 1978). Friedman proved that all stars are unstable, because of gravitational radiation. The unstable modes have angular dependence of the form $\exp(im\theta)$ where m (≥ 2) must be sufficiently large. The growth rate of the instability is proportional to $(c/v)^{2m+1}$, where v is a characteristic velocity.

On the other hand the unstable modes of the stellar axisymmetric systems considered by Vandervoort and Ipser are of the Dedekind type (ellipsoids with fixed boundaries in an inertial system, and with a strong internal circulation), i.e. they have $m = 2$. The instability appears for all angular velocities $\Omega > 0$. Therefore we may state that relativistic stellar systems are more unstable than the corresponding fluid systems.

(b) The second subject of interest is the dynamics of a system under the influence of a large central black hole. In very dense nuclei of clusters, or galaxies, one may have frequent collisions that may lead to the formation of a massive central black hole. Thus it is quite possible that the dynamics of the nuclei of some clusters, or galaxies, are dominated by a large black hole. This black hole should be in general of the Kerr type (rotating black hole).

The study of the orbits of particles and photons (geodesics) around Kerr black holes has received a great impetus after the discovery by Carter (1968) that such systems are completely integrable. Carter proved that the equations of motion of particles (or photons) around a Kerr black hole have one more integral of motion besides the classical ones (rest mass, energy and angular momentum). By using Carter's integral one can solve explicitly the equations of motion, using elliptic integrals (for a review see Sharp, 1979; see also Chandrasekhar, 1982, and Contopoulos, 1982). One can distinguish the following main types of orbits:

1) Orbits staying outside the ergosphere,

2) Orbits entering the ergosphere and coming out again,

3) Orbits staying always in the ergosphere,

4) Orbits entering the outer horizon of the black hole; such orbits do not come out again in finite coordinate time.

5) Orbits inside the inner horizon, which, however, are of no importance for the outside observer.

The relative position of the orbits with respect to the ergosphere is of interest in connection with the Penrose effect (Penrose, 1969). A particle in the ergosphere may split into one particle of negative energy and one of positive energy. The first particle enters the horizon, while the second one acquires a larger energy than the energy of the incoming particle. The energy excess is taken from the black hole itself, therefore it can be larger than the rest mass ($E = mc^2$) of the particle. The Penrose process has been invoked as a source of energy in the nuclei of galaxies. Although it is known now that the Penrose process is much less effective than previously thought (Bardeen et al., 1972; Chandrasekhar, 1982) the problem is of great theoretical and practical importance.

One problem of theoretical interest is whether negative energy particles can ever go out of the ergosphere. If this would happen, we might have closed time-like geodesics, which would lead to causality violation. However it can be proved that if we impose the condition that the coordinate time and the proper time increase together ($dt/d\tau > 0$) no negative energy orbit can go out of the ergosphere. However, this condition cannot be derived from the local dynamics of the splitting of a particle in the ergosphere.

Among the orbits around black holes of special interest are the nearly circular orbits. In fact if a quasi-stationary stellar system is formed around the black hole such orbits should contain most of the mass. Such orbits are the counterparts of the familiar epicyclic orbits of classical galactic dynamics. They have two basic frequencies, the rotation frequency Ω, and the epicyclic frequency κ of radial oscillations.

We come now to the problem of the perturbations of a Kerr black hole.

The linearized problem of perturbed Kerr black holes has been solved recently by Chandrasekhar (1982). This problem involves the solution of 76 coupled differential equations containing 50 unknown functions. It seems almost miraculous that this problem has an explicit solution.

The perturbations depend on the time through factors of the form $\exp(i\sigma t)$, where σ is complex and can be considered as an eigenvalue of the system.

From now on the problem takes the familiar aspect of galactic dynamics. The perturbations can be considered as due to a density distribution of matter (and radiation) around the black hole, which we call the imposed density. On the other hand the same distribution produces deviations in the orbits of particles (or photons) around the black hole. Thus the study of the orbits around <u>perturbed</u> black holes gives the response density, which must be equal to the imposed density. This condition is the well known self-consistency equation of galactic dynamics

$$\rho^{\text{imposed}} = \rho^{\text{response}} . \qquad (1)$$

Its solution will give the eigenvalues of the problem, i.e. the allowed values of σ.

The perturbed epicyclic orbits can be written in the form

$$r = r_o + \Sigma F_{mn} \exp[i(m\Omega + n\kappa + \sigma)t] . \qquad (2)$$

This form is similar to that of the orbits of stars in a spiral galaxy, if we set

$$\sigma = -m(\Omega_s + i\Omega_i) , \qquad (3)$$

where Ω_s is the angular velocity of the spiral pattern and $m\Omega_i$ its growth rate.

If we solve now the collisionless Boltzmann equation we find the response density in the form

$$\frac{\exp[i(m\Omega + n\kappa + \sigma)t]}{m\Omega + n\kappa + \sigma} \qquad (4)$$

which contains the expression $(m\Omega + n\kappa + \sigma)$ in the denominator. If σ has a zero, or small, imaginary part, then $(m\Omega + n\kappa + \sigma)$ may be zero, or close to zero, i.e. we have a resonance. The values of Ω and κ depend on the radius r, therefore at particular radii we have particular resonances. The most important resonances are the Lindblad resonances, where $m/n = \mp 2$, or

$$\frac{\kappa}{\Omega - \Omega_s} = \pm 2 , \qquad (5)$$

and the particle resonance (or corotation) where $n = 0$, or

$$\Omega = \Omega_s . \qquad (6)$$

Such resonances play an important role in spiral and barred galaxies and also in the rings of Saturn. Namely, when such resonances appear, the perturbations are so large that gaps are formed in the distribution of matter, both in galaxies and in Saturn's rings. We expect therefore a similar behavior in the distribution of matter around black holes; e.g., if an accretion disk is formed, it should have the form of the rings of Saturn with several gaps here and there. The same should be the case for a cluster, or a galaxy, that is dominated by the central black hole.

A similar problem refers to the dynamics of a system dominated by a massive central body with axial symmetry. Such is the case of a dense elliptical galaxy. In such a system we can consider perturbations in the same way as in the case of the third integral of motion of galactic dynamics (Contopoulos, 1960). This problem has been treated recently in the post-Newtonian approximation by Spyrou and Varvoglis (1982), and can be considered as a first approach to the problems of relativistic galactic dynamics.

(c) The third type of problems of relativistic stellar dynamics refers to perturbations in an expanding Universe.

It is well known that the equations of motion in a Friedmann Universe are integrable. Thus, we can write explicitly the solutions for the motions of particles (and photons) in such a Universe.

A more general integrable case is given by the metric

$$ds^2 = dt^2 - R^2(t)\left[\frac{dr^2}{1 - \frac{2\mu}{r} - kr^2} + r^2(d\theta^2 + \sin^2\theta d\phi^2)\right], \tag{7}$$

which may represent a spherical condensation in an expanding Universe. This formula includes as limiting cases the Friedmann Universes if $\mu = 0$ and $k = -1$, 0 or +1.

The geodesics in such a metric are plane and we can take $\theta = \pi/2$. Thus the Lagrangian can be written

$$\mathcal{L} = \frac{1}{2}\left\{\dot{t}^2 - R^2\left(\frac{\dot{r}^2}{1 - \frac{2\mu}{r} - kr^2} + r^2\dot{\phi}^2\right)\right\}, \tag{8}$$

where dots mean derivatives with respect to the affine parameter τ. The conserved angular momentum is

$$R^2 r^2 \dot{\phi} = L \tag{9}$$

and we have also the Eulerian equation

$$\frac{d}{d\tau}\left(\frac{\partial\mathcal{L}}{\partial\dot{r}}\right) = \frac{\partial\mathcal{L}}{\partial r}, \tag{10}$$

that gives

$$\frac{d}{d\tau}\left(\frac{R^2\dot{r}}{1 - \frac{2\mu}{r} - kr^2}\right) = \frac{R^2\dot{r}^2(kr - \mu/r^2)}{(1 - \frac{2\mu}{r} - kr^2)^2} + \frac{L^2}{R^2 r^3}. \tag{11}$$

Thus we find

$$\frac{R^4 \dot{r}^2}{1 - \frac{2\mu}{r} - kr^2} = \Gamma^2 - \frac{L^2}{r^2} , \tag{12}$$

where Γ is a constant, and we derive

$$R^2 \dot{r} = \Gamma \left\{ \left(1 - \frac{B^2}{r^2} \right) \left(1 - \frac{2\mu}{r} - kr^2 \right) \right\}^{\frac{1}{2}} , \tag{13}$$

where $B = L/\Gamma$.

On the other hand R is assumed to be given by the Friedmann equation with zero cosmological constant

$$\left(\frac{dR}{dt} \right)^2 = \frac{q}{R} - k . \tag{14}$$

Thus, R is a given function of t; hence, the above equations can be solved to give r and ϕ implicitly as functions of t.

Then the problem of perturbations is dealt with as in the case of a black hole. If we consider, in particular, nearly circular motions, we can define the rotational end epicyclic frequencies Ω and κ. Then we consider perturbations proportional to $\exp(i\sigma t)$, and find a response of the form (4). Thus we can search for eigenvalues σ, that define self-consistent solutions of the linearized Einstein equations and collisionless Boltzmann equations in the same way as in the case of a central black hole.

This type of research is still in its first stages. However, it is a useful approach if we want to understand the results of the N-body calculations in an expanding Universe of the kind presented to us by the movies of R. Miller and others during this Symposium.

REFERENCES

Bardeen, J.M., Press, W.H. and Teukolsky, S.A.: 1972, Astrophys J. 178, p.347.
Carter, B.: 1968, Phys. Rev. 174, p.1559.
Chandrasekhar, S.: 1982, "The Mathematical Theory of Black Holes" (in press).
Contopoulos, G.: 1960, Z. Astrophys. 49, p.273.
Contopoulos, G.: 1982, preprint.
Friedmann, J.L.: 1978, Commun. Math. Phys. 62, p.247.
Friedmann, J.L. and Shutz, B.F., 1978, Astrophys. J. 222, p.281.
Ipser, J.P.: 1975, IAU Symposium 69, p.423.
Lynden-Bell, D. and Wood, R.: 1968, Monthly Notices Roy. Astron. Soc. 138, p.495.

Penrose, R.: 1969, Riv. Nuovo Cimento 1, p.252.
Sharp, N.A.: 1979, Gen. Rel. Gravitation 10, p.659.
Spyrou, N. and Varvoglis, H.: 1982, Astrophys. J. 255, p.674.
Vandervoort, P.O. and Ipser, J.R.: 1982, Astrophys. J. 256, p.497.

Discussion

Vignato: Is it possible by Lindblad resonance theory to take into account nonmonotonic density distribution in a spherically symmetric system?

Contopoulos: Yes, the theory can take care of all cases.

McCrea: Why should causality not be violated?

Contopoulos: This is a philosophical question and I will not answer it here. But we may discuss it privately and I have a few things to tell you.

SPHERICAL AND TOROIDAL LOCAL BLACK HOLES

BASILIS C. XANTHOPOULOS
Department of Physics, University of Crete,
Iraklion, Crete, Greece.

Abstract. The local black holes describe physical situations involving a black hole surrounded by a finite vacuum region and then by matter and fields. The stationary and axisymmetric local black holes belong into two classes, the spherical and the toroidal ones, depending on the topology of their horizon. For the static black holes their metric tensors are given explicitly in terms of Legendre polynomials. For the stationary local black holes the problem is formulated interms of the Ernst potential of the rotational Killing field and the appropriate asymptotic conditions on the horizon are determined.

1. LOCAL BLACK HOLES

A black hole, being the final configuration of a burnt out and settled down massive star, is described in General Relativity by a stationary spacetime. A stationary isolated black hole - i.e., a spacetime representing a single black hole - has to be axisymmetric[1] as well and its horizon must have the topology of the sphere S^2 . In fact the two parametric family of the Kerr[2] solutions of the Einstein equations describes the most general[3-5] stationary isolated black hole (We shall not be concerned in this paper with electrically charged black holes). To study in General Relativity physical phenomena involving isolated black holes, we have to use approximation methods. For instance, in all studies[6-7] of the fall of particles into, or the scattering of waves by, a black hole we treat the falling matter or the scattered field as a perturbation, i.e., we treat them as test particles or test fields which move in, or are propagated in, the spacetime of the black hole, without actually affecting the geometry itself.

The concept of the local black holes[8] represents the first step in an effort to describe exactly in General Relativity physical phenomena involving a black hole surrounded by matter and fields. Since it is reasonable to assume that the matter that was in the immediate neighborhood of the black hole has been "eaten up" by the hole, we demand that the black hole is surrounded, for a finite distance from its horizon, by a vacuum region. No matter or any other field but the gravitational field exists in this vacuum region. However, matter and other fields may exist (and generally do exist) farther away from this vacuum region; it is only de-

G. O. Abell and G. Chincarini (eds.), Early Evolution of the Universe and Its Present Structure, 425–430.

manded that their density and strength fall off sufficiently fast away from the hole so that they represent an isolated system with a black hole at the center.

The general theory of the static and axisymmetric local black holes has been recently put forward by Geroch and Hartle[8]. In particular they have shown that the horizon of these black holes is homeomorphic either to the sphere S^2 or the torus $S^1 x S^1$; these two classes will be referred to as the spherical and the toroidal black holes, respectively. Besides indicating how to construct such solutions, Geroch and Hartle have investigated their global structure, their thermodynamic behavior and their evolution with the emission of Hawking radiation. Certain examples of local black holes had appeared in the literature[9-11] before the work of Geroch and Hartle.

2. THE STATIC AND AXISYMMETRIC LOCAL BLACK HOLES

The construction of the spacetime of the local black holes proceeds in two steps[8]. In the first we are concerned with the immediate (the vacuum) neighborhood of the black hole, where the actual difficulty is to satisfy the vacuum Einstein equations compatibly with the requirement that they admit a smooth event horizon. The second step then will be to match this solution (of the Einstein equations) with an exterior solution with sources, which will represent the region far away from the hole. All the studies of the local black holes are concerned, so far, with the first step. For the static and axisymmetric case in particular the metric tensors which describe the vacuum neighborhoods of the spherical and the toroidal black holes have been obtained explicitly by Chandrasekhar[7] and Xanthopoulos[12] respectively. Here we give only the results.

Spherical black holes[7]: The line element is

$$ds^2 = \frac{(\eta-1)}{(\eta+1)} e^S (dt)^2 - \frac{m^2(\eta+1)}{\eta-1} e^{\sigma-S}(d\eta)^2 -$$

$$-m^2(\eta+1)^2 e^{-S} \{(1-\mu^2)(d\varphi)^2+(1-\mu^2)^{-1}e^{\sigma}(d\mu)^2\} \quad , \qquad (1)$$

where $S=S(\eta,\mu)=\Sigma\, A_k P_k(\eta)P_k(\mu)$ and the function $\sigma=\sigma(\eta,\mu)$ is determined from S by the equations

$$\frac{(\eta^2-\mu^2)}{(\eta^2-1)(1-\mu^2)}\sigma_{,\eta} = \frac{2\eta}{\eta^2-1} S_{,\eta} - \frac{2\mu}{\eta^2-1} S_{,\mu} - \mu\, S_{,\eta} S_{,\mu} +$$

$$+ \frac{\eta}{2(\eta^2-1)} \{(\eta^2-1)S_{,\eta}^2 + (\mu^2-1)S_{,\mu}^2\}, \qquad (2)$$

$$\frac{(\eta^2-\mu^2)}{(\eta^2-1)(1-\mu^2)} \sigma_{,\mu} = \frac{2\mu}{1-\mu^2} S_{,\eta} + \frac{2\eta}{\eta^2-1} S_{,\mu} + \eta\, S_{,\eta} S_{,\mu} +$$

$$+ \frac{\mu}{2(1-\mu^2)} \{(\eta^2-1)S_{,\eta}^2+(\mu^2-1)S_{,\mu}^2\}. \qquad (3)$$

The P_k's are Legendre polynomials and the constants A_k are subject to the condition $\Sigma A_{2k+1} = 0$. The mass of the hole is m, its horizon is the surface $\eta=1$, the surface area of the horizon is $16\pi m^2 e^{\alpha}$ and the surface gravity on the horizon is $(4\pi e^{\alpha})^{-1}$, where $\alpha=-\Sigma A_{2k}$. (All the summations are from zero to infinity).

Toroidal black holes[12]: The line element is

$$ds^2 = \frac{1}{4}(\eta^2-1)(1-\mu^2)e^{-S}(dt)^2 - 4m^2 e^{S}(d\varphi)^2 - $$
$$- 4m^2(\eta^2-\mu^2)e^{\sigma-S}\{(\eta^2-1)^{-1}(d\eta)^2+(1-\mu^2)^{-1}(d\mu)^2\}, \quad (4)$$

where $S=S(\eta,\mu)=\Sigma B_k P_k(\eta)P_k(\mu)$ and the function $\sigma=\sigma(\eta,\mu)$ is determined from S by the equations

$$\sigma_{,\eta} = -\frac{\eta(\mu^2-1)}{2(\eta^2-\mu^2)}\{(\eta^2-1)S_{,\eta}^2+(\mu^2-1)S_{,\mu}^2\} + \frac{\mu(\eta^2-1)(\mu^2-1)}{\eta^2-\mu^2}S_{,\eta}S_{,\mu} \quad (5)$$

$$\sigma_{,\mu} = \frac{\mu(\eta^2-1)}{2(\eta^2-\mu^2)}\{(\eta^2-1)S_{,\eta}^2+(\mu^2-1)S_{,\mu}^2\} - \frac{\eta(\eta^2-1)(\mu^2-1)}{\eta^2-\mu^2}S_{,\eta}S_{,\mu}. \quad (6)$$

The P_k's are Legendre polynomials and the constants B_k are subject to the conditions $\Sigma B_{2k+1}=0$, $\Sigma B_{2k}\dot{P}_{2k}(1)=0$, and $\Sigma B_{2k+1}\ddot{P}_{2k+1}(1)=0$, where the dots denote differentiations. The mass of the hole is m, the surface area of the horizon is $16\pi m^2 e^{\alpha}$ and the surface gravity on the horizon is $(4me^{\alpha})^{-1}$, where 2α equals to the constant value of the function σ on the horizon.

Geroch and Hartle[8] have shown that the above solutions can be smoothly continued to asymptotically flat non-vacuum solutions. It should be mentioned, however, that no non-vacuum continuation has been explicitly constructed so far corresponding to some physically interesting distribution of matter.

3. THE STATIONARY CASE

Contrary to the static black holes, which are non-rotating, the stationary and axisymmetric black holes are uniformly rotating. Unfortunately, the relevant stationary axisymmetric vacuum Einstein equations are quite complicated. For instance, in the static case a combination of the components of the metric tensor satisfies a single linear equation; in the stationary case, instead, we have to deal with a non-linear system of partial differential equations. In particular we choose a coordinate chart which covers a neighborhood of the horizon and we determine the asymptotic behaviors of the coefficients of the metric tensor which guarantee the existence of a smooth horizon. We find that a formulation of the problem based on the Ernst potential[13] associated with the rotational Killing field is the most appropriate. Finally, we determine a simple necessary condition which distinguishes the spherical and the toroidal holes.

A suitable expression for the line element of the general stationary axisymmetric spacetime is

$$ds^2=+e^{2\nu}(dt)^2-e^{2\psi}(d\varphi-\omega dt)^2-e^{2\mu_2}(dx^2)^2-e^{2\mu_3}(dx^3)^2, \tag{7}$$

where t and φ are the time and the azimuthal angle and the scalars ν,ψ,ω,μ_2 and μ_3 are functions of the two remaining spatial coordinates x^2 and x^3. The Einstein equations for the stationary and axisymmetric case can be found, for instance, in reference 7 equations 7,8,14,15, and 16 of chapter VI. We note that eq. 14, compatibly with which the gauge condition should be imposed[14], is the same in both the static and the stationary cases. We can choose the gauge, therefore, exactly as in the static case. Since the argument is presented in detail in ref. 12, we here give only the conclusions: We impose the gauge conditions

$$e^{\mu_3-\mu_2}=(\eta^2-1)^{1/2}(1-\mu^2)^{-1/2}, \quad e^{\psi+\nu}=m(\eta^2-1)^{1/2}(1-\mu^2)^{1/2} \tag{8}$$

for which the line element (7) becomes

$$ds^2=m(\eta^2-1)^{1/2}(1-\mu^2)^{1/2}\{e^{\nu-\psi}(dt)^2-e^{\psi-\nu}(d\varphi-\omega dt)^2\} -$$
$$-(\eta^2-1)^{1/2}(1-\mu^2)^{1/2}e^{\mu_2+\mu_3}\{(\eta^2-1)^{-1}(d\eta)^2+(1-\mu^2)^{-1}(d\mu)^2\}. \tag{9}$$

Here m is a constant, the range of $\mu=x^3$ is in $(-1,+1)$, and the horizon is the surface $\eta=x^2=1$.

The limit on the horizon of the determinant of the spacetime metric is

$$G_{(4)} = -2m^2(1-\mu^2)\lim\{(\eta-1)\, e^{2(\mu_2+\mu_3)}\} \tag{10}$$

and the limit on the horizon of the corresponding determinant of the induced metric is

$$G_{(2)} = 2m \lim\{(\eta-1)\, e^{\psi-\nu}\, e^{\mu_2+\mu_3}\}. \tag{11}$$

For the metric (9) to be well behaved on the horizon the limits $G_{(4)}$ and $G_{(2)}$ should be finite and non-zero. Therefore, we should also have that

$$\lim\{(\eta-1)^{1/2}e^{\psi-\nu}\} \text{ and } \lim\{(\eta-1)^{1/2}\, e^{\mu_2+\mu_3}\} \tag{12}$$

are finite and non-zero. We observe that we have obtained the same asymptotic conditions as in the static case[12].

By glancing on themetric (9) one can immediately see that the first of the conditions (12) is equivalent to the condition that the squared norm of the rotational Killing field is finite and non-zero on the horizon. This observation suggests to use the formulation of the stationary axisymmetric Einstein equations interms of the Ernst potential[13-14] associated with the rotational Killing field. Hence, instead of the variables $\psi-\nu$ and ω we consider

$$\Psi=(\eta^2-1)^{1/2}(1-\mu^2)^{1/2}\, e^{\psi-\nu} \tag{13}$$

and Φ defined by the equations

$$(\eta^2-1)\Phi_{,\eta}=\Psi^2\omega_{,\mu} \quad\text{and}\quad (\mu^2-1)\Phi_{,\mu}=\Psi^2\omega_{,\eta}\quad, \tag{14}$$

as the basic variables of the problem. In terms of the complex Ernst potential $Z=\Psi+i\Phi$ the vacuum Einstein equations read

$$(\mathrm{Re}Z)\{[(\eta^2-1)Z_{,\eta}]_{,\eta}+[(1-\mu^2)Z_{,\mu}]_{,\mu}\}=(\eta^2-1)Z_{,\eta}^2+(1-\mu^2)Z_{,\mu}^2 . \quad (15)$$

In addition, instead of $\mu_2+\mu_3$, we consider the M defined by

$$e^{\mu_2+\mu_3}=(\eta^2-\mu^2)(\eta^2-1)^{1/2}(1-\mu^2)^{-1/2}\ \Psi^{-1}e^M \quad (16)$$

as an equivalent variable. Interms of these variables the line element (9) takes the form

$$ds^2= m\{(\eta^2-1)(1-\mu^2)\Psi^{-1}(dt)^2-\Psi(d\varphi-\omega dt)^2\} -$$
$$-(\eta^2-\mu^2)\Psi^{-1}e^M\{(\eta^2-1)^{-1}(d\eta)^2+(1-\mu^2)^{-1}(d\mu)^2\} . \quad (17)$$

Note that, the second of the asymptotic conditions (12) is that e^M should be finite and non-zero on the horizon.

Turning now to the remaining of the Einstein equations-namely, equations 8 and 16 of chapter VI of the reference 7- we obtain, after a lengthy calculation, that M is determined from Ψ and Φ by the equations

$$2M_{,\eta}=(1-\mu^2)(\eta^2-\mu^2)^{-1}\{\eta A-2\mu(\eta^2-1)B\}$$
$$2M_{,\mu}=(\eta^2-1)(\eta^2-\mu^2)^{-1}\{\mu A+2\eta(1-\mu^2)B\}, \quad (18)$$

where

$$A=\Psi^{-2}\{(\eta^2-1)(\Psi_{,\eta}^2+\Phi_{,\eta}^2)+(\mu^2-1)(\Psi_{,\mu}^2+\Phi_{,\mu}^2)\}$$
$$B=\Psi^{-2}(\Psi_{,\eta}\Psi_{,\mu}+\Phi_{,\eta}\Phi_{,\mu}). \quad (19)$$

Note that by using equation (15) and the assumption that $\lim\Psi$ is finite and non-zero we can conclude that Φ is smooth in a neighborhood of the horizon. Hence A and B are bounded in a neighborhood of the horizon and the second of equations (18) implies that M is constant on the horizon. Therefore the second of the asymptotic conditions (12) is satisfied as a consequence of the first asymptotic condition.

Finally we obtain an additional necessary condition on the scalar Ψ by applying the Gauss-Bonnet[15] theorem. The induced on the horizon metric from the spacetime metric is

$$d\tau^2=mg(d\varphi)^2+g^{-1}e^M(d\mu)^2 \quad (20)$$

where $g=g(\mu)=\Psi(1,\mu)$ and M now stands for the constant value of the function $M(\eta,\mu)$ on the horizon. The scalar curvature of the metric (20) is easily found to be $R=-\ddot{g}e^{-M}$, where the dots denote differentiations. Then the Gauss-Bonnet formula $4\pi\chi=\int R dV$ gives that

$$2\chi m^{-1/2}\ e^{M/2}=\dot{g}(-1)-\dot{g}(+1) . \quad (21)$$

In the expression (21) χ is the Euler number[16] of the horizon of the black hole, which is a topological invariant; $\chi=2$ for a spherical and $\chi=0$ for a toroidal horizon. We conclude therefore that g has to satisfy the condition

$$\dot{g}(-1)-\dot{g}(+1)=\begin{cases} 4m^{-1/2}\, e^{M/2} & \text{for a spherical black hole.} \\ 0 & \text{for a toroidal black hole.} \end{cases} \tag{22}$$

We summarize the conclusions of this section: The stationary and axisymmetric local black holes should be searched among the smooth solutions of the Ernst equation (15) whose real part Ψ of the complex potential Z is finite and different from zero on the horizon $\eta=1$ and it satisfies, in addition, the condition (22). The metric tensor is given by equation (17) where ω and M are determined from Ψ and Φ via the equations (14) and (18).

REFERENCES

1. Hawking, S.W.: 1972, Commun. Math. Phys. 25, pp. 152-166.
2. Kerr, R.P.: 1963, Phys. Rev. Lett. 11, pp. 237-8.
3. Israel, W.: 1968, Phys. Rev. 164, pp. 1776-9.
4. Carter, B.: 1972, Phys. Rev. Lett. 26, pp. 331-3.
5. Robinson, D.C.: 1975, Phys. Rev. Lett. 34, pp. 905-6.
6. Misner, C.W., Thorne, K.S , and Wheeler, J.A.: 1973, "Gravitation", W.H. Freeman and Company, San Francisco.
7. Chandrasekhar, S.: 1982, "The mathematical theory of black holes", Oxford at the Clarendon Press.
8. Geroch, R., and Hartle, J.B.: 1982, J. Math. Phys. 23, pp. 680-92.
9. Israel, W., and Khan, K.A.: 1964, Nuovo Cim., 33, pp. 331-44.
10. Mysak, L.A., and Szekeres, G.: 1966, Can. J. Phys. 44, pp. 617-
11. Peters, P.C.: 1979, J. Math. Phys. 20, pp. 1481-5.
12. Xanthopoulos, B.C.: 1982, "Local toroidal black holes that are static and axisymmetric", Proc. R. Soc. Lond. (submitted).
13. Ernst, F.J.: 1968, Phys. Rev. 167, pp. 1175-8.
14. Chandrasekhar, S.: 1978, Proc. R. Soc. Lond. A358, pp. 405-20.
15. Hicks, N, J.: 1971,"Notes on differential geometry", Van Nostrand Reinhold Company, London.
16. Steenrod, N.: 1951, "The topology of fibre bundles", Princeton University Press.

Discussion

Novikov: Is the toroidal black hole stable against the small perturbations? I guess it may be unstable.

Xanthopoulos: It is unknown.

POSSIBLE CONTRACTION OF THE MEMBERS OF THE BINARY PULSAR PSR 1913+16 AND ITS ASTROPHYSICAL CONSEQUENCES

N. Spyrou
Astronomy Department, University of Thessaloniki,
Thessaloniki, Greece

ABSTRACT. It is proposed that the small difference between the observed and the theoretically predicted decrease of the orbital period of the Binary Pulsar PSR 1913+16 is not due to the insufficiency of the quadrupole formula and can be attributed to a mass-energy loss due to the contraction of the binary's members. Assuming that the pair's primary is a typical, noncontracting pulsar, is in favour of a slowly contracting, neutron-star companion, thus limiting the member's radii to at most 25 km and 28 km, respectively. The primary's computed total absolute luminosity is in excellent agreement with the observed upper limit of its X-ray and optical luminosities. Moreover, the companion's slow contraction rate implies that its present total absolute luminosity presents a maximum at wavelengths characteristic of X-rays. Finally, it suggests that if the energy-loss remains constant, the duration of the contraction phase will be of the order of 10^8 y and the final radius about 25 km.

Perhaps the most interesting feature of the Binary Pulsar PSR 1913+16 is the observed decrease of the orbital period P_b of the pulsar, the system's visible primary, by a well-determined amount:

$$(-\dot{P}_b/P_b)_{obs.} = (1.1 \pm 0.2) \times 10^{-16} \ s^{-1}$$

where a dot denotes total time-derivative (Taylor et al. 1979). This decrease is explained as a decay of the binary due to the emission of orbital energy and angular momentum in the form of gravitational radiation. The corresponding result predicted theoretically via the quadrupole formula (Peters and Mathews 1963) is:

$$(-\dot{P}_b/P_b)_{pred.} = 0.8 \times 10^{-16} \ s^{-1}$$

differing from the observationally determined value by an extra amount:

$$(-\dot{P}_b/P_b)_{ext.} = 0.3 \times 10^{-16} \ s^{-1}$$

G. O. Abell and G. Chincarini (eds.), Early Evolution of the Universe and Its Present Structure, 431–436.

approximately equal to its 24% and about 1.5 times larger than the quoted uncertainty of the observational data.

The above discrepancy is very close to the uncertainty of the observations, and so it is probable that more accurate, future observations will confirm the validity of the quadrupole formula in this case in spite of the expressed skepticism concerning this validity in general (Ehlers et al. 1976). On the other hand, the quadrupole formula has not been generalized in the case of a realistic binary so that to include the internal characteristics of the members and moreover post-Newtonian corrections to the Keplerian orbits, always assumed for the member's motion (see, however, Epstein and Wagoner 1975).

Here we present the motivation and the main results of an independent method for explaining the above discrepancy, which proved to be in favour of an active neutron-star companion, provided that the primary is treated as a typical noncontracting pulsar. Details and proofs will be given elsewhere (Spyrou, to be published). Thus, we consider that the discrepancy is not due to the insufficiency of the quadrupole formula. Instead, treating the members as realistic, extended bodies, and not simply as idealized point masses, we attribute the discrepancy in the value of $-\dot{P}_b/P_b$ to a change of the internal characteristics of the members.

It is obvious that we have to relate, somehow, the members' internal characteristics with their orbital motion. These characteristics, however, do not enter the dynamical laws of the orbital motions of the members provided by the Newtonian theory of gravity, usually applied. As the required generalization of these dynamical laws, we use the result of the recently developed post-Newtonian celestial mechanics for realistic binaries (Spyrou 1981 a,b; Caporali and Spyrou 1981). In this context each member is treated as a perfect-fluid body and the basic parameter describing it is its total mass-energy or inertial mass, m, defined as:

$$m = \overline{m} + c^{-2} E$$

where $\overline{m}$ and E are the body's rest mass and Newtonian total self energy and c is the velocity of light in vacuum. This theory has already been applied in the case of the Binary Pulsar (Spyrou 1981b) for determining the inertial masses, as well as the invariant rest masses under different assumptions, according to which both members do not contract, but rather they are in stable hydrodynamical equilibrium (constant and negative self energies). Moreover, most of this theory's dynamical laws, like, e.g., the orbital energy and orbital angular momentum per unit reduced mass, h and ℓ, as well as the generalized laws of Kepler and mass-function, have functional forms similar to their Newtonian counterparts.

In applying these results for explaining the extra decay of the Binary Pulsar, we shall let the inertial masses change due to the change of the self energies. (After all, we know that the Binary Pulsar has a nonvanishing luminosity.) Moreover, we shall demand that the purely orbital parts, h and ℓ, of the total energy and angular momentum, do not change, so that orbital energy and angular momentum losses in the form of gravitational waves be excluded. Then it can be shown that any change $\dot{M}$ of the total mass-energy will induce the following extra changes in the semimajor axis, a, eccentricity, e, and orbital period, P_b, of the relative orbit:

$$\dot{P}_b/P_b = \dot{a}/a = (e^{-2} - 1)^{-1} (\dot{e}/e) = \dot{M}/M .$$

In the case of the Binary Pulsar, these relations reduce to:

$$\dot{M}/M = -0.3 \times 10^{-16} \text{ s}^{-1} = \dot{a}/a, \quad \dot{e}/e = -0.4 \times 10^{-16} \text{ s}^{-1} ,$$

showing that the extra decay ($\dot{a} < 0$, $\dot{e} < 0$) can be attributed to a certain decrease of the total mass-energy ($\dot{M} < 0$) due to both members.

For the evaluation of the quantity $\dot{m}/m$ for a member, we assume that this member is a homogeneous, spherically symmetric and uniformly rotating collapsed object whose interior is described by the Fermi-Dirac statistics for noninteracting particles of degenerate matter. Under these assumptions, it can be show that:

$$\frac{\dot{m}}{m} = C_R \frac{\dot{R}}{R} - \frac{2E_{kin}}{\bar{m}c^2} \frac{\dot{P}}{P}$$

where R, P and E_{kin} are the member's radius, period of axial rotation and kinetic energy of rotation, respectively, and the coefficient C_R, which is a known function of the member's internal characteristics, for a contracting star must be positive.

The coefficient C_R is always negative for typical white dwarfs, showing that isolated (as far as changes of their interior are concerned) white dwarfs cannot contract (and according to current theories of stellar evolution they, as well as isolated neutron stars, cannot expand). So if the companion is a white dwarf, the mass-energy loss is due solely to the visible pulsar's contraction. The same is true for a dead pulsar or a black hole companion. The coefficient C_R, however, for isolated neutron stars, can be either negative or positive, and so in the case of an active neutron-star companion, the mass-energy loss, in general, is due to both members' contraction.

In the case of the Binary Pulsar PSR 1913+16, the white dwarf and dead pulsar companions are ruled out on evolutionary grounds concerning the system itself (Srinivasan and van den Heuvel 1982), and so the companion can be either a neutron star or a black hole. In speculating about the unseen companion's nature, we recall that the visible pulsar, on the basis of its short pulse period (Smarr and Blandford 1976) and its weak dipole, magnetic field strength (Srinivasan and van den Heuvel 1982), is believed to be the older member of the binary. So it seems very probable that the major part of the mass-energy loss is due to the companion's contraction, and this rules out the possibility of a black hole companion, because, obviously, a black hole cannot emit energy, at least classically. Since, moreover, the companion is believed to be the younger member of the binary, it, most probably, is a rotating neutron star. This, along with the close proximity of the inertial masses of the members, implies that the slow-down rate of the companion cannot differ drastically from typical values, as is the primary's slow-down rate (Manchester and Taylor 1977). This slow-down rate solely, however, is not enough for explaining the mass-energy loss. So finally we are left with a rotating slowing-down and contracting neutron-star companion, which is responsible for almost all of the proposed mass-energy loss.

In view of the above, we adopt here the point of view according to which the visible pulsar does not contract and that the mass-energy loss is entirely due to the contraction of the companion neutron star. Moreover, we shall assume that the mass-energy densities of both members are in the range of the currently accepted densities of stable pulsars, namely, between $10^{13.40}$ gr.cm^{-3} and $10^{15.80}$ gr.cm^{-3} (Misner, Thorne and Wheeler 1972). This assumption simply means that the proposed contraction rate, if any, must be very small. In this way it is proved that the condition $C_R = 0$ for the noncontracting primary pulsar, along with the constancy of its rest mass, limit the pulsar's present radius and mass-energy density to about 25 km and $10^{13.44}$ gr.cm^{-3}, respectively. (The corresponding radius for the lowest possible density $10^{13.40}$ gr.cm^{-3} is about 30 km.) Moreover, it can be shown that the positivity of the coefficient C_R, necessary for a contracting pulsar, implies that its self-energy is negative, $E < 0$. This again is in favour of a slow contraction and shows that the proposed contraction is to be considered as a relic of the supernova explosion, from which the pulsar was formed. In other words, the major part of the contraction occurred during the supernova explosion, and after that the contraction continues in a very slow rate, the pulsar passing through various stages of stable (quasistationary) hydrodynamical equilibrium ($E < 0$) until it reaches the final stage of no contraction ($C_R < 0$). From a physical point of view, this slow appraoch to the final stable state is more preferable than its instantaneous settlement after the supernova explosion, and shows that the gravitational collapse could continue even after the formation of the neutron star.

In the case of the quasistationary, companion pulsar, we arbitrarily assume that its mass-energy is equal to the lowest possible density of stable pulsars, $10^{13.40}$ gr.cm^{-3}. Then the positivity of the coefficient C_R and the constancy of the companion's rest mass limit its present radius to about 28 km, while its radius and density at the final stable state of no contraction are found to be 25 km and $10^{13.55}$ gr.cm^{-3}, respectively. We notice that the difference of 5 km between the two radii is more pronounced that the corresponding difference of 3 km in the case of the companion pulsar. This could mean that in the case of the primary and little more massive pulsar the contraction has proceeded in the past in a faster way, and has decelerated to an almost zero value. This conclusion is again in favour of an older pulsar primary.

From the estimated radius and the observationally known slow-down rate of the primary pulsar, we find that its total absolute luminosity, interpreted as (minus) the rate of change of its total self-energy is:

$$L_1 = 4.96 \times 10^{35} \text{ erg s}^{-1}$$

This result is in excellent agreement with the observed upper limits of the primary's X-ray and optical luminosities (Davidsen et al. 1975):

$$L_{1x} < 3 \times 10^{35} \text{ erg s}^{-1} \qquad L_{1v} < 0.17 \times 10^{35} \text{ erg s}^{-1}$$

and shows that its energy emission is practically all in the X-ray range. The corresponding absolute temperature is $T_1 = 2.9 \times 10^6$ °K, showing that the emitted radiation presents a maximum at a wavelength about 10 Å, characteristic of the X-rays of energy of approximately 1.2 KeV.

In the case of the companion pulsar from the known values of the inertial masses, the radius and the coefficient C_R and assuming a slow-down rate $\dot{P}_2/P_2$ of the same order of magnitude as the primary's one, the contraction rate is estimated equal to:

$$\frac{\dot{R}_2}{R_2} = - 0.06 \times 10^{-16} \text{ s}^{-1}$$

Moreover, the companion's absolute luminosity consists of two parts, the first of which, L_{2P}, depends only on the slow-down rate, while the second, L_{2R}, depends only on the contraction rate. These two parts are estimated equal to:

$$L_{2P} = 6.27 \times 10^{35} \text{ erg s}^{-1} \qquad L_{2R} = 1.42 \times 10^{38} \text{ erg s}^{-1}$$

The corresponding absolute temperatures are 3.04×10^6 °K and 1.22×10^7 °K showing that the emitted energies present their maxima at wavelengths 9.5 Å and 2.4 Å, respectively, characteristic of X-rays. The total luminosity, 1.43×10^{38} erg s^{-1}, corresponds to an absolute temperature 1.23×10^7 °K and presents its maximum at a wavelength 2.4 Å (X-rays of energy 5.24 KeV).

Finally, some further consequences of the companion pulsar's contraction hypothesis are the following: i) Provided that the mass-energy-loss rate remains constant, the duration of the contraction phase will be about 4×10^8 y; ii) The primary's spin-down time is 2.1×10^8 y and it is equal to its slow-down time and to twice its cooling rate; iii) The companion's inertial mass at the final stable state will be 0.49 $m_\odot$, approximately 40% of its present value; iv) Although the companion's final rest-mass density will increase (by about 38%) due to the continuous squeezing of the baryons, the final inertial-mass density will reduce (by about 37%), due to the continuous loss of the star's self-energy; v) The companion's number of baryons is about 6×10^{58} (that of one solar mass pulsar being of the order 10^{57}).

In conclusion, the physically reasonable pulsar contraction hypothesis explains in an independent and very satisfactory way the visible pulsar's X-ray emission and other observed properties, and along with the presently accepted evolutionary data on the Binary Pulsar is in favour of an active neutron-star companion.

REFERENCES

Caporali, A., Spyrou, N. 1981, Gen. Rel. Grav., 13, 689.
Davidsen, A., Margon, B., Liebert, J., Spinrad, H., Middleditch, J., Chanan, G., Mason, K.O., and Sanford, P.W. 1975, Ap. J. (Letters), 200, L19.
Ehlers, J., Rosenblum, A., Goldberg, J.N., and Havas, P. 1976, Ap. J. (Letters), 208, L77.
Epstein, R., and Wagoner, R.V. 1975, Ap. J., 197, 717.
Manchester, R.N., and Taylor, J.H. 1977, Pulsars, Freeman and Co., San Francisco.
Misner, C.W., Thorne, K.S., and Wheeler, J.A. 1972, Gravitation, Freeman and Co., San Francisco.
Peters, P.C., and Mathews, J. 1963, Phys. Rev., 131, 435.
Smarr, L.L., and Blandford, R. 1976, Ap. J., 207, 574.
Spyrou, N. 1981a, Gen. Rel. Grav., 13, 473.
Spyrou, N. 1981b, Gen. Rel. Grav., 13, 487.
Srinivasan, G., and van den Heuvel, E.P.J. 1982, Astron. Astrophys., 108, 143.
Taylor, J.H., Fowler, L.A., and McCulloch, P.M. 1979, Nature, 277, 437.

Discussion

Rees: Bertoth, Carr and I have recently been considering what pulsar timing data may tell us about an important cosmological phenomenon: the possible presence of a stochastic background of long-wavelength gravitational waves, produced either in the initial instants of the big bang, or by (for instance) Population III massive objects at large redshifts. These waves would generate a very small time-varying redshift (i.e., change in ratio of clock rates) between a pulsar and the Earth. Limits on the "timing noise" or ordinary pulsars set interesting limits on waves with periods of 1 - 100 years. However, the best limits on periods 10 - 10^4 years are potentially offered by the orbital behavior of the binary pulsar. If we assume that the intrinsic secular behavior is given by the Landau-Lifshitz formula, this can be predicted with an accuracy which is already $\sim 10^{-11}$ parts per year, and which can be improved at least to 10^{-12}. The absence of any discrepancy between observed and predicted orbital decay would then imply that gravitational waves with periods in the band 10 - 10^4 years constitute less than 10^{-4} to the density parameter Ω.

OBSERVATIONAL TESTS OF BARYON SYMMETRIC COSMOLOGY

F. W. Stecker, Laboratory for High Energy Astrophysics
NASA/Goddard Space Flight Center, Greenbelt, MD. 20771,
U.S.A.

To the gods alone belongs it never to be old or die. But all things else melt with all-powerful time....

Sophocles

1. INTRODUCTION

With the advent of grand unified theories (GUTs) has come the concept (among others) that baryons (protons, etc.) can decay by changing into leptons ("Diamonds are not forever.") and *vice versa*, baryonic matter can be created from the thermal blackbody radiation in the early universe (provided, of course, that the hot big-bang model is basically correct). Using this concept, models have been suggested to generate a *universal* baryon asymmetry, with the consequence that no important amount of antimatter would be left in the universe at the present time (see, e.g. Langacker 1981 and references therein). These models have been motivated by observational constraints on antimatter, at least in our little corner of the universe (Steigman 1976). However, some of these constraints have been shown to be overrestrictive (Stecker 1978, Allen 1981) and an alternative model, also based on GUTs, has been suggested which maintains matter-antimatter (i.e., baryon) symmetry on a *universal* scale, but results in separate "fossil domains" of clusters of matter galaxies and clusters of antimatter galaxies. The basic physics argument regarding the choice between a baryon symmetric and an asymmetric cosmology hinges on the manner in which CP violation occurs in nature (or GUTs) at the temperature when the matter (antimatter) excesses are produced from the blackbody radiation. If the CP violation is spontaneous, it will arise with random sign changes in causally independent regions (Brown and Stecker 1979, Senjanović and Stecker 1980) leading to separate regions of matter and antimatter excesses. The creation of these excesses subsequent to a de Sitter phase arising from a GUT first order phase transition can result in fossil domains of astronomically

G. O. Abell and G. Chincarini (eds.), Early Evolution of the Universe and Its Present Structure, 437–445.

relevant size (Sato 1981). The general scenario is shown in Fig. 1. Details of the theory have been discussed and reviewed by the author recently (Stecker 1981, 1982). I will concentrate here on possible observational clues that large amounts of antimatter exist elsewhere in the universe.

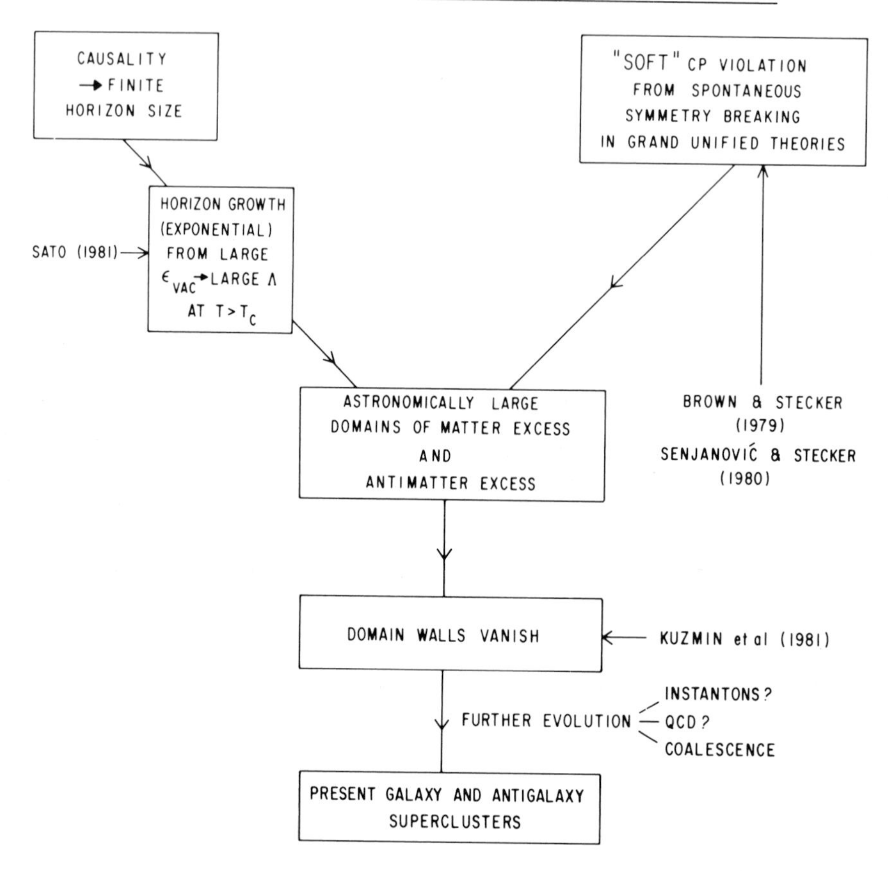

2. THE COSMIC γ-RAY BACKGROUND RADIATION

One of the most significant consequences of baryon symmetric big-bang cosmology lies in the prediction of an observable cosmic background of γ-radiation from the decay of π^0-mesons produced in nucleon-antinucleon annihilations. This is also a most encouraging aspect of this cosmology, since it satisfactorily explains the observed energy spectrum of the cosmic background γ-radiation as no other proposed mechanism does (with the possible exception of hypothetical point sources).

For high redshifts z, when pair production and Compton scattering become important, it becomes necessary to solve a cosmological photon transport equation in order to calculate the γ-ray background spectrum. This integro-differential equation takes account of γ-ray production, absorption, scattering, and redshifting and is of the form

$$\frac{\partial}{\partial t} + \frac{\partial}{\partial E}\left[-EH(z)\ \right] = \ (E,z) - \kappa_{AB}(E,z) + \int_{E}^{\varepsilon(E)} dE' \ \kappa_{sc}(E,z) \quad (E;\ E')\ dE' \qquad (1)$$

where

$$(E,z) \equiv (1+z)^{-3}\ I(E,z)$$

$$(E,z) \equiv (1+z)^{-3}\ Q(E,z)$$

and

$$\frac{\partial}{\partial t} = -\ (1+z)\ H\ (z)\ \frac{\partial}{\partial z}\ ,$$

$$H(z) = H_{o}(1+z)\ (1+\Omega z)^{1/2}$$

The second term in eq. (1) expresses energy loss from the redshift effect. The third term is the γ-ray source term from $p\bar{p}$ annihilation primarily into π^{o}s. The absorption term is from pion production and Compton interactions with electrons at high z and the scattering integral puts back Compton scattered γ-rays at lower energies E<E'.

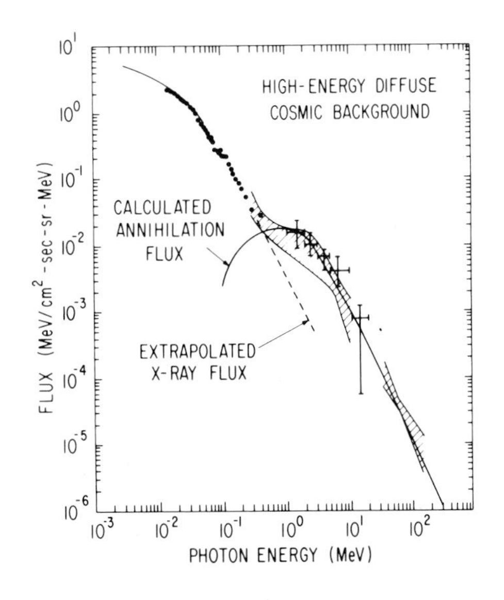

Fig. 2. The Cosmic γ-Ray Background Spectrum: Theory and Observational Data.

Fig. 2 shows the observational data on the γ-ray background spectrum. The dashed line is an extrapolation of the X-ray background component. The calculated annihilation spectrum (Stecker 1978) is also shown. The excellent agreement between the theory and the data is apparent. Other attempts to account for the γ-ray background radiation spectrum by diffuse processes give spectra which are inconsistent with the observations, generally by being too flat at the higher energies.

In Fig. 2 the spectrum is shown as an energy flux. The "bump" in the energy range of 1-10 MeV stands out clearly and can be used as <u>prima facie</u> evidence that a new spectral component dominates in this region. To illustrate this quantitatively, one may note that the energy flux in the 10-100 keV

X-ray background is $\sim 2\times10^{-5}$ erg cm^{-2} s^{-1} sr^{-1}. Using the same units, a power law extrapolation of the X-ray component, as shown in the figure, would give an energy flux of only $\sim 5\times10^{-9}$ in the 1-10 MeV range whereas the observed flux in this range is $\sim 2\times10^{-7}$, a factor of $\sim$40 higher! The observational data, in order of increasing energy, are from Marshall, et al. (1980), Rothschild, et al. (1982), Trombka, et al. (1977), Schönfelder, et al. (1980) and Fichtel et al. (1978). (The data of Fichtel, et al. contain a component of galactic γ-radiation which causes a flattening at the higher energies.)

It is possible that the γ-ray background is made up of a superposition of point sources. However, since only one extragalactic source has been seen at energies above $\sim$1 MeV, this remains a conjecture. Such a hypothesis must be tested by determining the spectral characteristics of extragalactic sources and comparing them in detail with the characteristics of the background spectrum. It presently appears, e.g., that Seyfert galaxies may have a characteristic spectrum which cuts off above a few MeV, so that they could not account for the flux observed at higher energies.

3. ANTIMATTER IN THE COSMIC RADIATION

Measurements of cosmic-ray antiprotons can give us important information about cosmic-ray propagation and also provide a test for primary cosmological antimatter. Buffington, et al. (1981), observing at energies well below the secondary cutoff, appear to see a signal of primary antiprotons. Data on $\bar{p}$ fluxes at higher energies (Bogomolov, et al. 1979, Golden, et al. 1979) give measured values a factor of 4-10 above the fluxes expected for a standard "leaky box" type propagation model with the primaries passing through $\sim$ 5 g/cm^2 of material (Stecker, et al. 1981 and references therein). In fact, the $\bar{p}$ flux integrated over the observed energy range is $\sim$ 7 times the expected flux. But what is particularly striking is that the flux observed by Buffington, et al. (1981) in the 150-300 MeV range is orders of magnitude above what is expected (see Fig. 3).

The reason that standard secondary $\bar{p}$ production models give a very low flux in the 150-300 MeV energy range is easily understood and is a basic feature of the relativistic kinematics (Gaisser and Levy 1974), viz., antiprotons with less that $\sim$ 1 GeV energy must be produced backward in the cms of the collision, and those with energy as low as 150-300 MeV must be produced by cosmic-ray protons significantly above threshold. Since the cosmic-ray proton energy spectrum falls off steeply with energy, the secondary $\bar{p}$ flux has a natural low-energy cutoff. This leaves two explanations for the cosmic-ray $\bar{p}$'s: (1) they are primary, or (2) they are secondary and have undergone significant deceleration. For case (1) we would expect that the $\bar{p}/p$ is independent of energy as observed ($\bar{p}/p$ = (3.2 $\pm$ 0.7) x 10^{-4}), consistent with the primary hypothesis.

It can be easily demonstrated that solar modulation effects will not produce the deceleration required by the secondary hypothesis to

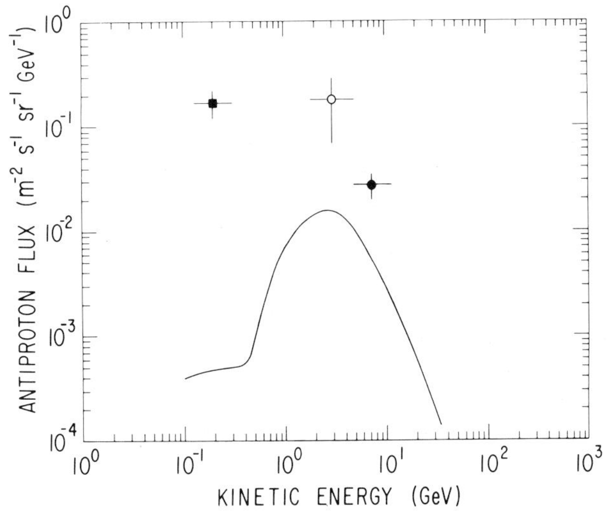

Fig. 3. Cosmic ray antiproton fluxes: Data and predictions from the standard propagation model with energy losses (Protheroe 1982).

account for the 150-300 MeV flux (Stecker, *et al.* 1981). Stecker *et al.* (1981, 1982) have argued that an extragalactic primary flux from antimatter active galaxies could supply a $\bar{p}$ flux with a ratio $\bar{p}/p \sim 5 \times 10^{-4}$, the protons being overwhelmingly galactic in origin (Stecker 1975). They have further argued that the lack of cosmic-ray $\bar{\alpha}$'s at present detection levels can plausibly be accounted for by spallation and photodisintegration in the core of these sources. They estimate spallation and photodisintegration times of $\tau_{sp} \sim 0.2 - 6 \times 10^4$ yr and $\tau_{pd} \sim 3 \times 10^8$ yr in these sources. Finally, it is predicted that the $\bar{\alpha}/\alpha$ ratio should be determined by $\bar{\alpha}$ acceleration in normal antimatter galaxies and that the resultant flux should be in the range $5 \times 10^{-6} \lesssim \bar{\alpha}/\alpha \lesssim 5 \times 10^{-5}$. It is also estimated that extragalactic cosmic-rays can reach us by diffusion from distances of up to 500 Mpc (Stecker, *et al.* 1982).

Other possible explanations for the cosmic-ray $\bar{p}$ flux have recently been reviewed by Protheroe (1982). These alternatives appear to have serious problems. Production of $\bar{p}$'s through n-$\bar{n}$ oscillations gives a flux orders of of magnitude below the observed flux (at best). Galactic primordial black holes are quite *ad hoc*. Suggestions for secondary generation and deceleration of $\bar{p}$'s in galactic cosmic-ray sources have energetics problems.

4. FUTURE TESTS

We have seen how the γ-ray and cosmic-ray $\bar{p}$ observations can be accounted for by a baryon symmetric cosmology. We have also seen how measurements of $\bar{\alpha}$'s in the cosmic radiation can provide a future test. Suggestions to look for $\overline{Fe}$ have also been made recently (Ahlen, et al. 1982).

Several suggestions have been made for using high-energy neutrino astronomy to look for antimatter elsewhere in the universe (Learned and Stecker 1980; Berezinsky and Ginzberg 1981, Brown and

Stecker (1982). These suggestions are all based on the fact that cosmic ray pp and pγ interactions favor the secondary production on π^+'s over π^-'s, whereas for $\bar{p}\bar{p}$ and $\bar{p}\gamma$ interactions the situation is reversed. The subsequent decay of the pions results in equal amounts of ν_μ's and $\bar{\nu}_\mu$'s of almost equal energies. However, π^+ decay leads to ν_e production, whereas π^- decay leads to $\bar{\nu}_e$ production. A production mechanism of particular importance in this context because of its large inherent charge asymmetry, involves the photoproduction of charged pions by ultrahigh energy cosmic rays interacting with the universal 3K blackbody background radiation. The most significant reactions occur in the astrophysical context principally through the Δ resonance channel.

There is a significant and potentially useful way of distinguishing ν_e's from $\bar{\nu}_e$'s, namely through their interactions with electrons. The $\bar{\nu}_e$'s have an enhanced cross section (resonance) through formation of weak intermediate vector bosons such as the W^-. For electrons at rest in the observer's system, the resonance occurs for cosmic $\bar{\nu}_e$'s of energy $M_W^2/2m_e = 6.3 \times 10^3$ TeV for $M_W \simeq 80$ GeV.

The cosmic and atmospheric fluxes for $\bar{\nu}_e$'s, based on cosmic ray production calculations have been given by Stecker (1979). Assuming that there is no significant enhancement in the flux from production at high redshifts, the integral $\bar{\nu}_e$ spectrum from $\gamma\bar{p}$ interactions is expected to be roughly constant at 10^{-18} to 10^{-17} $\bar{\nu}_e$'s cm^{-2} sr^{-1} up to an energy of $\sim 2 \times 10^7$ TeV, above which it is expected to drop steeply. It is expected that the largest competing background flux of $\bar{\nu}_e$'s will be prompt $\bar{\nu}_e$'s from the decay of atmospherically produced charmed mesons. A cosmic $\bar{\nu}_e$ signal may be heavily contaminated by prompt atmospheric $\bar{\nu}_e$'s at the W resonance energy. The cosmic flux is expected to dominate the higher energies so that the existence of higher mass bosons B^- may be critical to any proposed test for cosmic antimatter using diffuse fluxes (Brown and Stecker 1982). An acoustic deep underwater neutrino detector may provide the best hope for testing for cosmic antimatter by studying the diffuse background neutrinos. The practical threshold for such devices appears to be in the neighborhood of $10^3 - 10^4$ TeV (Bowen and Learned 1979). One gains much in looking for higher mass resonances at higher energies. Acoustic detectors of effective volume $\gg 10$ km^3 (10^{10} tons) may be economically feasible and event rates of $\sim 10^2 - 10^4$ yr^{-1} may be attained in time.

Indirect future observational tests involve studies of primordial He (Stecker 1980) and distortions in the high frequency side of the microwave background radiation (Stecker and Puget 1973). Observation of angular fluctuations in the 100 MeV γ-ray background radiation using the Gamma Ray Observatory satellite could play a key role in determining whether the flux is from point sources or more diffuse "ridges" as predicted by baryon symmetric cosmology.

There is an intriguing connection between the "cell structure" of galaxy clustering presented in many papers at this symposium and the large scale fossil domain structure predicted by our baryon symmetric cosmology. It is obvious that future theoretical studies exploring this relationship could be of significant value.

5. REFERENCES

Ahlen, S. P., Prince, P. B., Salamon, M. H. and Tarlé, G., 1982, Astrophys. J. 260, 20.
Allen, A. J., 1981, Mon. Not. R. Astron. Soc. 197, 679.
Bogomolov, E. A., et al., 1978, Proc. 16th Intl. Cosmic Ray Conf., Kyoto 1, 330.
Bowen, T. and Learned, J. G., 1979, Proc. 16th Intl, Cosmic Ray Conf., Kyoto 10, 386.
Brown, R. W. and Stecker, F. W., 1979, Phys. Rev. Lett. 43, 315.
Brown, R. W. and Stecker, F. W., 1982, Phys. Rev. D26, 373.
Buffington, A., Schindler, S. M. and Pennypacker, C. R., 1981, Astrophys. J. 248, 1179.
Fichtel, C. E., Simpson, G. A., and Thompson, D. J., 1978, Astrophys. J. 222, 833.
Gaisser, T. K., and Levy, E. H., 1974, Phys. Rev. D10, 1731.
Golden, R. L., et al., 1979, Phys. Rev. Lett. 43, 1196.
Kuzmin, V. A. et al., 1981, Phys. Lett. 105B, 167.
Langacker, P., 1981, Phys. Rpts. 72, 185.
Learned, J. G. and Stecker, F.W., 1980, Proc. Neutrino 79 Intl. Conf., Bergen Norway (ed. Haatuft, A. and Jarlskog, C.) 2, 461.
Marshall, et al., 1980, Astrophys. J. 235, 4.
Protheroe, R. J. 1982, in Proc. NATO Adv. Study Inst., Erice, Italy, Reidel Pub. Co., Dordrecht, p. 119.
Rothschild, R. E., et al., 1982, UCSD SP-82-23, to be published.
Sato, K., 1981, Phys. Lett. 99B, 66.
Schönfelder, V., et al., 1980, Astrophys. J. 240, 350.
Senjanović, G. and Stecker, F. W., 1980, Phys. Lett., 96B, 285.
Stecker, F. W., 1975, Phys. Rev. Lett. 35, 188.
Stecker, F. W., 1978, Nature 273, 493.
Stecker, F. W., 1979, Astrophys. J. 228, 919.
Stecker, F. W., 1980, Phys. Rev. Lett. 44, 1237.
Stecker, F. W., 1981, Proc. Tenth Texas Symp. on Relativistic Astrophys. Ann. N.Y. Acad. Sci. 375, 69.
Stecker, F. W., 1982, Proc. 1981 Oxford Intl. Symp. on Progress in Cosmology (ed. A. W. Wolfendale), Reidel Pub. Co., Dordrecht, 1.
Stecker, F. W., Protheroe, R. J. and Kazanas, D., 1981, Proc. 17th Intl. Cosmic Ray Conf., Paris 9, 211.
Stecker, F. W., Protheroe, R. J. and Kazanas, D., 1982, to be published.
Stecker, F. W. and Puget, J. L., 1973, in Gamma Ray Astrophysics (ed. F. W. Stecker and J. I. Trombka) U. S. Gov't. Printing Off., Washington, 381.
Steigman, G., 1976, Ann. Rev. Astron. Astrophys. 14, 339.
Trombka, J. I., et al., 1977, Astrophys. J. 212, 925.

Discussion

Shafer: Perhaps it should be pointed out that there is more than one way to draw the dotted line in your graph of the X-ray spectrum. While it is true that a thermal component, or a broken power law, would contribute little to the flux at energies > 100 keV, there are other components observed in the X-ray regime that might be significant contributors at the higher energies. In particular, active galactic nuclei contribute only about 20% of the 2 - 10 keV flux, but are characterized by a fairly flat spectrum. If this spectrum is continued out to higher energies, as is observed to be the case for a few of the brightest examples, then they could be responsible for almost all the flux above 100 keV. The observational data to directly test this, particularly in the MeV range, is admittedly meager, but it should not be thought that annihilation radiation is the only explanation of the background above 100 keV. Indeed, there may be room for a significant excess from annihilation, even if active galactic nuclei dominate the flux.

Salpeter: Does the angular distribution of the observed γ-rays give any evidence for boundaries between clusters and anti-clusters?

Stecker: There are no data as yet on this point, but it is hoped that the planned gamma-ray observatory (GRO) satellite may obtain data on the angular distribution at $\geq$ 100 MeV, corresponding to low-redshift annihilation radiation near the boundaries of clusters or superclusters.

Smoot: I would like to comment on the antiproton evidence for antimatter. The low-energy point of Andy Buffington et al., which appears so far above the predicted galactic flux of antiprotons produced by high-energy cosmic rays, is not necessarily very high when the adiabatic deceleration in the solar cavity is taken into account. The evidence for primordial antimatter is not strong.

Stecker: Calculations of solar modulation effects, as done by Protheroe and also reported in our forthcoming paper on this subject, show that solar modulation effects do not reduce the discrepancy by an acceptable amount.

van der Laan: While you clearly can compute the shape of the annihilation contribution to the γ-ray background spectrum, its amplitude is subject to many parameters. You do not claim to have fit the data by a calculated curve, do you?

Stecker: The magnitude is a function of the surface area of the boundary regions and therefore the size of the fossil domains. Puget and I gave estimates a while back arguing that supercluster or large cluster size for the final dimensions would be consistent both with galaxy formation theory and the magnitude of the flux, but this calculation is, of necessity, very rough.

Wright: Don't you need a fairly special amount of inflation to get from the horizon scale at CP breaking to a scale between superclusters and the Hubble radius?

Stecker: Yes.

PARTICLE PHASE TRANSITIONS CAN PREVENT AN INITIAL COSMOLOGICAL SINGULARITY

S.A. Bludman
Department of Physics, University of Pennsylvania
Philadelphia, Pennsylvania 19104

The present universe has small vacuum energy density (cosmological constant) and small spatial curvature. Therefore, before symmetry breaking, it had to have huge positive vacuum energy density, significant curvature, and low entropy. Because the energy positivity condition was not then satisfied, there need not have been an initial singularity (Hot Big Bang). If the universe is spatially open or the initial entropy high enough, then before the elementary particle phase transition, it was indeed an inflecting universe with an initial singularity. But if the universe is spatially closed and of low initial entropy, it needed to be a curvature-dominated bouncing universe, with a minimum size and maximum temperature (Tepid Little Bang). The closure of the present universe will therefore determine whether or not there was an initial singularity.

Motivated by the successful unification of electromagnetic and weak interactions, theoretical physicists have proposed various grand unification theories (GUTs) of electroweak and strong interactions at energies above 10^{14} Gev. In these theories generally, the Weinberg parameter is naturally fixed, neutrinos have a small mass, and domain walls and magnetic monopoles are created at elementary particle symmetry breaking.

In the present universe, elementary particle symmetry is hidden or seemingly "broken" by the presence of an order parameter $\sigma \sim 10^{14}$ Gev. In the early hot universe ($T > T_c$), this order vanished ($\sigma = 0$) so that the presently hidden symmetry was unbroken. Although some of the above implications of GUTs may be observed in relatively low-energy terrestrial laboratories, the most dramatic consequences appear in the hot and dense very early universe. This happens because as the universe expands and cools below T_c the vacuum (elementary particle ground state) undergoes a transition from unbroken to broken symmetry with the release of a huge latent heat $\varepsilon_V(0) - \varepsilon_V(\sigma) \approx 1/8\ B\sigma^4$. The ordered phase has lower energy $\varepsilon_V(\sigma)$ than the disordered phase, $\varepsilon_V(0)$. This is why the ordered phase is the stable phase at $T < T_c$.

In the present universe, the energy density of empty space (cosmological constant $\Lambda = 8\pi G/3\ \varepsilon_V$) is observed to be very small or zero. Precisely because of this empirical fact, the early universe must

G. O. Abell and G. Chincarini (eds.), Early Evolution of the Universe and Its Present Structure, 447–451.

have had a huge vacuum mass density $\rho_V \equiv \varepsilon_V(0)/c^2$. As a consequence of accepted particle and gravitational physics, there had to be a huge cosmological constant in the very early universe, precisely where its gravitational effects are most important.

THE PRESENT BROKEN-SYMMETRY UNIVERSE

The present universe has vanishing energy density $\varepsilon_{vac}(\sigma) = 0$ and is described, in the large, by the Friedman equation:

$$H^2 \equiv \left(\frac{\dot{R}}{R}\right)^2 = \frac{8\pi G\rho}{3} - \frac{kc^2}{R^2} \tag{1}$$

where the constant space curvature is described by $k = +1, 0, -1$ (closed, critical density, open universe). The radiation-dominated Friedman equation depends on only one parameter:

$$a_R^2 \equiv \frac{8\pi G}{3} \varepsilon_R R^4 = 2\pi(3/4\ a_2')^{1/3} S^{4/3} m_{PL}^{-2} \tag{2}$$

where $\varepsilon_R = a_s T^4$ is the radiation energy density, $S = 4/3\ a_s T^3 R^3$ $4/3\ a_s^{1/4}(\varepsilon_R R^4)^{3/4}$ is the total entropy, and $m_{PL}^{-1} \equiv G^{1/2} = 1.6 \times 10^{-33}$ is the Planck length. (We use units $h = c = k_B = 1$ so that the radiation constant $= \pi^2/30$ and, allowing for three flavors of two-component neutrinos, $a_s = [\pi^2/30]\ g_s = 0.64$). The total entropy is, in the absence of dissipations, the only strictly conserved quantity.

Ever since $R > R_x \equiv (\varepsilon_R R/\varepsilon_M)_o \sim 10^{-3} R_o$, the universe has been matter-dominated so that defining:

$$a_M \equiv \frac{8\pi G}{3} \varepsilon_M R^3 \approx 10^{-29} a_R^{3/2} m_{PL}^{1/2}, \tag{3}$$

$$\dot{R}^2 = \left(\frac{a_R}{R}\right)^2 + \frac{a_M}{R} - k. \tag{4}$$

Because the present scale $R_o > 10^{28}$ cm and temperature $T_o \sim 3K$, $RT > 10^{28}$, $S > 10^{87}$ and $a_R > 10^{56} m_{PL}^{-1}$, $a_M > 10^{57} m_{PL}^{-1}$. Defining $\Omega \equiv \rho/\rho_{CR} \equiv \rho(3H^2/8\pi G)$, Eq. (4) can be rewritten:

$$1 - \Omega^{-1} = \frac{3k}{8\pi G\rho R^2} = k\left[\left(\frac{a_R}{R}\right)^2 + \left(\frac{a_M}{R}\right)\right]^{-1}. \tag{5}$$

At present, $\Omega \sim 1$ and the closure of the universe is practically impossible to determine because the curvature term in Eq. (4) is and always has been so small relative to the radiation and matter energy density terms. This flatness or huge total entropy suggests huge dissipation in the early universe. Although this huge dissipation might be gravitational or hydrodynamic in origin, in the rapidly expanding early universe the GUTs' symmetry breaking leads naturally to extreme supercooling and entropy generation before the phase transition is completed. This happens because after supercooling the temperature rebounds almost up to T_c, the universe has meanwhile been expanding exponentially by a factor $> e^{100}$, so that the entropy increases by a factor $> 10^{87}$. Because the universe within the present horizon has expanded from a very small (presumably homogeneous) patch, the presently observable universe is homogeneous and practically devoid of magnetic monopoles or domain walls. Any huge dissipation would explain the flatness of the present universe. The exponential growth, brought about by the drawn-out first-order phase transition in the expanding universe, explains homogeneity of the present universe and the absence of magnetic monopoles.

The inflationary scenario predicts that $\Omega = 1$ to a very high accuracy (10^{-9}) and that galaxies are formed out of a scale-invariant spectrum of initial fluctuations. Both predictions are very hard to test. Dynamical observations show that, for mass clustered on the scale of superclusters, $\Omega \sim 0.2$, but neutrinos, photinos, gravitinos or other collisionless particles that do not cluster on this scale could make Ω larger. It is therefore important to push up the lower bound on Ω, since if Ω exceeds unity significantly, the inflationary scenario would be disproven.

THE UNIVERSE BEFORE THE SYMMETRY-BREAKING PHASE TRANSITION

Before the symmetry-breaking phase transition, the universe was dominated by the hugh vacuum energy density $\varepsilon_V(0)$ so that, in place of Eq. (4):

$$\dot{R}^2 = \frac{8\pi G}{3}(\rho_R + \rho_v)R^2 - k = \left(\frac{a'_R}{R}\right)^2 + \left(\frac{R}{b}\right)^2 - k \tag{6}$$

$$\ddot{R} = -\frac{4\pi G}{3c^2}(\varepsilon + 3p)R = -\frac{a'^2_R}{R} + \frac{R}{b^2}, \tag{7}$$

$$a'^2_R \equiv \frac{8\pi G}{3}\varepsilon_R R^4 = 2\pi(3/4\ a'_s)^{1/3}\, S'^{4/3}\, m_{PL}^{-2} \tag{8}$$

$$b^{-2} \equiv \frac{8\pi G}{3}\varepsilon_v = (7.0 \times 10^9\ \text{Gev})^2 = (3.5 \times 10^{23}\ \text{cm}^{-1})^2$$

The primes refer to values before symmetry breaking so that we expect $S' \sim 1 \sim a'_R m_{PL}$. The vacuum energy density now defines a second length scale $b = 3 \times 10^{-24}$ cm. The ratio of these two scales in the symmetry-unbroken universe, $\alpha \equiv (2 a'_R/b)^2$, is a measure of the radiation energy or total entropy before the GUTs' phase transition.

AVOIDING THE INITIAL COSMOLOGICAL SINGULARITY

From Eq. (7) there is an inflection in R(t) at $R = (a'_R b)^{1/2}$, when $\varepsilon_R = \varepsilon_V$. From Eq. (6), $\dot{R}$ can vanish only if both $k = +1$ (closed universe) and $\alpha \leq 1$ (low entropy initial state). Thus, there are two possibilities for homogeneous isotropic radiation-dominated universes with positive vacuum energy:

1) If $k = -1$ (open universe) and/or $\alpha > 1$ (relatively high entropy initial state), there can be no extremum in R. We have an inflectional universe with a singular origin (Hot Big Bang) at which the radiation density $\varepsilon_r \gg \varepsilon_v$. This is followed by a short coasting phase, followed by vacuum-dominated exponential growth during which the universe cools down, finally undergoing the phase transition to the present broken-symmetry universe. This inflectional model, in which the curvature term was insignificant in the early universe, is the one advocates of the inflationary scenario usually have in mind.

2) If $k = +1$ (closed universe) and $\alpha \equiv \sin^2\phi \leq 1$ (low entropy initial state), then we have a bouncing universe with an extremum in R(t)

$$R^2_{max,min} = \frac{b^2}{2}(1 \mp \cos\phi), \quad R_{max} R_{min} = \frac{b^2}{2}\alpha$$

where the curvature and energy terms cancel. Such a low entropy is reasonable before the GUTs' phase transition. If there were a singular origin, R(t) would stay less than $R_{max} = b \sin \phi/2$ and the universe would have remained very small and very hot, never becoming our present universe; we must reject this possibility. There is, however, another reversing solution, $R(t) \geq R_{min} = b \cos \phi/2$ in which $T < T_{max} \sim m_{PL} \sin \phi/2$. There is no initial singularity because the radiation energy never dominated the vacuum energy. Summarizing: If the universe is open and began radiation-dominated, there had to be an initial singularity. If it is closed and began not too hot, there was no initial singularity. We call this latter solution the Tepid Little Bang in contrast to the Hot Big Bang. If the inflationary scenario is correct, $\Omega = 1 \pm (10^{-9})$, so that it will be practically impossible to distinguish observationally between an open and closed universe. For those who prefer to maintain Einstein's gravitational theory, however, the Tepid Little Bang enjoys the theoretical advantage of avoiding the initial singularity which would otherwise signal the incompleteness of Einstein's theory.

Especially interesting is the critical case $\alpha = 1$ for which $R_{max} = R_{min} = b/\sqrt{2} \equiv R_E$. Then, at $R = R_E = (3/2\pi B)^{1/2} m_{PL}/\sigma\text{-}2$, the radiation and vacuum energies balance precisely so that $\dot{R} = R = 0$. (This fine-tuning may be a consequence of some supergravity theory or new cosmological principle.) This is the original Einstein static universe that is known to be gravitationally unstable. If, at $t = -\infty$, it begins to expand from $R = R_E$, it slowly goes into exponential expansion until the phase transition takes place and we enter into the conventional closed universe, with the usual baryosynthesis, nucleosynthesis and galaxy synthesis scenarios. If the universe is closed and began with the critical low entropy, this Tepid Little Bang avoids the initial singularity which would otherwise signal the breakdown of conventional gravitational theory.

Fig. 1. The Tepid Universe, which is gravitationally unstable, expands exponentially from finite size at $t = -\infty$ and undergoes phase transition to the present universe.

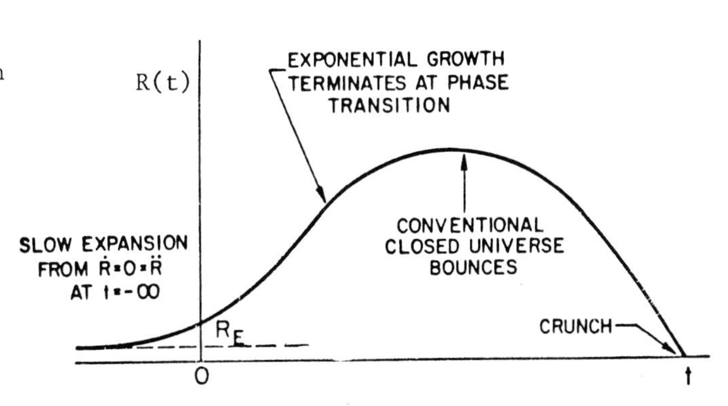

Discussion

Shklovskii: The Anthropic principle (Dicke et al.) is very important.

Peacock: Do you think there is any prospect that GUTs will eventually explain why the cosmological constant is presently zero (or very small)? If a value $\Lambda = 0$ has to be assumed ad hoc, the theory of the phase transition is seriously incomplete.

Bludman: Hopefully, some future theory will explain the fact that, at present, $\Lambda = 0$. Accepting this empirical fact, symmetry-unbreaking requires an original cosmological constant of the sign and magnitude needed to avoid an initial singularity.

Lukash: The model you suggested begins with the Einstein unstable model, which, for this reason, cannot be continued to $t = -\infty$. Is that not the case?

Bludman: Because the Einstein universe is unstable, if it begins to grow, it enters exponential growth slowly. In contrast with the Hot Big Bang, the time before the phase transition is infinitely long.

ACCELERATION AND DISSOLUTION OF STARS IN THE ANTIBANG

E.R. Harrison
University of Massachusetts, Amherst, MA 01003, U.S.A.

If the universe is spatially closed, and the simplest cosmological models are valid approximations, then in $\sim 10^{11}$ years the universe will recollapse into an antibang. (A "bang" is explosive, whereas an "antibag" may be considered implosive.) In this note I shall add one or two details to the seminal work of Rees (1969).

In the collapse phase, galaxies merge and tend to vanish when the radiation background has a temperature $T_o \simeq 300$ K. Stars with an initial velocity of $v_o \simeq 10^3$ km s^{-1} now accelerate, and if their motion is free, then $\gamma\beta T_o = \beta_o T$, where $\beta = v/c$, $\gamma = (1-\beta^2)^{-1/2}$. As shown by Rees, the stars become relativistic when $T \simeq 10^5$K, and their collisions with one another are negligible.

Let us now take into account the drag force on a spherical body of mass M, radius R, moving through the background radiation of energy density $u = aT^4$. The force is:

$$F = \pi R^2 u\gamma^2 \beta\{\frac{4}{3} + (1 + \frac{1}{3}\beta^2) \frac{2GM}{R^2c^2} \ell n \frac{D}{R}\} , \tag{1}$$

where the second term in curly brackets is the rate of momentum to distant photons due to small-angle deflections, and D is a cutoff length. For bodies large compared with black holes, this second term is negligible. On integrating the equation of motion, we now have:

$$\gamma\beta = \beta_o \times e^{-\alpha(x^2-1)} , \tag{2}$$

where $x = T/T_o$, and

$$\alpha = 2M_\gamma/M , \tag{3}$$

with $M_\gamma = \pi R^2 L_o u_o/c^2$ equal to the photon mass in a cylinder of radius R and the Hubble length at T_o. For (BD) black dwarfs we have $\alpha \sim 10^{-13}$, for (WD) white dwarfs $\alpha \sim 10^{-17}$, and for (NS) neutron stars $\alpha \sim 10^{-22}$. From equation (2) we find γ has a maximum value:

G. O. Abell and G. Chincarini (eds.), Early Evolution of the Universe and Its Present Structure, 453–455.

$$\gamma_m = \beta_o (2\alpha e)^{-1/2} \tag{4}$$

at $T_m = T_o(2\alpha)^{1/2}$, and hence $\gamma_m \sim 10^4$, 10^6, 10^9 for BD, WD, NS, respectively, and the corresponding temperatures are $T_m \sim 10^9$, 3×10^{10}, 3×10^{13}. It is unlikely that these relativistic speeds will be attained because of i) star dissolution, and ii) electron pair production (particularly for WD and NS).

Dissolution due to mass loss is a difficult calculation. In the temperature range of interest, it appears that the incident energy flux due to radiation ($\sim u\gamma^2 c$) is greater than that due to gas ($\sim \rho_g \gamma^2 c^3$). Avoiding complications such as "limb skimming" and hydrogen burning, and taking account only of atmospheric boiling, the dissolution time scale is:

$$t_{dis} \sim \pi R^2 \gamma^2 u/c \; . \tag{5}$$

In the temperature range of interest $(\beta_o^{-1} < x < \alpha^{-1/2})$,

$$\gamma = \frac{T}{T_o} \beta_o \; , \tag{6}$$

and dissolution occurs therefore on a time scale corresponding to the temperature:

$$T_{dis} \sim T_o \beta_o^{-1/2} \alpha^{-1/4} \; . \tag{7}$$

We find: $T_{dis} \sim 10^7$K, $\gamma \sim 10^2$ for BD; $T_{dis} \sim 10^8$, $\gamma \sim 10^3$ for WD; and $T_{dis} \sim 10^9$K, $\gamma \sim 10^4$ for NS.

The increase in specific entropy due to the acceleration of stars is:

$$\frac{\Delta s}{s} \sim \frac{n_b \; \gamma m_b c^2}{n_\gamma kT} \; ,$$

and using equation (1), it is seen:

$$\frac{\Delta s}{s} \sim \frac{n_b}{n_\gamma} \frac{\beta_o m_b c^2}{kT_o} \sim 3 \times 10^7 \frac{n_b}{n_\gamma} \; , \tag{8}$$

where n_b/n_γ is the ratio of baryon and photon densities. If the ratio of densities lies in the range $10^{-8} - 10^{-9}$, the entropy increase is significant, as was foreseen by Rees.

REFERENCES

Rees, M.J. 1969, The Observatory, 89, pp. 193-198.

Discussion

Kazanas: I would like to comment that a very simple calculation can show that if the recollapsing phase of the universe is a time-reversed version of its expansion, i.e., the relation between the scale factor and time is that given during the expansion: $R \sim t^{1/2}$, the mass of any black hole diverges at a finite temperature given by $T \sim 10^{12}(M/M_{\odot})^{-1/2}$ K. This is consistent with the view that the time's arrow is in the direction of increasing entropy, since a homogeneous model has a very low gravitational entropy compared with that of a black hole, which is the maximum entropy state for a collection of particles.

I consider this single argument to really impose very serious problems on models of an oscillating universe.

INFLATIONARY UNIVERSE, PRIMORDIAL SOUND WAVES AND GALAXY FORMATION

V.N. Lukash and I.D. Novikov
Space Research Institute, Moscow

I. Observational data and requirements to theory

At present cosmology experiences a qualitatively new state of its development. Up to now cosmological theory successfully extrapolated to the past the Universe we observe now. This extrapolation was reliable up to the temperatures $T \sim 10^{10}$K. The endeavors to advance nearer to the singularity imply that even cardinally new hypotheses in the elementary particle physics (e.g. Hagerdorn's hypothesis) brought about the relatively insignificant variations in cosmology—the Universe expansion rate obeyed the power law as before, the main features of the cosmological model (homogeneity, singularity, the existence of horizone, etc.) remained unchanged.

The great success of the first stage of cosmology development was a formulation of the initial conditions which should have been prescribed from the very beginning of the cosmological expansion and resulted in the present state of the Universe. These conditions of the early Universe and their relation to the observable world are given below.

1. Spatial homogeneity and isotropy of the cosmological expansion. This property follows from reliably determined large-scale isotropy and homogeneity of the observed world (data on relic radiation, distribution of galaxy clusters, abundance of He^4 and so on) which indicate that the Universe expanded in the past in accordance with the Friedmann model. Non-Friedmannian beginning of the expansion attended by isotropization would result in the observed effects which have not been detected so far (isotropization problem, see [1,2] etc.). The fact that the physical conditions are the same in the causally nonrelated space regions, filled up with radiation and matter, cannot be explained in the context of classical cosmology (horizon problem).

2. Primordial cosmological perturbations. The developed structure of the observed Universe (galaxies and their clusters) on scales of $\sim$0.01 horizon and less along with a high degree of homogeneity and isotropy on larger scales (see 1) indicate that the observed structure of the Universe appeared as a result of the development of small perturbations—deviations from isotropy and homogeneity—which should exist in the beginning of the expansion. At the radiation-dominated stage, when pressure is one-third of the energy density ($p = \frac{\epsilon}{3}$), the small primordial perturbations are the cosmological sound waves of constant amplitude. For these perturbations to increase to unity in the

G. O. Abell and G. Chincarini (eds.), Early Evolution of the Universe and Its Present Structure, 457–462.

epoch after hydrogen recombination their amplitude at the same $p = \frac{\epsilon}{3}$ should be of the order of 10^{-4} on scale of clusters. This is much more than the statistical or quantum density fluctuations. The absence of the large amount of the primordial black holes and of large-scale fluctuations of the relic radiation suggests that the primordial perturbation spectrum is close to the flat one [3].

3. Baryon asymmetry. The observed world consists of matter, the noticeable presence of anti-matter in the Universe has not been detected. Qualitatively a number of baryons (at the hot stage—their excess over antibaryons) is convenient to characterize relative to relic photons, the number of which is strictly determined in the hot Universe. This ratio—one baryon per $\sim 10^8$ photons—is conserved in the process of expansion, insignificantly varying at the epoch of annihilation of electron-positron and other pairs at the hot stage of expansion. Such a high value of specific entropy evidences the hot past of the Universe when the matter was ultrarelativistic with the equation of state $p = \frac{\epsilon}{3}$.

4. Mean density of the matter and spatial flatness of the Universe. The density of the visible matter in the Universe is less than the critical one by the factor of ~ 30. Other data show the possible presence in space of invisible matter. The fact that the mean density of the matter is close to the critical one in 15 billion years after the beginning of expansion evidences that the three-dimensional space of the Universe on the horizon scale is practically flat.

5. Absence of Λ-term. Physical vacuum of all the fields has no gravity.

The modern second stage of the cosmological theory development is aimed to explain these five puzzles of the early Universe. This became possible due to the advances of the elementary particle physics. Spontaneously broken gauge theories (see reviews [4,5,6] and others) realized in a simple mathematical form the idea of unification of physical interactions at very high energies and predicted the presence of phase transitions with breaking of the ground state of quantum fields in the process of cosmological expansion.

Not long before a typical phase transition the Universe expansion is determined by the density of a symmetric "false" vacuum whose energy then converts partly to usual matter. At this stage the Universe expands exponentially fast and for several tens of cosmological times the Planck scale can inflate up to very large sizes, e.g., to the present visible horizon and even more. During such an expansion all random (chaotic) unhomogeneities and curvatures which could take place at the beginning are ironed, as at this stage the effect of the vacuum is equivalent to the presence of a large Λ-term, homogeneous in space and as if preserving the future homogeneity, isotropy and flatness of the Universe. At this stage the world approaches the part of de Sitter's world, therefore the inflationary stage is often called de Sitter's.

So, the existence of de Sitter's expansion stages in the early Universe with the successive decrease of the value of Λ solves in principle problems 1, 4, and 5 [7]. On the other hand grand unified theories, predicting baryon-and CP-nonconservation intereactions with energies $10^{14} - 10^{17}$ GeV, explain successfully question 3 on the origin of baryon asymmetry by the decay of heavy leptons and Higgs' bosons (see, e.g., reviews [8,9]). To make this baryon number generation possible, de Sitter's expansion stage should have occurred before the moment the world had a temperature of the grand unification energy for the last time. The existence of such rather stretched exponential expansion stage could have also explained the absence of superheavy monopoles which were formed during the grand unification symmetry breaking since these monopoles would get beyond the visible horizon nowadays [10].

Before discussing the main point of our paper—generation of the primordial cosmological perturbations—it should be mentioned that de Sitter's stage near the cosmological singularity could appear not only due to Higg's breaking which does not effect the gravitational field (the metric is classical and its symmetry does not change) but also, in principle, due to quantum gravitational effects [11,12].

This problem is closely connected with the problem of the cosmological singularity and, to our point of view, only a future theory of quantum gravity can solve it.

II. Generation of primordial perturbations

During phase transition at the exponential expansion stage unhomogeneities form spontaneously, it would be desirable to identify them with those primordial (see 2). The resulting spectrum is flat—this property of de Sitter's stage is inherent in both gravitational waves [13] and potential perturbations [14,18]. The question about potential fluctuations amplitude is more problematic since the latter depends on the phase transition kinetics.

It was assumed in the first versions of the theory for the simplest model of Higg's scalar field that the first order phase transition to the stage of a new phase ($\phi \neq 0$) results at once in the formation of "true" vacuum bubbles (see [4-6]). In this case practically all released energy goes to the walls of nucleated bubbles which, expanding and colliding, form large unhomogeneities contradicting the observations.

In a new version of the theory [10] that differs by a special choice of the effective potential determining the type of the ϕ-field nonlinearity, the phase transition kinetics is such that a formed bubble continues expanding exponentially and can expand up to the size larger than the Universe observed. In this version 1, and 4 initial condition problems and also the absence of the noticeable number of monopoles and primordial black holes are naturally explained. The termination of de Sitter's stage and the transfer to the radiation-dominated expansion stage are ensured by a decay of Λ-term inside the bubble with the formation of the "true" vacuum and thermal particles. A rigorous account of this transition stage depending on the type of the effective potential, makes it possible to calculate the value of potential perturbations generated in the bubble. In the new inflationary Universe scenario for the simplest phase transition model the amplitude of potential perturbations also turns out to be too high to be consistent with observations [15-18].

The amplitude of potential fluctuations created during the phase transition on scales less than the cosmological horizon (mind, that the bubble inflates with the velocity of light) is directly connected with the choice of the effective potential of Higgs' field considered. In the real Universe the symmetry of physical fields was apparently more complex, during the transition fields could interact intensively, the effective potential could have quite another form than that investigated up to now. Evidently, there may exist theories with acceptable effective potentials (see, e.g., [19]). Thus, one of the basic problems is to build a theory which does not result in fairly large density fluctuations and baryon charge of the Universe.

In this paper we would like to pay attention to perturbations the generation of which does not depend on details of phase transitions, but is rather provided by the existence of the intermediate de Sitter stage. These perturbations are spontaneously generated from the point-zero fluctuations of metric through the parametric amplification effect on scales more or of the order of the cosmological horizon [14] (the latter grows as fast as a scale factor at the de Sitter stage). To determine these perturbations correctly we need some assumptions about the matter since matter and metric perturbations of potential type are strictly coupled [20]. Independently of the choice of a field theory, the matter in such large scales can be described as a perfect fluid with the effective equation of state $p = p(\epsilon)$.

The perturbations of this gravitating fluid are relativistic sound waves described by the gauge invariant 4-scalar $q(t, x)$ [14,21]:

$$\Box_\beta \bar{q} = U\bar{q},$$

where $ds^2 = dt^2 - a^2(dx^2 + dy^2 + dz^2)$ is the metric of background Friedmann model, $a = a(t)$ is the scale factor, $\Box_\beta = \frac{\partial^2}{\partial\eta^2} - \beta^2(\frac{\partial^2}{\partial x^2} + \frac{\partial^2}{\partial y^2} + \frac{\partial^2}{\partial z^2}$ is the acoustic d'Alambertian operator, $\eta = \int \frac{dt}{a}$ is the conformal time, $\beta = (\frac{dp}{d\epsilon})^{\frac{1}{2}}$ is the sound velocity, $U = (\xi a)^{-1}\frac{d^2}{d\eta^2}(\xi a)$ is the sound effective potential reflecting nonstationarity of the Universe, $\xi = \frac{(1+p/\epsilon)^{\frac{1}{2}}}{2\beta}$, $\bar{q} = \xi a q$. In the short-wave asymptotic q has a meaning of hydrodynamic velocity potential. For large scales being of interest for us q is mainly metric perturbations.

At the hot evolutionary stage when the Universe expansion is determined by relativistic particles, q-field proves conformally coupled: $U = 0$. At this stage, using the procedure of secondary quantization, the vacuum of q-field can be determined, which on large scales corresponds to quantum metric potential perturbations.

The possibility of spontaneous birth of phonons, quanta of the q-field, comes due to the fact, that sound effective potential differs from zero at de Sitter stage, $U \neq 0$. This effect works as a typical parametric amplification. The resulting perturbation spectrum somewhat differs from the flat one [22,23]: it damps in a direction of large scales inversely with perturbation size and has a characteristic maximum with the exponential cut in a direction of short-waves, the perturbation amplitude in the maximum is of the order of the temperature squared (in Planck units), at which the Universe was at the beginning of the first de Sitter stage.

For these perturbations could be primordial to form galaxies, the spectrum maximum should be within the mass range from $10^3 M_\odot$ to $10^{15} M_\odot$. This requires the total duration of all de Sitter stages from 30 to 60 cosmological times. To ensure the required perturbation amplitude in the scales of clusters the first phase transition should occur not later than in $\sim 10^4$ Planck times, that corresponds to the energy of $\sim 10^{17}$ GeV. This figure does not yet contradict to the estimate of the density of gravitational waves, produced due to this transition, which will affect the large-scale isotropy of relict radiation [13,24,25]. The induced quadrupole relict anisotropy will be of the same order of magnitude as the density perturbation amplitude in the spectrum's maximum.

1. A.G. Doroshkevich, V.N. Lukash, I.D. Novikov, *Zh. Eksp. Teor. Fiz.* **64**, 1457 (1973).
2. C.B. Collins, S.Q. Hawking, *Astroph. J.* **180**, 317 (1973).

3. Ya.B. Zeldovich, *Mon. Not. R. Astron. Soc.* **160**, 1L (1972).
4. S. Dimopoulos, Proc. of IAU Symposium 104 "Early Evolution of the Universe and Its Present Structure," Crete, Greece, Aug 30-Sep 2 (1982).
5. K. Sato, Proc. of IAU Symposium 104 "Early Evolution of the Universe and Its Present Structure," Crete, Greece, Aug 30-Sep 2 (1982).
6. S. Bludman, Proc. of IAU Symposium 104 "Early Evolution of the Universe and Its Present Structure," Crete, Greece, Aug 30-Sep 2 1982 , (1983) p. 447.
7. A.H. Guth, *Phys. Rev.* **D23**, 347 (1981).
8. A.D. Dolgov, Ya.B. Zeldovich, *Rev. Mod. Phys.* **53**, 3 (1981).
9. M. Turner, Proc. of IAU Symposium 104 "Early Evolution of the Universe and Its Present Structure," Crete, Greece, Aug 30-Sep 2 (1982).
10. A.D. Linde, *Phys. Lett.* **108B**, 389 (1982).
11. A.A. Starobinski, *Phys. Lett.* **91B**, 99 (1980).
12. L.P. Grishchuk, Ya.B. Zeldovich, Proc. of 2nd Seminar "Quantum Gravity," Moscow, Oct 13-15 (1981).
13. A.A. Starobinski, *Pis'ma Zh. Eksp. Teor. Fiz.* **30**, 719 (1979).
14. V.N. Lukash, *Zh. Eksp. Teor. Fiz.* **79**, 1601 (1980).
15. J.M. Bardeen, Proc. of Nuffield Workshop on the Very Early Universe (<1 sec), Cambridge, June 21-July 9 (1982).
16. A.H. Guth, Proc. of Nuffield Workshop on the Very Early Universe (<1 sec), Cambridge, June 21-July 9 (1982).
17. S.W. Hawking,Proc. of Nuffield Workshop on the Very Early Universe (<1 sec), Cambridge, June 21-July 9 (1982).
18. A.A. Starobinski, Proc. of Nuffield Workshop on the Very Early Universe (<1 sec), Cambridge, June 21-July 9 (1982).
19. P.J. Steinhardt, Proc. of Nuffield Workshop on the Very Early Universe (<1 sec), Cambridge, June 21-July 9 (1982).
20. E.M. Lifshitz, *Zh. Eksp Teor. Fiz.* **16**, 587 (1946).
21. V.N. Lukash, *Pis'ma Zh. Eksp. Teor. Fiz.* **31**, 631 (1980).
22. D.A. Kompaneets, V.N. Lukash, I.D. Novikov, *Astron. Zh.* **59**, 424 (1982).
23. V.N. Lukash, I.D. Novikov, Proc. of Nuffield Workshop on the Very Early Universe (<1 sec), Cambridge, June 21-July 9 (1982).
24. L.P. Grishchuk, *Zh. Eksp. Teor. Fiz.* **67**, 825 (1974).
25. A.V. Veryaskin, V.A. Rubakov, M.V. Sazhin, submitted to *Phys. Lett. B* (1982).

Discussion

B. Jones: Are there any dissipative processes, occurring perhaps when the X-bosons decouple, which could inhibit or damp the growth of the fluctuations which you created at earlier times?

Novikov: We are interested in the fluctuations, the scale of which is much greater than the horizon at the early universe. These perturbations are not influenced by any matter effects, in particular by dissipative processes.

Bonometto: I would like to understand why the equation for sound wave propagation contains a term violating conformal invariance during the exponential expansion, while this term is not there before and after. In other terms, a characteristic scale enters the equation during the inflationary phase. What is this scale?

Novikov: The effective potential of sound waves $U = \alpha''/\alpha$ = (time scale to the minus two) is exactly equal to zero, when the universe expands linearly in the conformal time $\alpha \sim \eta$ (it takes place for the $p = \varepsilon/3$ equation state), and it appears non-zero when the expansion deviates from the linear law, e.g., when it is exponential. A characteristic scale here is the scale of the horizon at the beginning of the inflationary stage. If the latter is provided by GUTs, then the characteristic scale turns out to be the GUT scale, because the temperature at the beginning of the exponential stage is of the order of the "vacuum" density, α = (scale factor) (1 + pressure/density)$^{1/2}$ /2 (sound velocity).

PHASE TRANSITIONS OF COSMOLOGICAL VACUUM AND PRIMORDIAL BLACK HOLES

N.S. Kardaskhev, I.D. Novikov
Space Research Institute, USSR Academy of Sciences

We propose to discuss here the possibility of the origin and existence of primordial black holes and their expected parameters in the scenarios of the early Universe in which vacuum phase transitions play a decisive role.

The possibility of the origin of primordial black holes was advanced in a paper by Zeldovich and Novikov (1966) and later in a paper by Hawking (1971) and was successfully developed subsequently by others.

The earlier work dealt with the investigation of the physical meaning of the cosmological constant as a manifestation of vacuum properties and was initiated by Zeldovich (1967) and Sakharov (1967). Papers by Kirzhnitz (1972), Kirzhnitz and Linde (1972), Linde (1974) and Weinberg (1974) initiated the development of the theory of phase transitions with spontaneously broken symmetry, which according to the Grand Unified Theories should have occurred at early stages of the universal expansion. This trend then had a brilliant development. The essential stage was the scenario of the universal evolution with a term caused by quantum effects advanced by Starobinsky (1980), and the scenario of the universal evolution with a Λ-term in the Grand Unified Theories advanced by Guth (1981) and Linde (1981). We now believe that we are close to the solution of many important problems: the problem of the beginning of the cosmological expansion (origin of the universe in the context of supersymmetry theories), the isotropy and homogeneity of the Universe at large scales due to the causal connection between the parts in the earlier period of exponential expansion, the nearness of the present density to the critical one (flat three-dimensional space), and the appearance of inhomogeneities essential for forming stars and galaxies. One of the most complicated problems of the developing theory is the concrete definition of the phase transition process. Although much remains to be worked out, a qualitative picture of the universal expansion at the earlier stages seems to be as follows (Figure 1): In the process of expansion, the universe vacuum was initially in the state of so-called false vacuum with high energy density. Now we can single out three states.

<u>The first state</u> corresponds to the unified gravitational, strong, weak and electromagnetic interactions (the temperature was close to the Planck mass: $T_1 = \sqrt{\hbar c/G} = 2.2 \times 10^{-5}$ g $= 1.2 \times 10^{19}$ GeV $= 1.4 \times 10^{32}$ K).

G. O. Abell and G. Chincarini (eds.), Early Evolution of the Universe and Its Present Structure, 463–468.

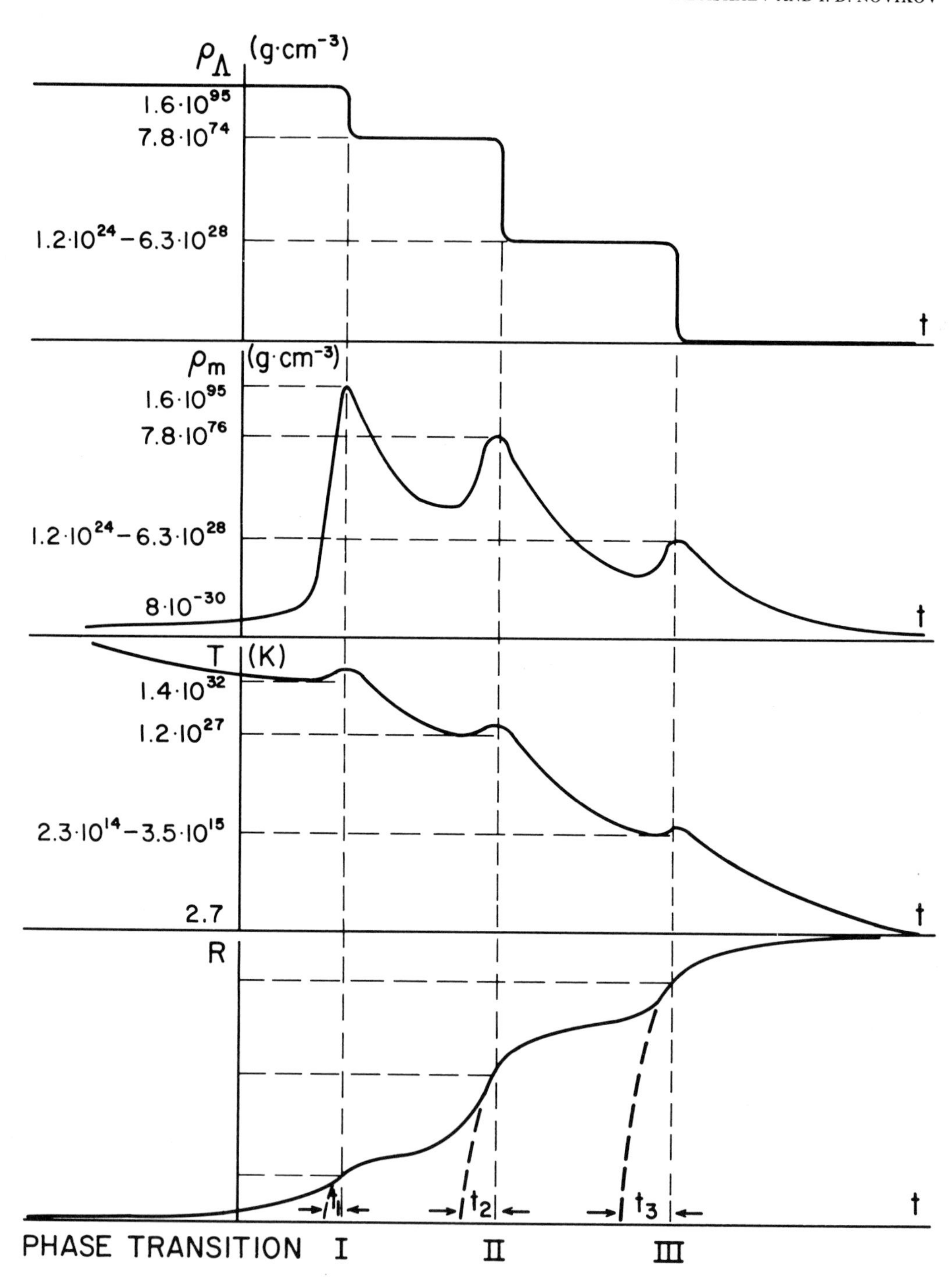

Figure 1. Expected evolution in the early big bang with three phase transitions. ρ_Λ is the vacuum energy density, ρ_m is the matter density, T is the particle temperature, and R is the scaling facto

The vacuum energy density at this temperature and possibly at higher temperatures $\rho_1 = \pi^2 N(kT)^{-4}/30\hbar^3 c^5 \simeq 1.6 \times 10^{95}$ g/cm^3, where N is the number of various types of elementary particle. We adopt N = 100 for all calculations. This state corresponds to a cosmological constant $\lambda_1 = 8\pi G\rho_1 = 2.7 \times 10^{89}$ s^{-2}, $\Lambda_1 = \lambda_1/c^2 = 3.0 \times 10^{68}$ cm^{-2}. After the transition to the new vacuum state we can assume that the expansion obeys the law of the common cosmological model of a flat world, where the main density of vacuum energy is transformed to that of relativistic matter, heated to approximately the same temperature, and a small part of vacuum energy corresponds to the next metastable state. It can be assumed that during the transition from one vacuum state to another some small part of the matter could transform to the black hole state (see, for example, Sato 1981).

From the relation for Friedmann models at relativistic stages, $\rho_1 = 3/32\pi G t^2$, we determine $t_1 = 1.7 \times 10^{-45}$ s, which characterizes a time interval of density variation, and provides an estimate of a possible characteristic mass of appearing primordial black holes at this stage: $M_1 = 4/3\ \pi(ct_1)^3\rho_1 \simeq 8.4 \times 10^{-8}$ g. Such holes should rapidly evaporate, or even fail to form in general because of quantum effects. In the process of further expansion, the matter density and temperature fall according to a power law, and at some time the Λ-term of the next metastable vacuum state becomes decisive for another expansion.

The second state is the unified strong, weak and electromagnetic interactions; $T_2 = 10^{14}$ GeV $= 1.2 \times 10^{27}$ K, $\rho_2 = 7.8 \times 10^{74}$ g/cm^3, $\lambda_2 = 1.3 \times 10^{69}$ s^{-2}, $\Lambda_2 = 1.4 \times 10^{48}$ cm^{-2}; $t_2 = 2.4 \times 10^{-35}$ s, and $M_2 = 1.2 \times 10^3$ g are determined, respectively. The lifetime of such black holes due to the Hawking process is: $t_H = 8.8 \times 10^{-27} M_2^3 = 1.5 \times 10^{-17}$ s.

After this phase transition, the vacuum energy density once again transforms to hot matter, it expands, and the Λ-term of the new metastable state again becomes decisive.

The third state is the unified weak and electromagnetic states. $T_3 \approx g^{-1/2} \simeq 300$ GeV $\approx 3.5 \times 10^{15}$ K (g is the weak interaction constant), $\rho_3 = 6.3 \times 10^{28}$ g/cm^3, $\lambda_3 = 1.1 \times 10^{23}$ s^{-2}, $\Lambda_3 = 120$ cm^{-2} $t_3 = 2.7 \times 10^{-12}$ s, $M_3 = 1.4 \times 10^{26}$ g, $Z_3 = 9.2 \times 10^{15}$.[1] According to a paper by Guth and Weinberg (1980), the transition to the spontaneous phase variation occurs for $T_3^* = K(\alpha)(2-\alpha)^{1/4}\sigma$, $K(\alpha)$ smoothly varies from 0.081 to 0.087 as α varies from 2 to 0. We take $\alpha \simeq 1$. The value $\sigma = (\sqrt{2}g)^{-1/2} = 246$ GeV is experimentally determined, from which $T_3^* = 20$ GeV $= 2.3 \times 10^{14}$ K. Then $\rho_3^* = 1.2 \times 10^{24}$ g/cm^2, $\lambda_3^* = 2.0 \times 10^{18}$ s^{-2}, $\Lambda_3^* = 2.2 \times 10^{-3}$ cm^{-2}, $t_3^* = 6.1 \times 10^{-10}$ s, $M_3^* = 3.1 \times 10^{28}$ g, and $Z_3^* = 6.0 \times 10^{14}$. The masses obtained for black holes (of the order of the mass of the Moon and terrestrial planets) have very weak Hawking glow and should have been preserved up to the present. Apparently, after the third phase transition (although maybe it is not so), the true vacuum forms (its energy density and the Λ-term are equal to zero), and the expansion is close to the Einstein-de Sitter flat model, with all main properties of the existing theory of the large explosion. The kinetics and even the duration of

[1] $Z_3 = R_0/R_3 = (0.5\ N)^{1/2}\ T_3/T_0 = 9.2 \times 10^{15}$ is the compression measure of the Universe (redshift), $T_0 = 2.7$ K is the modern temperature of the cosmological background.

each phase of the vacuum "boiling" have not quite been cleared up yet. is possible that the number of transitions is much more than three, but is also possible that they all merge together forming a common region stretched over temperatures. Probably, however, the conclusion is valid that only in the latter case (Z_3) do primordial black holes form that are capable of existing up to the present. One possibility is that these black holes represent a hidden mass giving the Universe a density close the critical one. During the subsequent evolution, the accretion process scarcely changes their mass (Novikov and Polnarev 1980). Under our assumption of the age of the Einstein-de Sitter model, $t_0 = 10^{10}$ years, $H_0 = 2/3t_0 = 65$ km/s, the density, $\rho_0 = 3H_0^2/8\pi G = 8 \times 10^{-30}$ g/cm^3, is completely determined by primordial black holes. Then, for the moment t, their contribution to the density is $\rho_{BH} = \rho_0 Z_3^3 = 6.2 \times 10^{18}$ g/cm^3, and $\rho_{BH}/\rho_3 = 10^{-10}$; i.e., primordial holes can constitute an insignificant part of density at this moment. The important peculiarity of primordial black holes is the fact that all these objects cannot have relativistic velocities even during the initial epoch (though the opposite possibility should be investigated). The theory of nonlinear perturbations in the gas from stars or black holes can be used, resulting in the formation of black hole clusters (Doroshkevich, Zeldovich 1974; Peebles 1974). The cluster mass is $M_c = M_3 Z^{6/7}$, where Z is the beginning of the nonlinear stage.

With a uniform distribution of primordial black holes in space, their density will be $n_0 \simeq 2.5 \times 10^{-58}$ to 5.6×10^{-56} cm^{-3}, the distance to the nearest hole is $n_0^{-1/3} = 1$ to 15 pc. If it is assumed that the hidden mass in the Galaxy near the sun is $\rho_0 \simeq 0.05\ M_\odot/pc^3$ (Faber and Gallagher 1979), the black hole density is $n_0 \simeq 1.1 \times 10^{-52}$ to 2.4×10^{-5} cm^{-3} and the distance to the nearest hole is $n_0^{-1/3} = 2.3 \times 10^3$ to 1.4×10^4 a.u.

In conclusion, it should be mentioned that the presence of a large number of black holes with masses in the interval considered does not contradict observations. Figure 2, taken from Carr (1979), shows that just in this interval of masses the contribution of black holes providing the critical density of the Universe is possible.

REFERENCES

Carr, B.Y. 1978, Comments on Astrophysics, 7, 161.
Doroshkevich, A.G., Zeldovich, Ya.B. 1975, Astrophys. Space Sci., 35, 55
Faber, S.M., Gallagher, J.S. 1979, Ann. Rev. Astron. Astrophys., 17, 135
Guth, A.M., Weinberg, E.J. 1980, Phys. Rev. Lett., 45, 1131.
Guth, A.M. 1981, Phys. Rev., D23, 347.
Hawking, S.W. 1971, M.N.R.A.S., 152, 75.
Kirzhnitz, D.A. 1972, Zh. Exp. Theor. Phys. Pis'ma, 15, 745.
Kirzhnitz, D.A., Linde, A.D. 1972, Phys. Lett., 42B, 471.
Kirzhnitz, D.A., Linde, A.D. 1974, Zh. Exp. Theor. Phys., 67, 1263.
Linde, A.D. 1982, Phys. Lett., 108B, 389.
Novikov, I.D., Poluarev, A.G. 1980, Astron. Zh., 57, 250.
Peebles, P.J.E. 1974, Astrophys. Space Sci., 31, 403.
Sakharov, A.D. 1967, Doklady Akad. Nauv. SSSR, 177, 70.
Sato, K. 1981, Progr. Theor. Phys., 66, 2287.

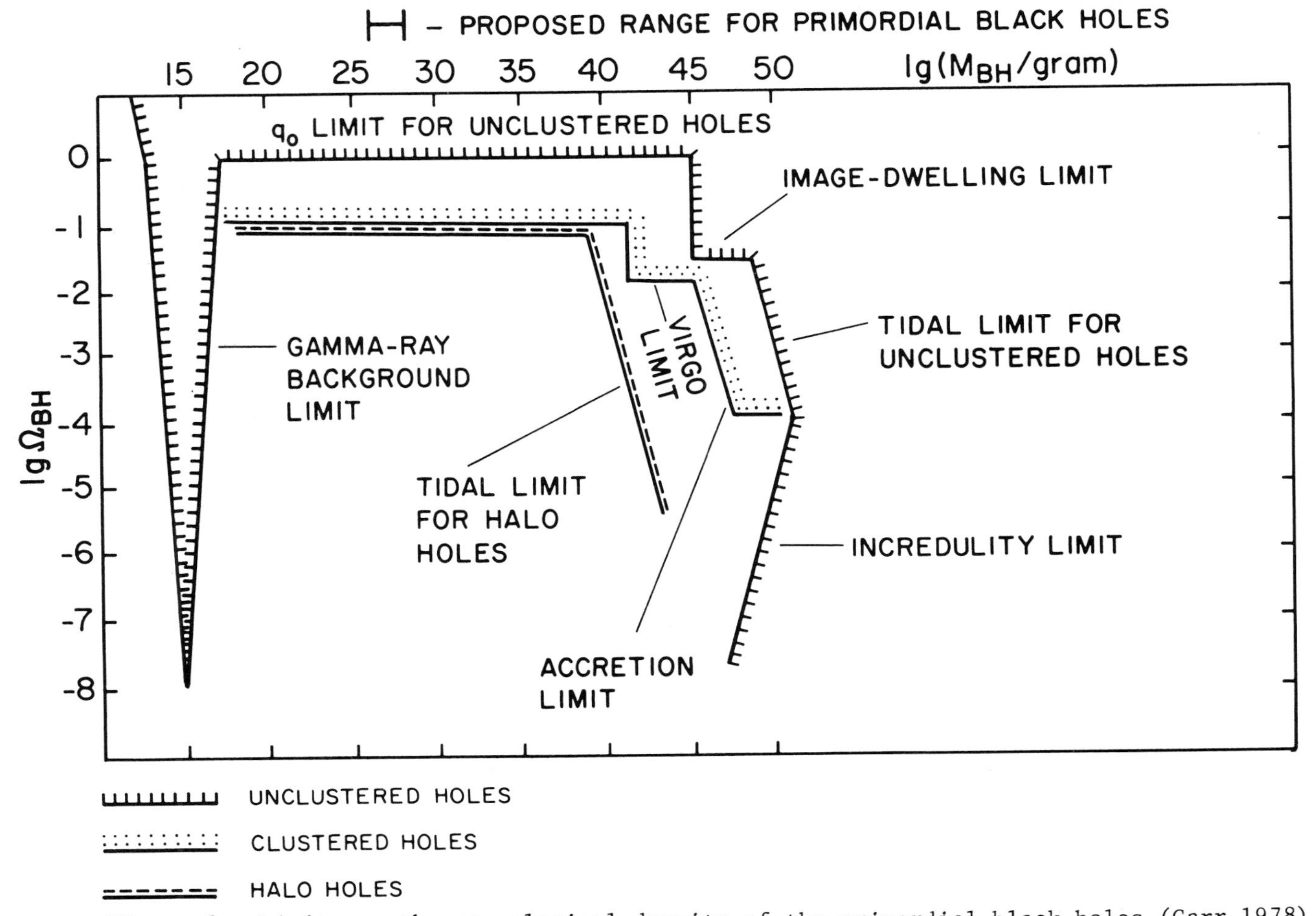

Figure 2. Limits on the cosmological density of the primordial black holes (Carr 1978).

Starobinsky, A.A. 1980, Phys. Lett., 91B, 99.
Weinberg, S. 1974, Phys. Rev., D9, 23357.
Zeldovich, Ya.B., Novikov, I.D. 1966, Astron. Zh., 43, 758.
Zeldovich, Ya.B. 1967, Zh. Exp. Theor. Phys. Pis'ma, 6, 883.

Discussion

Rees: You have, I think, assumed that the black holes start to cluster immediately after they form. However, for reasons discussed in papers by (among others) Guyot and Zeldovich, and Meszaros, this clustering will not start until the end of the radiation-dominated phase (i.e., $T \simeq 10^5\ \Omega°K$). Taking this into account, it will be hard to evolve a cluster of even a stellar mass.

Kardashav: We agree.

UNIFIED GAUGE THEORIES AND GALAXY FORMATION

G. Lazarides
The Rockefeller University, New York, New York 10021, U.S.A.

Abstract. Grand Unified strings may provide us with the primordial density fluctuations needed for galaxy formation. The properties and the cosmological evolution of such strings are discussed. A Grand Unified Theory with strings is constructed.

Grand Unified Theories (GUTs) of strong weak and electromagnetic interactions (Georgi and Glashow 1974) predict that the universe, as it cools down after the big bang, undergoes a series of phase transformations during which the original unifying gauge symmetry is reduced in stages down to the zero temperature symmetry $SU(3)_c \times U(1)_{em}$. During some of these transitions topological objects may be produced (Kibble 1980). These objects may be points, lines or surfaces where the higher temperature phase is preserved by a topological conservation law. They are called magnetic monopoles, strings or domain walls respectively.

We will concentrate on the strings because, as suggested by Zel'dovich (1980), they may be of great cosmological importance. They have the right properties needed to produce the primordial density fluctuations that lead to galaxy formation. Consider a gauge group G which breaks down to H by the vacuum expectation value (VEV) of a scalar field ϕ (for GUTs, $M \sim 10^{15}$GeV, and g is the gauge coupling),

$$G \xrightarrow{\langle\phi\rangle \sim (M/g)} H \, ,$$

To decide whether this breaking leads to string production, we must look at the topology of the vacuum manifold V=G/H. Strings are produced iff the fundamental group $\pi_1(G/H)$ is non-trivial, i.e., iff there exist loops in G/H that cannot be continuously deformed down to a point. Such loops are called homotopically non-trivial. A string is a tube of thickness $d \sim M^{-1} \sim 10^{-29}$cm. Outside this tube $\langle\phi\rangle \varepsilon$ G/H. As one describes a loop around the string, $\langle\phi\rangle$ describes a non-trivial loop in G/H. This guarantees the topological stability of the string. On the string axis, $\langle\phi\rangle = 0$ and the symmetry G is unbroken. The string carries an energy per unit length $\sigma \sim M^2/\alpha \sim 10^{20}gm/cm(\alpha = g^2/4\pi)$.

G. O. Abell and G. Chincarini (eds.), Early Evolution of the Universe and Its Present Structure, 469–471.

At a cosmic time $t \sim 10^{-35}$ sec or at a critical temperature $T_c \sim M/g \sim 10^{15}$GeV, G breaks down to H and a network of superheavy strings is produced. Initially, the scale of this network(i.e., the mean distance between two neighbouring strings) is $\sim M^{-1}$. It, then, grows very rapidly and soon becomes of the order of the particle horizon t and remains so thereafter (Kibble 1980). We, thus, essentially have one string piece per horizon volume $\sim t^3$. This produces a density fluctuation $\delta\rho/\rho \sim \rho_s/\rho_r$, where ρ_r is the radiation energy density and ρ_s the energy density due to strings. This fluctuation has the right magnitude $\sim 10^{-3}$ for galaxy formation. $\delta\rho/\rho$ remains constant until decoupling at $t_d \sim 10^{12}$ sec; only the scale of the fluctuation grows as t. After matter domination and decoupling $\delta\rho/\rho$ grows as $t^{2/3}$, becomes ~ 1 at $t \sim 10^{16}$ sec, and galaxies are formed.

A very simple and elegant example (Kibble, Lazarides and Shafi 1982) of a string producing theory is based on the gauge group Spin(10). The symmetry breaking goes as follows:

$$\text{Spin}(10) \xrightarrow[\underline{126}]{M_s} SU(5) \times Z_2 \xrightarrow[\underline{45}]{M_x} [SU(3)_c \times SU(2)] \otimes U(1)$$

$$\times Z_2 \xrightarrow[\underline{10}]{M_w} SU(3)_c \otimes U(1)_{em} \times Z_2.$$

At the first stage of symmetry breaking the unbroken group contains a discrete factor $Z_2 = (1,-1)$contained in the centre of Spin(10) which is a Z_4 subgroup generated by $i\Gamma^0$. Here $\Gamma^0 = i^5\, \Gamma^1\Gamma^2 \ldots \Gamma^{10}$ is the "chirality operator" and Γ^i's are the generalized Dirac Matrices in 10 dimensions. One can show that π_1 (Spin(10)/SU(5) x Z_2) = Z_2. Thus a superheavy string network is produced during this phase transition. The strings remain unaffected down to T=0 since the Z_2 never breaks.

This work is partially supported by the Department of Energy, Grant No. DE AC02-82ER 40033.B000.

References:

Georgi,H., and Glashow, S.L.: 1974, Phys. Rev. Lett. 32, pp.438

Kibble, T.W.B.: 1980, Phys. Rep. C67, pp. 183

Kibble, T.W.B., Lazarides, G., and Shafi, Q.: 1982, Phys. Lett. 113B, pp. 237

Zel'dovich, Ya.B.: 1980, Mon. Not. Roy. Astron. Soc. 192, pp. 663.

Discussion

Contopoulos: 1) What is the relation between your strings and the superclusters we discussed here? 2) Can the monopoles you are considering be of galactic size?

Lazarides: 1) Since the strings act as seeds for galaxy formation, there may be a relation between these gauge theory strings and the linear structures in the universe. This deserves further investigation. 2) No, the monopoles are actually of microscopic size. Their core has a radius of order 10^{-29} cm.

Segal: Why is $i\gamma_5$ representative of an element of Spin (10)?

Lazarides: The center of Spin (10) is a four-element group generated by $h = \exp(i\ \pi/2\ \sigma_{12})\cdots\exp(i\pi/2\ \sigma_{9,10}) \ \varepsilon$ Spin (10), where $\delta_{ij} = 1/2i[\Gamma^i, \Gamma^j]$. It is trivial to show that h is equal to $i\Gamma^0$, where $\Gamma^0 = i^5\Gamma^1\cdots\Gamma^{10}$ in the "chirality" operator in ten dimensions. It is also easy to see that $[h, \sigma_{ij}] = 0$.

Bond: Are the strings absolutely stable, and if so, how would they interact with matter?

Lazarides: Strings are absolutely stable against any local disturbance. This is seen by the topological argument in my talk.

Strings have a negative tension whose magnitude is equal to the string mass per unit length. Because of this, their gravitational field has the following "peculiar" properties: It does not affect slowly moving nonrelativistic particles, but it does cause deviation of light rays.

Hogan: What are the most realistic observational tests of this picture?

Lazarides: Strings produce deviation of light rays. Tests based on this are the most realistic.

PHYSICS OF THE BI-PARTITION OF THE UNIVERSE

Evry SCHATZMAN
Observatoire de Nice, B.P. 252 Nice-Cedex 06007 FRANCE

Using already tested physics of matter-antimatter interaction, an attempt is made of a check of the bi-partition model of the Universe of Fliche, Souriau and Triay (1982b). No inconsistency is found.

Fliche, Souriau and Triay (1982b) have considered an expanding closed Universe with:

$$dt = (hH_0)^{-1} A^{-1/2} R\, dR, \quad A = \Lambda R^4 - K R^2 + \Omega R$$

$$\Lambda = 1 + K - \Omega, \quad \Omega = 0.06 \pm 0.01, \quad K = 0.20 \pm 0.01$$

suggesting a bi-partition of the Universe, half matter and half anti-matter, separated by a zone of absence of QSOs. Schatzman (1982) has tested the model with respect to the following points: i) Delay of recombination due to gamma-ray production at the interface. Recombination takes place for $z \cong 600$ at the interface; ii) Thickness of the region where ionization excess prevents recombination from taking place. When $z = 1200$, the thickness of the ionized region measured at $z = 0$ is about 100 megaparsecs, in agreement with the estimate of Fliche *et al.* (1982a); iii) Production of gamma rays above 35 Mev at Earth. The result of Fichtel *et al.* (1978), $5.7 \cdot 10^{-5}$ photons cm^{-2} sr^{-1} s^{-1} can be interpreted by the present annihilation of matter and antimatter along the zone of absence of QSOs (from $z = 0.9$, $\alpha = 17^h44^m$, $\sigma = -5°55$ to $z = 30$ [at the opposite pole] [extrapolated]), the fit between observations and model being obtained for $\Omega h^2 = 0.1$, which is the right order of magnitude

Bibliography

Fichtel, C.E., Simpson, G.A., Thomson, D.J. 1978, *Ap. J.*, 222, 833.
Fliche, H.H., Souriau, J.M., Triay, R. 1982a, *Astron. Astrophys.*, 108, 256.
——————————— 1982b, preprint.
Schatzman, E. 1982, Cours à l'Ecole de Goutelas, preprint.

G. O. Abell and G. Chincarini (eds.), Early Evolution of the Universe and Its Present Structure, 473–474.

Discussion

Cristiani: Do you expect any effect on spectra of QSOs at larger red-shifts, say, any particular kind of absorption lines?

Schatzman: The missing zone has been found by Fliche et al. by condensing large clouds which define it. As far as I remember, they are visible through absorption lines.

CAUSAL STRUCTURE OF THE EARLY UNIVERSE

J. Richard Gott, III
Princeton University Observatory

The standard big bang model has always had a couple of fundamental problems: the universe's matter excess and the isotropy of the cosmic microwave background radiation. Recent developments in particle physics now point to answers to these problems and lead to new models for the causal structure of the early universe. One interesting possibility is the formation of multiple bubble universes. In such models there exist event horizons and the thermal background radiation can be produced by the Hawking mechanism.

Grand Unified Theories (GUT's) now provide an explanation for the universe's matter excess. If the cosmic blackbody radiation was initially pure thermal radiation with exactly equal amounts of matter and anti-matter at an initial temperature higher than $T \sim 10^{14}$ Gev then the GUT's now show that the observed baryon excess $n_b/n_\gamma \sim 10^{-9}$ can be produced from CP violations at later epochs (Weinberg 1979, Toussaint et al. 1979).

The other development of importance for cosmology in particle physics is the idea of "false vacuum" epochs in the early universe where $P = -\rho$ and the universe expands exponentially. Such false vacuum epochs may occur at the Planck density $T \sim 10^{19}$ Gev as proposed by Brout, Englert and Spindel (BES) (1979), Gott (1982) and Zeldovich (1981) or at the GUT transition $T \sim 10^{14}$ Gev as proposed by Guth (1981). By allowing more time for different regions to come into causal contact these models can all solve the isotropy problem. We shall discuss each of these models and their causal structure.

Figure 1 shows a Penrose diagram of an open, negatively curved, $k = -1$ standard big bang cosmology (see Hawking and Ellis 1973 for description). χ is a spacelike coordinate and η is a timelike coordinate, light travels at 45°. The big bang singularity at $t = 0$ is shown as a serrated line. Spacelike and timelike infinity i^0, i^+ are shown as well as future null infinity $\mathscr{I}^+$. A typical hypersurface

G. O. Abell and G. Chincarini (eds.), Early Evolution of the Universe and Its Present Structure, 475–481.

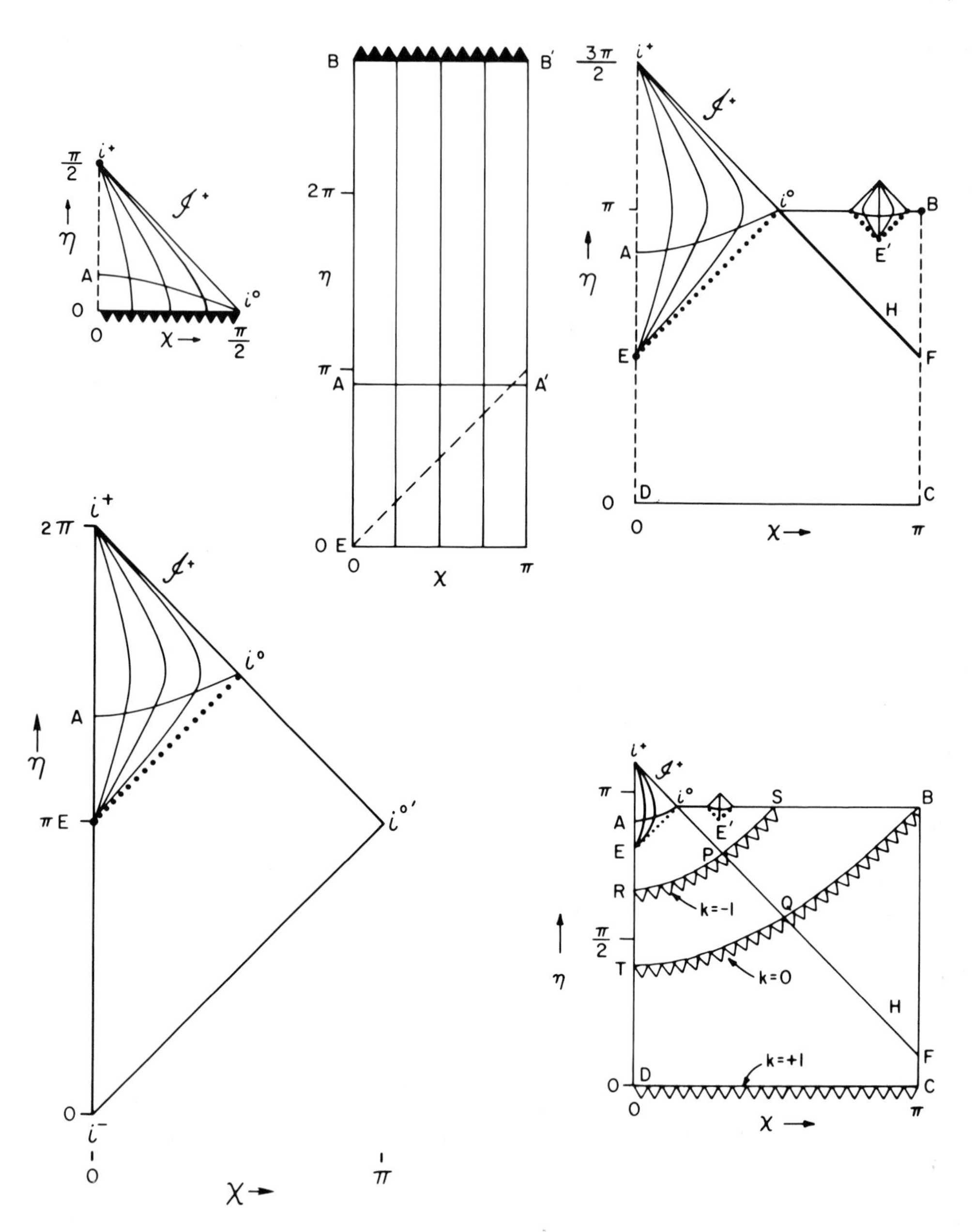

Figure 1. Top left: Standard big bang model, also Guth model. Bottom left: Brout, Englert, Spindel model. Top center: Zeldovich model. Top right: Gott model. Bottom right: Gott model with a finite past de Sitter phase.

at time $t = \tau$ is shown as Ai^0, the timelike worldlines of co-moving observers are shown as lines connecting the singularity to i^+. This is a big bang model that expands forever, it is only the conformal map projection that makes it appear contracting. Our galaxy's world line from $t = 0$ to $t = \infty$ is $\chi = 0$ from $\eta = 0$ to $\eta = \pi/2$. Our current position at $t = t_0$ is at $\eta = \eta_0$. Let Ai^0 for the moment stand for the epoch of recombination. Let A^1 be the intersection of the past light cone of ($\eta = \eta_0$ and $\chi = 0$) and the line Ai^0, i.e. $\eta(A^1) + \chi(A^1) = \eta_0$. In the standard big bang model $\eta_0 > 2\eta(A^1)$ so that when we look back to recombination we see a region so large that it could not have been causally connected in the past.

The Guth model at early times is radiation dominated: $P = 1/3\ \rho$ and $\rho \propto a^{-4}$; when $\rho = \rho_c$ it enters a "false vacuum" phase where $P = -\rho_c$, $\rho = \rho_c$ and $a \propto \exp(t/t_{ex})$. During this phase $T_{\mu\nu} = -\rho_c g_{\mu\nu}$, and the geometry approximates a piece of deSitter space. Finally it exits the deSitter phase and becomes a standard radiation dominated big bang model again: $P = 1/3\ \rho$, $\rho \propto a^{-4}$. (Linde 1981). The Penrose diagram of the Guth model is exactly the same as the standard big bang model. If the deSitter phase lasts $\tau > 65 t_{ex}$ then the recombination epoch Ai^0 is pushed upward so that $\eta_0 < 2\eta(A^1)$ and the region we can see is causally connected. But if we wait long enough we will eventually expect to see a chaotic universe just as in the big bang model.

The BES model proposes that one starts with an empty Minkowski space with $\rho = 0$. There is a quantum tunnelling event that within the future light cone of an event E produces an open $k = -1$ universe with a deSitter $\rho = \rho_c$ phase, finally leading at the epoch Ai^0 to a standard radiation dominated $P = 1/3\ \rho$ phase. In this model $t = 0$ at E and $a \sim t_{ex} \sinh(t/t_{ex})$ for small t. The Penrose diagram of the BES model is shown in Figure 1. The region $Ai^0 i^+$ in the diagram is equivalent to the same region in the standard big bang model (Fig. 1). Now however note that for any η_0 in the range 0 to π, $\eta_0 < 2\eta(A^1)$ so the observer always sees uniform background radiation no matter how long he waits. The entire hypersurface Ai^0, at which the phase transition occurs, is causally connected to the event E.

Gott (1982) and Zeldovich (1981) have proposed models intermediate between the Guth and BES models that simply begin the universe with a deSitter state. A complete deSitter space may be embedded as the hyperboloid $X^2 + Y^2 + Z^2 + W^2 - V^2 = r_0^2$ in a 5 dimensional space with metric $ds^2 = -dV^2 + dX^2 + dY^2 + dZ^2 + dW^2$ where $T_{\mu\nu} = -\rho_c g_{\mu\nu}$, $\rho_c = 3/8\pi r_0^2$, $P = -\rho_c$. This solves Einstein's field equations with zero cosmological constant. In the Zeldovich model there is a phase transition to $P = 1/3\ \rho$ along the 3-sphere $V = V_0 > 0$. This leads to creation of one closed positively curved $k = +1$ universe. The Penrose diagram of a complete deSitter space is a square $0 \leqslant \chi \leqslant \pi$, $0 < \eta < \pi$. $\eta = 0$ corresponds to $t = -\infty$, $V = -\infty$; $\eta = \pi$ corresponds to $t = +\infty$, $V = +\infty$; $\chi = 0$ and $\chi = \pi$ correspond to opposite poles where $X = Y = Z = 0$. The Zeldovich model is given by a rectangle $0 \leqslant \chi \leqslant \pi$, $0 < \eta \leqslant \eta_{BB}$, (Figure 1). There is a big crunch singularity at $\eta_{BB'} < 3\pi$ and the phase transition occurs at $\eta = \eta_{AA'} < \pi$. The phase transition is acausal because one can not find any event in the deSitter space which can causally communicate with all of $\eta = \eta_{AA'}$, $0 \leqslant \chi \leqslant \pi$. The Zeldovich model can solve the observed isotropy of the cosmic background but if one waits long enough one will eventually see chaotic conditions. Zeldovich does not believe that the deSitter space existed for all time in the past, he thinks it likely that it originated from singular initial conditions.

The Gott model starts with the same initial deSitter space but the phase transition to $P = 1/3\ \rho$ occurs along a 3-hyperboloid given by $W = W_0 > r_0$ and $V > 0$. This creates an open negatively curved universe with $k = -1$ and $\Omega < 1$ which will continue to expand forever. The Penrose diagram for the Gott model is shown in Figure 1. The phase transition occurs along Ai^0. The region Ai^0i^+ can be mapped directly into the corresponding region of a standard big bang model (Fig. 1). This is a low density bubble universe in a high density deSitter space.

The Gott model contains an event horizon (boundary of past of $\mathcal{I}^+$) shown by the line H in Figure 1. None of the other models we have discussed have an event horizon. Hawking (1974) has shown that in the black hole case the existence of an event horizon leads to production of pure thermal radiation. The GUT's theories now allow us to start the universe with pure thermal radiation so Hawking radiation is ideal. DeSitter phases have been introduced in order to explain the isotropy of the cosmic background radiation. But one of the most famous properties of a complete deSitter space is that it has event horizons and is filled with Hawking radiation at a temperature of $T = 1/2\pi r_0$ (Planck units) [c.f. Gibbons and Hawking (1977)]. In fact the total energy momentum tensor for the Gibbons and Hawking thermal state has the required $P = -\rho$ form in a curved deSitter space because of trace anomalies (Gott (1982), deWitt (1982), Page (1982)). A self-consistent state is achieved at a density of $\rho_c \sim 135/2N$ (Planck units) where N is the number of spin states (Gott 1982). The diagram shows how another bubble universe (starting at E') can form behind the event horizon which is causally disconnected from our universe. The causal structure depends on the structure of deSitter space at late times and does not require that the deSitter space exist since $t = -\infty$. It can begin with singular or chaotic initial conditions. As an example of this consider a Guth type model in which the deSitter phase ends not with a smooth transition but by breaking into Gott type bubble universes (Fig. 1): $k = -1$ bubble universes can form from $k = -1$, $k = 0$ and $k = +1$ initial Guth type initial models. In each case consider only the appropriate singularity and ignore any parts of the diagram to the past of it. The event horizon for our universe is shown (H) as well as another universe beyond it.

The detailed bubble formation outcome depends on several factors. Let ε be the probability per unit four volume r_0^4 of forming an event

E which creates a bubble. If the deSitter space has existed since $t = -\infty$ isolated bubble universes form if ε is infinitesimal. Only a few isolated bubble universes are allowed in this case. If a large number are created they will run into each other and destroy the event horizons and create one frothy chaotic universe. This will also occur if ε is finite. If however the deSitter space is created in the finite past the situation is as follows. If ε is infinitesimal then a finite but arbitrarily large number of bubble universes form each surrounded by an event horizon. If $0 < \varepsilon < \varepsilon_{CRIT}$ (Guth (1982)) has shown $5.8 \times 10^{-9} < \varepsilon_{CRIT} < 0.24$.) then an infinite number of separate bubble cluster universes form. A bubble cluster universe is a bubble with bumps on it and smaller bumps on those bumps ad infinitum so that it looks like a fractal snowflake. A bubble cluster universe is quite acceptable because it is surrounded by its own event horizon and its background radiation stays uniform until times much later than the present if ε is sufficiently small. If $\varepsilon > \varepsilon_{CRIT}$ then the bubbles coalesce to fill the space and create an unacceptable frothy chaotic structure with many wormholes and event horizons (Kodama et al. 1981).

These new developments allow a much greater freedom in the initial conditions of the universe.

REFERENCES

Brout, R. Englert, F. and Spindel, P. 1979, *Phys. Rev. Lett.*, **43**, 417.

DeWitt, B.S. 1979, In General Relativity And Einstein Centennial Survey (eds. Israel, W. and Hawking S.W.) (Cambridge University Press).

Gibbons, G.W. and Hawking, S.W. 1977, *Phys. Rev. Lett.*, **D15**, 2738.

Gott, J.R. 1982, *Nature*, **295**, 304.

Guth, A.H. 1981, *Phys. Rev. Lett.*, **D23**, 347.

Guth, A.H. 1982, preprint.

Hawking, S.W. and Ellis, D.F.R. 1973, The Large Scale Structure of Space-Time (Cambridge University Press).

Hawking, S.W. 1974, *Nature*, **248**, 30.

Linde, A.D. 1981, preprint.

Page, D. 1982, preprint.

Kodama, H., Sasake, M., Sato, K. and Maeda, K. 1981, *Prog. Theor. Phys.*, **65**, 2052.

Touissant, D., Treiman, S.B., Wilczek, F. and Zee, A. 1979, *Phys. Rev. Lett.*, **D19**, 1036.

Weinberg, S. 1979, *Phys. Rev. Lett.*, **42**, 850.

Zeldovich, Ya. B. 1981, *Sov. Astron. Lett.*, **7(5)**, 323.

Discussion

Novikov: The probability of the origin of the Friedman universe in the de Sitter universe is constant for all points. The de Sitter model has an infinite past. It means that all space of the de Sitter universe is transformed into the new stage and different Friedman universes collide. Do you agree?

Gott: Yes.

Paal: The isotropy of the cosmic background radiation is uaually described as a consequence of the homogeneity and isotropy of the observable universe. In the standard Big Bang cosmology, this description implies that one has to accept six space-like symmetries in a very large cosmic region, the different parts of which are not yet causally connected. Such symmetries cannot therefore be explained as a result of physical smoothing processes.

In order to avoid this unwanted situation and ensure causal connections, R. Gott proposes a model which uses de Sitter's spacetime, i.e., ten symmetries corresponding to the constant four-dimensional curvature. This means presupposing more regularity to explain less regularity, which is not an obvious progress. The really satisfactory solution would be to start with chaos and deduce regularity. Such an attempt was made, e.g., by Misner, his "mixmaster model"; however, it failed to work without very special and artificial assumptions and has not therefore been accepted as a real physical explanation. Similarly, the different ad hoc constructions of unconventional cosmologies have not clarified so far the mystery of the isotropy of the microwave background radiation.

Gott: I agree that this is a case of broken symmetry where the O(4,1) symmetry of the de Sitter space is broken by bubble formation to leave each bubble with the O(3,1) symmetry of an open universe. However, the de Sitter false vacuum state which is characterized by a T^{μ}_{ν} proportional to the metric is one which can arise naturally from a variety of chaotic or singular initial conditions. As the de Sitter exponential expansion phase sets in, it forgets its initial conditions on an expansion timescale. So starting with random or chaotic initials conditions, this leads at late times to a de Sitter geometry with its associated symmetries.

Segal: As you may be aware, the Einstein universe covariantly contains all the universe you describe, and yields isotropy without any relatively exotic and complex features such as instantaneous symmetry breaking and Hawkins radiation. Therefore, why not just use the Einstein universe?

Gott: The Einstein static universe has no expansion and no event horizons.

Contopoulos: The "mystery of the microwave isotropy" is a special case of the much more general problem: "Why are the physical laws everywhere the same in the universe?" which is extremely difficult to answer. For example, in an open universe, we cannot have a causal relation between all its parts, whatever the law we adopt for the initial expansion.

Gott: That's not correct. In the model I have proposed (Nature, 1982), as well as in the Brout, Englert, Spindel model, the universe is open and yet all its parts have had a causal relation with each other. In my model, for example, from $t = 0$ to $t = \tau \sim 69t_{ex}$, the expansion law is given by $a(t) = t_{ex} \sinh(t/t_{ex})$ where t_{ex} is the expansion timescale. The comoving distance traveled by a photon is $d\eta = dt/a$ and since $a \sim t$ as $t \to 0$, the integral of $d\eta$ from $t = 0$ to $t = \tau$ is logarithmically divergent. So even two observers who are separated by an arbitrarily large comoving distance at the present epoch have had time to communicate with each other since $t = 0$. Another way to see this is to note that in the Penrose diagram the entire negatively curved open universe hypersurface at epoch $t = \tau$ (Ai°) is causally connected to the event E.

Contopoulos: I notice that most unconventional theories today are much less unconventional than the ones that were fashionable ten or twenty years ago. They depart from the standard big bang only at times prior to the Planck time (10^{-43} s) or the GUT time (10^{-37} s).

R^2 GRAVITY AND THE STRUCTURE OF THE UNIVERSE

Kenneth Brecher
Department of Astronomy
Boston University
Boston, MA 02215

The "standard" hot big bang model accounts for the expansion of the Universe, the existence of the microwave background radiation, and the mass fraction of the light elements up to ^4He. It does not account for the high degree of isotropy and homogeneity of the Universe in the large, nor of the existence of structure (galaxies, clusters) on smaller scales. Other problems, such as the lepton to baryon ratio, the preponderance of matter over antimatter, and the "coincidences" of dimensionless ratios of several fundamental physical and cosmological "constants" also lie outside of the "standard" model at present.

A solution of the large scale homogeneity and isotropy problems, as well as the so-called "flatness" problem, has been proposed recently involving extreme supercooling during a phase transition associated with spontaneous symmetry breaking in Grand Unified Theories (Guth, 1981). This and related solutions are based on a direct application of the Einstein field equations for General Relativity $G_{\mu\nu} = -8\pi G T_{\mu\nu} - \Lambda g_{\mu\nu}$. The new feature of such theories is due to the introduction of a cosmological constant Λ at high densities due to the GUTs. Such theories do not, however, alleviate the problem of the origin of galaxies.

We propose a solution to both the large and small scale structure problems based on a modification of the left hand side of the field equations (Frenkel and Brecher, 1982). By starting with a gravitational field Lagrangian $R + AR^2$ (where A is a new dimensional constant), we derive a new set of field equations, consistent with all current weak field gravitational tests, which modify the cosmological solutions to General Relativity at early times when the high order terms predominate. Such solutions are without a particle horizon (with scale factor $a(t) \propto t$ as $t \to 0$). Furthermore, perturbations in both the density and metric tend to grow much more rapidly than in unmodified General Relativity.

REFERENCES

Frenkel, A. and Brecher, K.: 1982, Phys. Rev. D 26, pp. 368 - 372.
Guth, A.H.: 1981, Phys. Rev. D 23, pp. 347 - 356.

G. O. Abell and G. Chincarini (eds.), Early Evolution of the Universe and Its Present Structure, 483–484.

Discussion

Trimble: Presumably, you chose to add the particular term AR^2 to your Lagrangian because it was the simplest. Is there any more rhyme or reason to it than that?

Brecher: Reason, yes. In order to remove singularities in physical theories, adding a term with higher derivatives often does the trick. Rhyme? Yes, too. I present it below, in hopes that it will inspire Furgeson

UNIFIED FIELD THEORY

By Tim Joseph

In the beginning there was Aristotle,
And objects at rest tended to remain at rest,
And objects in motion tended to come to rest,
And soon everything was at rest,
And God saw that it was boring.

Then God created Newton,
And objects at rest tended to remain at rest,
But objects in motion tended to remain in motion,
And energy was conserved and momentum was conserved and matter was conserved,
And God saw that it was conservative.

Then God created Einstein,
And everything was relative,
And fast things became short,
And straight things became curved,
And the universe was filled with inertial frames,
And God saw that it was relatively general, but some of it was especially relative.

Then God created Bohr,
And there was the principle,
And the principle was quantum,
And all things were quantified,
But some things were still relative,
And God saw that it was confusing.

Then God was going to create Furgeson,
And Furgeson would have unified,
And he would have fielded a theory,
And all would have been one,
But it was the seventh day,
And God rested,
And objects at rest tended to remain at rest.

Tim Joseph works for a nutrition program affiliated with Cornell University. He also writes science fiction.

THE STRONG EQUIVALENCE PRINCIPLE AND ITS VIOLATION

V. M. Canuto and I. Goldman
NASA, Goddard Institute for Space Studies
2880 Broadway, New York, New York 10025

Abstract. In this paper, we discuss theoretical and observational aspects of an SEP violation. We present a two-times theory as a possible framework to handle an SEP violation and summarize the tests performed to check the compatibility of such violation with a host of data ranging from nucleosynthesis to geophysics. We also discuss the dynamical equations needed to analyze radar ranging data to reveal an SEP violation and in particular the method employed by Shapiro and Reasenberg.

I. The Strong Equivalence Principle, SEP

The Strong Equivalence Principle(1), demanding that local physics (described in a local lorentzian coordinate system) be the same anywhere and at anytime in the universe, assumes that the outcome of local experiments is independent of cosmological influence. This is equivalent to complete decoupling of local physics from the global structure of the universe, which can therefore influence local systems by determining only boundary conditions, i.e. the background space-time structure at large distances. In this sense, the SEP allows only for a very restricted realization of "Mach's principle".

If the SEP holds, the different dynamical physical fields of nature must be coupled in a way that is independent of cosmology, i.e. by coupling coefficients that are constant. On the other hand, if the SEP is violated on a cosmological time scale, there must exist an underlying dynamics connecting the cosmological structure with local physics. Before discussing how to construct a theoretical framework that incorporates an SEP violation, we shall consider an aspect of an SEP violation that has direct observational consequences.

Consider Einstein's equations and electrodynamics. Both dynamical theories are not invariant under scale transformations. Because of this property, each of them defines its own unit of time (i.e. a clock) which serves as a basis for an intrinsic system of units. (On the contrary, equations that are scale invariant, like Maxwell equations for photons, cannot give rise to a clock). Since there exist several non-scale invariant dynamics in nature, each defining its own clock and units, the question arises as to how do the different clocks relate to one another.

The SEP demands that all clocks be equivalent, i.e. that the ratio

G. O. Abell and G. Chincarini (eds.), Early Evolution of the Universe and Its Present Structure, 485–492.

of the intrinsic units of time be constant. On the other hand, if the SEP is violated, the couplings coefficients among different fields are cosmology dependent and so are the ratios of the periods of the various clocks, i.e. one faces the possibility of non-equivalent systems of units.

A minimal form of an SEP violation is one in which all microscopic units are equivalent (e.g. atomic clocks, nuclear clocks and weak interaction clocks have constant ratios of their periods), but gravitational units are different, a choice consistent with the fact that gravity is the only interaction important on a cosmological scale. Taking into consideration the observed homogeneity of the universe, one may restrict the SEP violation to depend only on the time coordinate of the comoving cosmological coordinate system. One may therefore write

$$ds_E = \beta_a(t)ds_a \; , \qquad \beta_a(t_o) = 1 \qquad (1)$$

where ds_E is a physical space-time line element measured with gravitational clocks, (e.g. a planet revolving a star), while ds_a is the corresponding line element measured with atomic clocks. The function β_a epitomizes our ignorance of how the two different dynamics couple: a constant β_a implies that the two systems of units are equivalent, a non-null $\dot{\beta}_a$ implies a violation of the SEP. (In what follows, EU and AU will be used to denote gravitational or Einstein units and atomic units respectively).

Presently available lunar data[2] suggest $\dot{\beta}_a \sim 10^{-11}$ yr^{-1} and radar ranging data[3] to the inner planets provide an upper limit $|\dot{\beta}_a| < 10^{-10}$yr^{-1}.

II. Frameworks that Incorporate an SEP Violation

Let us turn to the formulation of a framework that allows for an SEP violation of the order of H_o. One may follow two alternatives.

1) Local approach. β_a is assumed to be a dynamical field coupled locally to matter sources. This approach, adopted for example by Brans-Dicke[4], has two limitations:

a) Since solar system data[5] require $\omega > 500$, it follows that $\dot{\beta}_a \leqslant 10^{-13}$ yr^{-1} since the BD theory predicts[6] $\dot{\beta}_a \sim H_o/(1 + \omega)$. Therefore, with a BD-type of theory, it is not possible to accommodate both solar system data as well as $\dot{\beta}_a \sim H_o$.

b) From a physical point of view, one would expect in gravitational units the trajectories of macroscopic objects to be described by geodesics i.e. that gravity be described by a metric theory[1] in the system of units defined by its own intrinsic clock. However, this is not so in the BD framework, where gravity is described by a metric theory not in gravitational units but rather in atomic units, which however belong to a different dynamics.

2) Global approach[7]. Since there is no a priori reason why one should restrict an SEP violation to be described by a local space-time field, it is conceivable that β_a in eq. (1) is not coupled to local matter sources but rather represents an average over the global distribution of matter in the universe at any given time. One may of course try to construct a complete theory to predict how β_a is related to the global structure of the universe. However, a great deal can be learned even from the use of a framework in which β_a is treated phenomenologically, i.e. its dynamics is not specified; β_a enters such a framework as an external quantity

much as for example viscosity and diffusion coefficients enter the classical hydrodynamic equations. Such phenomenological framework has been meaningfully employed to limit the allowed degree of violation of the SEP using geological, astrophysical and cosmological data[8,11,12,13].

III. Breaking the Symmetry between Atomic and Gravitational Clocks.

The main objective of any theory that tries to incorporate an SEP violation is that of achieving a symmetry breaking between AU and EU, which, according to the SEP, are indistinguishable systems of units. To be more specific, one must construct gravitational and atomic clocks whose periods P and p respectively, satisfy the following symmetry-breaking conditions

$$\text{EU:} \qquad P_E = \text{constant} \qquad p_E \sim \beta_a \tag{2}$$

$$\text{AU:} \qquad p_a = \text{constant} \qquad P_a \sim \beta_a^{-1} \tag{3}$$

The construction of such gravitational and atomic clocks was presented in [7]. The most crucial ingredients are the equations of motion for micro and macroscopic objects, which must clearly be different from one another in order for conditions (2) and (3) to be satisfied. To achieve our goal, it is convenient to introduce a mathematical formalism whereby equations are first expressed in a general system of units. This operation does not introduce any physics and is merely a convenient mathematical tool. In [7], it was shown that in general units, the equations of motion of an object with mass μ (macro, M or microscopic, m) is given by ($\Delta^{\alpha\nu} = u^\alpha u^\nu - g^{\alpha\nu}$).

$$u^\alpha_{;\nu}\, u^\nu + \frac{(\mu\beta^{2-g})_{,\nu}}{(\mu\beta^{2-g})}\,\Delta^{\alpha\nu} = \frac{e}{\mu}\, u^\nu F^\alpha_\nu \tag{4}$$

Here, the function β defines the general system of units to which we have arrived starting from EU, taken to be our fiducial system, and performing the transformations

$$L_E = \beta L, \qquad G_E = G\beta^g, \qquad M_E = M\beta^{1-g} \tag{5}$$

where g is a constant number. The details of the mathematical solutions of (4) were presented in [7]. Here, we shall follow a different approach. Taking the φ component of (4), we obtain for a circular orbit, the relation

$$\Pi/(r^2\mu\beta^{2-g}) = \text{constant} \tag{6}$$

where r is the radius distance and Π the period. Eq (6) is valid in any units and for any mass μ. Let us now consider the following argument. In AU, the period of an atomic clock and the dimension of it, must clearly be constant and so must the ratio r/Π. The same argument must hold true in EU for a gravitational clock. We therefore conclude that the dimensionless ratio r/Π must remain constant for either clock in <u>any</u> units. This changes eq. (6) to

$$\Pi \sim 1/(\mu\beta^{2-g}) \tag{7}$$

Applying (7) to a macroscopic object (μ ≡ M, Π ≡ P) and a microscopic one (μ = m, Π = p) and imposing (2) - (3), we obtain

$$\begin{array}{llll} \text{EU}, \ \beta \equiv 1 & P_E \sim M_E^{-1} & \text{i.e.} & M_E = \text{constant} \\ & p_E \sim m_E^{-1} & & m_E \sim \beta_a^{-1} \\ \text{AU}, \ \beta \equiv \beta_a & p_a \sim m_a^{-1}\beta_a^{g-2} & & m_a\beta_a^{2-g} = \text{constant} \\ & P_a \sim M_a^{-1}\beta_a^{g-2} & & M_a\beta_a^{1-g} = \text{constant} \end{array} \quad (8)$$

To proceed further, we must define the AU and EU more precisely. It is physically reasonable to characterize these units and the nature of th physical clocks defining them, by requiring that microscopic masses be constant in AU and that macroscopic masses be constant in EU, i.e. the results (8) will be supplemented by the two conditions (the second of (9) is already contained in (8))

$$\text{AU:} \quad m_a = \text{constant} \qquad\qquad \text{EU:} \quad M_E = \text{constant} \quad (9)$$

which, together with (8) imply that the parameter g must be 2. We therefore conclude that the symmetry breaking conditions (2) and (3) can be met provided the following conditions are satisfied

$$\begin{array}{lllll} \text{AU:} & \beta = \beta_a, & m_a, e_a = \text{const.}, & M_a \sim \beta_a, & G_a \sim \beta_a^{-2} \\ \text{EU:} & \beta = 1, & m_E \sim \beta_a^{-1}, & e_E = \text{const.} & M_E, G_E = \text{const.} \end{array} \quad (10)$$

It is important to stress the fundamental role played by (9) in achieving a symmetry breaking.

IV. Baryons and Photons

From (10), it follows that the number of particles N must vary like

$$N \sim \beta_a \quad (11)$$

independently of the system of units, as indeed expected for N is a pure number. We therefore conclude that, within the present framework, a violation of the SEP can be realized provided the number of particles varies with time. This in turn implies that in AU, many-body thermodynami relations will also depend on β_a[8], a result that can be visualized in terms of a "viscosity-like" role played by β_a. In fact, while at the one body level, say Dirac equation, β_a does not appear, it does so at the many-body level, as a consequence of the fact that the conservation laws are changed, since in AU the Einstein equations have changed[9]. It is in fact only in EU that Einstein equations are assumed to be unaltered

As for photons, it has recently been shown[8] that the number of photons N_γ is constant, independently of the value of g. This, together with the value g = 2, are sufficient to allow the construction of a photon theory in which all the photon relations remain unaltered, in spit of a possible violation of the SEP.

This result leads us to suggest a scenario in which β_a was constant during the radiation dominated era. The argument is as follows. The function β_a represents a cosmological influence on local physics. The dynamics of β_a is determined by the global structure of the universe whos dynamics is governed by the energy density (and pressure) of matter and radiation. Since matter is affected by β_a, while radiation is not, one may expect that during the matter dominated era the rate of variation of β_a to be comparable with the rate of expansion $\dot\beta_a/\beta_a \sim 1/t$, i.e. $\beta_a \sim t^{-n}$ $n \sim 1$. However, during the radiation era, such a variation must have bee considerably smaller $\dot\beta_a/\beta_a \ll 1/t$, since the dynamical role played by matter was negligible. This is in agreement with the results of a recent numerical study of nucleosynthesis[10] showing that in order to explain the present abundance of light elements, $\beta_a/\beta_a \ll 1/t$.

V. Compatibility Tests

The compatibility of an SEP violation has been tested against a host of geophysical, astrophysical and cosmological data. The global result is that a variation of the type $\beta_a \sim t^{1/2}$ during matter dominated era and $\beta_a \sim$ constant during the radiation era, is consistent with the following set of tests: 1. Nucleosynthesis[8], 2. The 3K black-body radiation[8], 3. The m vs. z relation for QSOs, radio galaxies and elliptical galaxies[11], 4. Angular diameters for radio galaxies[11], 5. The isophotal angles for optical galaxies[11], 6. The N(S) vs. S relation for radio galaxies[12], 7. Ages of stars and globular clusters[8], 8. The early temperature of the Earth[8], 9. The Earth's paleoradius[13]

VI. Testing the SEP by Radar Ranging.

A planet orbiting the Sun is a gravitational clock. Therefore, by tracking the motion of such a planet by means of atomic clocks, one compares directly gravitational vs. atomic clocks and can therefore determine the value of $\dot{\beta}_a$. The observations consist of measuring with precision atomic clocks the round trip time of radio beams sent from earth and bounced back by the planet surface. A series of such measurements have been analyzed over the years by Shapiro and Reasenberg[3]. Since these measurements and their analysis predated by several years the appearance of the two-times theory presented here, it may be instructive to analyze the mathematical framework employed by SR vis á vis the present framework.

Let us begin with Newton's equations in their standard form, i.e. in dynamical units (subindex E)

$$\frac{d^2 \vec{x}_E}{dt_E^2} = - \frac{G_o M}{x_E^3} \vec{x}_E \tag{12}$$

where G_o is a constant. With the solutions of these equations, one can construct the "range" $\rho_E(t_E)$, i.e. half the round trip time,

$$\rho_E(t_E) = (x_{1E}^2 + x_{2E}^2 - 2x_{1E}x_{2E} \cos \theta_{12})^{1/2} \tag{13}$$

which is a quantity of direct observational interest. Let us now perform a transformation from dynamical units x_E and t_E to the corresponding atomic ones x and t, by writing in general

$$x_E(t_E) = \Lambda_1(t)x(t), \qquad dt_E = \Lambda_2(t)dt, \tag{14}$$
$$\Lambda_k(t) = 1 + \lambda_k t + \ldots \qquad t_E = t + 1/2\lambda_2 t^2 + \ldots$$

where $t = t_E = 0$ denote the initial time of the ranging data Substitution in (12) yields (up to first order in λ_1, λ_2)

$$\frac{d^2 x^k}{dt^2} = - \frac{G_o M}{x^3} x^k - (2\lambda_2 - 3\lambda_1) \frac{G_o M}{x^3} x^k t - (2\lambda_1 - \lambda_2) \frac{dx^k}{dt} \tag{15}$$

At the same time, the perturbation in the range $\delta\rho \equiv \rho(t) - \rho_E(t)$ can easily be evaluated using $\rho(t) = \Lambda_1^{-1}(t)\rho_E(t_E)$. The result is

$$\delta\rho = -\lambda_1 t\rho_E(t) + \frac{1}{2}\lambda_2 t^2\dot{\rho}_E(t) \quad (16)$$

Defining the finite differences $\Delta t \equiv t_2 - t_1$, $\Delta t_E = t_{2E} - t_{1E}$, as well as $2t = t_1 + t_2$, we obtain

$$\Delta x_E = (1 + \lambda_1 t)\Delta x, \qquad \Delta t_E = (1 + \lambda_2 t)\Delta t \quad (17)$$

which give the transformation laws for finite lengths and time intervals. With these results, we shall investigate two choices for λ_1 and λ_2.

A) $\lambda_1 = \lambda_2 \equiv \lambda_A$. This corresponds to requiring the same scaling for lengths and times, i.e. velocities remain invariant under scaling, whereas angular momenta per unit mass, J, do not, since the dimensions of J are $\sim$ (length)^2/time. Eqs. (15) and (16) now become

$$\frac{d^2x^k}{dt^2} = -\frac{G_oM}{x^3}x^k + \lambda_A\frac{G_oM}{x^3}x^k t - \lambda_A\frac{dx^k}{dt} \quad (18)$$

$$\delta\rho = -\lambda_A t\rho_E(t) + \frac{1}{2}\lambda_A t^2\dot{\rho}_E(t)$$

The first of (18) is precisely the form of Newton's equation derived from the ingeodesic equations in atomic units, proposed several years ago[9]. To make the identification complete, we must call $\lambda_A \equiv \dot{\beta}_a/\beta_a$. Because of the presence of the "viscosity term" $\sim dx^k/dt$ in (18), the angular momentum J cannot be conserved in atomic units.

B) $\lambda_1 = 1/2\ \lambda_2 \equiv \lambda_B$. This corresponds to requiring that J remains constant under scaling, i.e. that lengths and times scales differently. With this choice eqs. (15) and (16) become

$$\frac{d^2x^k}{dt^2} = -\frac{G_oM}{x^3}x^k - \lambda_B\frac{G_oM}{x^3}x^k t \quad (19)$$

$$\delta\rho = -\lambda_B t\rho_E(t) + \lambda_B t^2\rho_E(t)$$

Let us note that the viscosity term has disappeared and therefore J may be constant. The first of Eqs. (19) are exactly the relations employed by SR in their analysis.

Now the question is: is the choice $\lambda_1 = 1/2\ \lambda_2 \equiv \lambda_B$ acceptable? While there may be physical situations in which times and lengths scale differently, in the experimental situation at hand, ranging from planets, the units of length and time are not independent, distances being in fact defined as round trip times. It follow that only the choice $\lambda_1 = \lambda_2$ is acceptable, i.e. eqs. (19) are not compatible with the present two-times theory and the accompanying assumption that in gravitational units Newton's equation holds unchanged.

Given the unphysical nature of the scaling $\lambda_1 = \lambda_2/2$, what can one say about eqs. (19) from the theoretical and experimental point of view?

Let us first consider the theoretical aspect. One may not want to adopt the present two-times theory and the assumption that eq. (12) is valid in EU. In that case, the derivation of (19) through scaling would no longer exist. For example, the Brans-Dicke-type theories do not adopt eq. (12) in EU, rather they demand that in AU Newton's law should read

$$\frac{d^2 \vec{x}}{dt^2} = - \frac{G(t)M}{x^3} \vec{x} \tag{20}$$

i.e., (19) with $\lambda_B = (\dot{G}/G)_o$. Therefore, the starting equation in the SR analysis can be said to belong to the Brans-Dicke-like theoretical frameworks. However, because of the arguments given in IIa, one may expect that $(\dot{G}/G)_o << H_o$.

Let us now consider the experimental situation. Suppose that a particular set of ranging data is used for which the quadratic term in $\delta\rho$ dominates over the linear one (e.g. the Mercury data). If so, eq. (18), representative of the present two times theory, and (19), representative of the BD-type theories, give

$$\delta\rho \simeq \frac{1}{2} \dot{\rho}_E \lambda_A t^2, \qquad \delta\rho \simeq \dot{\rho}_E \lambda_B t^2 \tag{21}$$

i.e. two expressions which are equally good to detect whether the coefficient of the quadratic term (independently of how we call it) is non-null, a result which would signal an SEP violation.

This conclusion is in accord with the claim made by SR sometime ago[14] (on the basis of a similar but less general analysis), that the goal of first "smoking out" the presence of an SEP violation could be achieved in a "theory independent" fashion.

The differentiation between the present two times theory and the BD-like theories can be carried out if one employs the Viking data, since in that case the short time interval (since 1976) together with the long period of Mars, make the linear term in $\delta\rho$ important. The Viking data are at present analyzed by SR with this aspect in mind; it is hoped that the results will be soon forthcoming.

Acknowledgements. It is a pleasure to thank Drs. I. I. Shapiro and R. D. Reasenberg for interesting discussions.

References

1. Will, C. M., in "General Relativity", (eds. Hawking, S. W. and Israel, W. (Cambridge Univ. Press, New York, 1979). See also: Thorne, K. S., Lee, D. L. and Lightman, A. P., Phys. Rev. D7, 3563 (1973).
2. For a summary of the results see: Van Flandern, T. C., Ap. J. 248, 813 (1981).
3. Reasenberg, R. D. and Shapiro, I. I., in "On the Measurements of Cosmological Variations of G", (ed. by L. Halpern), (Univ. Presses of Florida, Gainsville, Fla. 1978).
4. Dicke, R. H., "The Theoretical Significance of Experimental Relativity", (Gordon and Breach, Science Publ., N. Y., N. Y., 1964).
5. Reasenberg, R. D., Shapiro, I. I., Macneil, P. E., Goldstein, R. B., Breidenthal, J. C., Brenkle, J. P., Cain, D. L., Kaufman, T. M., Konarek, T. A., and Zygielbaum, A. I., Ap. J. Lett. 234, L219 (1979).
6. Weinberg, S., "Gravitation and Cosmology: principles and applications of the General Theory of Relativity", pp. 628-629 (Wiley, N.Y. 1972).
7. Canuto, V. M. and Goldman, I., Nature 296, 709 (1982).
8. Canuto, V. M. and Goldman, I., 1982 Nature (submitted).
9. Canuto, V. M., Adams, P.J., Hsieh, S.-H. and Tsiang, E., Phys. Rev. D16, 1643 (1977).

10. Rothman, T. and Matzner, R., Ap. J. 257, 450 (1982).
11. Canuto, V. M., Hsieh, S.-H. and Owen, J.R., Ap. J. Suppl. 41, 263 (197
12. Canuto, V. M. and Owen, J. R., Ap. J. 41, 301 (1979).
13. Canuto, V. M., Nature 290, 739 (1981).
14. Shapiro, I. I. and Reasenberg, R. D., private communication (1982).

Discussion

Spyrou: Could you please comment on the systematic post-Newtonian uncertainties and errors you mentioned concerning the solar system?

Goldman: In analyzing the planetary radar ranging data, one employs a post-Newtonian model of the gravitational dynamics of the solar system. The model contains parameters such as solar and planetary masses, planetary orbital elements, post-Newtonian gravity parameters, and in our case also, $\dot{\beta}_a$. These parameters are fitted so as to minimize the residuals between the computed and measured range, as a function of time.

Prior to the use of the Viking Mars lander, the main source for uncertainty in $\dot{\beta}_a$ determination came from unknown topography of the planetary surfaces from which the radar waves were reflected. With the Viking lander there is a systemic uncertainty due to the modeling of the asteroid belt. Another limitation to the accuracy stems from possible correlations of $\dot{\beta}_a$ with other parameters; e.g., an error in Mars' semimajor axis can, for a short observing time, mimic a $\dot{\beta}_a$ effect. Nevertheless, it is estimated that the Viking data will provide an accuracy of $\sim$ 1 part in 10^{11} yr^{-1}. good enough to detect a cosmological effect of magnitude $\sim H_0$.

THE ROLE OF PARTICLE PHYSICS IN COSMOLOGY AND GALACTIC ASTRONOMY

D.W. Sciama
International School for Advanced Studies, Trieste, Italy, International Centre for Theoretical Physics, Trieste, Italy and Department of Astrophysics, Oxford University, UK.

ABSTRACT

A non-technical introduction is given to (a) the inflationary universe, (b) the production of baryon asymmetry by GUTs, (c) the possible role of massive neutrinos and (d) the possible role of the massive photinos and goldstinos of broken supersymmetric theories.

INTRODUCTION

The organizers have asked me, in this final lecture, not to summarize the Symposium but to give a broad review for astronomers of the role of particle physics in cosmology and galactic astronomy. Recent developments have shown that the possibilities here are very great, and that many observable features of the universe, and even of galaxies including our own, may have their origin either in processes occurring extremely close to the big bang, say within 10^{-35} seconds; or in effects arising from the most recent developments in particle physics, such as Grand Unified Theories (GUTs) and supersymmetry. All these matters are still very speculative. Nevertheless, I believe that the time has now come for astronomers to take these possibilities seriously, and clearly our organizers feel the same way! One problem is that modern particle physics and quantum field theory are highly elaborate and technical subjects which do not lend themselves to simple exposition. In consequence, most astrophysicists are unfamiliar with them. On the other hand, most particle physicists do not know the intricacies of modern astrophysics. Nature herself faces no such problems, of course, and all one can do is to stumble hopefully in her wake. This review is written in the further hope that it will encourage more astronomers to join the stumblers.

G. O. Abell and G. Chincarini (eds.), Early Evolution of the Universe and Its Present Structure, 493–506.

QUANTUM GRAVITY

The quantum gravity epoch presumably occurred at the Planck time $(hG/C^5)^{1/2} \sim 10^{-43}$ seconds. Unfortunately, not much progress has been made recently in finding a satisfactory quantum theory of the gravitational field (see, for example, Isham, Penrose and Sciama 1981). Many people now pin their hopes on supergravity and its unification with particle theories such as GUTs. Severe technical problems still remain to be solved in this programme, however, and the outcome is not clear. For the astronomer the importance of quantum gravity is that (a) it may eliminate the classical singularity at t = 0, thereby leading to a clearer view of the initial conditions in the hot big bang; (b) it may lead to the formation of particle-antiparticle pairs of various species, whose annihilation could have provided the initial heat needed to generate the hot big bang; (c) the same pair production process might have equalized the expansion rates of the universe in different directions, so accounting, in part, at least, for the presently observed high isotropy of the Hubble constant and of the 3^oK background.

I think that it is fair to say that not much progress has been made with these problems in recent times. This is in strong contrast to the remaining topics in this review, all of which have been transformed, or even come into existence, in the last few years.

THE INFLATIONARY UNIVERSE

This refers to a remarkable scheme designed to solve the so-called horizon and flatness problems (although, of course, whether the scheme is really valid should be decided eventually from first principles). Let us begin by considering what these problems are.

The Horizon Problem

The 3^oK background is observed to be isotropic to better than 1 part in 10^3, yet, according to standard theory, if one looks at the background in opposite directions the points observed on its last scattering surface would lie nearly one hundred horizon lengths apart (one horizon length corresponding to the largest region in causal contact with itself in terms of processes propagating with the speed of light). Thus, there would not have been time for transport processes to even out the temperature over the whole of the last scattering surface, even if they had started operating close to the big bang itself. It follows that either the rather precise equality of temperatures was imposed as an initial condition in the big bang, or some non-standard mechanism intervened to enlarge the horizon or to eliminate it. As we shall see, the inflationary universe could provide such an enlargement mechanism.

The Flatness Problem

The mean density of the universe to-day is probably within a factor of ten of the critical density. An exact equality of these densities would represent a sort of equilibrium point for Robertson-Walker models in the sense that this equality would then hold at all times (Einstein-de Sitter model). But this equilibrium is unstable in the sense that a deviation between the two densities would increase with time. For example, if the present density were one tenth of the critical value, then at a temperature of 1 Mev it would have been within one part in 10^5 of critical, while at a temperature of 10^{14} Gev it would have been within one part in 10^{49}. Such "fining tuning" looks unnatural, and suggests that either the density is indeed exactly critical, for a reason still to be discovered, or that some non-standard mechanism intervened to drive the present value of the density close to the critical value. Again, we shall see that the inflationary universe claims to provide such a mechanism.

To see what would be involved in a solution to the flatness problem, let us take a look at the standard Friedmann equation for an expanding universe

$$H^2 = \frac{\dot{R}^2}{R^2} = \frac{8\pi G\rho}{3} - \frac{k}{R^2} \tag{1}$$

in the usual notation. If the curvature k were zero, we would have the Einstein-de Sitter model with

$$\frac{8\pi}{3} \frac{G\rho_c}{H^2} = 1 \quad ,$$

giving the critical density ρ_c in terms of the Hubble constant H. In the radiation dominated case, which has

$$\rho \propto T^4 \propto \frac{1}{R^4} \quad ,$$

the resulting differential equation for R is trivial to solve and yields

$$R \propto t^{1/2} \, .$$

Now, our present task is to prevent the k term in (1) from unduly dominating the ρ term for large R (note that the present epoch corresponds to about 10^{60} Planck times - the Planck time being the only natural time scale defined in a radiation dominated universe). However, in the standard theory precisely the opposite occurs, that is, the k term is suppressed at small R. This familiar result arises because, with $\rho \propto 1/R^4$ and the k term $\propto 1/R^2$, clearly the ρ term must dominate for sufficiently small R. This is the basis for the usual discussion of a radiation dominated universe near $t = 0$.

How then can we arrange for the k term to be small, or at least not unduly large, at large R? A brilliant answer to this question occurred to Guth in 1981. We shall approach his idea in a number of steps, drawing on our knowledge of the byways of cosmology, of which Guth tells me he was unaware. Some of us remember the steady state theory of the universe and in particular Hoyle's(1948) discussion of how the steady state could be reached asymptotically in the future, that is, for large R, under the driving action of the "C field." This is relevant to our discussion, because the asymptotic steady state (which in fact corresponds to the de Sitter form

$$R \propto e^{Ht} \;) \quad ,$$

indeed has a negligible k term, as we would require. The driving mechanism is based on the C field which, roughly speaking, has a constant energy density ρ_0 given by

$$8\pi G \rho_0 = \lambda \quad ,$$

where λ can be interpreted alternatively as the cosmical constant. A glance at (1) shows that with this part of ρ now independent of R, the k term does indeed become negligible at large R. When it is negligible the differential equation is again easy to solve, and one obtains the exponential form characteristic of the de Sitter metric. If the ρ and k terms start out comparable, then the asymptotic form is accurately reproduced in a few expansion time-scales 1/H, owing to the tendency of the scale factor R to increase exponentially fast.

A second reason for thinking that the steady state theory may be relevant to our problems is that an exactly de Sitter universe would have no particle horizon at all. This difference from the usual case has to do with the very different behaviours of R for small t in the exponential form, where it tends to a constant, and in algebraic forms like $t^{1/2}$, where it tends to zero.

We can thus try introducing a new idea which might solve both the flatness and the horizon problems. If there had been an early stretch of approximately steady state behaviour, say for 100 Hubble times, then R would have been inflated by e^{100}, the k term would have been suppressed by a large factor, unless it had originally been extremely dominant, and the horizon might become enlarged sufficiently to permit the causal propagation of transport processes across the presently observable universe.

The difficulty with this idea is giving the steady state stretch a sound physical foundation. The weak point of Hoyle's proposal (apart from the accumulating observational evidence against the steady state theory) was that his C field was introduced in an ad hoc manner, unrelated to basic physical theory. Of course, Hoyle was ahead of his time - Guth's proposal for interpreting the λ term as an energy density is based on GUTs, and came 33 years later!

Guth's basic idea is that a so-called false vacuum would have an energy density of this form. Let us first try and understand what a false vacuum is. A good analogy, often cited, is with a ferromagnet. Above its Curie temperature a ferromagnet, free of an external magnetic field, would have zero net magnetic moment - the directions of its spins would have no long-range order. As its temperature is lowered, a phase change occurs when the system passes through its Curie point. At this critical temperature a change of symmetry occurs, with the spins tending to line up in an ordered way, thereby giving the ferromagnet a net magnetic moment. This change of symmetry corresponds to a loss of isotropy - the spins choose a particular spatial direction in which to point. However, this loss of symmetry is only partial, in the sense that the energy of the system would be the same whatever the overall direction chosen by the spins (in the absence of an external magnetic field). Thus the actual direction chosen in any particular case has to do with the detailed history of that specimen and has no fundamental significance; isotropy is still present.

The final feature of this analogy which we shall need is that below the Curie point the unmagnetized state possesses more energy than the magnetized one. Indeed, that is why the transition to the magnetized state takes place. However, while the temperature is being lowered through the Curie point, supercooling may occur, with a consequent delay in the onset of the phase transition. The supercooled phase would then have an excess energy and this, as we shall see, is analogous to the energy of the false vacuum in Guth's hypothesis.

We now return to GUTs. These are generalizations of the electroweak theory (which unified the electromagnetic and weak interactions) involving also the strong interactions. (For technical accounts of these theories see Langacker (1981) and Ellis (1981)). At high temperatures the lowest energy state of the fields is highly symmetrical (like the unmagnetized state of a ferromagnet, which defines no preferred direction in space). As the temperature is lowered through a critical value ($\sim 10^{14}$ Gev) a phase change occurs in which the fields lose their high symmetry, in a manner analogous to the ferromagnet, that is, by choosing one amongst an infinite number of states of the same energy in an "accidental" way; each of these states defining a preferred direction in an abstract space associated with the basic fields of the theory. It seems likely that this phase change is of the first order(that is, possesses a latent heat) in which case supercooling would be expected to occur. Of course, in the application of these ideas which Guth has in mind, the reduction of the temperature is associated with the expansion of the universe, and the supercooling would persist for a time determined by kinetic considerations. When the supercooling terminates and the system makes a transition from the false vacuum to the true vacuum, the latent heat is released, thereby increasing the entropy of the universe. According to current calculations this entropy increase is very considerable, and should be expected to be responsible for the major part of the present $3^{o}K$ radiation background.

Guth's idea is that the energy density of the supercooled phase would have the form of a λ term which drives a "steady state stretch" for the duration of this phase. The existence of this stretch could then solve the horizon and flatness problems in the manner which we have already indicated. Guth's idea is widely regarded as highly plausible, but I must emphasize that it depends on an aspect of the theory which is not at all well understood. To see what is involved let us first consider the true vacuum state at temperatures below the critical value of 10^{14} Gev. This state picks out a preferred direction in the abstract space, but is not expected to do so in physical space-time. Accordingly, the energy momentum tensor $T_{\mu\nu}$ of this state must have the form $\lambda g_{\mu\nu}$, where $g_{\mu\nu}$ is the metric tensor and λ is a constant. This has just the same form as a cosmical term in Einstein's field equations. However, in the universe to-day, the cosmical constant is observed to be many powers of ten smaller than any "natural" value it might have in a GUT theory (other than zero). It is therefore presumed that either the present GUT value of λ is strictly zero, or some cancellation mechanism exists (perhaps related to further symmetry considerations) which ensures that the total value of λ is zero.

The idea then is that any cancellation mechanism gets "used up" in this way, so that since the false vacuum in the supercooled phase has higher energy than the true vacuum, this excess energy survives and is physically real. This real energy would also have the form of a λ term, and would be expected to have a GUT-like value (corresponding, say, to an energy density $\sim T^4$ with $T \sim 10^{14}$ Gev). This assumption, and the related question of the present value of the cosmical constant from a field-theoretic point of view are the least understood parts of Guth's ideas, and constitute perhaps the most important unsolved problems in the relationship of gravitation to the other forces of nature.

If we accept this reasoning, we can use the λ term from the supercooled phase to generate a stretch of approximately de Sitter type exponential expansion in which the k term becomes highly suppressed and the extent of the horizon grows considerably. The duration of this stretch, and the details of the transition from the false vacuum to the true one, depend on uncertain parameters of the GUT theory involved. In Guth's original version the transition occurred sporadically in small bubbles which later collided in a complicated way. This led to a number of difficulties. In the new version, the symmetry change at the phase transition is ascribed to a different mechanism (it is induced by so-called radiative corrections, in a manner first proposed by Coleman and Weinberg (1973)). This would eliminate many of the difficulties. In particular, the whole of our observable universe would now fall well inside a single bubble. Apart from eliminating the collision problem, this would have the advantage of avoiding an embarassingly large flux of magnetic monopoles which would be associated with bubble walls.

This improved version was discussed in great detail at a Workshop organized in Cambridge this summer by G.W. Gibbons and S.W. Hawking, the proceedings of which will be published by the Cambridge University Press. The general consensus at the Workshop was that the new version represents an important step forwards, but that the assumptions leading to the Coleman-Weinberg form of the potential involve too much fine tuning. A more natural theory is still needed. The possibilities here are wide open. For instance, the inflation might have occurred during the quantum gravity phase at the Planck time, or it might involve supersymmetry (a theory which we shall discuss later on). A further problem is that irregularities arising during the supercooled phase,which might later lead to galaxy formation, for the moment seem to be embarassingly large.

I have tried to keep a fair balance by highlighting the difficulties, but in my opinion, and that of many people at the Workshop, they are the kind of difficulties which we could reasonably expect to be resolved by new ideas. Their existence should not blind us to the remarkable potential of the inflationary hypothesis, which at a stroke could solve both the flatness and horizon problems, and provide us with a new conception of the large scale structure of the universe. Moreover it would change our attitude to the role of initial conditions in accounting for the present state of the universe. The reason for this is that, as we have already seen, when the false vacuum decays, its energy, as represented by the λ term, is released as latent heat. This heat would correspond to a black body radiation field at a temperature $\sim 10^{14}$ Gev, whereas when this heat is released the universe would have supercooled down to a much lower temperature, say 10^{8} Gev. Thus, very considerable reheating occurs and the universe would be "born again". In particular, as we have already mentioned, the 3°K background which we observe today, would have its main origin in this reheating.

I stress this last point, not only for its intrinsic interest to astronomers, but also because the restoration of a temperature of 10^{14} Gev would permit the operation of another GUT mechanism, which has been widely invoked to account for the presumed baryon asymmetry of the universe, and for the presently observed value of the ratio of baryon number to the photon number in the 3°K background. To this mechanism we now turn.

THE PRODUCTION OF BARYON ASYMMETRY IN THE UNIVERSE

We begin with the assumption that the universe at very high temperatures was baryon symmetric, with equal numbers of baryons and anti-baryons being thermally excited. As the universe cooled it is supposed that reactions occurred which led to the production of $1 + \varepsilon$ baryons for every anti-baryon. Eventually, the baryons and anti-baryons annihilated into photons, leaving of order ε baryons for every photon. Since at high temperatures the number of baryon pairs in thermal equilibrium would be of the same order as the number of photons, this would mean that today's value of n_b/n_γ would be of order ε. Thus, if $\varepsilon \sim 10^{-10} - 10^{9}$, we would have accounted for the observed number of baryons per photon in terms of the microscopic processes which determine ε.

An essential feature of this mechanism is that baryon number is not conserved (by a relative amount measured by the small quantity ε). Until recently, in particle physics baryon number was thought to be absolutely conserved, although it was recognized that this conservation law was on a less fundamental footing than, say, the conservation of charge. However, in the unification achieved by GUTs strongly interacting quarks and electroweak leptons are placed essentially on the same footing. Since baryons are made up of quarks, there would be no absolute barrier to their decay into leptons. The half-life of a proton, for example, turns out to be of order m_X^4/m_p^5, where m_X is the mass of the X boson which mediates the grand unified interaction. This mass is believed to be about 10^{14} Gev. (This is the characteristic energy scale for GUTs, at which, for example, strong, electromagnetic and weak coupling constants all have the same value.) The predicted half-life is thus about 10^{31} years. The present experimental lower limit is about 3×10^{30} years. There are a number of experiments now under way looking for proton decay. A few potential events have been found, but the consensus at the recent Paris meeting on elementary particle physics was that nothing definite can be said as yet, although the situation should be much clearer in a year's time.

Following an original analysis by Sakharov, various people have suggested that the X boson of GUTs could be used to produce a baryon asymmetry. In rough outline what is proposed is as follows (for a more careful,but still fairly general, discussion, see Weinberg (1982)). At temperatures in the early universe above 10^{14} Gev thermal equilibrium is expected to have prevailed, with equal numbers of baryons and anti-baryons, and of X and $\bar{X}$ being present. When the temperature dropped below 10^{14} Gev, there was insufficient thermal energy to replace the X and $\bar{X}$ which disappear by decaying into B and $\bar{B}$, but with $B = (1 + \varepsilon)\bar{B}$. (This imbalance is a consequence of the CP non-invariance of the X boson's interactions.) As we have seen, this imbalance could then account for the present baryon asymmetry if $\varepsilon \sim 10^{-9} - 10^{-10}$. Attempts to calculate ε in detail have not led to very definite results, but a value of this order certainly falls within the range of possibilities.

So far I have described this idea as it was originally proposed, before Guth's inflationary universe had been introduced. In Guth's scheme the re-heating produced when the false vacuum decayed was so extensive that it would have reduced the prevailing baryon photon ratio to utterly negligible proportions. It is therefore an important feature of this scheme that after the re-heating the temperature was restored to 10^{14} Gev, so that the baryon asymmetry mechanism as described above could have then come into operation.

IMPLICATIONS OF NEUTRINO MASSES

Since many varieties of GUT, except admittedly the simplest one, lead to non-zero rest masses for neutrinos, it seems appropriate to consider here the possible role of such neutrinos in cosmology and galactic astronomy. I reviewed this subject at the recent Vatican conference, the proceedings of which have now been published (Sciama 1982a) and have reported on the ultraviolet aspects elsewhere in this symposium (1983). I will therefore be brief and will concentrate mainly on some recent developments.

We can link up this discussion with our earlier considerations by following Guth in his conclusion that the universe today possesses essentially the critical density. If we further assume that $\lambda = 0$, then the universe must be close to the Einstein-de Sitter form, and astronomical estimates of its age would indicate that the Hubble constant must be close to 50 km.sec.$^{-1}$ Mpc^{-1} (corresponding to an age $\sim 13 \times 10^9$ years).

To determine the contribution of massive neutrinos to the density of the universe we note that at high temperatures they would have been in thermal equilibrium, through weak interactions of the type $\nu + \bar{\nu} \leftrightarrow e^- + e^+$. For left-handed Majorana neutrinos these interactions become too slow to maintain equilibrium when the temperature has dropped to $\sim$ 1-3 Mev, at which time the neutrinos would have become decoupled. This detail is important because electron pairs permanently annihilate somewhat later (at $T \sim 1/2$ Mev), so that the resulting photons would have fed the 3°K background, but not the neutrinos. As a result, if the neutrinos were relativistic at decoupling, their present concentration n_ν would be suppressed below that of the photon concentration in the 3°K background ($\sim$ 400 cm^{-3}) by a factor of order 4. If the neutrinos were non-relativistic at decoupling ($m_\nu C^2 > kT_d$) they would be further suppressed by previous annihilation, but we shall not consider this case here. In summary, for $m_\nu < 1$.Mev we would have for each neutrino type

$$n_\nu \sim 100 \text{ cm}^{-3} .$$

If the neutrinos are essentially responsible for the critical density we would then require that

$$\sum m_\nu \sim 25 \text{ eV} \quad , \tag{2}$$

where the sum is over the different neutrino types. In practice, of course, one type (the tau neutrino?) might be much more massive than the others.

It is well known that the number of types N_ν, which were relativistic at decoupling is constrained by observations of the abundances of D, He^3, He^4 and Li^7 in comparison with the results of nucleosynthesis calculations for the hot big bang. The recent tendency has been for this constraint to become more stringent, and it is now widely considered that $N_\nu < 4$ (e.g. Barrow and Morgan (1982)). Since three neutrino types are already known (although the tau neutrino might be too massive to have been relativistic at decoupling) this is a remarkable result. We shall use it with greater force when we come to consider the photinos of supersymmetric theories in the next section.

It has been widely conjectured that such massive neutrinos might dominate galactic halos as well as the whole universe. A key role would be played here by the Liouville theorem, which would apply to neutrinos after decoupling (Tremaine and Gunn 1979). This theorem requires their phase space density following the motion to be constant in time, or if phase mixing is important in some process of violent relaxation, to be decreasing. Accordingly, their density ρ_ν in the galaxy today would have to be bounded as follows:

$$\rho_\nu \leq \rho_Q = \left(\frac{2\pi}{3}\right)^{3/2} \frac{m_\nu^4 v_0^3}{h^3} ,$$

where m_ν is the mass of one neutrino type, v_0 is its three-dimensional velocity dispersion (assuming a gaussian distribution) and h is Planck's constant. For example, if $m_\nu = 100$ eV and $v_0 = 100$ km sec^{-1}, then $\rho_Q = 10^{-24}$ gm. cm^{-3}, which turns out to be a typical galactic density.

Following Caldwell and Ostriker (1981) we might model the galactic halo by

$$\rho = \frac{\rho_0}{1 + \left(\frac{r}{a}\right)^2} .$$

This would yield the desired flat rotation curve for $r >> a$, but would allow for the gravitational influence of stars and gas for $r \lesssim a$. Their preferred galactic model then has

$$\rho_0 = 10^{-24} \text{ gm. cm}^{-3} ,$$

with an uncertainty of about a factor 2, and

$$a = 7.8 \text{ k pc} .$$

It is tempting to identify the required cut-off in the increase of ρ at small r with the approach of the neutrino density to the maximum permitted value ρ_Q. If we further assume that v_0 is independent of r, then its value is determined from the circular velocity v_c of the outer halo by $v_0 = \sqrt{3/2}\, v_c$. If we take $v_c \sim 220$ km sec^{-1}, we would have $v_0 \sim 270$ km sec^{-1}. The assumption that $\rho_0 = \rho_Q$ then leads to

$$m_\nu \sim 45 \text{ eV} \quad ,$$

assuming for simplicity that the mass of one neutrino type is dominant.

It is not quite clear whether this represents good agreement with (2) or whether the discrepancy of a factor ~ 2 should be taken seriously. We are, of course, assuming that phase mixing has not reduced appreciably the phase space density in the central regions of the galaxy; otherwise the discrepancy would be increased. However, Melott's (1982a,b) numerical simulations of neutrino collapse do suggest that the phase density is not much reduced in the central regions of the collapsed system.

In view of all the uncertainties, one should perhaps regard the two mass estimates, which are completely independent, as in remarkable agreement. However, the speculation (Sciama and Melott 1982), described in my other article in these proceedings, that photons emitted by galactic neutrinos are responsible for the high ionization stages represented in the galactic halo (and in the halos of other galaxies) by Si IV and CIV would require that $m_\nu \sim 100$ eV. This further increase of mass would correspond to a definite discrepancy and would lead to problems with the age of the universe (if $\lambda = 0$). These problems, however, could be relieved by replacing neutrinos with photinos, as we shall now see.

SUPERSYMMETRY AND MASSIVE PHOTINOS

GUTs do not yet lie at the end of the road, according to most particle physicists. This is partly because they still contain a large number of undetermined quantities, and partly because they present particular problems, such as the famous hierarchy problem: why are the important masses so widely spaced out, e.g. at 100 Gev (electroweak gauge particles, the W and Z bosons), 10^{14} Gev (the X bosons of GUTs), and 10^{19} Gev (the Planck mass)? This is particularly puzzling because one would expect interactions to drag, say, the masses of the 100 Gev bosons up to 10^{14} Gev, unless a miraculous cancellation occurs to many places of decimals - the fine tuning problem again.

In supersymmetric theories (for a technical review see Fayet and Ferrara (1977)) this miracle can occur by virtue of the supersymmetry itself. This new symmetry (which is the only one left unexploited by existing gauge theories) has the remarkable property of interrelating bosons and fermions, which thereby can occupy the same multiplet of particles. Thus, if supersymmetry were an exact symmetry, the electron, for example, would have a scalar partner of the same mass. Such a particle is known not to exist, and so one would have to suppose that in the real world the supersymmetry is broken, just as the GUT symmetry is broken at energies below 10^{14} Gev. The scalar partner of the electron could then have a much higher mass which could leave it unobservable at present. The energy at which supersymmetry is broken is not known, although we shall see that cosmology may provide some clues both to this and to other parameters of supersymmetric theories.

For our purposes the important new particles thrown up by supersymmetry are the photino, which is the spin 1/2 partner of the photon, and the goldstino, a spin 1/2 particle related to the breaking of the supersymmetry. These particles are important for at least two reasons. First of all they may have been sufficiently numerous when the temperature of the universe was 1 Mev to influence the time scale of the expansion and so the outcome of nucleosynthesis. Secondly, if they have masses of order tens or hundreds of electron volts, they could replace massive neutrinos in regard to the critical density for the universe, the dark matter in galactic halos, and the emission of ultraviolet photons.

These considerations are highly speculative, since there is as yet no experimental evidence in favour of supersymmetry. Nevertheless, the possibilities are intriguing, particularly because the argument goes both ways, that is, while aspects of supersymmetry would be important for cosmology, cosmological constraints would be important for supersymmetry. For example, if we are allowed only a fraction of an extra "neutrino type" by the nucleosynthesis argument, this could be achieved by supposing that photinos and goldstinos decoupled before muons and pions annihilated, since this would result in their further suppression relative to photons in the 3°K background. Such a decoupling requirement would have strong implications for the coupling constants of supersymmetric theories, and for their structural features generally (such as their capacity to solve the hierarchy problem) (Sciama 1982c). In addition the resulting suppression of the photino number density would permit 100 eV photinos to provide the critical density without running into an age problem for the universe (Sciama 1982b).

CONCLUSIONS

The reader may be appalled by the amount of speculation in this article. All I can say in my defence is that I find it hard to believe that it is <u>all</u> wrong and/or misleading. Even if only a small part of it is found to be on the right lines, we would still be witnessing the birth of imaginative new possibilities for our understanding of the universe, which will presumably leave their permanent mark on the growth of this understanding.

REFERENCES

Barrow, J.D. and Morgan, J.: 1982, to be published.
Bruck, H.A., Coyne, G.V. and Longair, M.S., Eds.: 1982, Astrophysical Cosmology, Pontifical Academy of Sciences.
Caldwell, J.A.R. and Ostriker, J.P.: 1981, Ap.J. 251, 61.
Coleman, S. and Weinberg, E.J.L 1973, Phys.Rev. D 1888.
Ellis, J.R.: 1981, Les Houches Summer School.
Fayet, P. and Ferrara, S.: 1977, Phys.Reps. 32C, 249.
Gibbons, G.W. and Hawking, S.W., Eds.: 1983, Cambridge Workshop on the Very Early Universe, Cambridge Unversity Press.
Guth, A.H.: 1981, Phys.Rev. D23, 347.
Hoyle, F.: 1948, Mon.Not.Roy.Astr.Soc. 108, 372.
Isham, C.J., Penrose, R. and Sciama, D.W., Eds.: 1981, Quantum Gravity 2; a Second Oxford Symposium, Oxford University Press.
Langacker, P.: 1981, Phys.Reps. 72, 185.
Melott, A.L.: 1982a, Phys.Rev.Lett. 48, 894; 1982b, Nature 296, 721.
Sciama, D.W.: 1982a, in Astrophysical Cosmology, Eds. Bruck, H.A., Coyne, G.V. and Longair, M.S., Pontifical Academy of Sciences, p.528; 1982b, Phys.Lett. 114 B, 19; 1982c, Phys.Lett. to be published; 1983, these Proceedings.
Sciama, D.W. and Melott, A.L.: 1982, Phys.Rev. D25, 2214.
Tremaine, S. and Gunn, J.E.: 1979, Phys.Rev.Lett. 42, 407.
Weinberg,S.: 1982, in Astrophysical Cosmology, Eds. Bruck, H.A., Coyne, G.V. and Longair, M.S., Pontifical Academy of Sciences, p.503.

Discussion

Schmidt: You stated that massive neutrinos could provide the Oort dark matter near the sun. Would they, in fact, provide all of it, or only part of it?

Sciama: They could provide all of it if neutrinos of low velocity dispersion form a thin disk, while those of higher velocity dispersion form a quasi-spherical halo. The total column density of neutrinos would be fixed by the observed rotation velocity of the galaxy. The phase space density constraint applied to the thin disk would then require the neutrinos (or, better, photinos) to have a higher rest-mass than is normally considered, say, 100 - 250 eV. This would be compatible with the critical density if the photinos were sufficiently suppressed by becoming decoupled sufficiently early, say, at a temperature $\sim$ 200 MeV.

McCrea: Why is there any ordinary matter? Why is the universe not <u>all</u> neutrinos? Do we exist simply because $\varepsilon \neq 0$?

Sciama: The ordinary matter which now exists arises, according to the theory I outlined in my talk, from the slight excess, ε, of baryons over antibaryons when X-bosons decay at a temperature $\sim 10^{14}$ GeV. Thus, on this view we do indeed simply exist because $\varepsilon \neq 0$ (and because galaxies, stars, planets, etc., came into being!).

Peacock: If we accept all this, then what do you see as the fundamental remaining cosmological problems?

Sciama: The audience can answer this as well as I can! Some problems which clearly remain are: the origin of galaxies, whether the universe is open or closed, how to deal with the initial singularity, etc., etc. We shall clearly need at least one more IAU symposium on cosmology.

Code: Apparently massive neutrinos or photinos can play similar roles in the cosmological scenario. Thus, either one or perhaps both are operative. You must, therefore, mean that you can provide some constraints or limits on the photino coupling constants from astrophysics and not a determination of the coupling constants.

Sciama: The photino coupling constant is constrained by the requirement that photinos decouple sufficiently early in the big bang that their number-density is suppressed relative to that of neutrinos. This suppression is required a) by the upper limit on the number of particle types permitted by the big bang nucleosynthesis argument and b) by the upper limit on the photino mass-density imposed by the age of the universe. Of course, this second limit depends on the rest-mass of the photino, for which only indirect arguments (both astronomical and from particle physics) exist at the moment.

G. Ellis: Is there some form of energy non-conservation during the exponential phase of the universe's expansion?

Sciama: There is no energy non-conservation. The argument is the same as used to be used in the early days of the steady state theory. The work done by the large negative pressure during the expansion reappears as energy-density which can then remain constant despite the expansion.

INDEX OF NAMES

INDEX OF SUBJECTS

IAU SYMPOSIUM No. 104

EARLY EVOLUTION OF THE UNIVERSE AND ITS PRESENT STRUCTURE

G. O. ABELL and G. CHINCARINI (EDS.)

In the area of cosmology there have recently been dramatic advances, both on the observational and theoretical fronts. Modern high-efficiency detectors have made possible extensive magnitude-limited redshift surveys, which have permitted observational cosmologists to construct three-dimensional maps of large regions of space. What seems to emerge is a distribution of matter in extensive, flat, but filamentary, and possibly interconnected superclusters, serving as interstices between vast voids in space. Meanwhile, theoretical ideas that were highly speculative a few years ago have begun to be taken seriously as possibly describing conditions in the very early universe. Brand new ideas, such as that of the inflationary universe, hold promise for solving outstanding observational, theoretical, and philosophical problems in cosmology. A new look at grand unified theories and concepts of supersymmetry have brought observational and theoretical cosmologists to a common meeting ground with modern particle physicists. These subjects provided the focal points for IAU Symposium No. 104, the Proceedings of which appear here.

Cover picture:
A computer simulation of galaxy clustering illustrates the filamentary nature of the structures that form in the "pancake" theory of galaxy formation. Surfaces of constant density are shown in an N-body model of a region about 100 Mpc in diameter. *(Courtesy, C. S. Frenk, S. D. M. White and M. Davis, University of California, Berkeley)*

D. REIDEL PUBLISHING COMPANY

DORDRECHT / BOSTON / LANCASTER